W · O · R · L · D
GEOGRAPHY

Herbert H. Gross, Ph.D.

Concordia Teachers College
River Forest, Illinois

Teacher Consultants

Roy C. Everson, Jr.
Waldo Junior High School
Aurora, Illinois

Myras Osman
Homewood-Flossmoor High School
Flossmoor, Illinois

Helyn E. Jones
Junior High School 118
New York, New York

James Reha
Horace Mann High School
Gary, Indiana

David Kernwein
Johnston High School
Austin, Texas

Don Valiska
Thornridge High School
Dolton, Illinois

Blase Markun
Gompers Junior High School
Joliet, Illinois

Moses Weintraub
Evander Childs High School
Bronx, New York

ALLYN AND BACON, INC., Newton, Massachusetts

Rockleigh, NJ   Atlanta   Warrensburg, MO   Dallas   Rancho Cordova, CA
London   Sydney   Toronto

Table of Contents

ISBN 0-205-09535-6
Printed in the United States of America
1 2 3 4 5 6 7 8 9 10   90 89 88 87 86 85

Map List

To the Student

Geography is the study of the relationship between earth's people and their physical environment. Ever since the first humans walked the earth, people have found ways to use their physical environment to build lives for themselves and opportunities for their children. Today's societies have all grown out of yesterday's choices—choices about how to use, tame, and sometimes even change the features of the earth in order for people to survive and prosper.

People all around the globe are still making these choices today. They are still seeking answers to the age-old question: How can we use our physical home, the earth, to support ourselves and future generations in as much comfort as possible? In the pages that follow, you will learn about the features of the earth and the natural forces that help create them. You will also learn how people in different parts of the earth are using the resources of their lands and waters to build unique and lasting societies.

The Atlas

Photographs, such as this one taken from space, give us some information about the earth. Maps also give information about the earth. The Atlas contains the maps you will use most often in this book. Study the different kinds of maps in the Atlas. Then you will always know where to find the map you need when you need it.

Mountain Peak

Plain

Source of River

Basin

Volcano

Mountain Range

Foothills

Waterfall

Lake

River

Reservoir

Valley

Divide

Basin

Dam

River

Canyon

Plateau

Branch or Tributary

Upstream →

← Downstream

River

Highland

Glacier

Cliff

Plain

Lowland Plain

Port

Bay

Marsh

Delta

Mouth of River

Harbor

Isthmus

Fiord

Inlet

Seacoast

Channel

Canal

Iceberg

Cape

Gulf

Sound

Strait

Island

Peninsula

Ocean or Sea

Reef

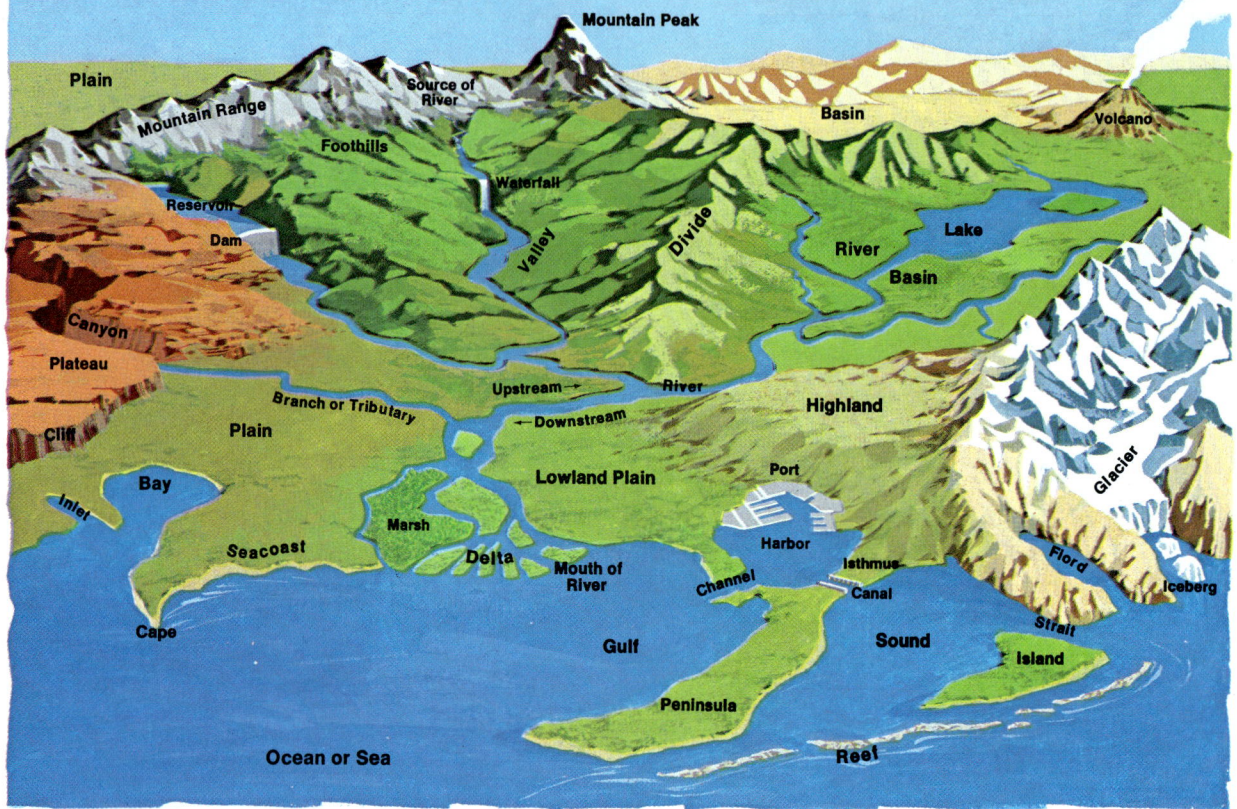

A Dictionary of Geographical Terms

altitude The height above sea level.

basin An area of land largely enclosed by higher land.

bay Part of a body of water that reaches into the land.

bed The bottom of a body of water, as riverbed or ocean bed.

branch A river or stream that flows into a larger river or stream.

canal A channel dug for irrigation or transportation.

canyon A deep, narrow valley with steep sides.

cape A point of land stretching out into a body of water.

channel A narrow body of water connecting two larger bodies of water; the deepest part of a waterway.

cliff A high, steep wall of rock.

coast Land next to a sea or ocean.

continent One of the seven largest bodies of land on the earth.

current The flow of a stream of water.

dam A structure built to stop or slow down the flow of water.

delta Land built up by deposits at a river's mouth.

desert A dry, barren area of land where few plants can grow.

dike A wall of earth or stone built by people to hold back flooding.

divide A height of land that separates river basins.

downstream The direction of a river's flow —toward its mouth.

drainage basin An area of land drained by a river and its tributaries.

elevation The height above sea level.

equator An imaginary line around the earth that is halfway between the North and South poles.

fiord A narrow inlet of the sea between steep banks or cliffs, usually formed by a glacier.

foothills Hills at the base of mountains.

glacier A large body of slowly moving ice.

gulf Part of a sea or ocean that reaches into the land.

harbor A sheltered place where ships may anchor safely.

highland An area of hills, mountains, or plateaus.

hill A raised part of the earth's surface with sloping sides, smaller than a mountain.

iceberg A floating mass of ice that has broken off from a glacier.

inland Away from the coast.

inlet A small part of a body of water that reaches into a coast.

island Land entirely surrounded by water, smaller than a continent.

isthmus A narrow strip of land that connects two larger bodies of land.

lake An inland body of water.

Map Symbols

Physical Features

- ～ River or stream
- 🟦 Lake
- - - - Ice pack
- ᐃᐃ ᐃ Mountains
- ▲ + Mountain peak
- 〰 Swamp or marsh
- 🟩 Grassland
- 🟫 Woodland
- 🟨 Desert

Cultural Features

- — ‧ — ‧ — International boundary
- — ‧ — ‧ — National boundary
- — — — International date line
- ▬▬▬▬▬ Trust territory boundary
- ⌒⌒⌒ Continental boundary
- ⊗ National capital
- ★ State or province capital
- ⬤ Large city
- ● ● Other cities
- ▭ National park

- 🟥 Reserve or reservation
- 🟦 Canal
- ▮ Dam
- ++++++++ Railroad
- ●●●●●● Pipeline
-)(Bridge
- Ⓐ ✈ Airport
- ▬▬▬ Main road
- ═══ Secondary road
- ─── Other road

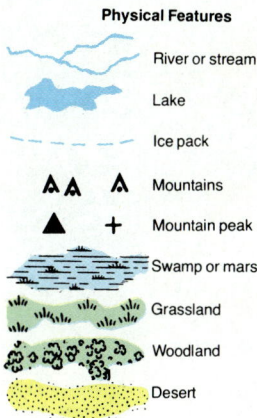

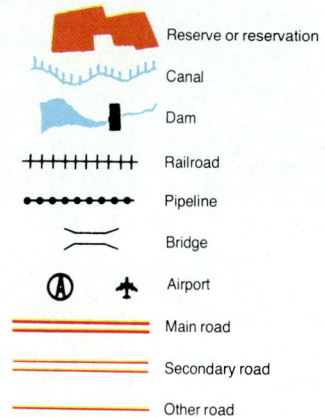

latitude The distance in degrees north or south of the equator.

longitude The distance in degrees east or west of the prime meridian.

lowland An area of low and usually level land.

map key An explanation of the meaning of the symbols on a map. The map scale is often included in the key.

map scale A measuring guide that shows what distance on the earth is represented by a certain measure on a map.

marsh An area of low, wet land where tall grasses grow.

mountain High, rocky land, usually with steep sides and a pointed or rounded top, higher than a hill.

mountain peak The pointed top of a mountain.

mountain range A chain of mountains.

mouth (of a river) The place where a river flows into a larger body of water.

North Pole The point on the earth that is farthest north.

oasis A place in a desert where water is found in a spring or a well, making the growing of crops possible.

ocean One of the four largest bodies of water on the earth.

ocean current A stream of water moving in a definite direction in the ocean.

pampa A grass-covered plain in South America.

peninsula A body of land almost surrounded by water.

plain An area of broad, level land.

plateau An area of mostly high, flat land.

port A harbor, town, or city where ships can load and unload their cargoes.

prairie A large plains region with tall grasses, usually found in the United States or Canada.

prime meridian An imaginary line on the earth's surface running through Greenwich, England, from the North Pole to the South Pole and used as the starting point from which degrees of longitude are measured.

rain forest A tropical woodland area where rainfall is heavy and branches of tall trees spread a covering layer over the top of the area.

rapids A part of a river, generally shallow, where the current moves swiftly over rocks.

reef A ridge of rock or sand at or near the surface of water.

reservoir A lake where water is stored for future use, often formed by a dam.

river A large stream of water flowing through the land.

river valley Low land through which a river flows.

savanna A level, tropical grassland with scattered trees.

sea A large body of water, usually salt water, partly or completely enclosed by land.

sound A wide channel connecting two bodies of water or an inlet between the mainland and islands.

source (of a river) The place where a river begins, usually in highlands.

South Pole The point on the earth that is farthest south.

steppe A vast area of dry, level grassland, usually found in Asia.

strait A narrow stretch of water that connects two larger bodies of water.

swamp An area of land, sometimes forested, that is always moist or soaked with water.

tide The regular rising and falling of the water of the ocean and waters connected with the ocean.

timber line A line in mountain regions above which trees do not grow.

tributary A river or stream that flows into a larger river or stream.

tropics The warm region lying on both sides of the equator.

tundra A cold, treeless plain in Arctic regions.

upstream The direction toward a river's source, opposite to its flow.

valley Low land between hills or mountains.

volcano A mountain formed of rock or ash thrown up from inside the earth.

waterfall Water falling over a steep drop in the land.

ARCTIC OCEAN

Ellesmere
Island

Thule

GREENLAND
(Denmark)

Norwegian

Jan Mayen I.
(Nor.)

80°

Beaufort
Sea

Banks
Island

Baffin
Bay

Point
Barrow
Barrow

Wrangel I.

U.S.S.R.

UNITED STATES

Victoria
Island

Great Bear
Lake

Port
Radium

Baffin
Island

Denmark Strait

ICELAND

70°

Nome

St.
Lawrence I.

ALASKA

Fairbanks

Yukon River

Arctic Circle

Great Slave
Lake

Godthaab

Reykjavik

Faeroes
(Den.)

Shetland Is.

Orkney Is.

North

60°

Bering Sea

Anchorage

Dawson

Whitehorse

Juneau

Hudson

Churchill

Bay

Julianehaab

Cape Farewell

UNITED

SCOTLAND

NORTH Sea

KINGDOM

ENGLAND

NO. IRELAND

London

Pribilof
Islands

Kodiak I.

Gulf of
Alaska

CANADA

Hudson Strait

Dublin

IRELAND

50°

Dutch Harbor

Aleutian Islands

Queen Charlotte
Islands

Vancouver
Island

NORTH

Edmonton

AMERICA

Calgary

Regina

Lake
Winnipeg

Winnipeg

Mackenzie River

Goose
Bay

Newfoundland I.

Gander

St. John's

Halifax

NORTH

English Channel

Bay of
Biscay

Paris

FRANCE

NORTH

PACIFIC

Seattle

Portland

Minneapolis

Great
Lakes

Missouri R.

Chicago

Detroit

Cleveland

Ottawa

Toronto

Quebec

Montreal

Boston

New York

Philadelphia

ATLANTIC

PORTUGAL

Lisbon

Madeira
Is.
(Port.)

Azores
(Port.)

Madrid

SPAIN

Strait of
Gibraltar

Tangier
Rabat
Casablanca

Algiers

40°

San Francisco

Salt Lake
City

Denver

St. Louis

UNITED STATES

Washington, D.C.

Norfolk

MOROCCO

ALGERIA

Los Angeles

Dallas

Atlanta

Mississippi R.

Bermuda
(Br.)

Canary Is.
(Sp.)

OCEAN

Houston

New
Orleans

Jacksonville

0°

PACIFIC

Tropic of Cancer

Monterrey

MEXICO

Guadalajara

Gulf of
Mexico

Miami

THE BAHAMAS

Havana

West
Indies

CUBA

DOMINICAN
REPUBLIC

Puerto
Rico
(U.S.)

OCEAN

CAPE VERDE

MAURITANIA

SENEGAL

Dakar

MALI

A

F

20°

Hawaiian Islands

Honolulu

HAWAII

UNITED STATES

Johnston I.
(U.S.)

Mexico City

Veracruz

BELIZE

GUATEMALA

EL SALVADOR

HONDURAS

NICARAGUA

JAMAICA

HAITI

Caribbean
Sea

Lesser
Antilles

Panama
Canal

GAMBIA

GUINEA-BISSAU

UPPER
VOLTA

GUINEA

SIERRA LEONE

IVORY
COAST

ACCRA

TOGO

BENIN

Niger R.

LIBERIA

Monrovia

OCEAN

Christmas I.

KIRIBATI

Central

COSTA RICA

PANAMA

America

TRINIDAD AND TOBAGO

Caracas

VENEZUELA

GUYANA

SURINAM

FRENCH GUIANA

10°

Medellín

Cali

Bogotá

COLOMBIA

Equator

Galápagos Is.
(Ecuador)

Quito

Guayaquil

ECUADOR

Manaus

Amazon R.

Belém

Fortaleza

0°

Marquesas
Islands
(Fr.)

PERU

Callao

Lima

Cuzco

SOUTH

AMERICA

BRAZIL

Recife

Ascension I.
(Br.)

10°

W. SAMOA

Society
Islands
(Fr.)

Tahiti
(Fr.)

Tuamotu
Arch.
(Fr.)

La Paz

BOLIVIA

Sucre

Brasília

Belo Horizonte

Salvador

SOUTH

St. Helena
(Br.)

20°

Pitcairn I.
(Br.)

Tropic of Capricorn

Easter I.
(Chile)

Antofagasta

PARAGUAY

Asunción

Paraná R.

Tucumán

Rio de Janeiro

São Paulo

Pôrto Alegre

30°

SOUTH

Juan Fernández Is.
(Chile)

Valparaiso

Santiago

Concepción

Rosario

CHILE

ARGENTINA

URUGUAY

Buenos Aires

Montevideo

Rio de la Plata

ATLANTIC

40°

PACIFIC

Strait of Magellan

Punta Arenas

Tierra
del Fuego

Cape Horn

Falkland
Islands
(Br.)

South Georgia
(Br.)

OCEAN

50°

OCEAN

Drake Passage

60°

10

Antarctic Circle

ANTAR

70°

ARCTIC OCEAN

UNION OF SOVIET SOCIALIST REPUBLICS

EUROPE

ASIA

AFRICA

CHINA

MONGOLIA

INDIA

NORTH PACIFIC OCEAN

INDIAN OCEAN

AUSTRALIA

NEW ZEALAND

ANTARCTICA

International Date Line

Arctic Circle

Tropic of Cancer

Equator

Tropic of Capricorn

Antarctic Circle

The World
Political Map

⊛ National Capitals
★ Other Capitals
• Other Cities

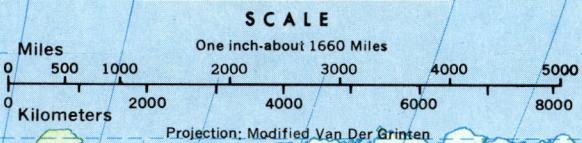

SCALE
One inch–about 1660 Miles

Miles
0 500 1000 2000 3000 4000 5000

Kilometers
0 2000 4000 6000 8000

Projection: Modified Van Der Grinten

11

© Follett Publishing Company

Polar Ice Pack

ARCTIC OCEAN

Ellesmere Island

Queen Elizabeth Islands

Greenland

Baffin Bay

Norwegian

Jan Mayen I.

Banks Island

Beaufort Sea

Victoria Island

Baffin Island

Davis Strait

Arctic Circle

Iceland

Wrangel I.

Point Barrow

Arctic Circle

Faeroe I.

Shetland Is.
Orkney Is.

Mackenzie River

Great Bear Lake

Hudson Strait

Cape Farewell

St. Lawrence I.

Bering Strait

Mt. McKinley (20,320)

Yukon River

Great Slave Lake

Hudson Bay

British Isles

North Sea

Bering Sea

Gulf of Alaska

NORTH

Lake Winnipeg

Pribilof Islands

Kodiak I.

Great Lakes

St. Lawrence R.

Newfoundland I.

English Channel

Aleutian Islands

Queen Charlotte Islands

AMERICA

Bay of Biscay

NORTH

Vancouver Island

Rocky Mountains

GREAT PLAINS

Missouri R.

Iberian Pen.

Atlas Mts.

NORTH

Columbia R.

Ohio R.

Appalachian Mts.

Azores

ATLANTIC

Strait of Gibraltar

Midway Is.

PACIFIC

Colorado R.

Mississippi R.

Bermuda

Madeira Is.

Hawaiian Islands

Baja Calif.

Rio Grande

Tropic of Cancer

Canary Is.

Tropic of Cancer

A F

Johnston I.

HAWAII

Gulf of Mexico

OCEAN

Niger R.

OCEAN

Yucatán Pen.

West Indies

Greater Antilles

Puerto Rico

Cape Verde Is.

Christmas I.

Equator

Galápagos Is.

Caribbean Sea

Lesser Antilles

Central America

Isthmus of Panama

Orinoco R.

Guiana Highlands

Amazon R.

Equator

Gulf of Guinea

SOUTH

Samoa Islands

POLYNESIA

SOUTH AMERICA

Brazilian Highlands

Ascension I.

Society Islands

Tahiti

Tuamotu Arch.

Titicaca

St. Helena

SOUTH

Andes

Paraná R.

Tropic of Capricorn

Pitcairn I.

Easter I.

Tropic of Capricorn

ATLANTIC

Juan Fernández Is.

Atacama Desert

Mt. Aconcagua (22,834)

Pampa

Río de la Plata

SOUTH

Patagonia

OCEAN

Strait of Magellan

Falkland Islands

PACIFIC

Tierra del Fuego

South Georgia

Cape Horn

Drake Passage

OCEAN

Antarctic Circle

Antarctic Circle

HEIGHT OF LAND

OVER 13,000 FEET
6,600 TO 13,000
3,300 TO 6,600
1,650 TO 3,300
650 TO 1,650
0 TO 650 FEET
BELOW SEA LEVEL

DEPTH OF WATER

0 TO 600 FEET
BELOW 600 FEET

12

Weddell Sea

ANTAR

135° 150° 165° 180° 165° 150° 135° 120° 105° 90° 75° 60° 45° 30° 15°

80° 70° 60° 50° 40° 30° 20° 10° 0° 10° 20° 30° 40° 50° 60° 70°

The World

Graphic-Relief Map

SCALE 1:105,000,000 or One Inch—About 1660 Miles

Miles
0 500 1000 2000 3000 4000 5000

Kilometers
0 2500 5000 7500

Projection Modified Van Der Grinten

13

© Follett Publishing Company

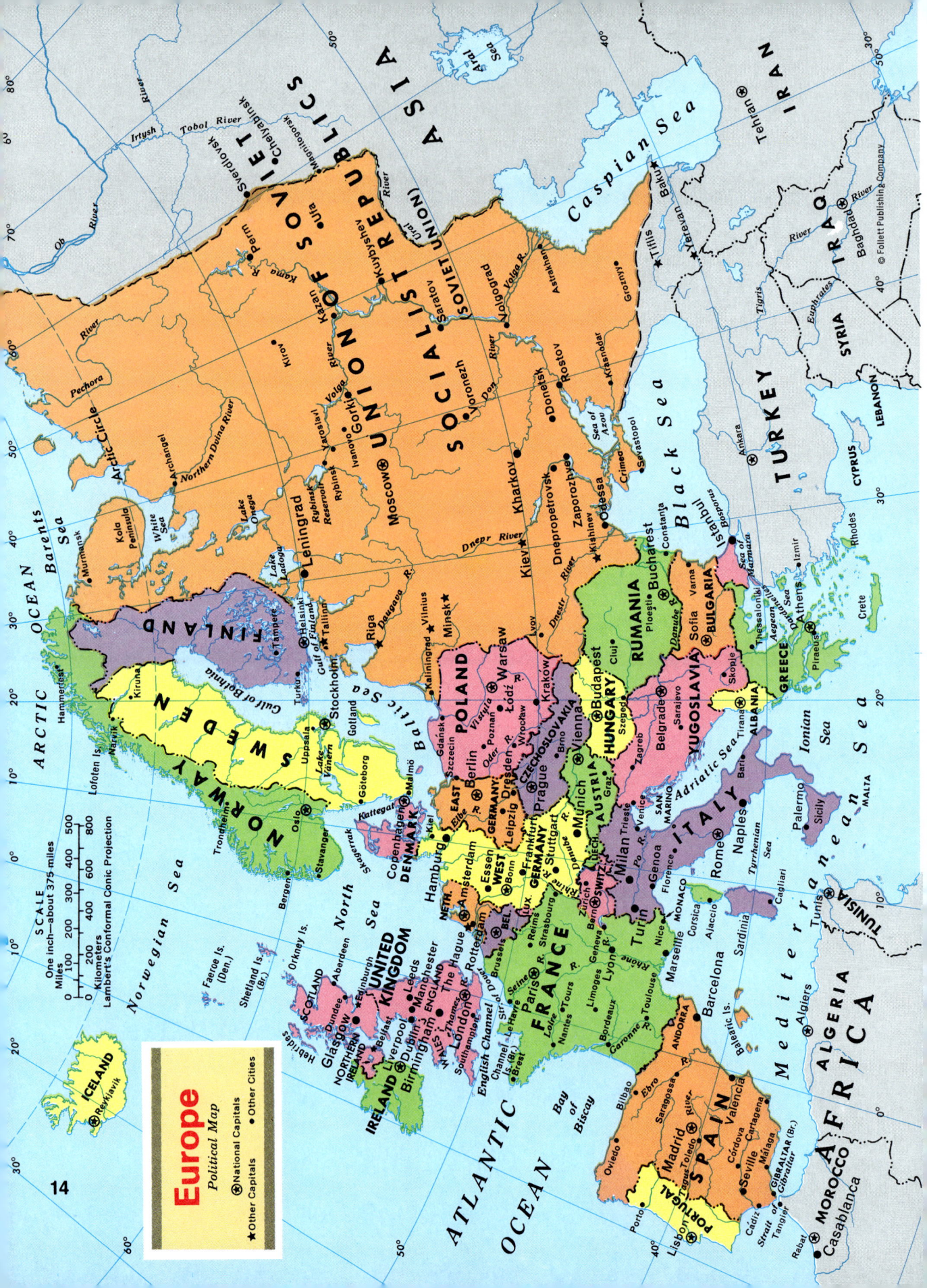

SOVIET
ASIA
IRAN
Irtysh
Tobol River
Sverdlovsk
Chelyabinsk
Magnitogorsk
Ob River
Pechora
River
UNION OF
Ural
Kama R.
Perm
Ufa
SOVIET UNION
Kazan
Kuybyshev
Volga River
SOCIALIST
Kirov
Saratov
Voronezh
Don River
Volgograd
Astrakhan
Caspian Sea
Tehran
Baku
Tiflis
Yerevan
Grozny
Krasnodar
TURKEY
IRAQ
Baghdad
Euphrates River
© Follett Publishing Company
SYRIA
LEBANON
Tigris
River

Ivanovo
Gorki
Yaroslavl
Rybinsk
Reservoir
Rybinsk
Arctic Circle
Archangel
Northern Dvina River
MOSCOW
Leningrad
REPUBLICS
UNION

ARCTIC OCEAN
Barents Sea
Murmansk
Kola Peninsula
White Sea
Lake Onega
Hammerfest
Lofoten Is.
Narvik
Kiruna
Lake Ladoga
Helsinki
Gulf of Finland
Tallinn
Tampere
Turku
FINLAND
Riga
Vilnius
Minsk
Daugava R.
Kaliningrad
Kharkov
Dnepr River
Kiev
Dnepropetrovsk
Zaporozhye
Odessa
Kishinev
Lvov
Sea of Azov
Crimea
Sevastopol
Rostov
Donetsk
Black Sea
Constanta
Istanbul
Bosporus
Ankara
CYPRUS
Rhodes

NORWAY
SWEDEN
Trondheim
Gulf of Bothnia
Uppsala
Lake Vänern
Stockholm
Göteborg
Oslo
Stavanger
Bergen
Skagerrak
Kattegat
Malmö
Gotland
Baltic Sea
Copenhagen
DENMARK
Kiel
Gdansk
Szczecin
Poznan
POLAND
Warsaw
Lódz
Vistula
Wroclaw
Oder R.
Kraków
CZECHOSLOVAKIA
Brno
Prague
Bratislava
HUNGARY
Budapest
Szeged
Cluj
RUMANIA
Bucharest
Ploesti
Danube R.
Sofia
BULGARIA
Varna
Skopje
YUGOSLAVIA
Belgrade
Sarajevo
Zagreb
Thessaloniki
GREECE
Athens
Piraeus
Aegean Sea
Crete
Tirana
ALBANIA
Ionian Sea

Norwegian Sea
Faeroe Is. (Den.)
Shetland Is. (Br.)
Orkney Is.
North Sea
Aberdeen
Edinburgh
Glasgow
Dundee
SCOTLAND
Belfast
NORTHERN IRELAND
IRELAND
Dublin
Liverpool
Manchester
Leeds
ENGLAND
WALES
Birmingham
UNITED KINGDOM
London
Thames R.
Southampton
English Channel
Channel Is. (Br.)
Brest
Hamburg
NETH.
Amsterdam
Rotterdam
The Hague
BEL.
Brussels
Elbe R.
EAST GERMANY
Berlin
Leipzig
Dresden
WEST GERMANY
Bonn
Frankfurt
Essen
Stuttgart
Munich
LUX.
Reims
Strasbourg
Paris
Seine R.
Le Havre
Cherbourg
FRANCE
Loire R.
Tours
Nantes
Limoges
Bordeaux
Garonne R.
Toulouse
Lyon
Rhône R.
Geneva
Bern
SWITZ.
Zürich
LIECH.
AUSTRIA
Vienna
Graz
Trieste
Venice
Milan
Turin
Genoa
Po R.
Florence
SAN MARINO
ITALY
Rome
Naples
Tyrrhenian Sea
Palermo
Sicily
MALTA
Cagliari
Sardinia
Corsica
Ajaccio
MONACO
Marseille
Bari
Adriatic Sea
Mediterranean Sea
Tunis
TUNISIA
ALGERIA
Algiers
AFRICA

Bay of Biscay
ANDORRA
Bilbao
Oviedo
Ebro River
Zaragoza
Barcelona
Balearic Is.
Valencia
Cartagena
SPAIN
Madrid
Toledo
Tagus River
Córdoba
Seville
Málaga
Cádiz
GIBRALTAR (Br.)
Strait of Gibraltar
Tangier
MOROCCO
Rabat
Casablanca
PORTUGAL
Lisbon
Porto
ATLANTIC OCEAN

ICELAND
Reykjavik

SCALE
One inch—about 375 miles
Miles
0 100 200 300 400 500
0 200 400 600 800
Kilometers
Lambert's Conformal Conic Projection

Europe
Political Map
⊛ National Capitals
• Other Cities
★ Other Capitals

14

Asia
Political Map
⊛ National Capitals
★ Other Capitals
• Other Cities

© Follett Publishing Company

International Date Line

PACIFIC OCEAN
ARCTIC OCEAN
ATLANTIC OCEAN
INDIAN OCEAN

UNION OF SOVIET SOCIALIST REPUBLICS (SOVIET UNION)

CHINA
MONGOLIA
Ulanbaatar
Peking
Harbin
Changchun
Shenyang
Tientsin
Tsingtao
Jinan
Taiyuan
Xi'an
Nanjing
Wuhan
Hangzhou
Nanchang
Changsha
Chengdu
Chongqing
Kunming
Canton
Fuzhou
Amoy
Shanghai
Lanzhou
Lhasa
Kashgar
Ürümqi

JAPAN
Tokyo
Yokohama
Nagoya
Kyoto
Osaka
Sapporo
Hokkaido
Honshu
Shikoku
Kyushu

NORTH KOREA
Pyongyang
SOUTH KOREA
Seoul
Pusan

TAIWAN
Taipei
HONG KONG
MACAO (Port.)

PHILIPPINES
Manila
Quezon City
Davao
Mindanao

VIETNAM
Hanoi
Ho Chi Minh City (Saigon)
Da Nang
LAOS
Vientiane
KAMPUCHEA
Phnom Penh
THAILAND
Bangkok
BURMA
Rangoon
Mandalay

MALAYSIA
Kuala Lumpur
SINGAPORE
BRUNEI
Bandar Seri Begawan
Kuching
Sumatra
Borneo
Celebes
Java
INDONESIA
Jakarta
Palembang

PAPUA NEW GUINEA
Port Moresby
AUSTRALIA

INDIA
New Delhi
Delhi
Agra
Lucknow
Kanpur
Varanasi
Calcutta
Nagpur
Hyderabad
Bombay
Poona
Bangalore
Madras
Madurai
Ahmadabad
Jaipur
NEPAL
Kathmandu
BHUTAN
Thimphu
BANGLADESH
Dacca
SRI LANKA
Colombo

PAKISTAN
Islamabad
Lahore
Karachi
AFGHANISTAN
Kabul
Herat

IRAN
Tehran
Meshed
Shiraz
Abadan
IRAQ
Baghdad
KUWAIT
SAUDI ARABIA
Riyadh
Mecca
Medina
BAHRAIN
QATAR
U.A.E.
Abu Dhabi
OMAN
Muscat
YEMEN
San'a
S. YEMEN
Aden
JORDAN
Amman
SYRIA
Damascus
LEBANON
Beirut
ISRAEL
Jerusalem
TURKEY
Ankara
Istanbul
Izmir
CYPRUS

Leningrad
Moscow
Gorky
Kuybyshev
Volgograd
Sverdlovsk
Chelyabinsk
Magnitogorsk
Omsk
Novosibirsk
Krasnoyarsk
Tomsk
Barnaul
Semipalatinsk
Karaganda
Alma-Ata
Tashkent
Samarkand
Aralsk
Tobolsk
Norilsk
Dikson
Murmansk
Archangel
Vorkuta
Khabarovsk
Vladivostok
Komsomolsk
Yakutsk
Chita
Irkutsk
Ulan Ude
Baku
Tbilisi
Yerevan
Rostov
Kharkov
Kiev
Odessa
Riga

EUROPE
UNITED KINGDOM
London
IRELAND
FRANCE
Paris
SPAIN
ITALY
Rome
Sardinia
Sicily
NORWAY
SWEDEN
FINLAND
DENMARK
NETH.
BEL.
GER.
SWITZ.
AUST.
POLAND
CZECH.
HUNG.
ROM.
YUGO.
BULG.
ALB.
GREECE
Berlin

AFRICA
LIBYA
EGYPT
Cairo
Alexandria
SUDAN
Khartoum
ETHIOPIA
Addis Ababa
SOMALIA
KENYA
Nairobi
TANZANIA
Dar es Salaam
Zanzibar

Mediterranean Sea
Red Sea
Black Sea
Caspian Sea
Arabian Sea
Bay of Bengal
South China Sea
Yellow Sea
Sea of Japan
Sea of Okhotsk
Bering Sea
Arctic Ocean
Persian Gulf
Coral Sea
Arafura Sea
Celebes Sea
Java Sea

SCALE
One inch—about 875 miles
Miles
0 200 400 600 800 1000
Kilometers
0 400 800 1200 1600
Bonne Equal-Area Projection

15

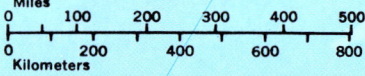

ARCTIC

North

Hammerfest

Lofoten Is.

Narvik

Kiruna

Norwegian Sea

Arctic Circle

ICELAND

Reykjavik ⊛

Trondheim

N O R W A Y

S W E D E N

Kjölen Mts.

Gulf of Bothnia

Tampere

Turku

Bergen

Oslo ⊛

Stavanger

Stockholm ⊛

Uppsala

Lake Vänern

Gulf

Faeroe Is. (Den.)

Shetland Is. (Br.)

Göteborg

Gotland

Baltic Sea

Skagerrak

Kattegat

DENMARK

Copenhagen ⊛

Malmö

Kaliningrad

Gdansk

Orkney Is.

Hebrides

SCOTLAND

Glasgow ●
North Channel

Aberdeen

Dundee

Edinburgh ★

North Sea

Kiel

Szczecin

P O L A N D

Poznan

Łódź

Warsaw ⊛

NORTHERN IRELAND
Belfast ★

IRELAND

Dublin ⊛

Irish Sea

Liverpool

UNITED KINGDOM

ENGLAND

Leeds

Manchester

Birmingham

London ⊛

Hamburg

Bremen

NETH.

Amsterdam ⊛

EAST

Berlin ⊛

GERMANY

Essen

Bonn ⊛

WEST GERMANY

Leipzig

Wrocław

Kraków

CZECHOSLOVAKIA

Prague ⊛

Brno

WALES

Southampton

Cork

St. George's Channel

Land's End

Thames R.

English Channel
Channel Is. (Br.)

Str. of Dover

Cherbourg

Rotterdam

Brussels ⊛

BEL.

LUX. ⊛

Frankfurt

Strasbourg

Stuttgart

Munich

Black Forest

Danube R.

Vienna ⊛

AUSTRIA

Graz

Budapest ⊛

HUNGARY

Szeged

R U

Le Havre

Paris ⊛

Seine R.

F R A N C E

Loire R.

Nantes

Tours

Limoges

Lyon

Geneva

Bern ⊛

Zürich

SWITZ.

LIECH.

Mt. Blanc (15,771)

Milan

Trieste

Turin

Venice

Po R.

Zagreb

Y U G O S L A V I A

Dinaric Alps

Belgrade ⊛

Sarajevo

Brest

Bay of Biscay

ATLANTIC OCEAN

Limoges

Bordeaux

Garonne R.

Toulouse

Pyrenees

ANDORRA

Ebro River

Saragossa

Nice

MONACO

Marseille

Genoa

Florence

SAN MARINO

Apennines

Adriatic Sea

Skopje

Tirana

ALBANIA

Pindus Mts.

GRE

Oviedo

Porto

Bilbao

PORTUGAL

Lisbon ⊛

S P A I N

Madrid ⊛

Toledo

Tagus River

Córdova

Seville

Cádiz

Málaga

Guadiana R.

Valencia

Cartagena

Barcelona

Balearic Is. (Sp.)

Corsica (Fr.)

Ajaccio

I T A L Y

Rome ⊛

Naples

Mt. Vesuvius (11,053)

Bari

Tyrrhenian Sea

Sardinia (It.)

Cagliari

Tangier

Strait of Gibraltar

GIBRALTAR (Br.)

Rabat ⊛

Casablanca

MOROCCO

Atlas

A F R I C A

Mountains

A L G E R I A

Algiers ⊛

Mediterranean Sea

Palermo

Sicily

Mt. Etna (4,190)

MALTA

Ionian Sea

TUNISIA

Tunis ⊛

16

HEIGHT OF LAND
OVER 13,000 FEET
6,600 TO 13,000
3,300 TO 6,600
1,650 TO 3,300
650 TO 1,650
0 TO 650 FEET
BELOW SEA LEVEL
DEPTH OF WATER
0 TO 600 FEET
BELOW 600 FEET

OCEAN
Barents Sea
Murmansk
Kola Peninsula
White Sea
Archangel
Pechora River
Ural Mountains
Ob River
Irtysh River

FINLAND
Helsinki
Tallinn
Lake Onega
Lake Ladoga
Leningrad
Rybinsk Reservoir
Northern Dvina River
Sverdlovsk
Tobol River
Perm
Chelyabinsk
Kirov
Kama R.
Magnitogorsk

UNION OF SOVIET

Rybinsk
Yaroslavl
Ivanovo
Gorki
Volga
River
Kazan
Ufa

Daugava River
Moscow
Kuybyshev

SOCIALIST REPUBLICS

ASIA

Vilnius
Minsk
Dnepr
Saratov
Voronezh
Ural River
(SOVIET UNION)

Aral Sea

Don
River
Volgograd
Caspian Depression

Volga River
Astrakhan

Kiev
River
Kharkov
Lvov
Dnestr River
Dnepropetrovsk
Donetsk
Rostov
Zaporozhye
Caspian Sea

Kishinev
Sea of Azov
Krasnodar
Odessa
Crimea
Groznyy
ANIA
Sevastopol
Mt. Elbrus (18,481)
Caucasus Mts.
Tiflis
Baku
Ploesti
Bucharest
Constanta
Black Sea
Surface 92 feet below sea level
Varna
Balkan Mts.
Sofia
BULGARIA
Bosporus
Yerevan

Istanbul
Sea of Marmara
Ankara
TURKEY
ASIA
Tehran
hessaloniki
Tigris
IRAN
Aegean Sea
Dardanelles
Izmir
Taurus Mts.
Euphrates River
Athens
Piraeus
SYRIA
ASIA
IRAQ
Crete (Gr.)
Rhodes (Gr.)
CYPRUS
Nicosia
Baghdad
LEBANON
17
© Follett Publishing Company

HEIGHT OF LAND

- OVER 13,000 FEET
- 6,600 TO 13,000
- 3,300 TO 6,600
- 1,650 TO 3,300
- 650 TO 1,650
- 0 TO 650 FEET
- BELOW SEA LEVEL

DEPTH OF WATER

- 0 TO 600 FEET
- BELOW 600 FEET

ATLANTIC OCEAN

Arctic Circle

Svalbard (Nor.)

Franz Josef Land (U.S.S.R.)

ARC

Barents Sea

Novaya Zemlya

Kara Sea

Murmansk

Archangel

Vorkuta

IRELAND

UNITED KINGDOM

London

NORWAY

SWEDEN

FINLAND

Leningrad

Tobolsk

PORTUGAL

Paris

FRANCE

NETH.

North Sea

DEN.

Baltic Sea

Riga

Moscow

Gorki

Perm

Sverdlovsk

Novos

SPAIN

SWITZ.

GER.

E. GER.

Berlin

POLAND

Kiev

Voronezh

Kuybyshev

Chelyabinsk

Omsk

MOROCCO

Pyrenees

ALPS

AUST.

CZECH.

Kharkov

Magnitogorsk

Semipalatinsk

Algiers

ITALY

Rome

YUGOSLAVIA

HUNG.

RUMANIA

Odessa

Rostov

Volgograd

Karaganda

TUNISIA

Sardinia (It.)

BULG.

Black Sea

Astrakhan

Aralsk

Lake Balkhash

ALGERIA

Sicily (It.)

ALB.

GREECE

Istanbul

Izmir

TURKEY

Ankara

Yerevan

Tiflis

Baku

Aral Sea

Syr Darya

Tashkent

Alma-Ata

LIBYA

Tripoli

Benghazi

Crete (Gk.)

CYPRUS

Beirut

SYRIA

LEB.

Baghdad

Tabriz

Caspian Sea

Samarkand

Meshed

Alexandria

Jerusalem

ISR.

Damascus

Amman

JORDAN

IRAQ

Tigris

Euphrates

Zagros Mts.

Tehran

Isfahan

Herat

AFGHANISTAN

Kabul

Hindu Kush

Kashgar

Cairo

EGYPT

Suez Canal

Abadan

KUWAIT

Shiraz

Mts.

Pamirs

Islamabad

Aswan

L. Nasser

Medina

SAUDI

Riyadh

BAHRAIN

QATAR

Persian Gulf

Abu Dhabi

Lahore

PAKISTAN

Him

NIGER

CHAD

Khartoum

Nile

Red Sea

Mecca

ARABIA

UNITED ARAB EMIRATES

Muscat

Karachi

Indus

Great Indian Desert

Delhi

New Delhi

NEPAL

NIGERIA

SUDAN

NORTH YEMEN

Sanaa

SOUTH YEMEN

OMAN

Ahmadabad

Agra

Lucknow

Kanpur

Varanasi

CENTRAL AFRICAN REPUBLIC

Addis Ababa

DJIBOUTI

Ethiopian Highlands

Aden

Gulf of Aden

Socotra (S. Yemen)

Arabian Sea

Bombay

Poona

INDI

Nagpur

Deccan

ZAIRE

ETHIOPIA

UGANDA

KENYA

SOMALIA

Laccadive Is. (India)

Plateau

Hydera

Western Ghats

Eastern Ghats

Madra

Zaire River

Lake Victoria

Nairobi

Equator

SEYCHELLES

MALDIVES

Bangalore

ANGOLA

TANZANIA

Zanzibar I.

Dar es Salaam

Madurai

SRI LAN

Colombo

ZAMBIA

18

INDIAN O

Mediterranean Sea

Sahara

AFRICA

UNION OF SO

(SO

A

Ob

Ural Mts.

Volga

Caucasus Mts.

Elburz Mts.

IRAN

Ili en

EUROPE

ITALY

ARCTIC OCEAN

Severnaya Zemlya

New Siberian Is.

Laptev Sea

East Siberian Sea

Ambarchik

Wrangel I.

St. Lawrence I. (U.S.)

Bering Sea

Aleutian Is. (U.S.)

International Date Line

Dikson

Tiksi

Lena R.

Verkhoyansk

Magadan

Kamchatka Peninsula

Petropavlovsk-Kamchatsky

Norilsk

Yenisei

Igarka

VIET SOCIALIST REPUBLICS

(VIET UNION)

Yakutsk

Sea of Okhotsk

Sakhalin

Tomsk

Krasnoyarsk

Stanovoy Mts.

Amur R.

Komsomolsk

Khabarovsk

Kuril Is. (U.S.S.R.)

birsk

Novokuznetsk

Lake Baikal

Irkutsk

Chita

Hailar

Harbin

Vladivostok

Sapporo

Hokkaido

Barnaul

Sayan Mts.

Ulan Ude

S I A

Sea of Japan

Honshu

SCALE

One Inch—About 650 Miles

Miles

0 200 400 600 800 1000

Kilometers

0 400 800 1200 1600

Altai Mts.

Ulaanbaatar

MONGOLIA

Gobi

Changchun

Shenyang

KOREA N.

Pyongyang

JAPAN

Tokyo

Yokohama

Ürümqi

Tian Shan

Anxi

Peking

Tientsin

Lüda

Seoul

KOREA S.

Kyoto

Pusan

Osaka

Nagoya

Shikoku

Ho R.

Huang Ho

Taiyuan

Jinan

Yellow Sea

Kitakyushu

Kyushu

Bonin Is. (Japan)

Altun Shan

Lanzhou

Taiyuan

Tsingtao

Shanghai

East China Sea

Ryukyu Is. (Japan)

PACIFIC

un

Lhasa

CHINA

Xi'an

Nanjing

Hangzhou

Tropic of Cancer

Mariana Is.

OCEAN

oyas

Mt. Everest (29,028)

BHUTAN

Katmandu

Chengdu

Yangtze River

Wuhan

Nanchang

Fuzhou

Chongqing

Changsha

Taipei

Ganges

Brahmaputra R.

Kunming

Canton

Amoy

Swatow

TAIWAN

Guam (U.S.)

Dacca

BANGLA DESH

cutta

HONG KONG (Br.)

MACAO (Port.)

Ki Jiang

Luzon

Philippine Sea

TRUST TERRITORY OF

Caroline Is.

THE PACIFIC ISLANDS (U.S.)

Mandalay

BURMA

Hanoi

VIETNAM

Hainan

ad

Bay of Bengal

Rangoon

Vientiane

LAOS

Da Nang

Manila

PHILIPPINES

Andaman Is. (India)

THAILAND

KAMPUCHEA

South China Sea

Iloilo

Mindanao

Bismarck Arch.

Nicobar Is. (India)

Bangkok

Phnom Penh

Ho Chi Minh City (Saigon)

Davao

Equator

Djajapura

PAPUA NEW GUINEA

BRUNEI

Bandar Seri Begawan

New Guinea

MALAYSIA

Moluccas

Celebes

INDONESIA

Port Moresby

Kuala Lumpur

Kuching

Borneo

East Indies

Arafura Sea

SINGAPORE

Sumatra

Palembang

Jakarta

Makasar

Timor

AUSTRALIA

Coral Sea

Java

CEAN

19

© Follett Publishing Company

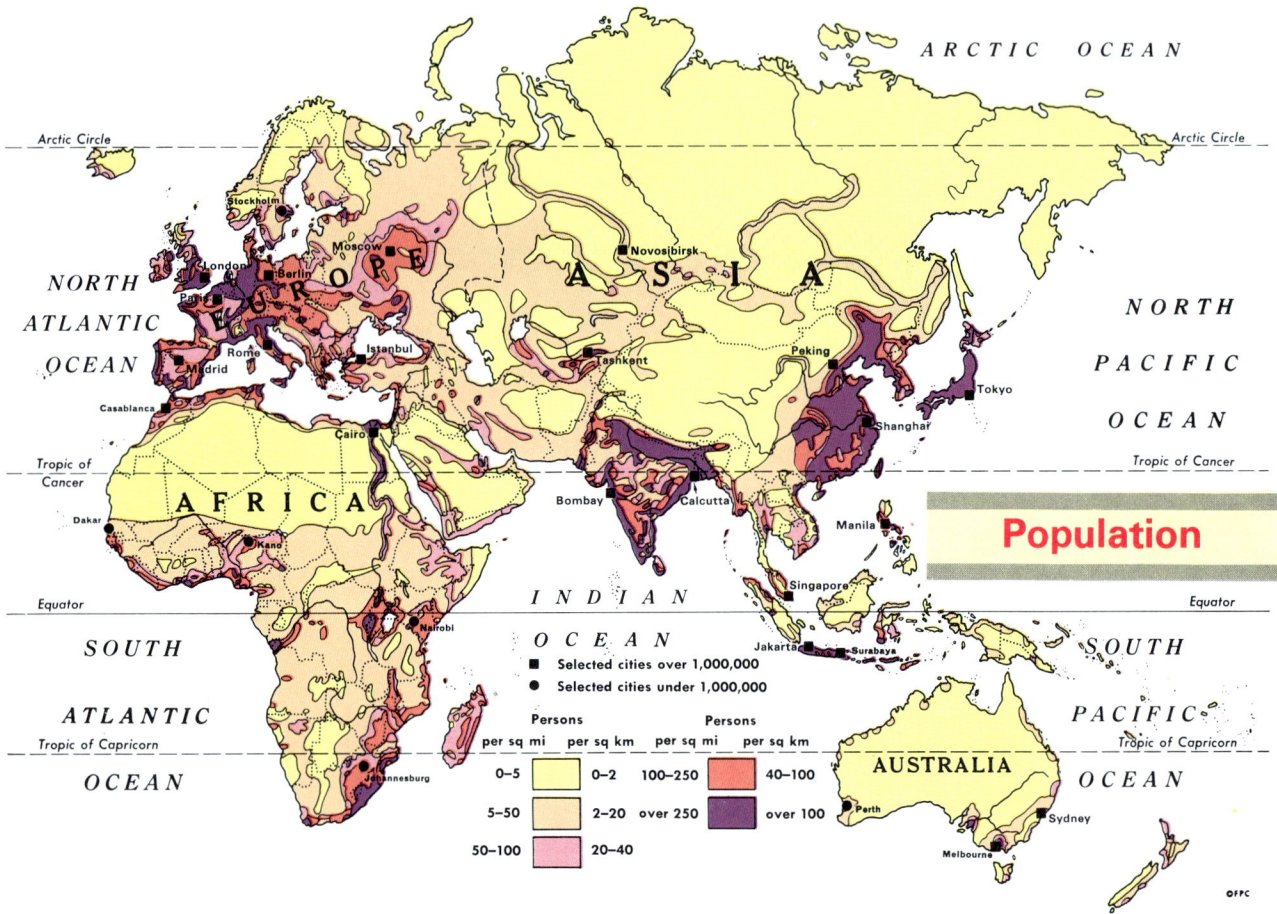

Population

Selected cities over 1,000,000

● Selected cities under 1,000,000

Persons per sq mi	Persons per sq km	Persons per sq mi	Persons per sq km
0–5	0–2	100–250	40–100
5–50	2–20	over 250	over 100
50–100	20–40		

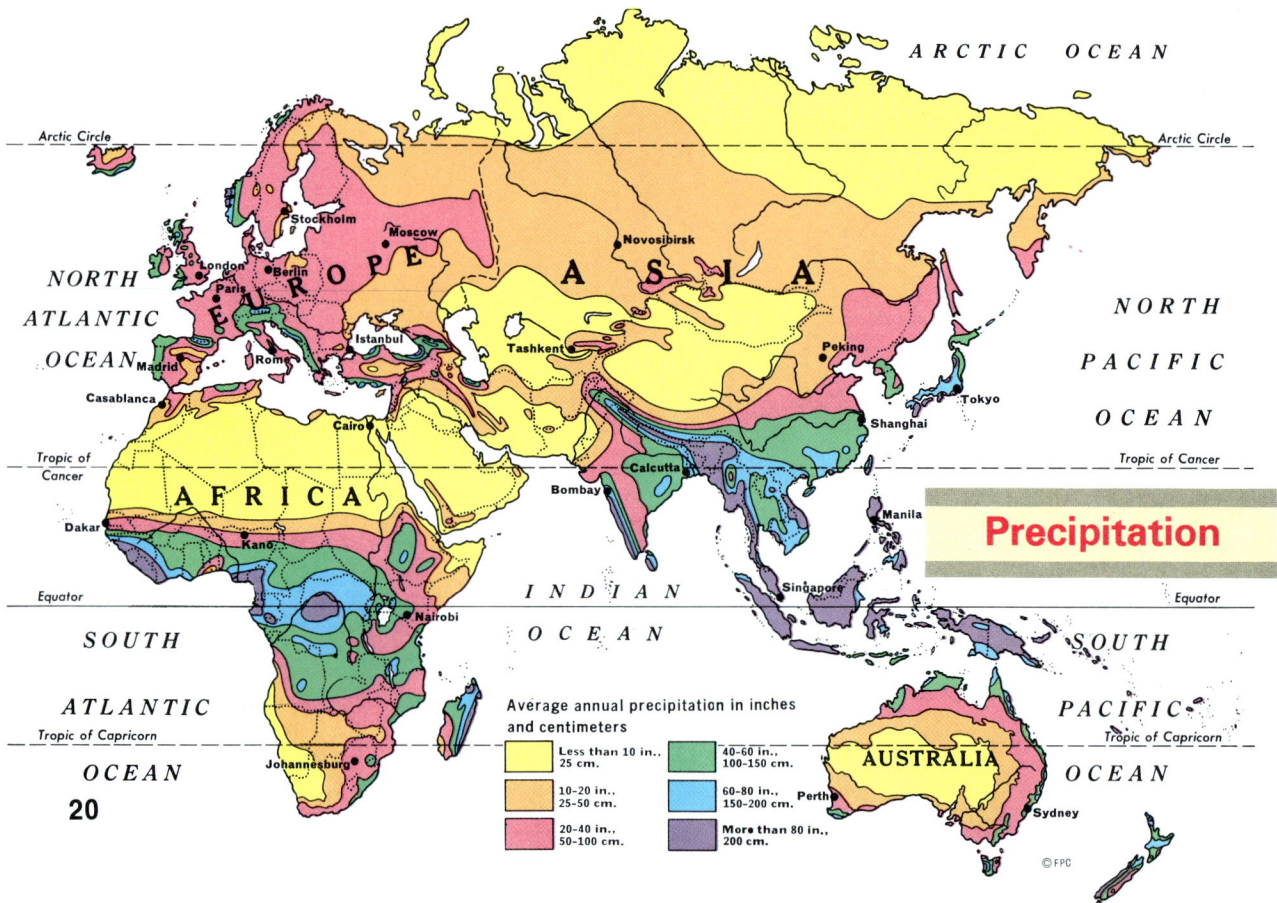

Precipitation

Average annual precipitation in inches and centimeters

- Less than 10 in., 25 cm.
- 10–20 in., 25–50 cm.
- 20–40 in., 50–100 cm.
- 40–60 in., 100–150 cm.
- 60–80 in., 150–200 cm.
- More than 80 in., 200 cm.

20

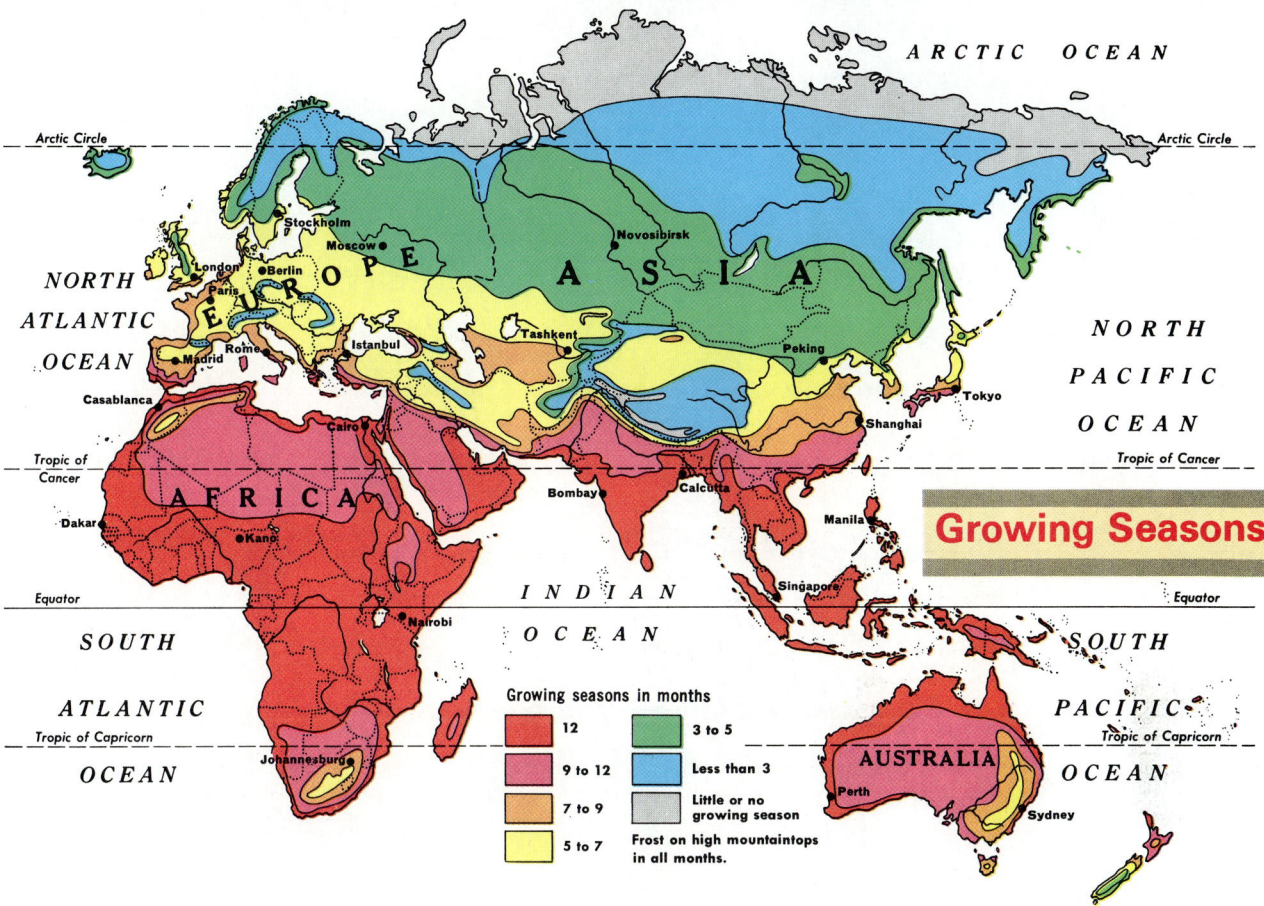

Growing Seasons

Growing seasons in months

- 12
- 9 to 12
- 7 to 9
- 5 to 7
- 3 to 5
- Less than 3
- Little or no growing season

Frost on high mountaintops in all months.

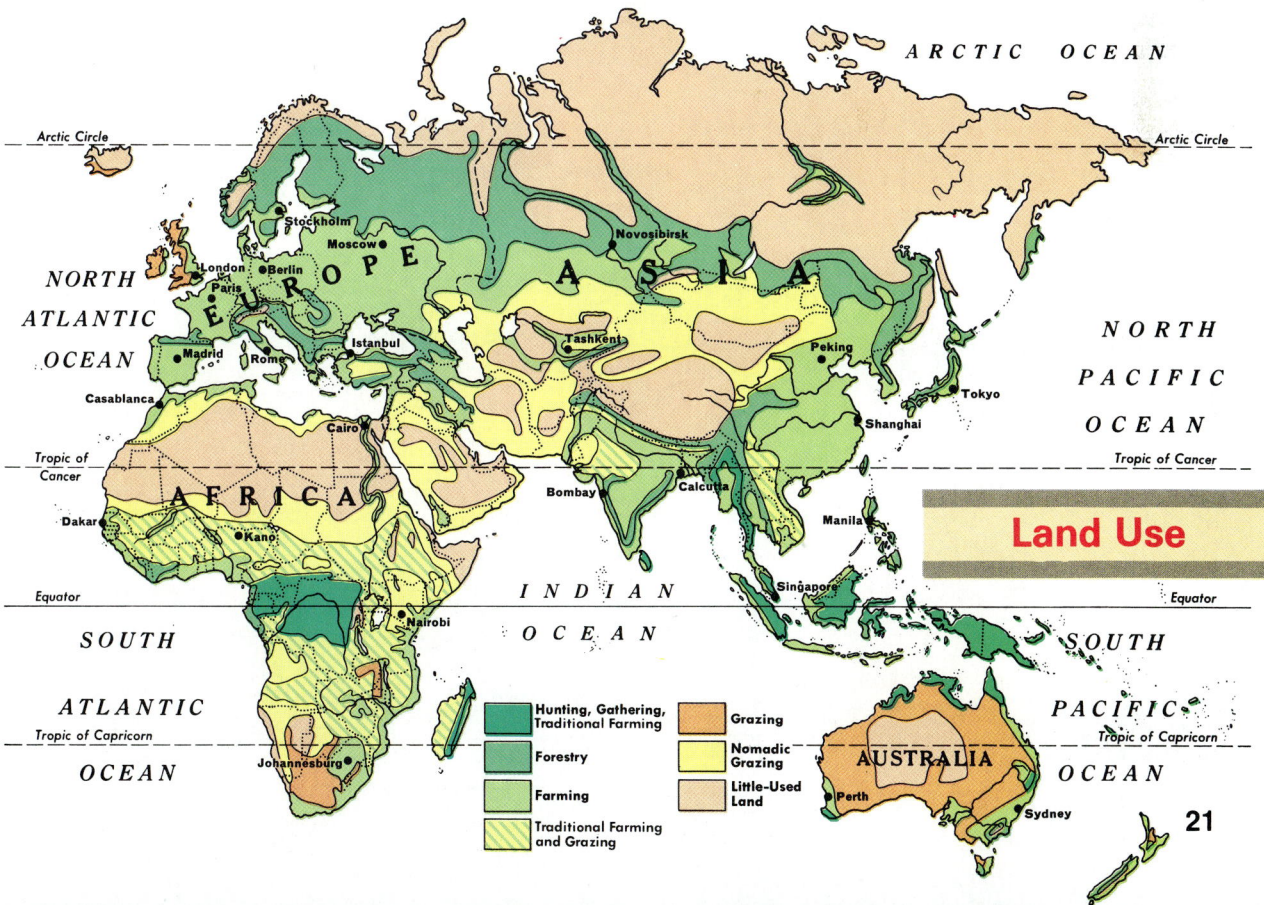

Land Use

- Hunting, Gathering, Traditional Farming
- Forestry
- Farming
- Traditional Farming and Grazing
- Grazing
- Nomadic Grazing
- Little-Used Land

21

Africa
Political Map
⊛ National Capitals
★ Other Capitals   • Other Cities

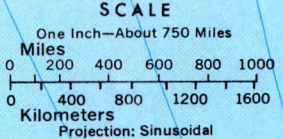

SCALE
One Inch—About 750 Miles
Miles
0 200 400 600 800 1000
Kilometers
0 400 800 1200 1600
Projection: Sinusoidal

22

ATLANTIC OCEAN

Mediterranean Sea

Black Sea   Caspian Sea   Aral Sea

EUROPE   ASIA

PORTUGAL SPAIN FRANCE ITALY GREECE TURKEY IRAN

MOROCCO ALGERIA TUNISIA LIBYA EGYPT

MAURITANIA MALI NIGER CHAD SUDAN

SENEGAL GAMBIA GUINEA-BISSAU GUINEA SIERRA LEONE LIBERIA IVORY COAST GHANA UPPER VOLTA TOGO BENIN NIGERIA CAMEROON CENTRAL AFRICAN REPUBLIC ETHIOPIA SOMALIA DJIBOUTI

EQUAT. GUINEA SAO TOME & PRINCIPE GABON CONGO ZAIRE UGANDA KENYA RWANDA BURUNDI TANZANIA

ANGOLA ZAMBIA MALAWI MOZAMBIQUE

NAMIBIA (SOUTH-WEST AFRICA) BOTSWANA ZIMBABWE SWAZILAND LESOTHO SOUTH AFRICA

MADAGASCAR

SAUDI ARABIA OMAN NORTH YEMEN SOUTH YEMEN UNITED ARAB EMIRATES BAHRAIN QATAR KUWAIT IRAQ SYRIA CYPRUS ISR JORDAN LEB

INDIAN OCEAN SEYCHELLES COMOROS

Tropic of Cancer   Tropic of Capricorn   Equator

© Follett Publishing Company

Prince Edward Is. (S. Afr.)   Crozet Is. (Fr.)   Kerguelen Is. (Fr.)

Africa
Graphic-Relief Map

⊛ National Capitals
★ Other Capitals • Other Cities

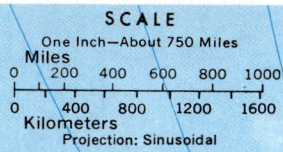

SCALE
One Inch—About 750 Miles
Miles
0 200 400 600 800 1000
Kilometers
0 400 800 1200 1600
Projection: Sinusoidal

HEIGHT OF LAND

OVER 13,000 FEET
6,600 TO 13,000
3,300 TO 6,600
1,650 TO 3,300
650 TO 1,650
0 TO 650 FEET
BELOW SEA LEVEL

DEPTH OF WATER

0 TO 600 FEET
BELOW 600 FEET

23

© Follett Publishing Company

© Follett Publishing Company

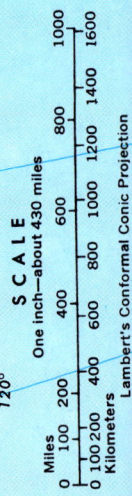

PACIFIC OCEAN

SOLOMON ISLANDS

Honiara ⊛
Guadalcanal

New Britain

PAPUA NEW GUINEA
NEW GUINEA

Port Moresby ⊛

VANUATU
Port-Vila ⊛

Loyalty Islands (Fr.)

New Caledonia (Fr.)

Norfolk Island (Austl.)

Tropic of Capricorn

NEW ZEALAND

North Island
North Cape
Auckland ●
Hamilton ●
Cook ⊛ Strait
Wellington
Christchurch ●
Mt. Cook (12,349)
Southern Alps
South Island
Dunedin ●
Invercargill ●

Tasman Sea

Coral Sea

Great Barrier Reef

Cairns ●
Townsville ●
Cloncurry ●

QUEENSLAND

Rockhampton ●
Bundaberg ●
Brisbane ★
Toowoomba ●
Ipswich ●

Great Dividing Range

Great Artesian Basin

Newcastle ●
Sydney ★
Wollongong ●
Canberra ⊛
Mt. Kosciusko (7,316)

NEW SOUTH WALES

Wagga Wagga ●
Broken Hill ●
Darling R.
Murray R.

Melbourne ★
Geelong ●
Ballarat ●

VICTORIA

Bass Strait

Launceston ●
Hobart ★

TASMANIA

Port Pirie ●
Adelaide ★
Port York Peninsula
Cape York
Torres Strait

Gulf of Carpentaria

Arnhem Land

Darwin ★

Arafura Sea

Banda Sea
Flores Sea
Savu Sea
Timor Sea

INDONESIA

Celebes
Sumbawa
Flores
Sumba
Timor
Bali
Lombok

24

NORTHERN TERRITORY

Alice Springs ●
Macdonnell Ranges

SOUTH AUSTRALIA

Lake Eyre (39 ft. below sea level)

WESTERN AUSTRALIA

Great Sandy Desert
Gibson Desert
Great Victoria Desert

Laverton ●
Meekatharra ●
Kalgoorlie ●

Great Australian Bight

AUSTRALIA

Hamersley Range

Broome ●
Carnarvon ●
North West Cape
Geraldton ●
Perth ★
Fremantle ●
Albany ●

INDIAN OCEAN

Australia and New Zealand

Graphic-Relief Map

⊛ National Capitals   ★ Other Capitals
● Other Cities

SCALE
One inch—about 430 miles

Miles 0 100 200 400 600 800 1000
Kilometers 0 100 200 400 600 800 1000 1200 1400 1600

Lambert's Conformal Conic Projection

HEIGHT OF LAND
OVER 13,000 FEET
6,600 TO 13,000
3,300 TO 6,600
1,650 TO 3,300
650 TO 1,650
0 TO 650 FEET
BELOW SEA LEVEL

DEPTH OF WATER
0 TO 600 FEET
BELOW 600 FEET

(longitude/latitude markings: 110°, 120°, 130°, 140°, 150°, 160°, 170°, 180°, 10°, 20°, 30°, 40°)

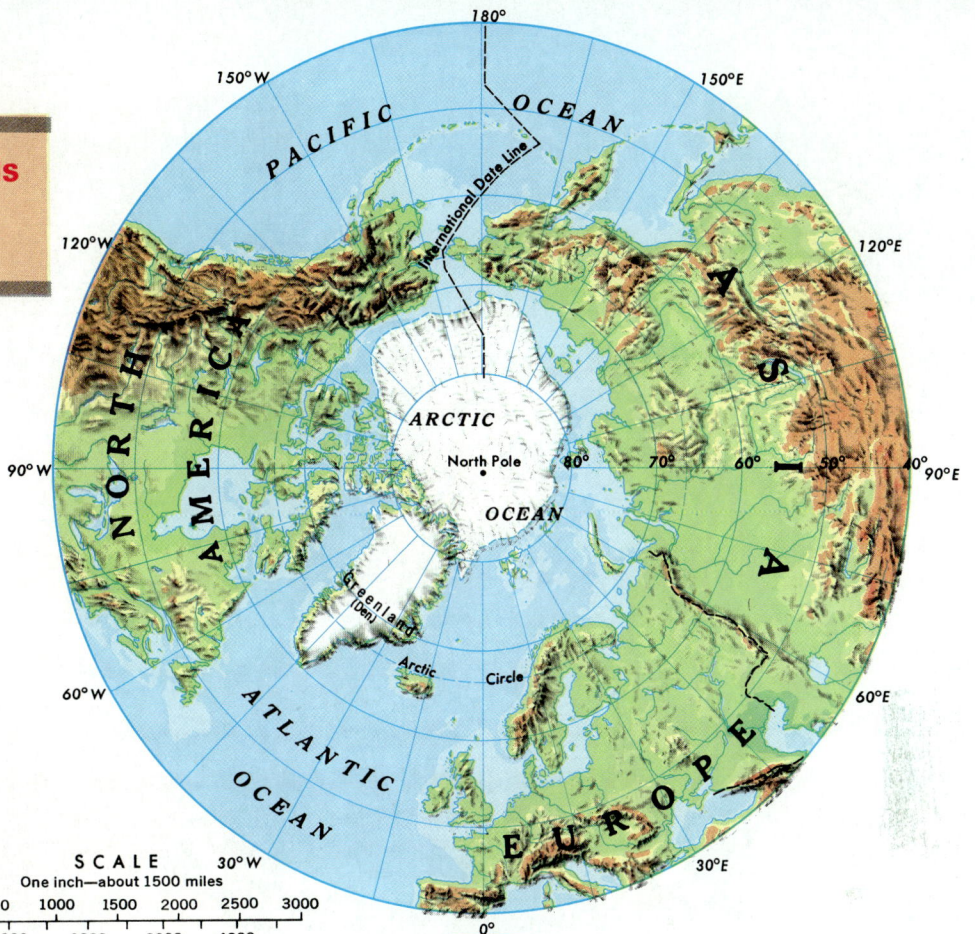

PACIFIC OCEAN

150°W — 180° — 150°E

International Date Line

120°W

NORTH AMERICA

ASIA

ARCTIC

90°W — North Pole — 80° — 70° — 60° — 90°E

OCEAN

50°

40°

Greenland (Den.)

120°E

Arctic Circle

60°W — 60°E

ATLANTIC OCEAN

EUROPE

30°W — 30°E

0°

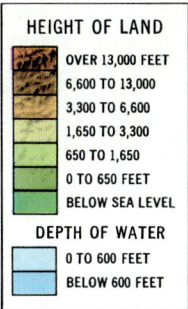

HEIGHT OF LAND

- OVER 13,000 FEET
- 6,600 TO 13,000
- 3,300 TO 6,600
- 1,650 TO 3,300
- 650 TO 1,650
- 0 TO 650 FEET
- BELOW SEA LEVEL

DEPTH OF WATER

- 0 TO 600 FEET
- BELOW 600 FEET

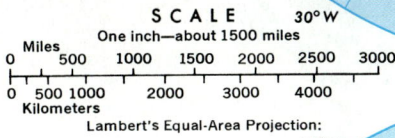

SCALE
One inch—about 1500 miles

Miles
0 500 1000 1500 2000 2500 3000

Kilometers
0 500 1000 2000 3000 4000

Lambert's Equal-Area Projection:

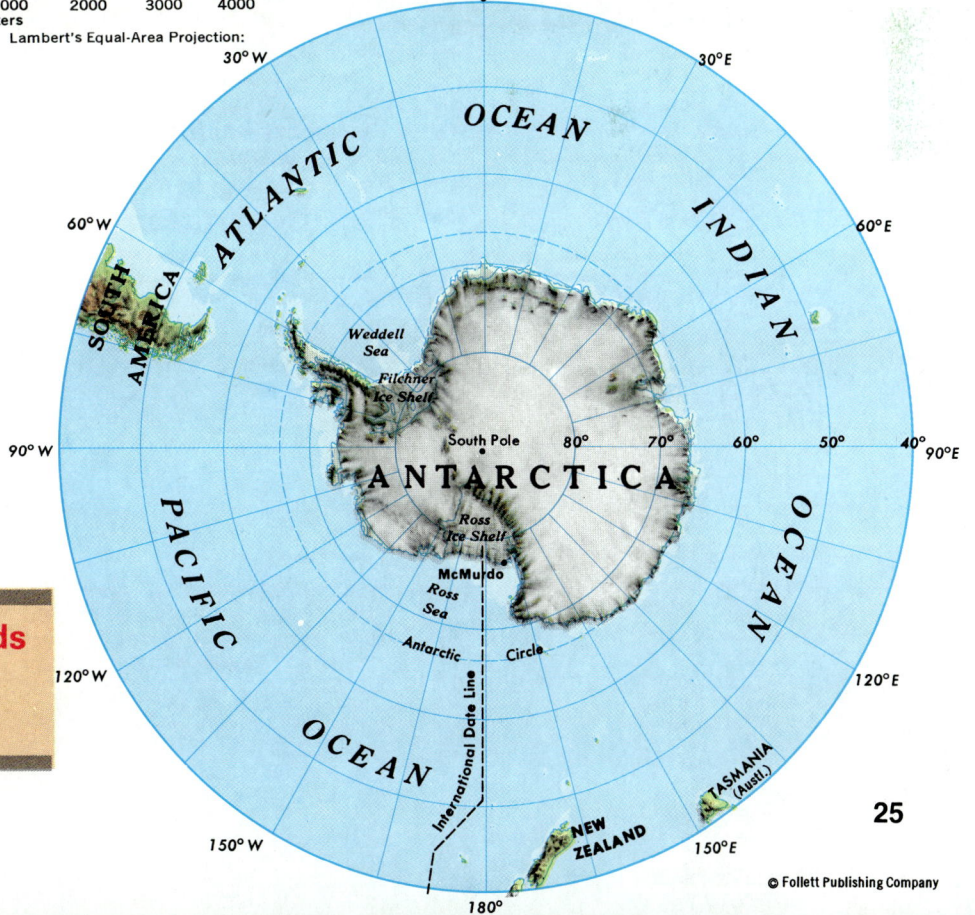

30°W — OCEAN — 30°E

ATLANTIC

INDIAN

60°W — 60°E

SOUTH AMERICA

Weddell Sea

Filchner Ice Shelf

90°W — South Pole — 80° — 70° — 60° — 50° — 40° — 90°E

ANTARCTICA

OCEAN

PACIFIC

Ross Ice Shelf

McMurdo Ross Sea

120°W — 120°E

Antarctic Circle

OCEAN

International Date Line

TASMANIA (Austl.)

150°W — 150°E

NEW ZEALAND

180°

25

© Follett Publishing Company

ASIA
U.S.S.R.

ARCTIC OCEAN

North Pole

GREENLAND
(Denmark)

ICELAND
Reykjavik

Denmark Strait
Arctic Circle

Bering Strait
Wrangel I.
St. Lawrence Island
Uelen
Bering Sea

Nome
Point Barrow
Barrow

Beaufort Sea

Banks I.
Ellesmere Island
Queen Elizabeth Islands

Baffin Bay

Thule

Cape Farewell
Godthaab
Julianehaab

ALASKA
Yukon River
Fairbanks
Dawson
Anchorage
Kodiak Island
Gulf of Alaska
Juneau
Whitehorse

Victoria Island

Baffin Island

Davis Strait

Frobisher Bay

Great Bear Lake
Yellowknife
Mackenzie R.
Great Slave Lake

Hudson Strait

Hudson Bay

Churchill

Schefferville

Newfoundland
Gander
St. John's

Queen Charlotte Islands
Peace R.
Lake Athabasca

CANADA

Vancouver Island
Vancouver
Victoria
Seattle
Fraser R.

Edmonton
Calgary
Regina
Winnipeg
Lake Winnipeg
Nelson R.

Gulf of St. Lawrence

Quebec
St. Lawrence R.
Montreal
Ottawa
Toronto
Hamilton
Buffalo

Halifax

Columbia R.
Great Falls
Portland
Snake R.
Boise

Duluth
L. Superior
L. Michigan
L. Huron
L. Ontario
L. Erie

Boston
Cape Cod

Minneapolis
Milwaukee
Detroit
Cleveland
New York

Great Salt Lake
Salt Lake City
Cheyenne

Omaha
Chicago
Pittsburgh
Baltimore
Philadelphia
Washington, D.C.

UNITED STATES

San Francisco

Denver
Kansas City
St. Louis
Cincinnati
Ohio R.

Norfolk
Cape Hatteras

San Luis Obispo
Las Vegas
Missouri R.
Wichita

Santa Fe
Oklahoma City
Memphis
Atlanta
Charleston

Los Angeles
San Diego
Phoenix
Salton Sea
Colorado R.
El Paso
Dallas
Birmingham
Savannah
Jacksonville

Ciudad Juárez
Rio Grande
San Antonio
Houston
New Orleans
Cape Canaveral

Bermuda I. (Br.)

Baja California

Monterrey
Gulf of Mexico
Miami
THE BAHAMAS
Tropic of Cancer

MEXICO
Tampico
Guadalajara
Mexico City
Veracruz
Puebla
Acapulco

Havana
CUBA
West Indies
HAITI
Port-au-Prince
DOMINICAN REP.
Hispaniola
Santo Domingo
Puerto Rico (U.S.)
San Juan

JAMAICA
Kingston

Yucatán Peninsula
BELIZE
Belmopan
GUATEMALA
Guatemala
San Salvador
EL SALVADOR
HONDURAS
Tegucigalpa
NICARAGUA
Managua
San José
COSTA RICA
PANAMA
Panama
Panama Canal

Caribbean Sea

Central America

SOUTH AMERICA

ATLANTIC OCEAN

PACIFIC OCEAN

North America
Political Map
⊕ National Capitals
★ Other Capitals
• Other Cities

SCALE
One inch-about 600 Miles

Miles
0 200 400 600 800
Kilometers
0 400 800 1200

26

© Follett Publishing Company

Projection: Lambert's Azimuthal

Equator

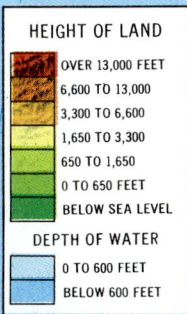

North America
Graphic-Relief Map

ASIA
U.S.S.R.

ARCTIC OCEAN

GREENLAND (Denmark)

ICELAND

Reykjavik

Denmark Strait

Godthaab

Cape Farewell

Baffin Bay

Baffin Island

Davis Strait

Queen Elizabeth Islands

Ellesmere Island

North Pole

Polar Ice Pack

Beaufort Sea

Point Barrow

Banks I.

Victoria Island

Bering Strait

Bering Sea

Wrangel I.

St. Lawrence Island

Uelen

Nome

Fairbanks

ALASKA

Yukon River

Mt. McKinley (20,300)

Alaska Range

Brooks Range

Anchorage

Whitehorse

Juneau

Gulf of Alaska

Kodiak Island

Alaska Pen.

Coast Mts.

Queen Charlotte Islands

Vancouver Island

Victoria

Seattle

Portland

Mt. Rainer (14,410)

Great Bear Lake

Mackenzie R.

Great Slave Lake

Lake Athabasca

Peace R.

CANADA

Edmonton

Calgary

Regina

Winnipeg

Lake Winnipeg

Nelson R.

Hudson Bay

Hudson Strait

Labrador Peninsula

Canadian Shield

Newfoundland

St. John's

Gulf of St. Lawrence

Halifax

Quebec

Montreal

Ottawa

Toronto

Hamilton

Buffalo

Boston

Cape Cod

New York

Philadelphia

Baltimore

Washington, D.C.

Norfolk

Cape Hatteras

Bermuda I. (Br.)

ATLANTIC OCEAN

Duluth

L. Superior

Minneapolis

Milwaukee

Detroit

Cleveland

Chicago

Pittsburgh

Cincinnati

L. Huron

L. Michigan

L. Erie

L. Ontario

St. Lawrence R.

Great Falls

Fraser R.

Columbia R.

Snake R.

Boise

UNITED STATES

Central Plains

Omaha

Missouri R.

Kansas City

St. Louis

Ohio R.

Cheyenne

Denver

Pikes Peak (14,110)

Great Salt Lake

Salt Lake City

Great Basin

Mt. Whitney (14,495)

Sierra Nevada

San Francisco

Las Vegas

Death Valley (282 ft. below sea level)

Salton Sea

Los Angeles

San Diego

Phoenix

Santa Fe

El Paso

Wichita

Oklahoma City

Memphis

Mississippi R.

Birmingham

Atlanta

Appalachian Mts.

Charleston

Savannah

Atlantic Coastal Plain

Dallas

Houston

San Antonio

Rio Grande

Ciudad Juárez

New Orleans

Gulf Coastal Plain

Cape Canaveral

Miami

THE BAHAMAS

Tropic of Cancer

Gulf of Mexico

Havana

CUBA

West Indies

HAITI

Port-au-Prince

DOMINICAN REP.

Hispaniola

Santo Domingo

Puerto Rico (U.S.)

San Juan

JAMAICA

Kingston

Caribbean Sea

Monterrey

Tampico

MEXICO

Guadalajara

Mexico City

Popocatepetl (17,887)

Puebla

Veracruz

Yucatán Peninsula

Acapulco

Central Plateau

W. Sierra Madre

E. Sierra Madre

Baja California

Gulf of California

PACIFIC OCEAN

Belmopan

BELIZE

GUATEMALA

Guatemala

San Salvador

EL SALVADOR

HONDURAS

Tegucigalpa

NICARAGUA

Managua

COSTA RICA

San José

PANAMA

Panama

Panama Canal

Central America

SOUTH AMERICA

Equator

HEIGHT OF LAND

OVER 13,000 FEET
6,600 TO 13,000
3,300 TO 6,600
1,650 TO 3,300
650 TO 1,650
0 TO 650 FEET
BELOW SEA LEVEL

DEPTH OF WATER

0 TO 600 FEET
BELOW 600 FEET

⊕ National Capitals
★ Other Capitals
• Other Cities

SCALE
One inch—about 600 Miles

Miles
0   200   400   600   800
Kilometers
0   400   800   1200

27
© Follett Publishing Company

Projection: Lambert's Azimuthal

Caribbean Sea

Central America

Panama Canal

Barranquilla
Cartagena
Aruba (Neth.)
Curaçao (Neth.)
Bonaire (Neth.)
Lesser Antilles
La Guaira
Port of Spain
TRINIDAD AND TOBAGO

Maracaibo
Barquisimeto
L. Valencia
Caracas
Ciudad Bolivar
Ciudad Guayana
Bucaramanga
VENEZUELA
Georgetown
Paramaribo
GUYANA
SURINAM
Cayenne
FR. GUIANA

Medellín
Manizales
Pereira
Bogotá
Cali
COLOMBIA
Pasto

Quito
ECUADOR
Guayaquil

Río Negro
Amazon River
Manaus
Santarém
Belém
São Luís
Fortaleza
Teresina
Natal
João Pessoa
Recife
Maceió
Salvador

Iquitos
Marañón R.
Ucayali R.
PERU
Trujillo
Cerro de Pasco
Lima
Callao
Machu Picchu (Ruins)
Cuzco
Puno
L. Titicaca
Arequipa
Arica
Iquique

Madeira R.
Tapajós R.
Tocantins R.
São Francisco R.

BRAZIL

Cuiabá
Brasília

BOLIVIA
La Paz
Cochabamba
Oruro
L. Poopó
Sucre
Potosi

Antofagasta
La Serena

Paraguay R.
PARAGUAY
Concepción
Asunción
Pilcomayo R.

Belo Horizonte
Volta Redonda
Campinas
São Paulo
Niterói
Rio de Janeiro
Santos
Curitiba

Tucumán
Salado R.
Paraná R.
Córdoba
San Juan
Mendoza
Santa Fe
Paraná
Rosario
Uruguay R.
Pôrto Alegre
Rio Grande
URUGUAY
Montevideo
Buenos Aires
La Plata
Río de la Plata

Viña del Mar
Valparaíso
Santiago
Concepción

CHILE
ARGENTINA

Colorado R.
Negro R.
Bahía Blanca
Mar del Plata

Puerto Montt

Comodoro Rivadavia

Falkland Islands (Br.)

Strait of Magellan
Tierra del Fuego
Punta Arenas
Cape Horn

South Georgia (Br.)

ATLANTIC OCEAN

PACIFIC OCEAN

Equator
Tropic of Capricorn

South America
Political Map

⊛ National Capitals ★ Other Capitals
● Other Cities

SCALE
One inch-about 520 Miles

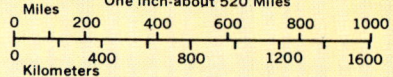

Miles
0   200   400   600   800   1000

Kilometers
0   400   800   1200   1600

Projection: Sinusoidal

© Follett Publishing Company

Caribbean Sea

Central
America

ATLANTIC

OCEAN

Lesser
Antilles

Aruba
(Neth.)
Curaçao
(Neth.)
Bonaire
(Neth.)

La
Guaira

Port of Spain
TRINIDAD AND TOBAGO

Barranquilla
Cartagena
Panama Canal

Maracaibo
Barquisimeto
Valencia
Caracas
Bucaramanga
Llanos
L.
Maracaibo
Orinoco R.
Ciudad
Bolívar

Georgetown
Paramaribo
Cayenne

Medellín
Manizales

Bogotá

VENEZUELA

Angel
Falls

GUYANA

SURINAM

FR.
GUIANA

Cali
COLOMBIA

Guiana
Highlands

Pasto

Río Negro

Equator

ECUADOR
Quito
Chimborazo
(20,561)
Cotopaxi
(19,347)
Guayaquil

Iquitos

Amazon R.
Manaus
Santarém
Basin
Tapajós R.

Belém
São Luís

Fortaleza
Teresina
Natal

Amazon
River

PERU
Trujillo

Ucayali R.
Marañón R.

Madeira R.

João Pessoa
Recife
Maceió

Huascarán
(22,205)
Cerro de Pasco

BRAZIL

Tocantins R.

Callao
Lima
Machu Picchu
(Ruins)
Cuzco

BOLIVIA

Plateau of
Mato Grosso

São Francisco R.

Salvador

El Misti
(19,110)
Arequipa

Lake
Titicaca
La Paz
Illimani (21,201)
Cochabamba
Arica
Oruro
Sucre
Iquique
Lake
Poopó
Potosí

Brazilian

Brasília

Belo
Horizonte

Cuiabá

Highlands

Paraguay R.

Tropic of Capricorn

Antofagasta

Gran
Chaco

Pilcomayo R.

PARAGUAY

Concepción

São Paulo
Curitiba
Santos

Rio de Janeiro

Atacama Desert

Asunción

Iguaçu Falls

CHILE

Salado R.

Tucumán

Paraná R.

Uruguay R.

Pôrto Alegre

Andes Mts.

Córdoba
San Juan
Aconcagua (22,834)
Valparaíso
Mendoza
Uspallata Pass
(12,650)
Santiago

Santa
Fe
Rosario

URUGUAY
Rio Grande

Concepción

Montevideo
Buenos Aires
Río de la Plata
La Plata

Pampa

ARGENTINA

Mar del Plata

Colorado R.
Negro R.
Bahía Blanca

Puerto Montt
Tronador
(11,660)

Patagonia

Gulf of San
Matías
(131 ft. below
sea level)

Gulf of
San Jorge

South America
Graphic-Relief Map

⊛ National Capitals   ★ Other Capitals
● Other Cities

SCALE
One inch—about 520 Miles

Miles
0   200   400   600   800   1000
0   400   800   1200   1600
Kilometers

Falkland Islands
(Br.)

HEIGHT OF LAND

OVER 13,000 FEET
6,600 TO 13,000
3,300 TO 6,600
1,650 TO 3,300
650 TO 1,650
0 TO 650 FEET
BELOW SEA LEVEL

DEPTH OF WATER

0 TO 600 FEET
BELOW 600 FEET

Strait of Magellan
Tierra del
Fuego
Punta Arenas

Cape Horn

South Georgia
(Br.)

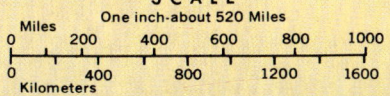

Projection: Sinusoidal

29

© Follett Publishing Company

PACIFIC OCEAN

POPULATION

LAND USE

Arctic Circle

Anchorage

Edmonton
Vancouver
Seattle
Winnipeg
Montreal
Toronto
Minneapolis
Detroit
New York
Philadelphia
Washington, D.C.
Chicago
St. Louis
San Francisco
Salt Lake City
Denver
Atlanta
Los Angeles
San Diego
Phoenix
Dallas
New Orleans
Miami
Houston

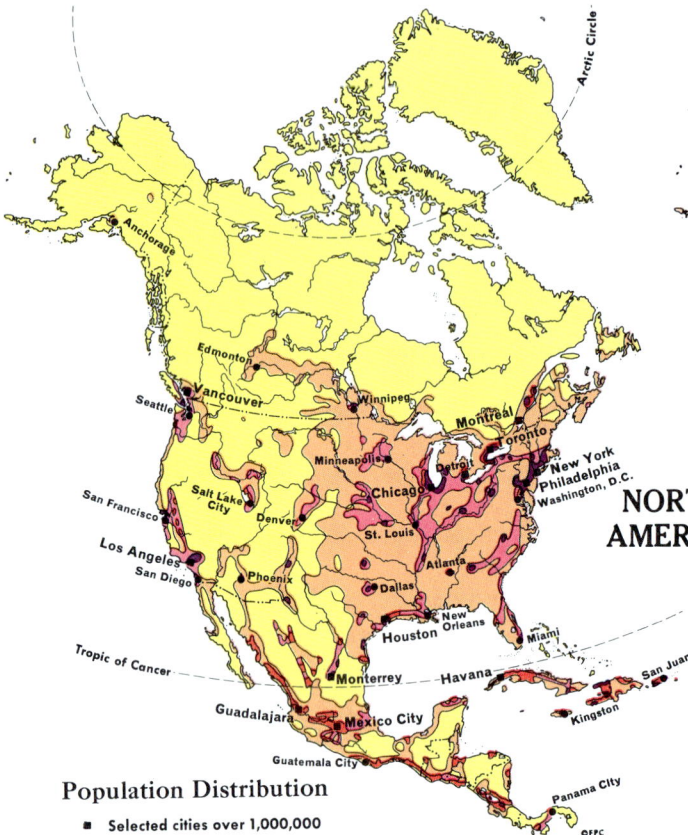

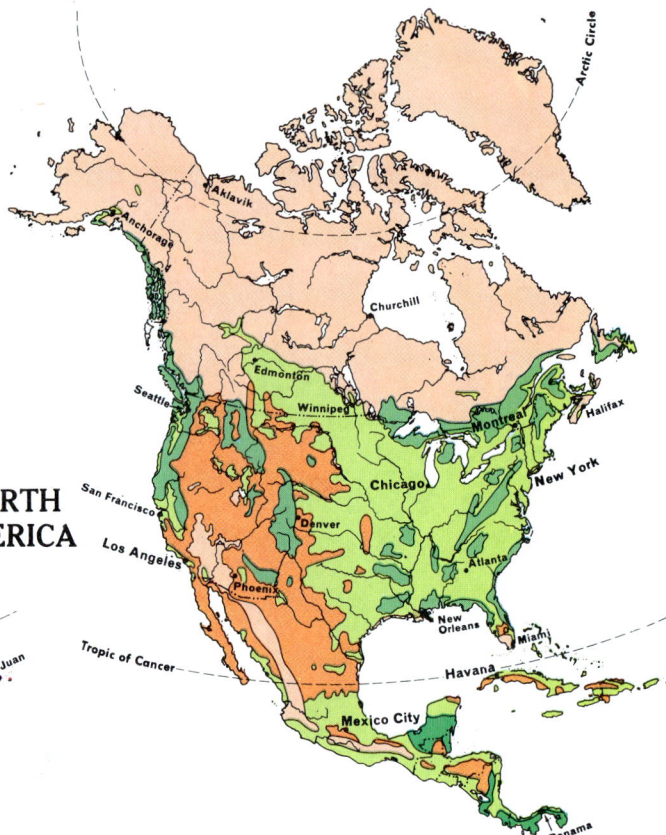

NORTH AMERICA

Tropic of Cancer

Monterrey
Havana
San Juan
Guadalajara
Kingston
Mexico City
Guatemala City
Panama City
©FPC

Population Distribution

- ■ Selected cities over 1,000,000
- ● Selected cities under 1,000,000

per sq mi	Persons per sq km	per sq mi	Persons per sq km
0–5	0–2	100–250	40–100
5–50	2–20	over 250	over 100
50–100	20–40		

Arctic Circle

Aklavik
Anchorage
Churchill
Seattle
Edmonton
Halifax
Winnipeg
Montreal
San Francisco
Chicago
New York
Denver
Los Angeles
Atlanta
Phoenix
New Orleans
Miami
Tropic of Cancer
Havana
Mexico City
Panama
©FPC

Land Use

■ Hunting, Gathering, Traditional Farming	▨ Traditional Farming and Grazing
■ Forestry	■ Grazing
■ Farming	■ Little-Used Land

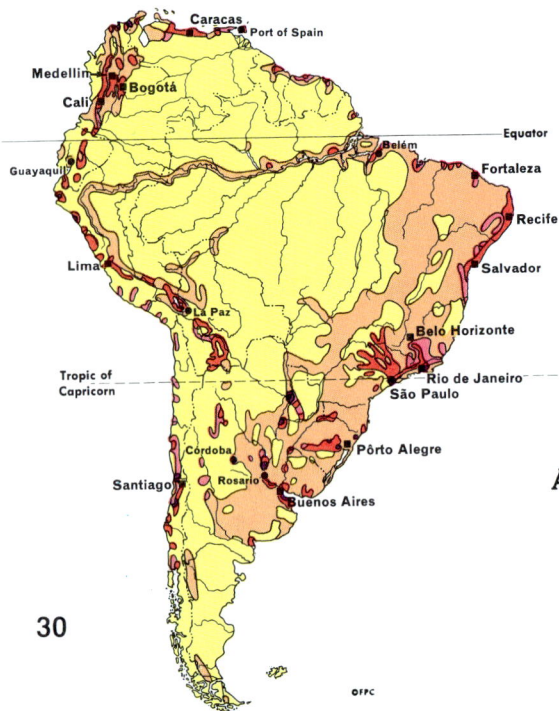

Caracas
Port of Spain
Medellin
Bogotá
Cali
Belém
Equator
Fortaleza
Guayaquil
Recife
Lima
Salvador
La Paz
Belo Horizonte
Tropic of Capricorn
Rio de Janeiro
São Paulo
Córdoba
Pôrto Alegre
Rosario
Santiago
Buenos Aires

Caracas
Paramaribo
Bogotá
Quito
Manaus
Belém
Equator
Recife
Lima
La Paz
Brasília
Asunción
Tropic of Capricorn
Rio de Janeiro
Córdoba
Santiago
Buenos Aires

SOUTH AMERICA

Punta Arenas
©FPC

PRECIPITATION

GROWING SEASONS

NORTH AMERICA

Arctic Circle

Aklavik · Anchorage · Churchill · Edmonton · Seattle · Winnipeg · Montreal · Halifax · San Francisco · Chicago · New York · Los Angeles · Denver · Phoenix · Atlanta · New Orleans · Miami

Tropic of Cancer

Havana · Mexico City · Panama

©FPC

Average Annual Precipitation
In Inches and Centimeters

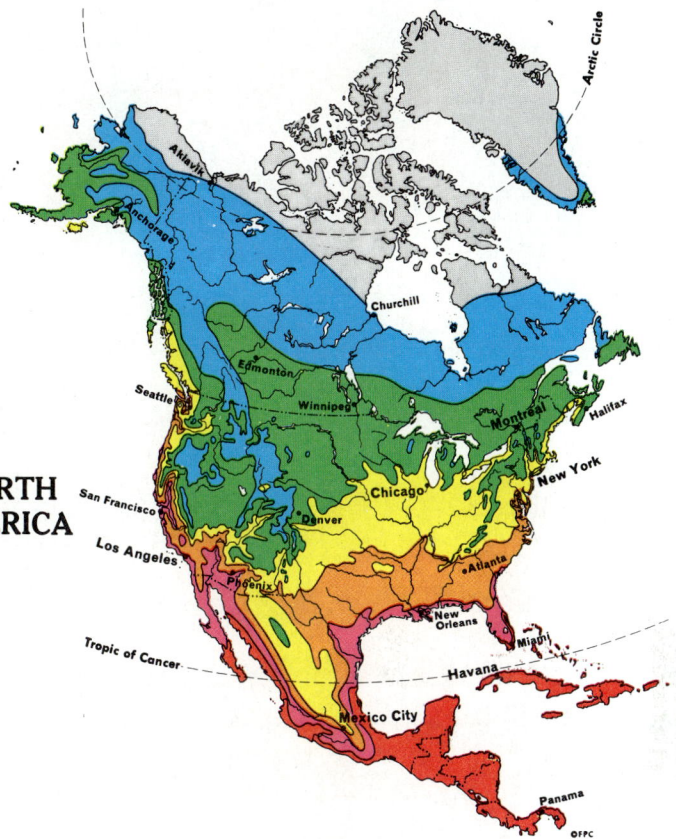

Less than 10 in., 25 cm.	40-60 in., 100-150 cm.
10-20 in., 25-50 cm.	60-80 in., 150-200 cm.
20-40 in., 50-100 cm.	More than 80 in., 200 cm.

Growing Seasons
In Months

12	5 to 7
9 to 12	3 to 5
7 to 9	Less than 3

Little or no growing season
Frost on high mountaintops in all months

SOUTH AMERICA

Caracas · Paramaribo · Bogotá · Quito · Manaus · Belém · Equator · Recife · Lima · La Paz · Brasília · Tropic of Capricorn · Rio de Janeiro · Asunción · Córdoba · Santiago · Buenos Aires · Punta Arenas

©FPC

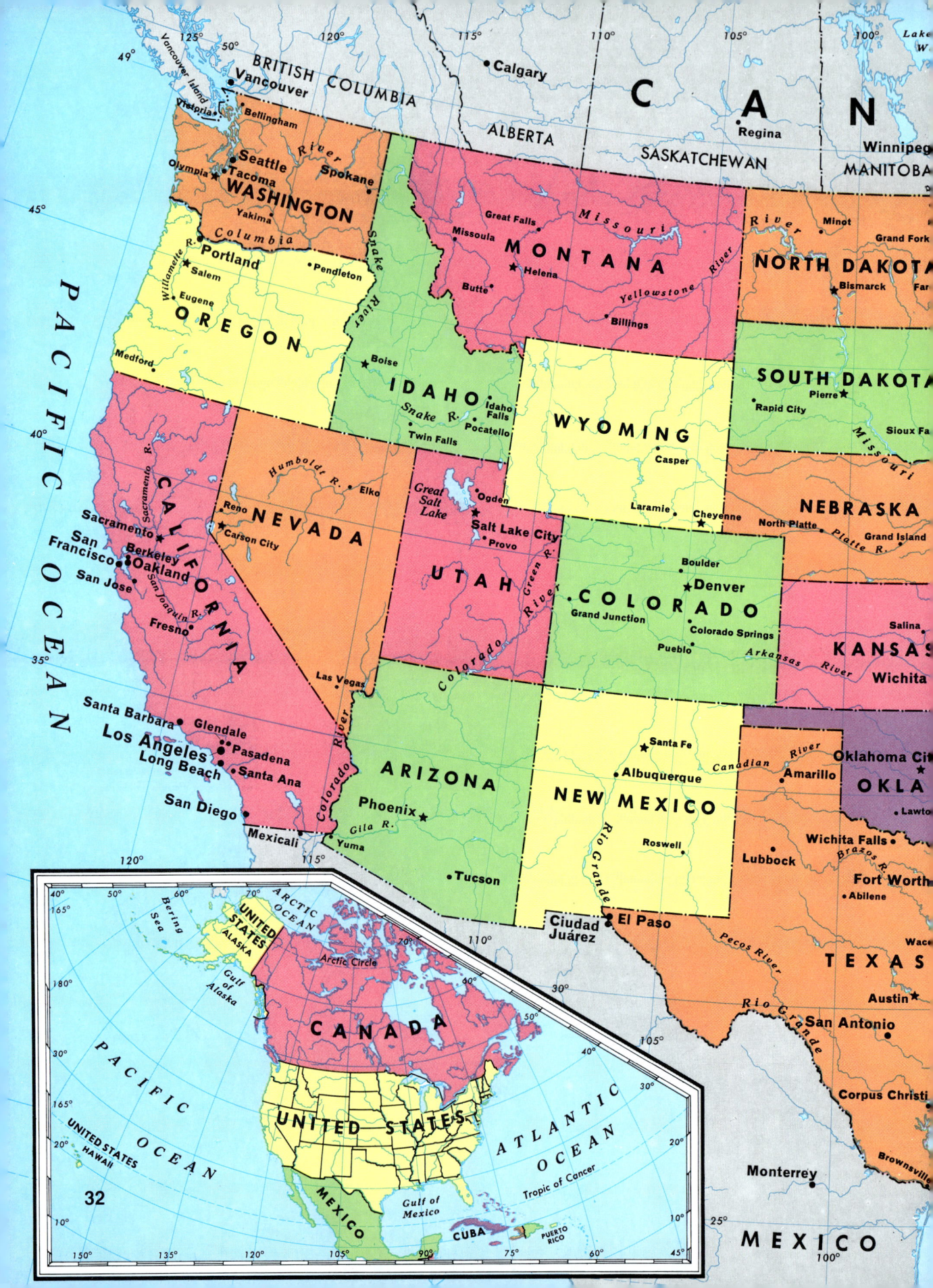

United States

Political Map

⊛ National Capitals   ★ State Capitals   ● Other Cities

SCALE

One inch—about 210 miles

Miles 0   100   200   300   400   500

Kilometers 0 100 200 300 400 500 600 700 800

ADA

ONTARIO

QUEBEC

NEW BRUNSWICK

Thunder Bay

Hibbing

Duluth

MINNESOTA

Minneapolis ★ St. Paul

Rochester

Sioux City

Waterloo

IOWA

Dubuque

Cedar Rapids

Davenport

Des Moines

Omaha

St. Joseph

Kansas City

Topeka

Jefferson City ★

MISSOURI

Springfield

Tulsa

Fort Smith

ARKANSAS

Little Rock ★

Pine Bluff

Greenville

MISSISSIPPI

Jackson ★

Shreveport

Meridian

LOUISIANA

Baton Rouge ★

Lake Charles

Beaumont

Houston

Galveston

New Orleans

Biloxi

Mobile

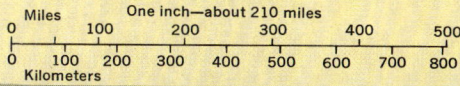

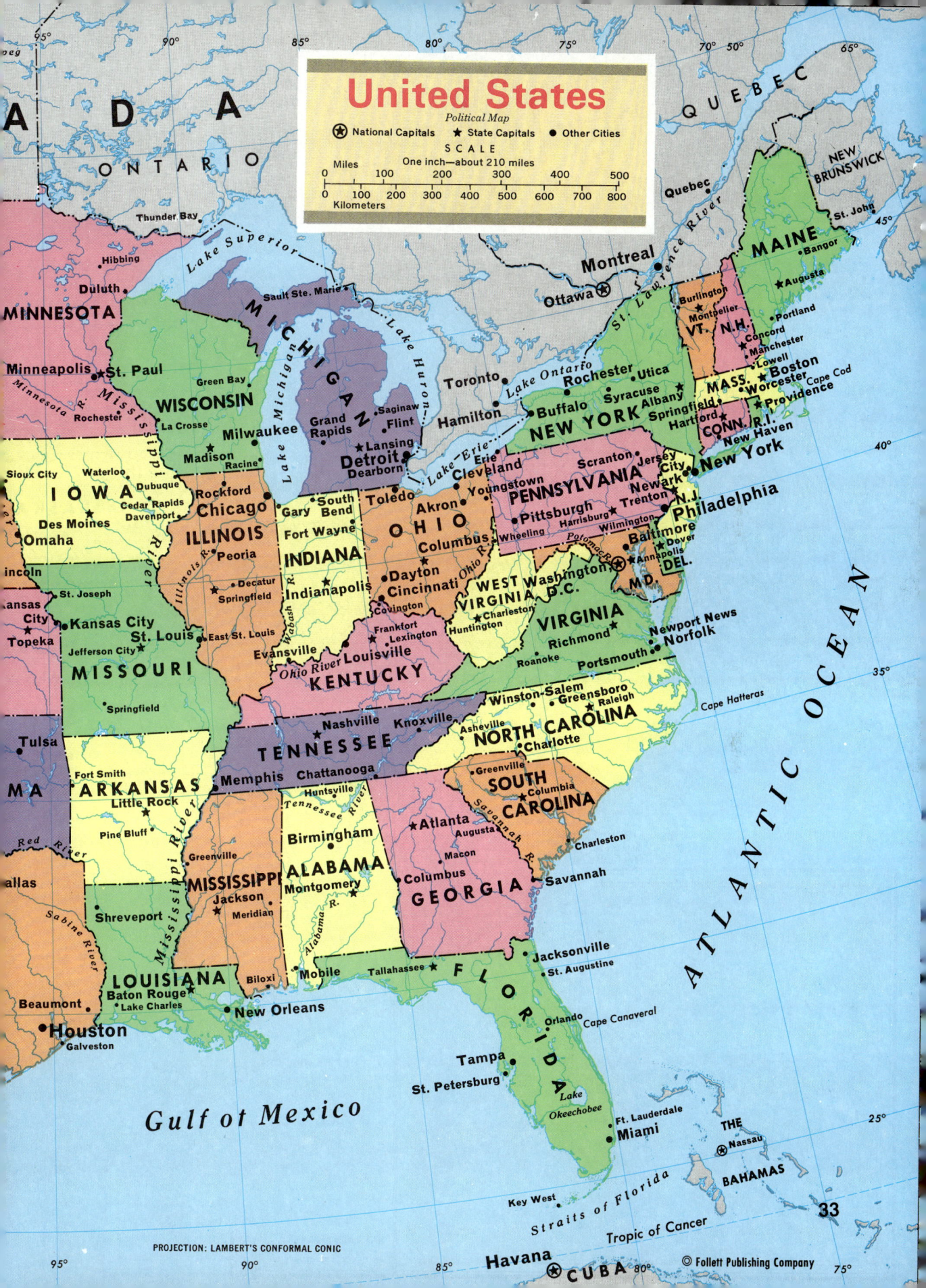

Gulf of Mexico

Lake Superior

Sault Ste. Marie

MICHIGAN

Lake Michigan

Lake Huron

WISCONSIN

Green Bay

Milwaukee

Madison ★

Racine

La Crosse

Rockford

Chicago

ILLINOIS

Peoria

Decatur

Springfield ★

Grand Rapids

Saginaw

Flint

Lansing ★

Detroit

Dearborn

South Bend

Gary

Fort Wayne

INDIANA

Indianapolis ★

Toledo

Akron

OHIO

Columbus ★

Dayton

Cincinnati

Covington

Lexington

Frankfort ★

Louisville

KENTUCKY

Evansville

Nashville ★

TENNESSEE

Knoxville

Chattanooga

Memphis

Huntsville

Birmingham

ALABAMA

Montgomery ★

Columbus

Tallahassee ★

FLORIDA

Jacksonville

St. Augustine

Orlando

Cape Canaveral

Lake Okeechobee

Tampa

St. Petersburg

Ft. Lauderdale

Miami

Key West

Straits of Florida

Tropic of Cancer

Havana

CUBA ⊛

THE BAHAMAS

Nassau ⊛

ATLANTIC OCEAN

Toronto

Hamilton

Lake Ontario

Lake Erie

Cleveland

Youngstown

PENNSYLVANIA

Pittsburgh

Wheeling

WEST VIRGINIA

Charleston

Huntington

Roanoke

VIRGINIA

Richmond ★

Harrisburg ★

Scranton

Jersey City

Newark

Trenton ★

N.J.

Philadelphia

Wilmington

Baltimore

Dover ★

DEL.

Annapolis ★

MD.

Washington D.C. ⊛

Portsmouth

Newport News

Norfolk

Cape Hatteras

Winston-Salem

Greensboro

Raleigh ★

Asheville

NORTH CAROLINA

Charlotte

Greenville

SOUTH CAROLINA

Columbia ★

Charleston

Atlanta ★

Augusta

Macon

Columbus

GEORGIA

Savannah

Buffalo

Rochester

Utica

Syracuse

NEW YORK

Albany ★

New York

Montreal

Ottawa ⊛

Quebec

St. Lawrence River

Burlington

Montpelier ★

VT.

N.H.

Concord ★

Manchester

Lowell

MASS.

Boston ★

Worcester

Springfield

Hartford ★

CONN.

R.I.

Providence ★

New Haven

Cape Cod

MAINE

Augusta ★

Bangor

Portland

St. John

Mississippi River

Minnesota R.

Illinois R.

Wabash R.

Ohio River

Tennessee River

Red River

Sabine River

Alabama R.

Savannah R.

PROJECTION: LAMBERT'S CONFORMAL CONIC

© Follett Publishing Company

33

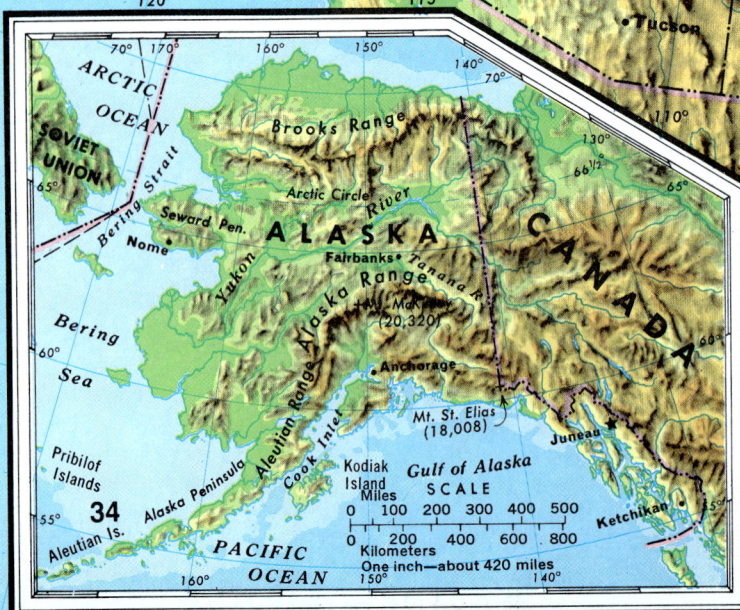
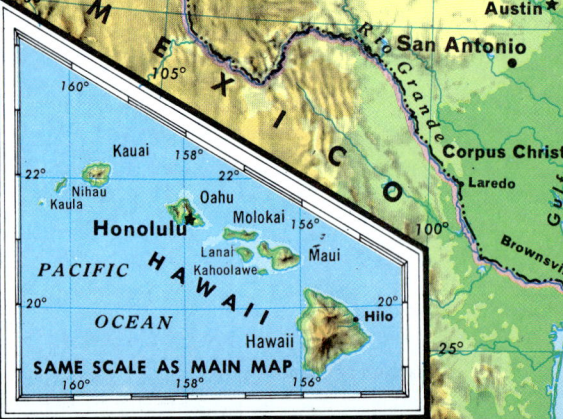

⊛ National Capitals   ★ State Capitals   • Other Cities

SCALE
Miles   One inch—about 210 miles
0   100   200   300   400   500
0   100   200   300   400   500   600   700   800
Kilometers

CANADA

ONTARIO

QUEBEC

NEW BRUNSWICK

Lake of the Woods

Thunder Bay

Hibbing

Duluth

MINNESOTA

Minneapolis   ★St. Paul

Minnesota R.

Sioux City

IOWA

Des Moines

Omaha

Lincoln

Kansas City

Topeka

Kansas City

Jefferson City   ★St. Louis

MISSOURI

Springfield

Ozark Plateau

Tulsa

Fort Smith

MA   ARKANSAS

Little Rock

Red River

allas

Sabine River

Shreveport

LOUISIANA

Baton Rouge

Houston

Galveston

Coastal Plain

Gulf of Mexico

Lake Superior

Soo Canals

Sault Ste. Marie

MICHIGAN

Green Bay

WISCONSIN

La Crosse

Milwaukee

Madison

Racine

Rockford

Chicago

Gary

ILLINOIS

Peoria

Springfield

Central Plains

Cedar Rapids

Davenport

Lake Michigan

Lake Huron

Flint

Grand Rapids

Lansing

Detroit

Fort Wayne

INDIANA

Indianapolis

Dayton

Cincinnati

Louisville

Frankfort

Lexington

KENTUCKY

Evansville

Paducah

Wabash R.

Ohio River

Nashville

TENNESSEE

Memphis

Tennessee River

Chattanooga

Birmingham

MISSISSIPPI

ALABAMA

Jackson

Meridian

Montgomery

Alabama R.

Mobile

Biloxi

New Orleans

Mississippi River

Toronto

Hamilton

Lake Ontario

Lake Erie

Erie

Toledo

Cleveland

Akron

OHIO

Columbus

Buffalo

Rochester

Syracuse

NEW YORK

PENNSYLVANIA

Pittsburgh

Wheeling

Harrisburg

WEST VIRGINIA

Charleston

VIRGINIA

Richmond

Roanoke

Knoxville

Appalachian Mountains

Blue Ridge Mountains

Piedmont

NORTH CAROLINA

Charlotte

Greenville

S. CAROLINA

Columbia

Atlanta

Augusta

GEORGIA

Columbus

Savannah River

Savannah

Charleston

Atlantic Coastal Plain

Cape Fear

Raleigh

Roanoke R.

Norfolk

Dismal Swamp

Chesapeake Bay

Cape Hatteras

Ottawa

Montreal

St. Lawrence River

Quebec

St. John

Bay of Fundy

MAINE

Augusta

Portland

VT.

N.H.

Concord

Montpelier

Burlington

Adirondack Mts.

White Mts.

Green Mts.

Mohawk R.

Albany

Catskill Mts.

MASS.

Boston

Springfield

Cape Cod

CONN.

Hartford

New Haven

R.I.

Providence

Long Island

New York

New York Bay

Newark

N.J.

Trenton

Hudson R.

Delaware R.

Susquehanna R.

Philadelphia

Baltimore

Dover

DEL.

Delaware Bay

Annapolis

Washington D.C.

MD.

Potomac R.

ATLANTIC OCEAN

FLORIDA

Okefenokee Swamp

Tallahassee

Jacksonville

Cape Canaveral

Tampa

St. Petersburg

Lake Okeechobee

The Everglades

Miami

Key West

Straits of Florida

Havana

CUBA

THE BAHAMAS

Nassau

Tropic of Cancer

ALASKA

HAWAII

UNITED STATES

ATLANTIC OCEAN

PACIFIC OCEAN

HEIGHT OF LAND
- OVER 13,000 FEET
- 6,600 TO 13,000
- 3,300 TO 6,600
- 1,650 TO 3,300
- 650 TO 1,650
- 0 TO 650 FEET
- BELOW SEA LEVEL

DEPTH OF WATER
- 0 TO 600 FEET
- BELOW 600 FEET

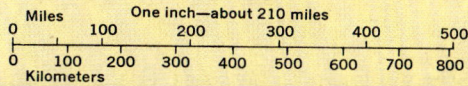

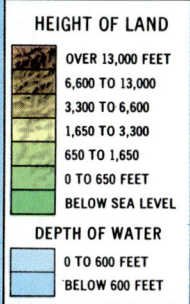

PROJECTION: LAMBERTS CONFORMAL CONIC

© Follett Publishing Company

POPULATION

Population Distribution

Persons per sq mi	per sq km
0–5	0–2
5–50	2–20
50–100	20–40
100–250	40–100
over 250	over 100

■ Cities over 1,000,000

● Selected cities under 1,000,000

Seattle
Portland
San Francisco
Los Angeles
San Diego
Salt Lake City
Phoenix
Denver
Minneapolis
Milwaukee
Chicago
Omaha
Kansas City
St. Louis
Oklahoma City
Dallas
San Antonio
Houston
New Orleans
Memphis
Birmingham
Atlanta
Cincinnati
Detroit
Cleveland
Pittsburgh
Buffalo
Boston
New York
Philadelphia
Baltimore
Washington, D.C.
Miami

Main Map
1 Inch–500 Miles
500 Miles–800 Kilometers

Alaska
1 Inch–1000 Miles
1000 Miles–1600 Kilometers
Anchorage

Hawaii
1 Inch–200 Miles
200 Miles–320 Kilometers
Honolulu

©FPC

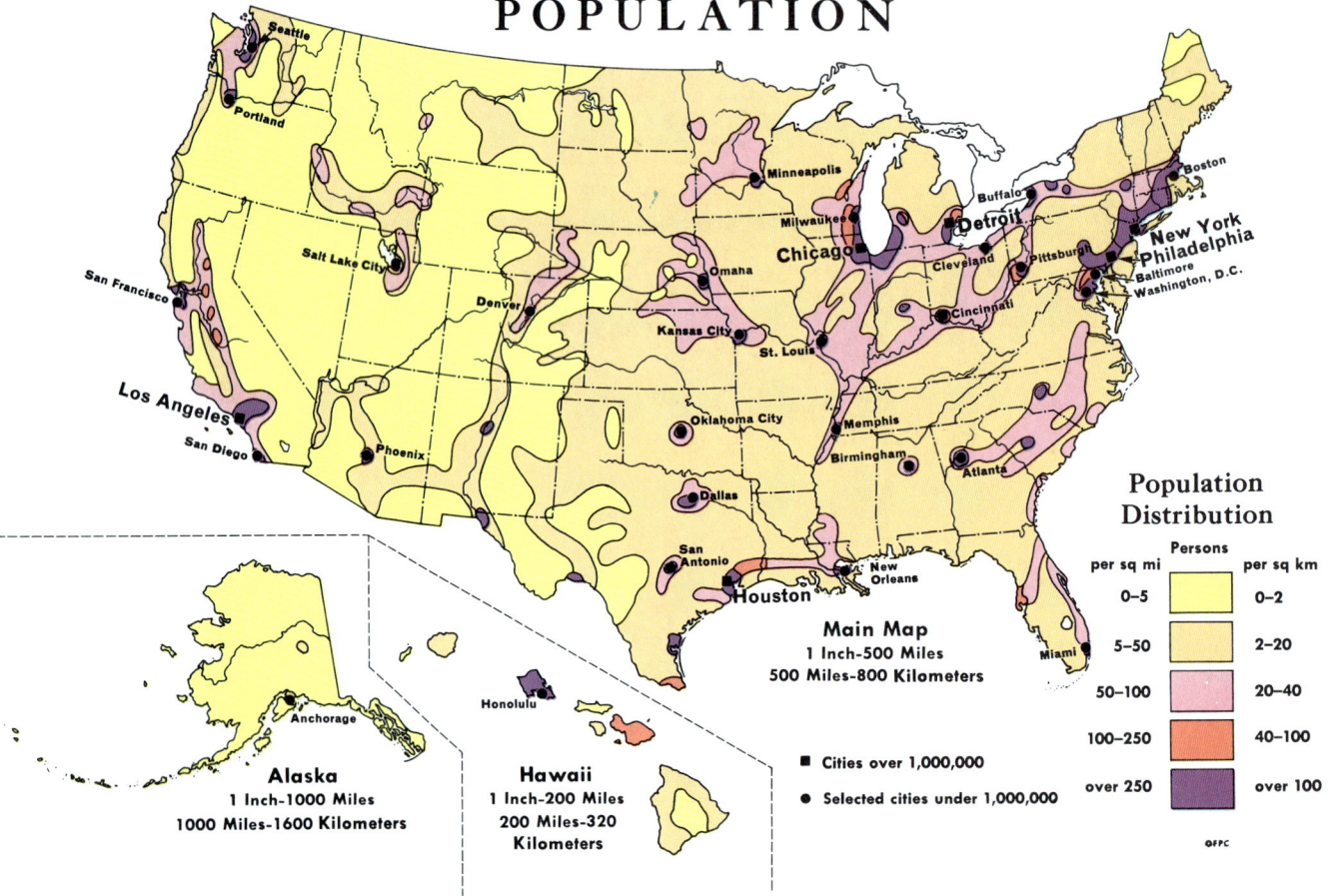

GROWING SEASONS

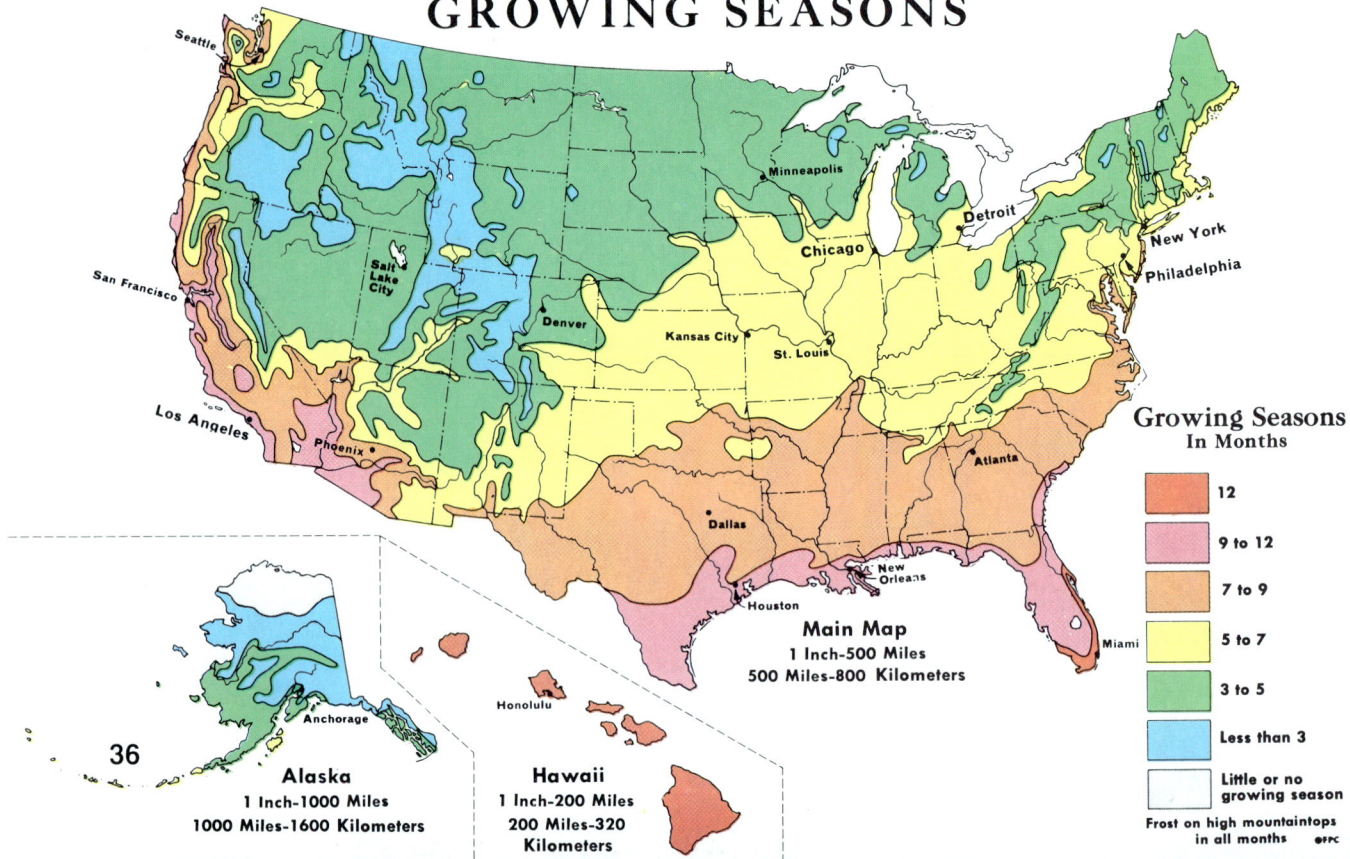

Growing Seasons
In Months

	12
	9 to 12
	7 to 9
	5 to 7
	3 to 5
	Less than 3
	Little or no growing season

Frost on high mountaintops in all months ©FPC

Seattle
San Francisco
Los Angeles
Phoenix
Salt Lake City
Denver
Minneapolis
Kansas City
St. Louis
Chicago
Detroit
Dallas
Atlanta
New Orleans
Houston
Miami
New York
Philadelphia

Main Map
1 Inch–500 Miles
500 Miles–800 Kilometers

36

Alaska
1 Inch–1000 Miles
1000 Miles–1600 Kilometers
Anchorage

Hawaii
1 Inch–200 Miles
200 Miles–320 Kilometers
Honolulu

LAND USE

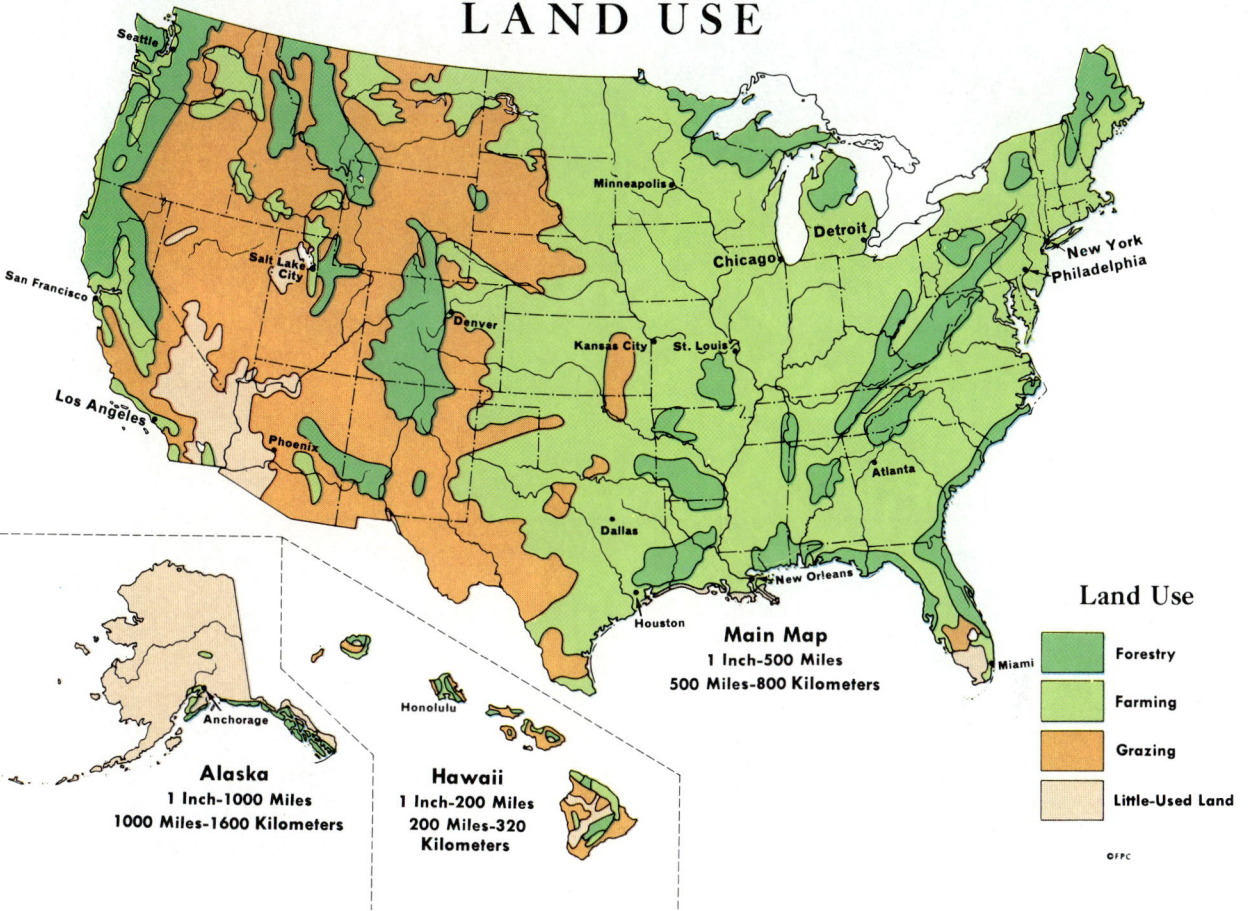

Seattle

Minneapolis

Detroit

Chicago

New York
Philadelphia

San Francisco

Salt Lake
City

Denver

Kansas City   St. Louis

Atlanta

Los Angeles

Phoenix

Dallas

New Orleans

Houston

Main Map
1 Inch–500 Miles
500 Miles–800 Kilometers

Miami

Honolulu

Anchorage

Alaska
1 Inch–1000 Miles
1000 Miles–1600 Kilometers

Hawaii
1 Inch–200 Miles
200 Miles–320
Kilometers

Land Use

🟩	Forestry
🟢	Farming
🟧	Grazing
⬜	Little-Used Land

©FPC

PRECIPITATION

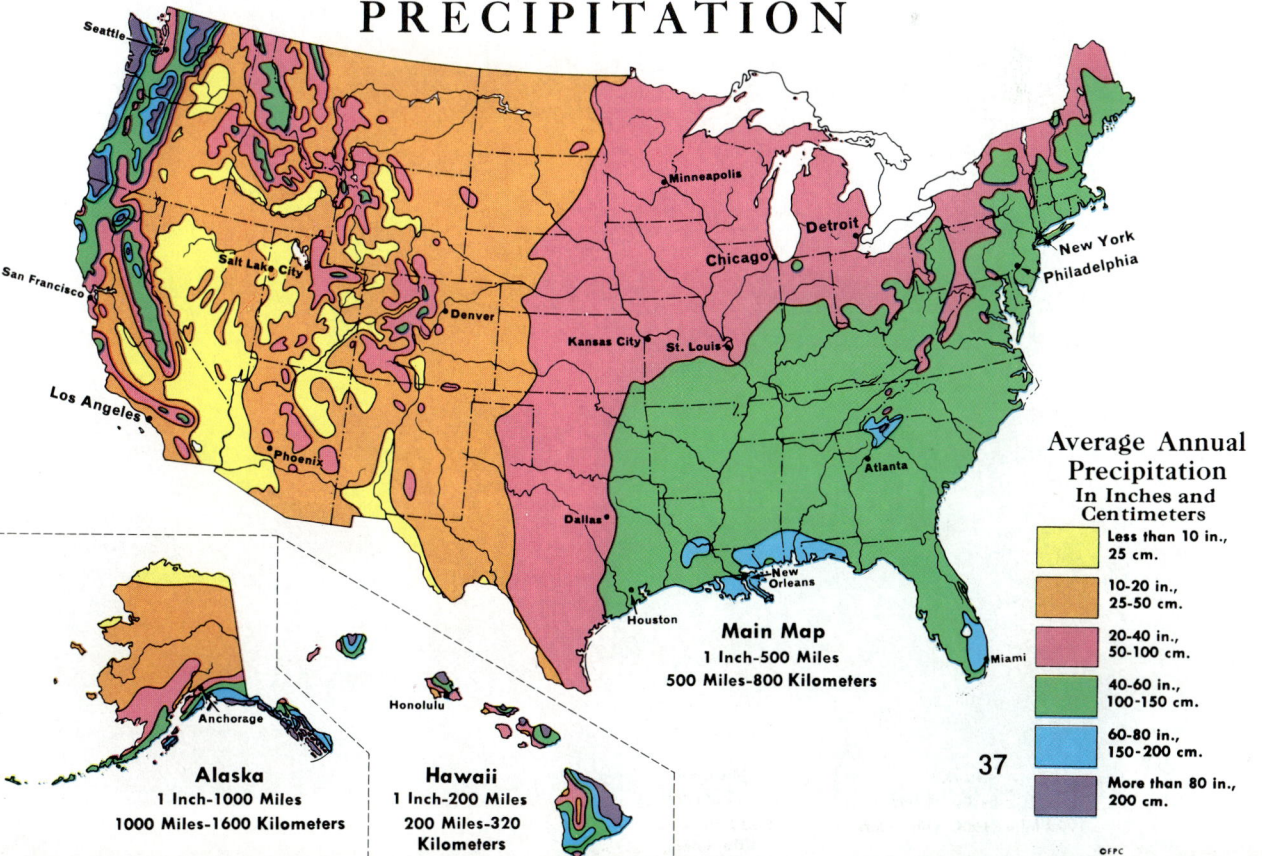

Seattle

Minneapolis

Detroit

Chicago

New York
Philadelphia

San Francisco

Salt Lake
City

Denver

Kansas City   St. Louis

Atlanta

Los Angeles

Phoenix

Dallas

New Orleans

Houston

Main Map
1 Inch–500 Miles
500 Miles–800 Kilometers

Miami

Honolulu

Anchorage

Alaska
1 Inch–1000 Miles
1000 Miles–1600 Kilometers

Hawaii
1 Inch–200 Miles
200 Miles–320
Kilometers

Average Annual Precipitation
In Inches and Centimeters

🟨	Less than 10 in., 25 cm.
🟧	10-20 in., 25-50 cm.
🟥	20-40 in., 50-100 cm.
🟩	40-60 in., 100-150 cm.
🟦	60-80 in., 150-200 cm.
🟪	More than 80 in., 200 cm.

37

©FPC

38

Part One

People and Their Global Environment

Part One of this book will introduce you to the realm of physical geography. It will explore the natural features of our earth and how they are distributed around the globe. It will help you understand how nature's never-ending processes help create global patterns of land surfaces, water distribution, air and weather, climate, soils, vegetation, and minerals. And it will show how the earth's patterns and features can affect people and their ways of life in all parts of the globe.

Part One will also show how people can influence and even change some features of their global home. Although the forces of nature are mighty enough to move mountains and fill seas, they are nonetheless delicately balanced. More and more, people are beginning to realize their power to upset this balance and threaten the very home that supports them. Physical geography, therefore, holds an important place in human studies. Through it, we can learn how to live in harmony with our global environment and preserve it for future generations.

Landsat satellite images—569 in all—were put together to make the map on the opposite page, showing the contiguous United States. Remote sensors in the satellite pick up reflected light from the earth's surfaces. The satellite sends this information to signal-receiving stations on earth, where the signals are changed to form images on film. The map on page 38 gives "false color" images. Green vegetation shows up in various shades of red. Clear, deep water is black, while water with sediments is bright blue. Colors for rocks and soils range from browns, yellows, and tans to shades of blue. Urban areas generally show up in blue-gray to bluish-black colors. White may be snow or cloud cover.

Unit 1 Geography—
Its Tools and Skills

Seeing the world from within—that is what these people are doing in a room shaped like a globe. Called a mapparium, this walk-in globe is in the Christian Science Publishing House in Boston, Massachusetts.

Lesson 1 Using Globes

Reading Focus

1. An understanding of the geographer's words is needed to understand the ideas in this lesson (and every other lesson). At the beginning of each lesson is a vocabulary list (see *Vocabulary Focus* below). The pronunciation and the definition for each term are given in the Glossary at the back of the book. Using the pronunciation key at the beginning of the Glossary, make sure you can pronounce each term correctly. Study the definition of each term. Ask questions if you do not understand any part of the definition. Become so familiar with each definition that you can put it into your own words.

2. As you read this lesson, look for answers to these key questions:
   a. How long have people been making and using globes?
   b. What is scale? Why is scale important on a globe?
   c. What is the global grid? What is the importance of latitude and longitude?
   d. What is a great circle? Are all lines on the global grid great circles? If not, which ones are?

Vocabulary Focus

sphere	*latitude*	*parallel*	*great circle*
scale	*longitude*	*meridian*	*hemisphere*
grid			

Globes have been useful tools of geographers since the time of the ancient Greeks. The first known globe was made around 150 B.C. By this time, the Greeks had discovered that the earth is a sphere. They had even figured out the circumference of, or distance around, the sphere. Their measurements were very close to the exact measurements of the earth's circumference.

History gives us only a general description of this first globe. It had four landmasses, or continents. The only landmass the Greeks really knew about was their own. The person who made the globe dreamed up the other three continents just to fill the empty space on the globe. He reasoned that everything in nature seems to have a balance. He figured that the earth's land and water bodies must also have a balance. He then separated the landmasses with a north-south ocean and an east-west ocean.

The oldest globe still preserved is in a museum in Germany. This wooden

First known globe

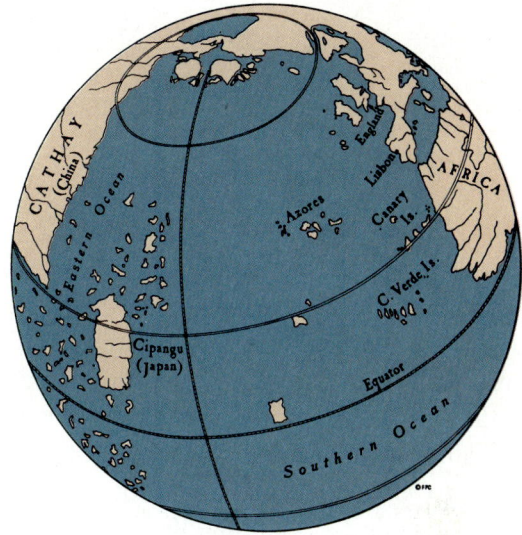
Oldest preserved globe

globe was made in 1492. It is 20 inches (51 cm) in diameter. The globe maker painted on the land and water bodies that he thought existed. What is interesting about this globe is that it shows the world as Columbus thought of it. It also shows that sometime between the time of the ancient Greeks and the time of Columbus, people had lost sight of the actual size of the earth.

The ancient Greeks had figured the circumference of the earth to be 24,000 miles (38,640 km). But at the time of Columbus, people thought it was only 14,000 miles (22,540 km). No wonder Columbus figured Japan would be only a little more than 3,000 miles (4,830 km) west of the Canary Islands.

After the 1500s, globe making became more and more exact as Europeans traveled and explored the world. Today, photographs from space, computers, and other instruments aid globe makers.

What Globes Show

Globes are useful because they show certain relationships about the earth's surface more accurately than maps. These relationships are scale and distance, size and shape of land and water bodies, and directions on the earth.

Scale and distance. The globe is a scale model of the earth. When things are drawn to scale, a smaller unit of measure stands for a larger unit of measure. One inch may stand for one foot. One inch may stand for one mile. One centimeter may stand for one meter. Or one meter may stand for one kilometer. Drawing or making things to scale produces a model that is an exact but smaller copy of the real thing.

In making a globe that is true to scale, globe makers start with the diameter of the sphere. The diameter is the distance through the center of the sphere. The

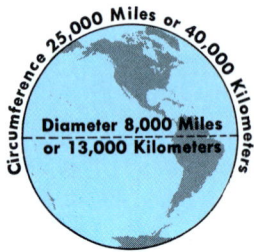

Circumference 25,000 Miles or 40,000 Kilometers
Diameter 8,000 Miles or 13,000 Kilometers

diameter of the earth is about 42,000,000 feet. A globe that has a diameter of one foot, then, is 42,000,000 times smaller than the actual size of the earth. On such a globe, the scale is 1:42,000,000. In other words, one foot on the globe stands for 42,000,000 feet on the earth. One inch on such a globe would stand for 42,000,000 inches on the earth. What would one centimeter on such a globe stand for?

To measure distances on a globe, notice the scale given on it. The scale may be stated in words. Or it may be shown by a line. On a globe with a 12-inch diameter, one inch stands for about 600 miles (960 km). On a 16-inch globe, one inch stands for about 500 miles (800 km). Check your classroom globe. What is its diameter? What is its scale?

A good way to measure distances on a globe is to use a piece of string. Place the piece of string between two points on the globe. Mark off the location of the two points on the string. Then measure the distance between the points on a ruler. On a 12-inch globe, suppose the distance between the two points is three inches. How many miles would separate the two points? How many kilometers?

Size and shape. Because it is true to scale, the globe accurately shows the size and shape of all land and water bodies in their true relationships to one another. When you look at Greenland on a globe, for example, its true shape is what you see. You can also accurately compare its size to any other land body.

Direction. Globes show the true compass direction from one point to any other point. North is always toward the North Pole. South is always toward the South Pole. When the globe is set so that the North Pole is at the top of the globe, east is to the right and west is to the left.

The directions *up* and *down* have their own meanings. *Up* is away from the center of the earth. *Down* is toward the center of the earth.

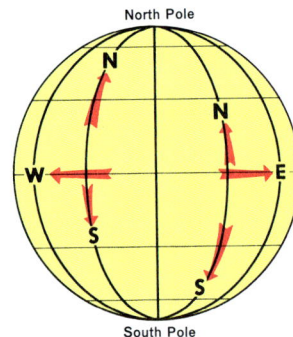

North Pole

N

N

W E

S

S

South Pole

The Global Grid

Geographers can find any place on a globe by locating it on a global grid. A grid is two sets of lines that cover the globe. One set of lines runs east and west. The other set of lines runs north and south. These lines are used to measure latitude and longitude. The unit for measuring latitude and longitude is the degree (° is the symbol for degree). A degree is further divided into minutes (′) and seconds (″).

Parallels of latitude. Parallels are the east-west lines of the global grid. Parallels on the global grid are also called lines of latitude. Parallels are always the same distance from one another. One degree of latitude is about 70 miles (112 km) from a degree of latitude north of it or south of it.

The starting point for measuring latitude is the equator. The equator is the east-west line that marks the midpoint between the North and South poles. Its latitude is 0°. Between the equator and the North Pole, there are 90° of latitude. There are also 90° of latitude between the equator and the South Pole. (Most globes do not show every line of latitude. They may show every ten, twenty, or even thirty degrees of latitude.) When giving the location of a place, geographers always state whether its latitude is north or south of the equator.

Meridians of longitude. Meridians are the north-south lines of the global grid. Each meridian, or line of longitude, meets its opposite meridian at the North and South poles. Together, the two meridians form a great circle. A great circle divides the earth into two equal parts. Unlike parallels, meridians are not the same distance apart everywhere on the globe.

Longitude has both a starting point and an ending point. The starting point for measuring longitude is the prime meridian. Its longitude is 0°. Directly opposite the prime meridian, on the other side of the globe, is the line that marks 180° of longitude. It is the ending point for measuring longitude. The area of the world to the east of the prime meridian up to 180° is east longitude. The area of the world west of the prime meridian up to 180° is west longitude. (Most globes do not show every line of longitude. They may show every fifteen, twenty, or thirty degrees of longitude.) When giving the location of a place, geographers always state whether the longitude is east or west of the prime meridian.

Using latitude and longitude. All places on the earth's surface can be located on a line of latitude. The latitude gives two facts about a place—its distance from the equator and its direction from the equator. Likewise, all places on the earth can be located on a line of longitude. The longitude gives two facts about a place—its distance and its direction from the prime meridian. Giving both the latitude and the longitude of a place marks its exact location.

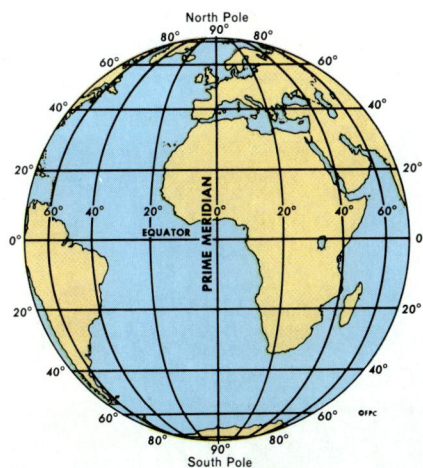

What two facts about a place does a line of latitude give? A line of longitude?

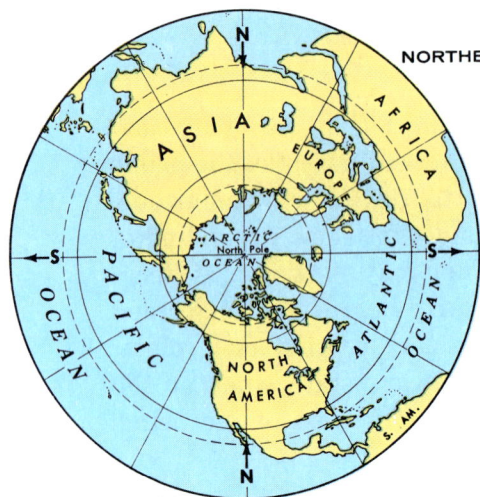

NORTHERN HEMISPHERE

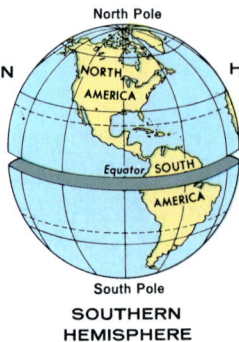

NORTHERN HEMISPHERE

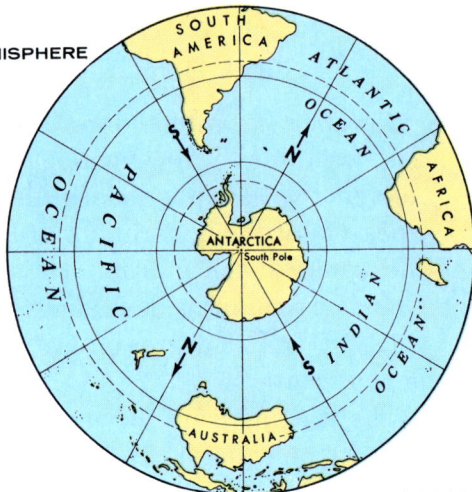

SOUTHERN HEMISPHERE

SOUTHERN HEMISPHERE

Hemispheres

Each of the great circles divides the earth into halves. Each half is called a hemisphere.

Northern and Southern Hemispheres. The equator is the only line of latitude that is a great circle. The equator divides the earth into the Northern and Southern Hemispheres. All of the land and water north of the equator are in the Northern Hemisphere. The land and water south of the equator are in the Southern Hemisphere. These hemispheres are shown on this page. Study them. Which continents are entirely in the Northern Hemisphere? Which continents are entirely in the Southern

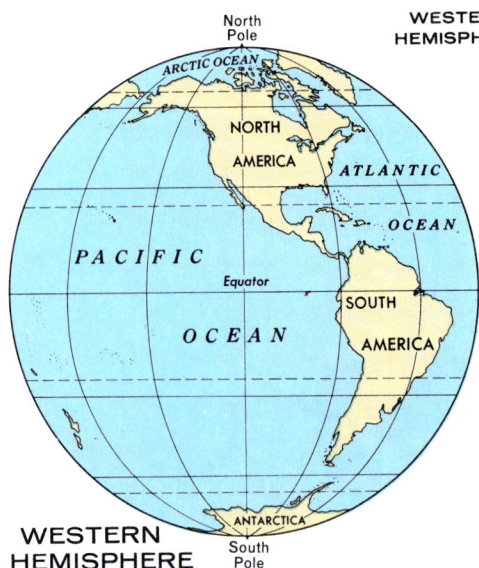

WESTERN HEMISPHERE

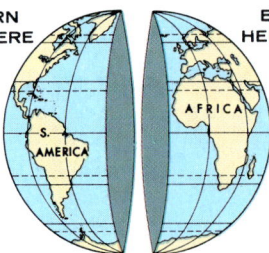

EASTERN HEMISPHERE

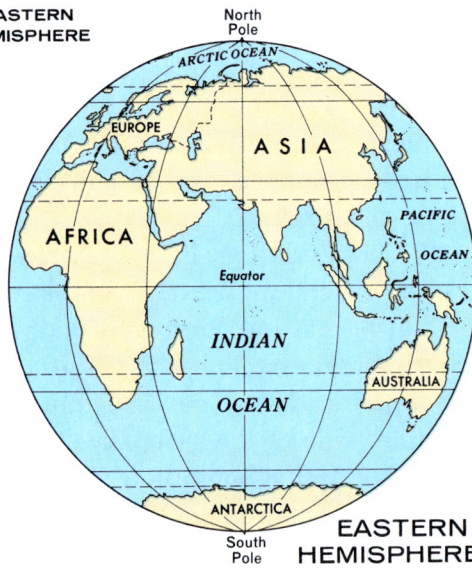

WESTERN HEMISPHERE

EASTERN HEMISPHERE

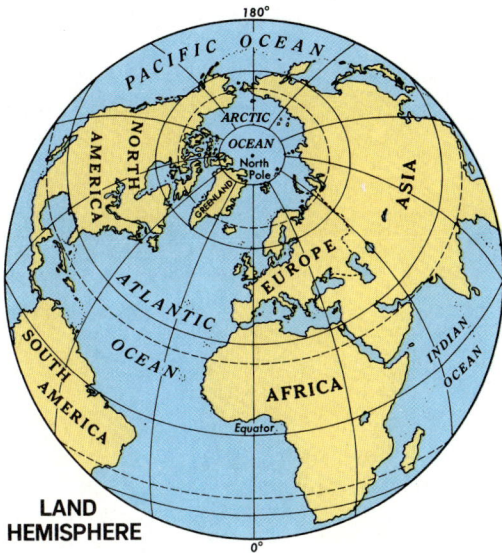

LAND
HEMISPHERE

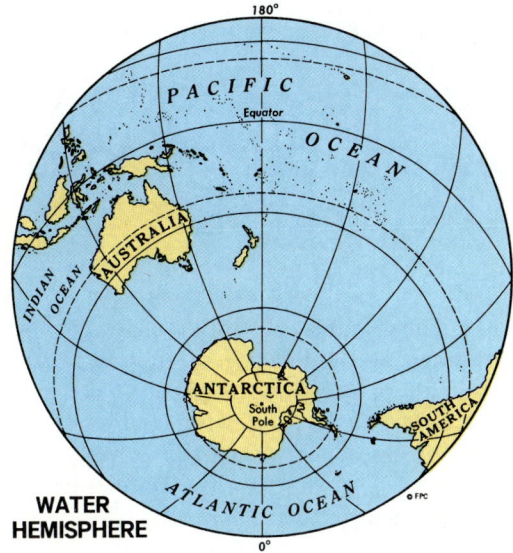

WATER
HEMISPHERE

Hemisphere? Which continents are partly in both? In which of these hemispheres do you live?

Eastern and Western Hemispheres. The boundaries between the Eastern and Western Hemispheres are generally drawn using 20° west longitude and 160° east longitude. Why do you think this great circle is used rather than the one formed by the prime meridian and the 180° meridian? Study the views of the Eastern and Western Hemispheres. Which continents are entirely in the Western Hemisphere? Which are entirely in the Eastern Hemisphere? Is there any continent that is in both the Eastern and the Western Hemispheres? In which of these hemispheres do you live?

Land and water hemispheres. Geographers sometimes divide the earth into land and water hemispheres. The land hemisphere is that half of the earth with the most land. The water hemisphere is that half of the earth with the most water.

Study the globe drawings that show the land and water hemispheres. Find these same hemispheres on a globe. The land hemisphere contains about 82 percent of all the earth's land. The center of the land hemisphere is near London, England. The center of the water hemisphere is near New Zealand.

Keeping Facts in Focus

1. What is a globe?
2. What is the definition of *north?* Of *south?* Of *up?* Of *down?*
3. What is the definition of *latitude?* Of *longitude?* Which are always the same distance apart—lines of latitude or lines of longitude?
4. What answers would you give to the key questions on page 42?

Working with Ideas

1. Your city, town, or village has a grid system much like the earth's. Explain how your local grid system works.
2. If you were to travel north four degrees of latitude from your school, in what state would you be? Near what city, town, or village would you be?

Lesson 2 Using Maps

Reading Focus

1. In some parts of this lesson, you will be asked to follow directions. Sometimes, following directions means only that you have to turn to a certain page to look at a map. Take the time to do it. Answer any questions your book asks. Following such directions will increase your understanding of the lesson. At other times, you may be directed to follow several steps. Always be sure you understand each step. Read all the steps through first. Then go back to the first step. Do it. Do the next step. Continue in this way until you have mastered all steps.

2. As you read this lesson, look for answers to these key questions:
   a. What advantages do maps have over globes? What disadvantages?
   b. How do symbols, map keys, and map scales help you to read maps?
   c. Why is it important to ask yourself what a map's purpose is?

Vocabulary Focus

symbol   *map key*   *graphic-relief map*   *political map*

A sketch drawn in the sand with a stick. Pieces of wood sewn on a piece of sealskin. Sticks tied together. These are some of the earliest forms of maps. Maps have been around for a very long time. People began to make a record of where they had been. A map could help them find their way back to a certain place. It could help them show others how to reach the same place—what directions to follow and how far they would have to travel. Maps could also help people describe the distant lands they had visited.

Comparing Maps and Globes

Maps have advantages that globes do not have. Maps can be folded. They can be easily carried or stored away when not in use. Globes cannot. Maps can show the whole world at one time. Globes cannot. When you look at a globe, you see only half of the world at one time. Maps can also show small parts of the world in great detail. Globes cannot. For a globe to show the kinds of details a map can show, it would have to be larger than most classrooms.

But maps have disadvantages that globes do not have. Globes are round, like the earth. They give a true picture of the shape and size of land and water bodies. They show true distances and true directions between two places.

Flat maps of a round earth. Maps, however, are flat. When the features of a round earth are put on a flat map, shapes of land and water bodies change. Sizes of these bodies change. When lines

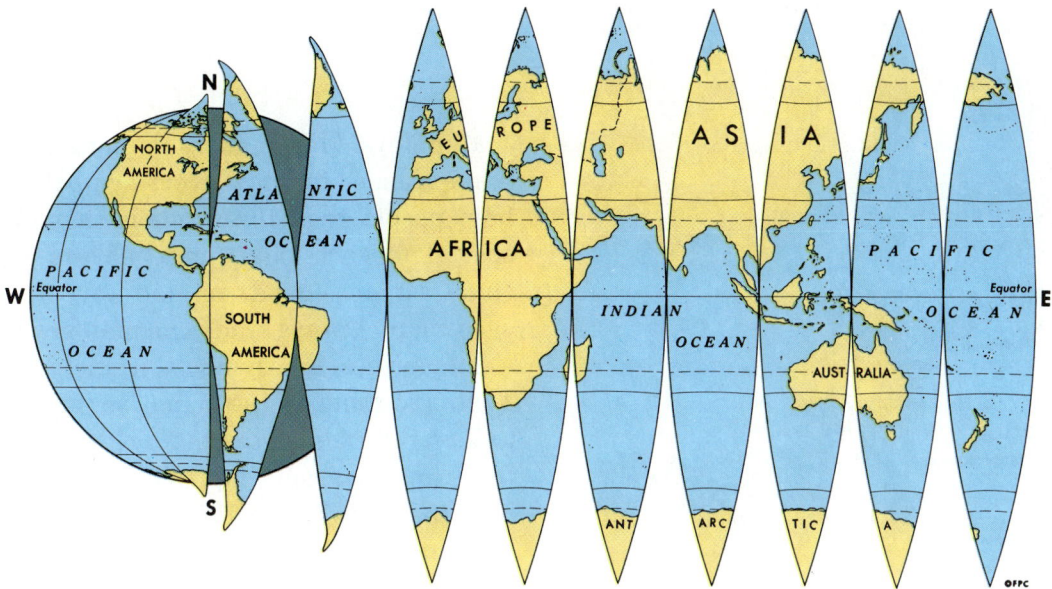

of longitude that follow the curve of the earth on a globe become straight lines on a flat map, distances between places change. Directions between places also change. You can see how all of these things happen by studying the illustration at the top of this page.

All of these four features—shape, size, distance, and direction—can be shown accurately on a globe. They cannot all be shown accurately on a map. A map may give a true picture of one or two of these features. But there will be some stretching or shrinking so that one or more of the features will not be true. Different kinds of maps show these features in different ways.

Great circle routes. A great circle is a line that divides the earth into two equal parts. Meridians meet to form great circles. The equator is a great circle. It is also possible to trace many other great circles on a globe. These would cut across the equator and the meridians at a slant. An arc, or part, of a great circle is the shortest distance between any two points. Pilots often fly great circle routes. They fly a shorter distance to reach their destinations in this way.

The map on page 49 shows the great circle route between Singapore and Panama City. It also shows that Panama City is mostly east of Singapore. A pilot could reach Panama City by flying east, the true compass direction. But the great circle route, which follows mostly a north-south direction, is shorter. By how many miles is it shorter? By how many kilometers?

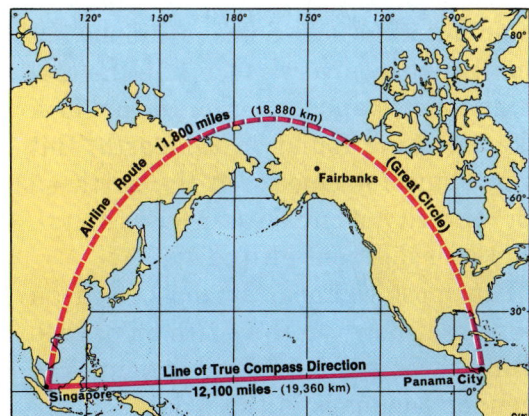

Transportation
- ——— Pan American Highway
- – – – Airlines
- ++++++++ Railroads
- ⊛ National Capitals

The airline route shown on the map is an arc of a great circle. On a globe, place a string between Singapore and Panama City, following the great circle route shown on the map. Extend the string all around the globe. Does it form a great circle? How do you know it is a great circle? You can also use the string method on a globe to check that the great circle route is shorter.

Reading Maps

Maps contain information, but that information is coded. Using maps skillfully means unlocking the code.

Symbols. Maps use symbols to present their information. Some symbols are small drawings very much like the real things they stand for. Others do not look like anything found on the earth. But they stand for objects on the earth—a dot for a city, a circled star for a capital, colors for different altitudes.

The most common symbols used on maps are shown in the Atlas on page 9. Study these symbols. Find examples of them on the maps in the Atlas.

Map keys. Map makers do not expect a map reader to know every symbol. To help people read the symbols that are special to each map, map makers include a map legend, or map key, with each map. Always study the map key of a map. Note the symbols it includes. Then find them on the map. Use them to unlock the code of the map. Now study the transportation map. What does the red line show? The broken blue line? What do the other symbols stand for?

Scale. Maps are drawn to scale, just as globes are. The map scale is usually a part of the map key. The scale tells the map user what distance on the earth is represented by a certain measure on the map. Each map has its own scale.

Scale can be shown in different ways. The scale may be stated in words: *One inch equals 500 miles* or *one centimeter equals 300 kilometers,* etc. A bar or line may also show scale. Maps in this text usually show scale both with words and with a bar or a line.

Look at the map on pages 10–11 in the Atlas. What is the scale of this map? What is the scale of the map on page 16?

The scale can be used to measure distances between places on the map. Take a piece of paper with a straight edge. Place the edge of the paper between the two points to be measured. Mark off each point on the paper. Now place the paper with its marked points on the map scale. Place the point on the left at zero. Note where the point on the right falls. If the scale is not long enough, keep moving the paper to the left, marking each end point until you have figured out the distance.

SCALE
One inch-about 400 Miles

Miles
0 100 200 300 400 500 600

0 200 400 600 800
Kilometers

SCALE
One inch-about 80 Miles

Miles
0 10 20 40 60 80 100 120 140

0 40 80 120 160 200
Kilometers

SCALE
One inch-about 6 Miles

Miles
0 2 4 6 8 10

0 4 8 12 16
Kilometers

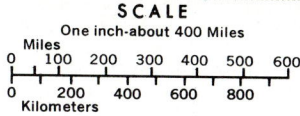

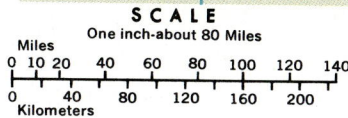

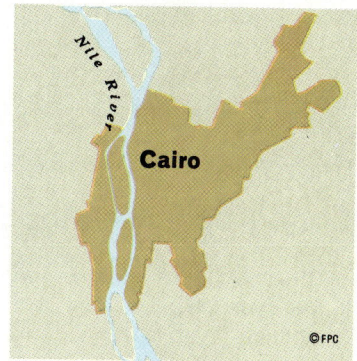

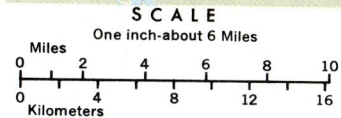

Note the boxed area on the map to the left. The middle map is an enlargement of this area, just as the map to the right is an enlargement of the boxed area on the middle map. On which map does one inch equal the shortest distance? Is this a large-scale or a small-scale map?

Practice finding distances using this method. Turn to page 14 of the Atlas. Locate the city of London in England. Then locate the city of Budapest in Hungary. Measure the distance between these two cities. How many miles do you measure between London and Budapest? How many kilometers? Now turn to the map on page 16 of the Atlas. Once again measure the distance between London and Budapest. Is there a difference in the two measurements?

The difference in the two measurements comes from the scale each map has. On which map does one inch cover a greater distance? This map has a smaller scale than the other one. Small-scale maps cannot tell the exact distance between two places. Always remember to use the word *about* when stating distances read from small-scale maps.

A large-scale map is one in which the distance covered by a unit (inch, centimeter, etc.) on the map stands for a short distance on the earth (a meter, a mile, a kilometer). Large-scale maps show many more features than small-scale maps. On large-scale maps, measurements can be much more accurate.

Some maps show scale in another way. On these maps, scale is shown by numbers. Such a map scale might read: 1:62,500. This means that one unit of measurement on the map stands for 62,500 units of the same measurement on the earth. The unit of measurement could be one inch, one centimeter, one meter, or even one matchstick. Many maps use this kind of scale.

Kinds of Maps

An important part of reading a map is deciding what the map's purpose is. Does the map indicate in some way what the land of a region looks like? If it does, it is a physical map. In this book, maps of this kind are labeled as graphic-relief maps. Does a map use different colors to show nations and their boundaries? If it does, it is a political map. Does the map show something else besides graphic-relief or the boundaries of nations or states? Then it is a special-purpose map.

Cartographers—Map Makers of the World

Throughout this book, you will find maps of all kinds. Every map—whether it is a globe drawing, a political map, a graphic-relief map, or a special-purpose map—is the work of one or more cartographers.

But a finished map needs more than the work of a cartographer. It also represents the work of many other people who have gathered the information shown on a map. Surveyors and aerial photographers gather information for land maps and oceanographers for water bodies. Weather observers gather reports on rainfall, temperature, etc. The list could go on and on, naming almost all kinds of scientists.

A good example of the work of a cartographer is the world graphic-relief map on pages 12–13 of the Atlas. The map maker chose a projection—one made by another cartographer—that would best show the shape and size relationships of land and water bodies. After tracing the global grid and the land and water shapes of the projection, the map maker then prepared the color design.

Still more decisions had to be made—the styles and sizes of the printed letters for different kinds and sizes of land and water bodies and the labeling of natural features. On this map, water bodies are labeled in italics. Land bodies are not. Oceans have a larger letter size than seas, continents a larger letter size than islands. Because the map's main purpose is to show natural features, no labels were given to countries, nor were cities located and labeled. If they were labeled, the largest cities and countries would have one size letter; smaller cities and countries would have another size letter.

A cartographer, then, is an artist and a technician. As an artist, a cartographer uses color and design to achieve a map's purpose. As a technician, a cartographer makes careful measurements to draw a map as accurately as its size will allow.

Many cartographers work for government agencies. They make maps of land and water areas, population maps, weather maps, maps showing income or education levels of people, county maps, state maps, navigation maps, flight maps, etc. Cartographers also work in industry, charting the locations of minerals and other raw materials. Those who work for publishers make the kinds of maps you see in magazines, in newspapers, and in this book.

Graphic-relief maps. A graphic-relief map uses shades of black to show land that is hilly or mountainous. Land that has no shading is nearly level land. A graphic-relief map may also use color as a symbol to indicate the height, or altitude, of the land above sea level. (Sea level is the average level of the sea where it meets the land.)

The diagram on page 53 shows how to read the key of a graphic-relief map.

Notice how the shading becomes darker as the land becomes more mountainous. Notice also the different colors and the range of altitudes each shows.

Notice that color is also used to show the depth of water. A lighter blue shows water less than 600 feet (180 m) deep. A darker blue shows water deeper than that. Turn to the graphic-relief map of the United States on pages 34–35 of the Atlas. How high above sea level is the

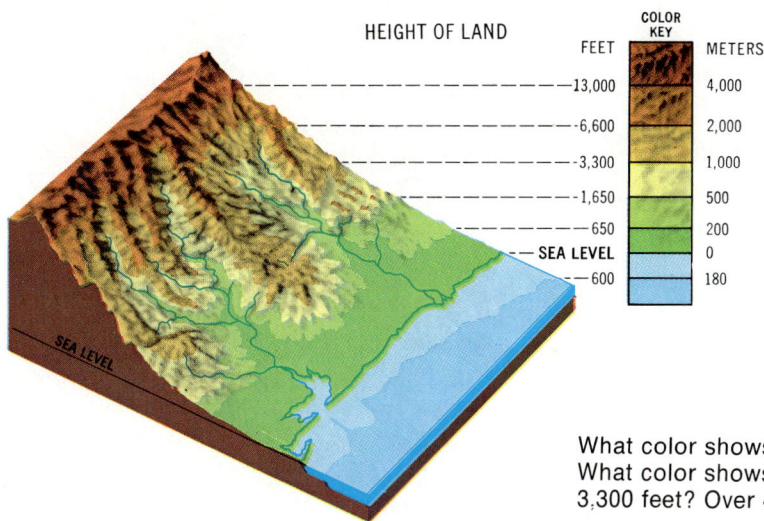

HEIGHT OF LAND

COLOR KEY

FEET	METERS
13,000	4,000
6,600	2,000
3,300	1,000
1,650	500
650	200
SEA LEVEL	0
600	180

What color shows land lower than 650 feet? What color shows land between 1,650 and 3,300 feet? Over 4,000 meters?

region where you live? Which is higher in altitude—the Sierra Nevada or the Coast Ranges?

Political maps. Political maps tell us more about the people on earth than they do about land and water. Many political maps tell us how people have divided the lands of the earth into countries. Some show how countries have been divided into states or provinces. Some even show how states have been divided into smaller parts, such as counties. Political maps also show where people have built towns and cities.

The colors on a political map are not really symbols as they are on a graphic-relief map. On a political map, colors simply make it easy to see political divisions.

Turn to the political map of South America on page 28 of the Atlas. Into how many nations is South America divided? Which country is the largest? Is Peru on the east or west coast of South America?

Turn to the symbols chart on page 9 of the Atlas. Notice that there are two symbols for political boundaries. What symbol shows a boundary between nations? What symbol shows a boundary between states or provinces?

Graphic-relief maps, too, give information about political boundaries. Does the graphic-relief map of the world on pages 12–13 show boundaries? What kind of boundaries does it show?

Special-purpose maps. These maps are just what the name suggests. They are maps used for special purposes. There are many different kinds of special-purpose maps. The best known is probably the road map. Others are used to show weather, population, land use, and climate patterns. Always study the key of a special-purpose map carefully. Use the maps on page 54 to tell which of the following statements are true:

1. Silver is mined in Kenya.
2. Kenya has two main ports.
3. Hokkaido receives between 40 and 60 inches of precipitation each year.

Economic Activities in Kenya

Farming areas
Manufacturing centers
Ports
Railroads
Cattle
Cement
Coffee
Corn
Gold
Millet
Salt
Sheep
Sisal
Tea

©FPC

Average Annual Precipitation
in inches and centimeters

20-40 in., 50-100 cm.
40-60 in., 100-150 cm.
60-80 in., 150-200 cm.
More than 80 in., 200 cm.

Hokkaido

Honshu

Shikoku

Kyushu

JAPAN

©FPC

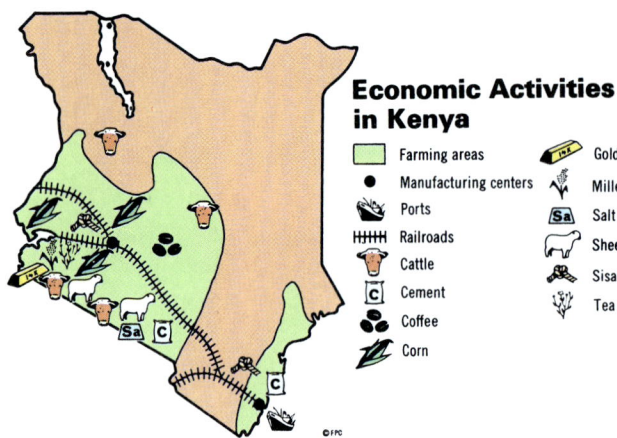

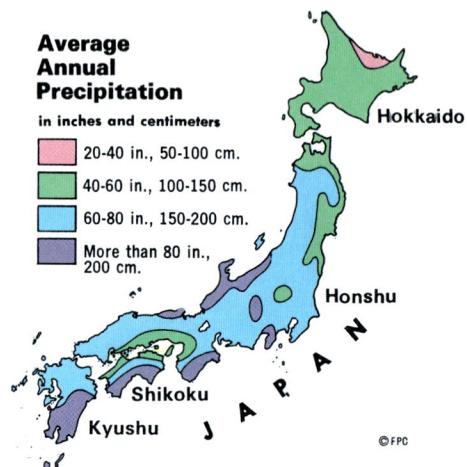

Keeping Facts in Focus

1. What is a symbol? Name three symbols found on most political maps.
2. What is a graphic-relief map? A political map? A special-purpose map?
3. What is a cartographer?
4. What answers would you give to the key questions on page 48?

Working with Ideas

1. In which occupations are people most likely to use maps on a daily basis?
2. Compare the world map on pages 10 and 11 with the world map on pages 12 and 13. Name three ways the first map differs from the second map. Name three ways the two maps are similar.

Map Projections

A map projection is a way of transferring the lines of the global grid and the shapes of land and water bodies from a globe to a flat sheet of paper.

There are many different kinds of map projections. Some of these are named after the map makers who invented them. Some are given names according to the ways they are drawn. Here are some widely used map projections:

Mercator's map. The Mercator map was first drawn in the 1500s. To draw his map, Mercator used an idea that many other map makers also use. He imagined a see-through, or transparent, globe. He imagined that a sheet of paper was wrapped around the globe, forming a tube, or cylinder. The paper touched the globe at all points along the equator.

Within the globe, at its center, was a light source. Mercator imagined how the light would cast shadows of the lines and shapes from the globe onto the paper. Then he used the rules of mathematics to draw the lines and shapes from the globe onto a flat sheet of paper.

Mercator's map stretches the lines of latitude and longitude. This stretching does several things. It keeps shapes nearly true, but not size or area. For example, a dime and a quarter have the same shape. The quarter, however, is larger than the dime. On a globe, as on the earth, South America is much larger than Greenland. But on Mercator's map, Greenland is much larger than South America. Their shapes, but not their sizes, are similar to the shapes they have on a globe.

More importantly, Mercator's map keeps directions true. In doing this, it gives sailors a way of plotting a true course at sea. Once sailors plot a straight line between two places, all they have to do to reach a destination is to cross each meridian at the same angle. Following this straight line lengthens their journey quite a bit. But it gets them where they want to go. Mercator's map is still used for sea travel today.

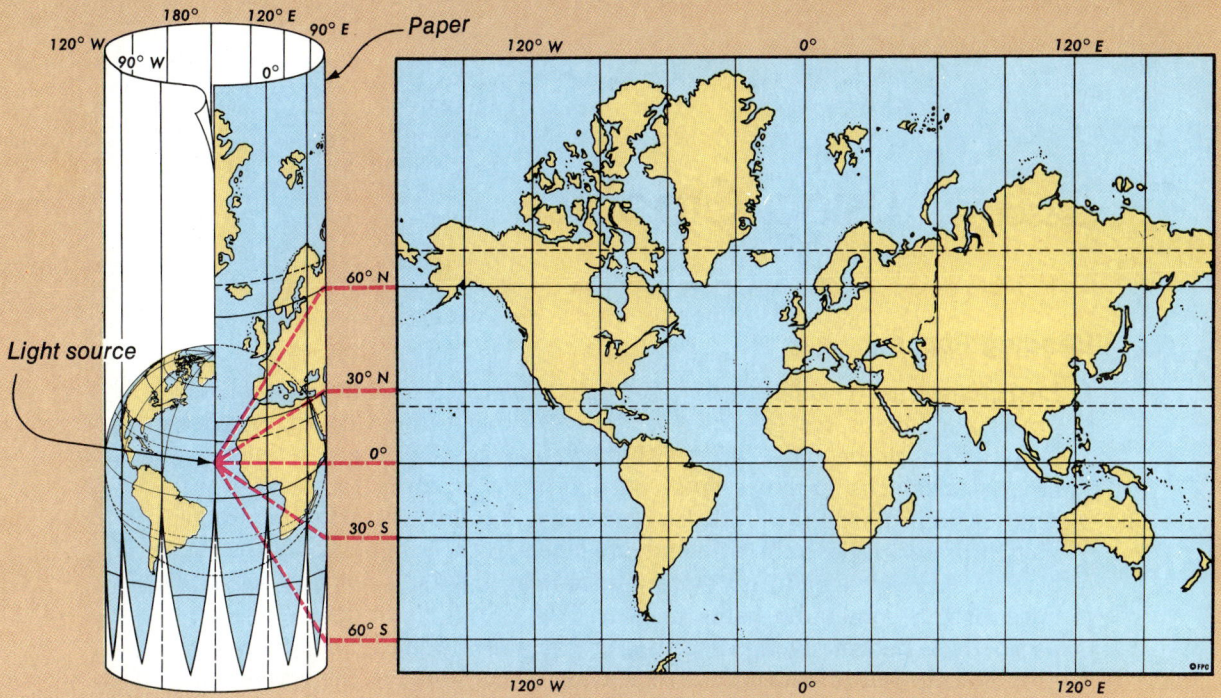

Mercator's Map

Directions are true.
Stretching toward poles.
Best used for ocean navigation.

Lambert's map. Lambert drew his map using Mercator's idea. Only Lambert changed the shape of the paper to a cone. He thought of the cone placed on the globe, much like a hat on a head. But he wanted the paper to touch the globe at two places. So he imagined that the paper cut through the globe.

What gets transferred to the paper is only a part of the world. This part is usually a country or a region in the middle latitudes. This projection keeps distance and direction true within two parallels, such as the ones shown on the drawing. This map also keeps shape fairly true within these parallels.

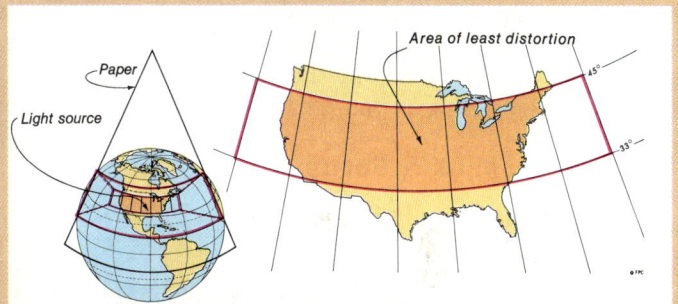

Lambert's Map

Distances and directions are fairly true in area of least distortion. Often used to show areas in middle latitudes.

Fixed-point projections. In this type of projection, a flat sheet of paper touches only one point on the globe. This fixed point becomes the center of the map. The fixed point may be the North Pole or the South Pole. It may even be a city. Maps with such a fixed point are called azimuthal (az′ə·məth′əl) projections.

Pilots use this kind of fixed-point projection to plot their courses to and from cities on different continents. They can draw a great circle route between the center point of the map and the cities that are their takeoff and landing points. Along this great circle route, distance and direction are true.

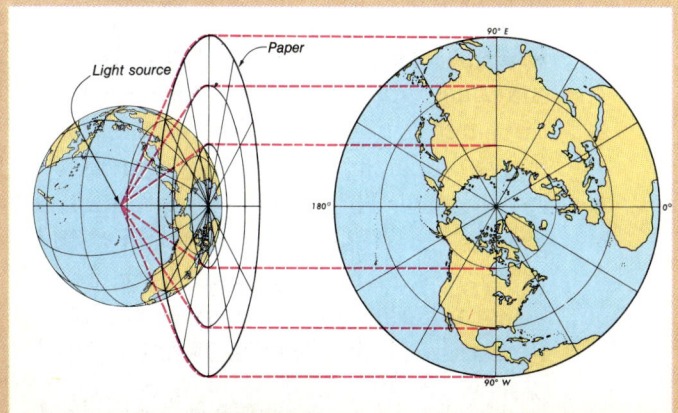

Fixed-Point Projection

Often used in flight navigation.

Lesson 3 Using Graphics

Reading Focus

1. Learning to use graphics—charts, graphs, diagrams, illustrations, photographs (and maps, too)—is the main theme of this lesson. Geographers use graphics to clothe bare facts with meanings by showing relationships among facts. The graphics must be read very carefully to discover the relationships among the facts. To read a graphic, first pay attention to the facts it presents. Read the title and any labels. If the graphic has a caption, be sure to read it and answer any questions it asks. Then ask yourself questions. Is something being measured? Are several things being compared? Is something being described or explained? Finally, how does this graphic help you to understand what is being discussed in the lesson?

2. As you read this lesson, look for answers to these key questions:
   a. What is a table?
   b. What are the different forms charts can take?
   c. What are the different forms graphs can take?
   d. What kinds of graphic tools are useful for showing how something is done or something is made?
   e. How do you "read" photographs and illustrations?

Vocabulary Focus

table	*diagram*	*vertical*
graph	*horizontal*	

Maps and globes are not the only tools that geographers use to tell people what the earth is like. Geographers use tables to list important facts. They use charts of all kinds. They use different kinds of graphs. They use diagrams to show how things are done or how things are made. Geographers also use photographs and illustrations to show us what certain places are like and to help us understand how people live in different parts of the world.

Tables

A table is a listing of facts. It usually has a title. The title helps to point out how the facts in the table are related. The facts may be a list of numbers, such as areas of countries, populations, or tons of coal mined. They may be a list of names, such as products imported or exported by a certain country.

Look at the table on page 57. It shows one way of organizing the countries of the world into regions. In the second

Regions of the World	Area in Square Kilometers
Antarctica	15,600,000
Asia	20,300,000
Europe and the Soviet Union	27,400,000
Latin America	20,500,000
North Africa and the Middle East	14,400,000
North America (United States and Canada)	21,500,000
Oceania and Australia	8,600,000
Sub-Saharan Africa	23,700,000
Total Land Area of World	152,000,000

part of this book, you will study these regions. What fact about each region does the table list?

Charts

There are many different kinds of charts. One kind of chart is very much like a table. It, too, shows facts. However, it usually shows more than one set of facts. Lines sometimes divide the sets of facts into rows and columns. Sometimes colors help to separate the sets of facts. Look at the National Profiles chart on pages 290–292. How many different sets of facts are included in this chart?

A second kind of chart makes use of illustrations or symbols. You can see an example of this kind of chart on this page. Use it to compare the resources of any two countries shown on the chart.

A third kind of chart is a time line. A time line shows a period of time and important events that took place within the time period.

A fourth kind of chart is called a flow chart. A flow chart may show how things are organized or related. It may show how decisions are made. Or it may show steps in a process.

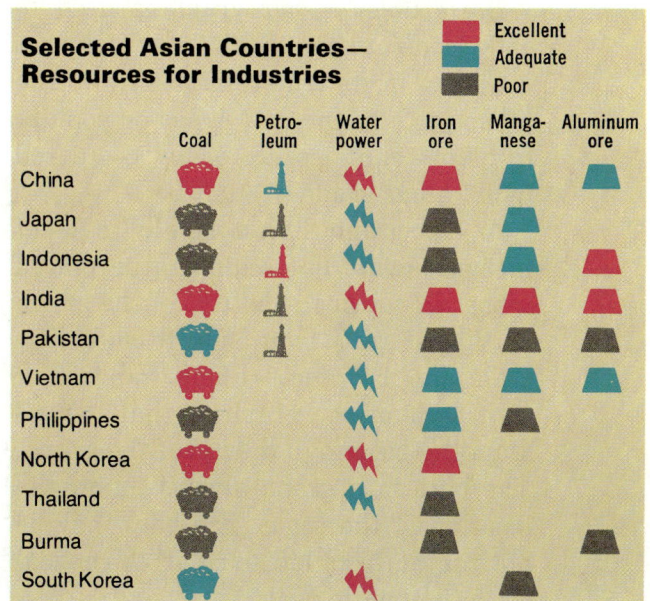

Selected Asian Countries— Resources for Industries

Legend: Excellent, Adequate, Poor

	Coal	Petroleum	Water power	Iron ore	Manganese	Aluminum ore
China	Excellent	Adequate	Excellent	Excellent	Adequate	Adequate
Japan	Poor	Adequate	Adequate	Poor	Poor	Poor
Indonesia	Poor	Excellent	Adequate	Poor	Adequate	Excellent
India	Excellent		Adequate	Excellent	Adequate	Adequate
Pakistan	Adequate	Adequate	Adequate	Poor	Poor	Poor
Vietnam	Excellent		Adequate	Adequate	Adequate	Poor
Philippines	Poor		Adequate	Adequate	Poor	
North Korea	Excellent		Excellent	Excellent		
Thailand	Poor			Poor	Poor	Poor
Burma	Poor				Poor	Poor
South Korea	Adequate		Adequate		Poor	

Land Area for Regions of the World in millions of square kilometers

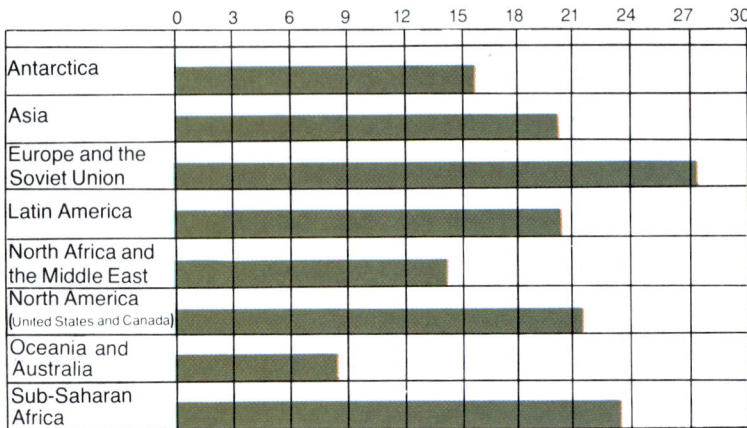

	0	3	6	9	12	15	18	21	24	27	30
Antarctica											
Asia											
Europe and the Soviet Union											
Latin America											
North Africa and the Middle East											
North America (United States and Canada)											
Oceania and Australia											
Sub-Saharan Africa											

Land Area for Regions of the World

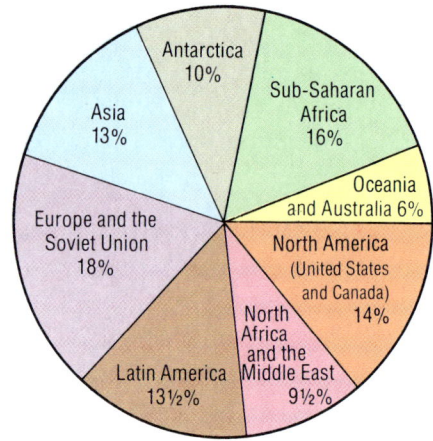

Antarctica 10%
Asia 13%
Sub-Saharan Africa 16%
Oceania and Australia 6%
Europe and the Soviet Union 18%
North America (United States and Canada) 14%
Latin America 13½%
North Africa and the Middle East 9½%

Graphs

A graph is a drawing that uses bars, circles, points and lines, or symbols to present facts and sets of facts. Bar graphs, circle or pie graphs, line graphs, and picture graphs are different forms of graphs. Any set of facts that deals with different amounts is easily shown on a graph.

Bar graphs. A bar graph is used to show comparisons. Each fact in the set is shown in its true relationship to the other facts in the set. Areas or populations of countries are easily compared using a bar graph.

A bar graph has a single scale of measurement. The scale can be placed along the top or the bottom of the graph so that it runs from side to side. The scale can be placed along the left or right side of the graph so that it runs from the bottom to the top of the graph. The bars in a graph always follow the position of the scale. The bars may be horizontal (from left to right) or vertical (from bottom to top).

Look at the bar graph on this page. It has the same information as the table on page 57. Look at the graph's scale. What does it measure? Compare the regions of the world. Which has the largest area? Which has the smallest? Which is larger—North Africa and the Middle East or Sub-Saharan Africa?

Like any tool, graphs cannot do more than they are made to do. Bar graphs seldom deal in exact figures. The graph maker may start with exact figures from a table. But then the graph maker rounds off the figures, especially if they are large numbers. Large numbers can be rounded off to different places—to the nearest ten, hundred, thousand, ten thousand, hundred thousand, million, or even billion. Read this figure: 5,384,289. This is how the number changes when it is rounded to different places:

nearest million	5,000,000
nearest 100 thousand	5,400,000
nearest 10 thousand	5,380,000
nearest thousand	5,384,000
nearest hundred	5,384,300
nearest ten	5,384,290

Have you figured out the rule for rounding off numbers? Try rounding off these numbers to each place in the list above: 9,748,278 and 3,264,826.

Now read the numbers from the bar graph on page 58. Look at the scale of measurement. It is in millions of square kilometers. Each division stands for 3 million square kilometers. Most graphs that deal with large numbers cannot show exact amounts. To read large numbers on graphs, always round off the numbers. For this graph, round the numbers to the nearest million. Now compare the readings you get from the bar graph with the numbers in the table on page 57. How close is your reading for each region's area to the number given in the table?

Pie graphs. A graph in the form of a circle is called a pie graph. A pie graph is useful for showing how parts of the whole are related to the whole. The circle, or "pie," stands for 100 percent. The sections, or "slices," of the pie show what percent each part is of the whole. Look at the pie graph on page 58. What percent of the world's land area is each region?

Percentages can also be shown in other forms besides pie graphs. For one example, see the bar graph below. What percent of India's people live in urban areas? In rural areas?

Monthly Average Temperatures for Ankara, Baghdad, and Beirut

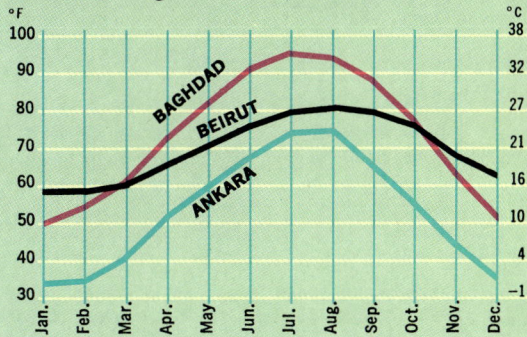

Urban and Rural Population of India

Urban 24%	Rural 76%

Line graphs. A graph can take still another form, using points and lines to connect the points. A line graph has both a horizontal and a vertical scale. This is because a line graph measures two things—amount and time or rate.

A line graph is often used to show a place's average temperature and precipitation. It may even show a set of facts for more than one place. Look at the line graph on this page. What does the graph measure? For how many cities does it measure it? Which of the three cities has the coldest temperatures? Which has the hottest? What is the average temperature of Ankara in May? During what two months are the average temperatures of Baghdad and Beirut about the same?

Picture graphs. This kind of graph uses symbols to stand for facts or sets of facts. Sometimes, the symbols are pictures very much like the real things they stand for. Sometimes, the symbols are abstract. They do not look like the real things they stand for.

59

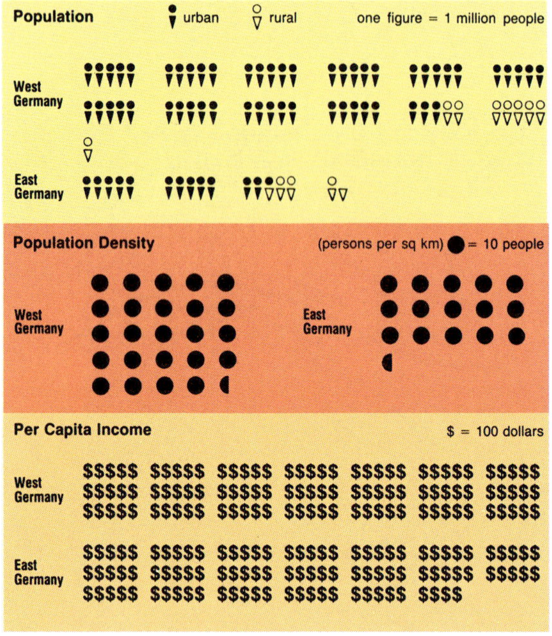
Population
urban
rural
one figure = 1 million people
West Germany
East Germany
Population Density
(persons per sq km) = 10 people
West Germany
East Germany
Per Capita Income
$ = 100 dollars
West Germany
East Germany

Look at the picture graphs on this page. What do the symbols in each graph stand for? What two countries are compared? Which of the two countries has a greater population? About how many people in East Germany live in rural areas? Which country has more people per square kilometer? On the average, how much does each person in West Germany earn a year?

Diagrams

Another tool that geographers use is the diagram, a drawing with labels. Most diagrams are used to show how something is made, what something looks like, or how something is done. Some diagrams show part of an object cut away to reveal the inside. Such cross section diagrams are useful for showing what something is made of or how something is done. The cross section diagram below shows the parts of a nuclear reactor and a heat exchanger. Follow the arrows to see how the splitting of atoms of uranium produces the heat that makes the steam used to generate electricity.

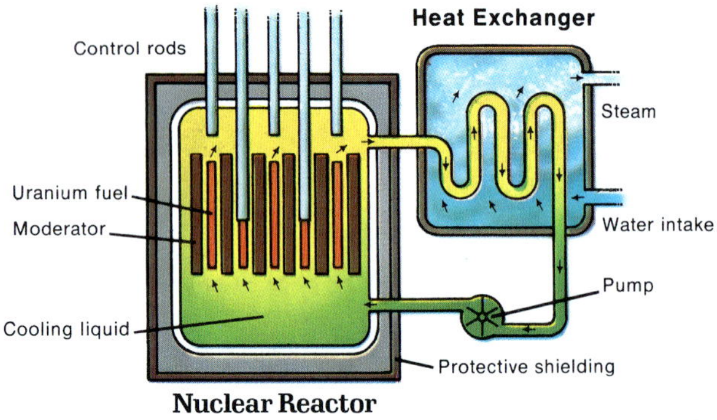
Heat Exchanger
Control rods
Steam
Uranium fuel
Moderator
Water intake
Pump
Cooling liquid
Protective shielding
Nuclear Reactor

Photographs

Photographs are very useful tools for bringing the faraway world close to hand. Looking at a photograph is almost like being at the place where it was taken.

Geographers look at photographs with trained eyes. They always ask questions about what the photographs show. You, too, must look at photographs with trained eyes. You must ask questions to get the most meaning from the photographs used in this text.

Earth scenes. One kind of photograph shows what different parts of the earth look like. Try to answer these questions when you see an earth scene:

1. Does the photograph show land? What kind of land does it show?
2. Does the photograph give any clues to the climate of the place? For example, does it show any kind of plant life?
3. Does the photograph give any clues about the people who live in the area and how they use the land?

A yak caravan in Nepal travels through a pass in the forbidding Himalayas, while well-tended farms lie in the valleys between ridges. Use the questions on pages 60-61 to gather more information from these photographs.

4. What makes this place different from the place where you live? What makes it similar?

Space-age scenes. Some photographs are taken from the air or from space. Airplane pilots and astronauts have taken some of them. Special cameras in orbiting satellites have taken others.

Space-age photographs are useful tools to geographers and other scientists. These photographs often show details that are not noticeable from the ground. Space-age photographs aid in weather forecasting. They aid in map making. Those taken with special film even show soil conditions in different parts of the world. You can find examples of space-age photographs on this page and on page 38.

This Landsat satellite image shows the southern part of Lake Michigan, the city of Chicago, Illinois, and surrounding metropolitan areas.

A village near Nigeria's Lake Kainji.

Culture scenes. People all over the world have the same basic needs: food, clothing, and shelter. But people in different parts of the world meet these needs in different ways. They have different customs and cultures. When you view culture scenes, such as the one above, ask yourself these questions:

1. Does the photograph give any clues about how the people in the area meet their needs?
2. What do the things shown—the homes and other buildings, the food, clothing, customs, or methods of transportation, etc.—tell about the environment the people live in?
3. Does the photograph give any clues about how the environment affects the way things are done?
4. What natural resources are the people using?
5. What machines or tools are the people using?
6. How is the way the people live different from the way you live? How is it similar?

Other scenes. You will see photographs in other places besides this book. Some will be historical scenes. Some will be scenes of the present in newspapers, magazines, or booklets. Others may be advertisements. Here are some additional questions to ask yourself about photographs:

1. What does the photograph show?
2. Does the photograph show only good things about a place or a group of people? Could the photograph be used to show that conditions are better or worse than they really are in that place?
3. Does the scene look as if it really happened that way, or is it a photograph of a posed scene?

Illustrations

Illustrations are drawings of people, places, and things. Illustrations often take the place of photographs. Illustrations are often used to show how things looked in the past. They often show how things were done in the past. They even show how people lived in the past. And sometimes illustrations are used to give people an idea of how things will be in the future. No matter what an illustration shows, it is usually based on what is known to be true, using evidence of the past or present. Illustrations should be "read" just as carefully as photographs. A trained eye will find many questions to ask about the land, the things, or the people shown in the illustrations. These questions will be similar to those asked about earth, space, or culture photographs.

Captions

Many graphics in this text have captions. Sometimes, the caption describes things in the graphic. Sometimes, the caption gives additional information. Many times the caption asks one or more questions.

Always read a caption carefully. Relate the information in it to the graphic. Answer any questions in the caption. Sometimes, the answer to a question will be found by reading the text. Sometimes, the answer is found in the graphic. And sometimes, a question asks you to think for yourself. It asks you to discover the answer by putting together information from several places in the text with the information you get from the graphic.

Keeping Facts in Focus

1. What are graphics?
2. Which kind of graph best compares amounts of things?
3. Which kind of graph best compares both amounts and time?
4. Which kind of graph can be used best to show how parts are related to the whole?
5. If a graph uses symbols to stand for real things, what kind is it?
6. What does a diagram usually show?
7. Name three kinds of photographs you will find in this text.
8. What answers would you give to the key questions on page 56?

Working with Ideas

Think of one way your city, town, or village has changed in the last ten years. What type of graphic would you use to best illustrate this change? Explain the reasons for your choice.

Lesson 4 The Earth and the Sun

Reading Focus

1. Previewing a lesson is a good study habit to follow. *Previewing* means "to see beforehand." In this case, it means to discover what the lesson is about before you read it. To preview a lesson, you can do these things. Read the key questions that are a part of every *Reading Focus.* Look at the headings used to divide a lesson into parts. Larger headings will give you an understanding of the main ideas. Smaller headings will give you an understanding of the ideas related to a main idea. Look at the graphics. Read their captions. Read the list of questions that begins and ends each lesson. Doing all of these things will help you see where you are going and how you will get there.

2. As you read this lesson, look for answers to these key questions:
   a. How does the sun affect the earth?
   b. In what ways does the earth move in space?
   c. Why do some parts of the earth have great seasonal changes?

Vocabulary Focus

solar system	axis	revolution	solstice
energy	rotation	orbit	equinox

No matter what day of the week it is, every day is a "sun day," a day when the sun shines on the earth. The sun, that great ball of hot gases in the sky, is a star. It is earth's closest star and the center of our solar system. The sun is the most important force acting on the earth.

The Powerful Sun

The power of the sun comes from its gases, which give off energy in the form of heat and light. These forms of energy, in one way or another, influence almost every activity that takes place on the earth.

The sun supports life. The earth is about 93,000,000 miles (149,000,000 km) from the sun. This is far enough away from the sun so that enough heat and light reach the earth. But it is close enough to give the earth the conditions it needs to support life.

The sun is a source of energy. The sun produces almost all the energy found on earth. Most of this energy comes from the sun in the form of heat and light. Some of this energy is held by the air that surrounds the earth. Some is held by the land. And some is held by the waters of the earth. Plants take light energy directly from the sun. They take other forms of energy from the air, the land, and the water. They use these forms of energy to grow. Animals and human beings use the energy stored in plants as food. Human beings make use of other forms of energy in other ways. You will learn more about energy in Unit 7.

The Moving Earth

For the earth, every day is also a "moving day." Every day, the earth moves. It spins on its axis. As the earth spins, it moves in a fixed path around the sun. These two movements—rotation and revolution—affect the earth and life on the earth.

The rotating earth. Every twenty-four hours, the earth rotates, or spins completely around, from west to east. The earth spins on its axis, an imaginary line that cuts through the earth from pole to pole. As the earth spins, the part of the earth that faces the sun has daylight. The part of the earth facing away from the sun has darkness.

The revolving earth. The earth also revolves, or follows a fixed path, around the sun. The path the earth takes around the sun is called its orbit. It takes the earth one year to make one complete orbit around the sun.

The tilted earth. One other fact is important to remember about the moving earth. This is that the earth's axis is tilted in relation to its yearly path around the sun. The axis stays tilted at a $23\frac{1}{2}°$ angle. This tilting of the earth's axis is the reason why seasons change in some parts of the earth. It is also the

reason why some parts of the earth receive more or fewer hours of daylight as the seasons change.

Understanding Earth-Sun Relationships

The graphics on page 66 will help you understand the relationships among the earth's movements, the tilt of its axis, and the amount of heat and light it receives from the sun.

Page 66 has several parts to it. The top part is a diagram that shows the earth's yearly path around the sun. It shows the earth's position, the tilt of its axis, and its daily rotation at four different times during the year. Each position is labeled to show the beginning of a season in the Northern Hemisphere. (In the Southern Hemisphere, winter begins on June 21 and summer on December 22. Spring begins on September 22 and fall on March 21. Always remember that seasons in the two hemispheres are not the same.)

Three globe maps make up the second part of the graphic. Each globe map shows the Western Hemisphere at the beginning of one of the seasons. It shows a half that is in daylight and a half that is in darkness. It shows what part of the earth receives the sun's direct rays on each date. It also shows the locations of three cities in the Western Hemisphere. What are these cities? (As you study these globe maps, remember that what is true for the Western Hemisphere is also true for the Eastern Hemisphere.) Take time to read the caption under each map.

March 21 (Vernal Equinox)

Equator

Direction of rotation causing day and night

Tropic of Cancer

June 21 (Summer Solstice)

EARTH'S YEARLY PATH AROUND THE SUN

Tropic of Capricorn

December 22 (Winter Solstice)

Equator

September 22 (Autumnal Equinox)

North Pole

Dawson — Arctic Circle
Tropic of Cancer
Equator — Quito
Tropic of Capricorn
Buenos Aires
Antarctic Circle

sun

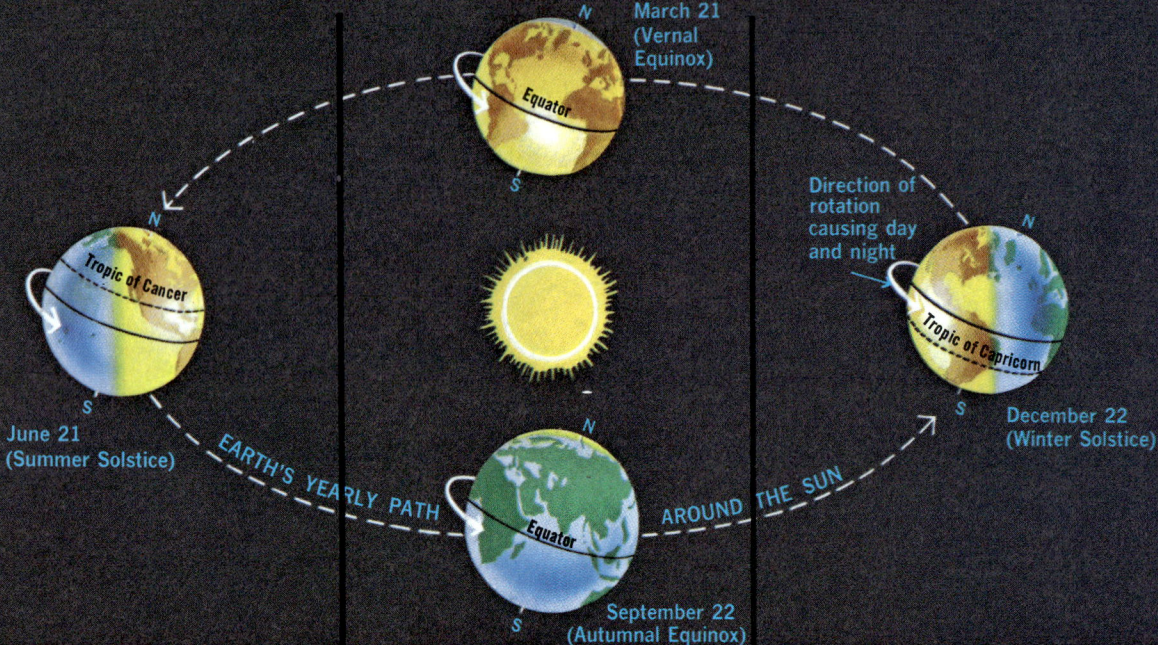

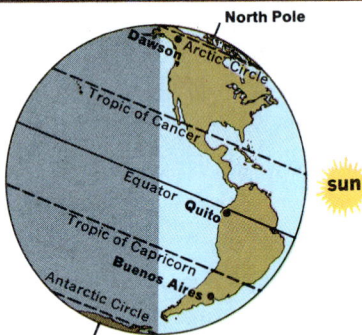

On June 21, the direct rays of the sun are over the Tropic of Cancer. The Northern Hemisphere is beginning summer; the Southern Hemisphere is beginning winter.

Hours of Daylight

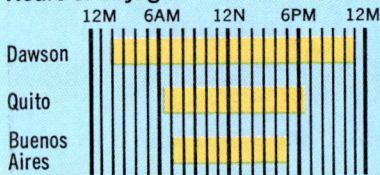

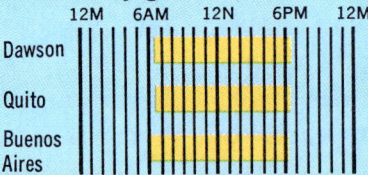

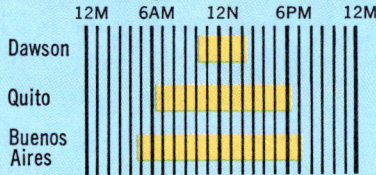

	12M	6AM	12N	6PM	12M
Dawson					
Quito					
Buenos Aires					

On June 21, the Northern Hemisphere has more hours of daylight than the Southern Hemisphere.

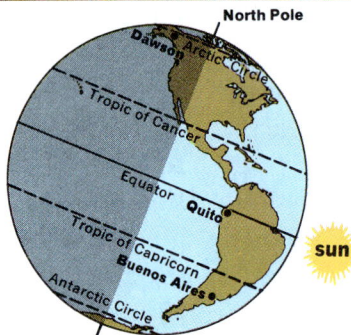

North Pole

Dawson — Arctic Circle
Tropic of Cancer
Equator — Quito
Tropic of Capricorn
Buenos Aires
Antarctic Circle

sun

On March 21 and September 22, the direct rays of the sun are over the equator. In the Northern Hemisphere, spring begins in March and fall in September.

Hours of Daylight

	12M	6AM	12N	6PM	12M
Dawson					
Quito					
Buenos Aires					

At the times of the two equinoxes, all places on earth receive 12 hours of daylight and 12 hours of darkness.

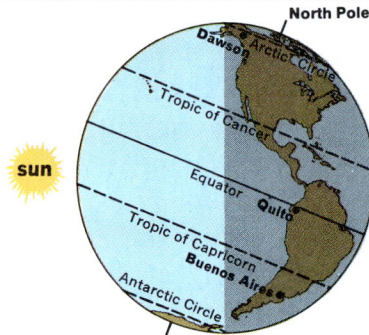

North Pole

Dawson — Arctic Circle
Tropic of Cancer
Equator — Quito
Tropic of Capricorn
Buenos Aires
Antarctic Circle

sun

On December 22, the direct rays of the sun are over the Tropic of Capricorn. The Northern Hemisphere is beginning winter, the Southern Hemisphere is beginning summer.

Hours of Daylight

	12M	6AM	12N	6PM	12M
Dawson					
Quito					
Buenos Aires					

On December 22, the Northern Hemisphere has fewer hours of daylight than the Southern Hemisphere.

Because the earth's surface is curved, the sun's rays do not strike all places on earth at the same angle. When rays of sunlight strike the earth directly, they cover a small area. The rays are very concentrated. They warm the earth's surface to a high degree. When rays strike the earth at a slanted angle, they cover a greater area. The heat is spread out. It is less concentrated. Which band of rays is less concentrated—A or B?

A

B

Below each globe map is a graph. Each graph shows how many hours of daylight each of the three cities has at each change of seasons. Take time to read the caption under each graph.

Have you noticed the heavy black lines dividing the top three parts of the graphic into three columns? You will understand earth-sun relationships more clearly if you study the graphics carefully from the top to the bottom of each column.

The summer and winter solstices. Starting with the column under June 21, use the globe map, the graph, and their captions to answer these questions. What season begins on this day in the Northern Hemisphere? What line of latitude receives the sun's direct rays? Is this line north or south of the equator? Is darkness covering the area beyond the Arctic Circle or the Antarctic Circle? According to the graph, which city receives the most hours of daylight? Is this city north or south of the equator? Which city receives the least hours of daylight? Is this city north or south of the equator?

Now move to the column under December 22. Study each part and any captions. Then answer the same questions asked for June 21. Note how the answers have changed.

A solstice is a time of the year when the direct rays of the sun are farthest north or farthest south of the equator. A solstice takes place on June 21 and on December 22. Which hemisphere has the greater number of hours of daylight on June 21? On December 22?

The busy harbor area of Bergen, Norway, is the center of a festival at which the people celebrate the arrival of the summer solstice with a bonfire.

The spring and autumnal equinoxes. The two equinoxes take place on March 21 and September 22. Study the three parts in the middle column. Notice that the sun's direct rays are over the equator. Notice that the hours of daylight and darkness are equal for all places on earth. An equinox is a time of the year when the direct rays of the sun strike the equator. How many hours of daylight does each place on earth receive at the time of an equinox? How many hours of darkness?

Temperature differences. The bottom diagram on page 66 helps to explain why some parts of the earth are warmer than others. Study the diagram. Read the caption carefully. Answer the question in the caption. Then answer these questions. Which would you expect to be warmer—the area between the Tropic of Cancer and the Tropic of Capricorn or the areas around the North Pole and the South Pole? Why?

Time Around the World

The way we keep time is related to rotation and revolution, the earth's movements around the sun. It takes the earth one year to revolve around the sun. Keeping time in years is easy—a year is the same everywhere on the earth. Keeping time in days and hours is not as easy. Because of rotation, the beginning of a day and the hourly times that follow are not the same everywhere on the earth.

Days and hours. In one day—one twenty-four-hour period—the earth rotates 360°. One hour is the length of time it takes for the sun's rays to pass over 15° of longitude. (You can check this for yourself by dividing 360, the total number of degrees of longitude, by 24, the number of hours in a day.)

Midnight and noon. A day begins and ends at midnight. People divide the day into two twelve-hour periods. One is from midnight to noon. These are the A.M. hours. The other period is from noon to midnight. These are the P.M. hours. But because midnight for any place occurs in darkness, noon (the time when the sun is highest in the sky) has long been used in figuring hours within a day.

Local time. Because of the earth's rotation, noon is always occurring someplace on earth. Start with the place where you live. When it is noon there, noon has already passed at a spot to the east. It has yet to occur at a spot to the west. Every four minutes, as the highest angle of the sun in the sky reaches another full degree of longitude, noon occurs at a spot farther to the west.

The system of keeping time by the sun is called local time. People used this system for hundreds of years. But changes came in the ways people lived. People formed nations and traded with other people in other nations. As people developed new and faster methods of communication and travel, they began to see that they needed a more fixed system of timekeeping. To meet this need, standard time was developed in 1884. Most nations in the world now use standard time.

Standard time. This system of timekeeping set up a model of 24 time zones. Each time zone covers 15° of longitude. Places within a time zone follow the local time of the line of longitude that passes through its center. One time zone was chosen as the starting point for figuring time within a day. The central line of longitude in that zone was used to determine noon for that zone. Time for any other zone would then be figured from that line, depending on how many zones east or west of the line it was. The prime meridian was chosen as the central line of the starting zone. Because the prime meridian passes through Greenwich (Gren'ich), a suburb of London, England, it is sometimes called the Greenwich meridian.

The line opposite the prime meridian (180° of longitude) became the line for determining midnight—the instant when one day ends and another begins. Crossing this line at any time takes a person into another day.

World time zones. But the model needed some changes to make it workable in everyday life. The time zones map on page 69 shows what changes have been made.

One change has to do with time zones. The boxes at the top and the bottom of the map show what time it is in the other 23 zones when it is noon at Greenwich. The times marked on the continents show how the time zones have been modified to fit the boundaries of nations, states, and cities. You will notice that some nations in western Europe and the Soviet Union have advanced their clocks one hour. Other nations have advanced their clocks a half hour. Still other nations follow local or irregular time.

A second change centers around the international date line—the line for determining the ending of one day and the beginning of another. Instead of following the 180° line of longitude, this line zigzags in certain places between land and water areas.

People who travel to distant places or make phone calls to people in other nations need to remember these things about time. Time in a zone to the east is always later than time in their home zone. Time in a zone to the west is always earlier. How much later or how much earlier depends on the number of zones between two places and whether nations have advanced their clocks or not.

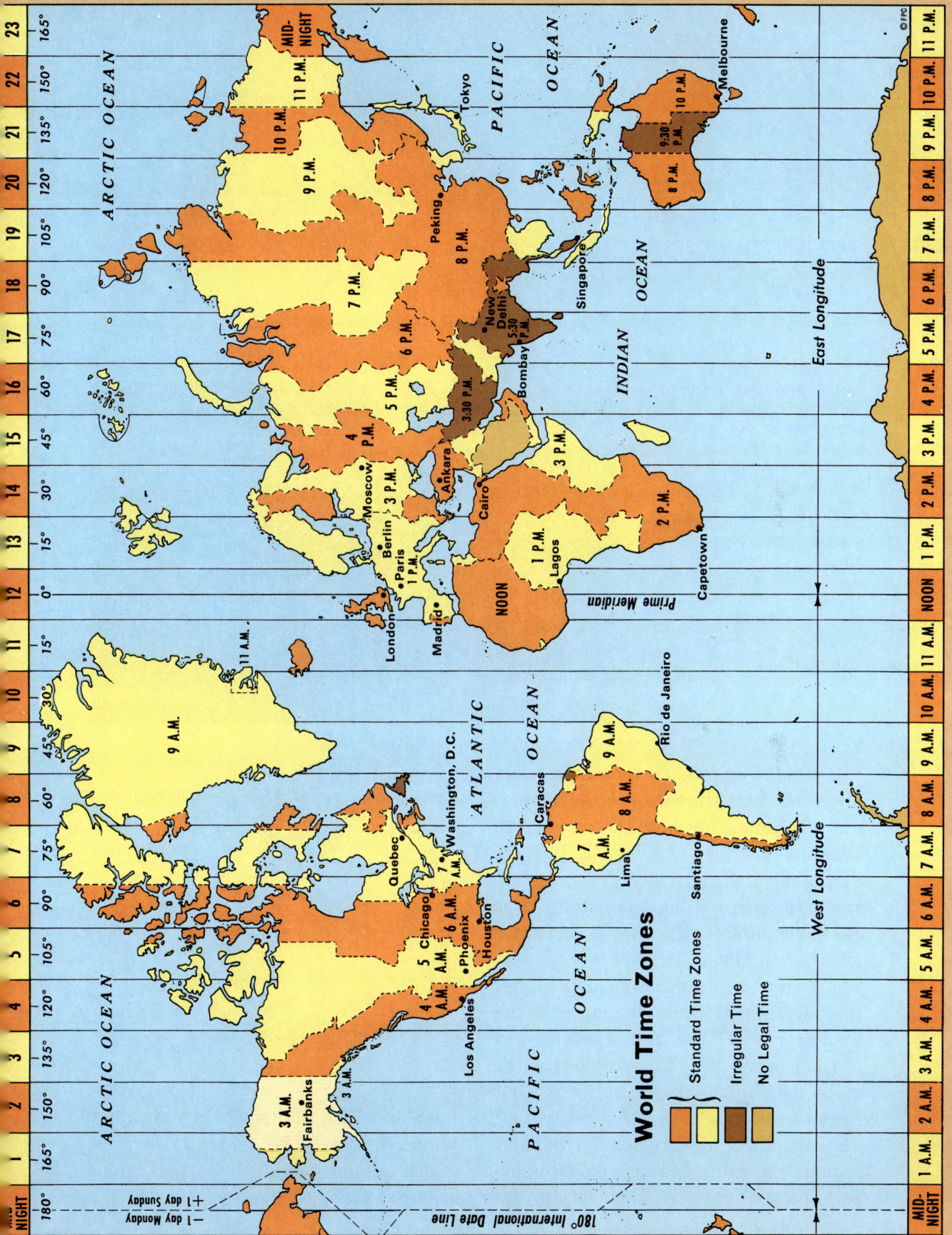

World Time Zones

Standard Time Zones

Irregular Time

No Legal Time

Top scale (left to right): 23 165° | 22 150° | 21 135° | 20 120° | 19 105° | 18 90° | 17 75° | 16 60° | 15 45° | 14 30° | 13 15° | 12 0° | 11 15° | 10 30° | 9 45° | 8 60° | 7 75° | 6 90° | 5 105° | 4 120° | 3 135° | 2 150° | 1 165° | NIGHT 180°

Right scale: 11 P.M. | 10 P.M. | 9 P.M. | 8 P.M. | 7 P.M. | 6 P.M. | 5 P.M. | 4 P.M. | 3 P.M. | 2 P.M. | 1 P.M. | NOON | 11 A.M. | 10 A.M. | 9 A.M. | 8 A.M. | 7 A.M. | 6 A.M. | 5 A.M. | 4 A.M. | 3 A.M. | 2 A.M. | 1 A.M. | MIDNIGHT

ARCTIC OCEAN

PACIFIC OCEAN

INDIAN OCEAN

ATLANTIC OCEAN

East Longitude

West Longitude

Prime Meridian

180° International Date Line

−1 day Monday   +1 day Sunday

MID-NIGHT   11 P.M.   10 P.M.   9 P.M.

MIDNIGHT

Melbourne 10 P.M.   9:30 P.M.   8 P.M.

Tokyo

Peking   8 P.M.

Singapore

New Delhi 5:30 P.M.   Bombay   5:30 P.M.

3:30 P.M.

7 P.M.   6 P.M.   5 P.M.   4 P.M.

Moscow 3 P.M.   Berlin   Ankara   3 P.M.   Cairo

Paris 1 P.M.   London   Madrid

Capetown   2 P.M.

Lagos 1 P.M.   NOON

9 A.M.

11 A.M.

Rio de Janeiro

Caracas

Washington, D.C.   7 A.M.

Quebec   8 A.M.   9 A.M.

Chicago 6 A.M.   Houston

Phoenix   5 A.M.

Los Angeles   4 A.M.

Lima 7 A.M.   Santiago

Fairbanks 3 A.M.   3 A.M.

9 A.M.

69

In the polar lands north of the Arctic Circle, a Laplander on skis watches a herd of reindeer foraging for food. In the tropics, a New Guinea family found this grassy patch above the rain forest a good place to make home.

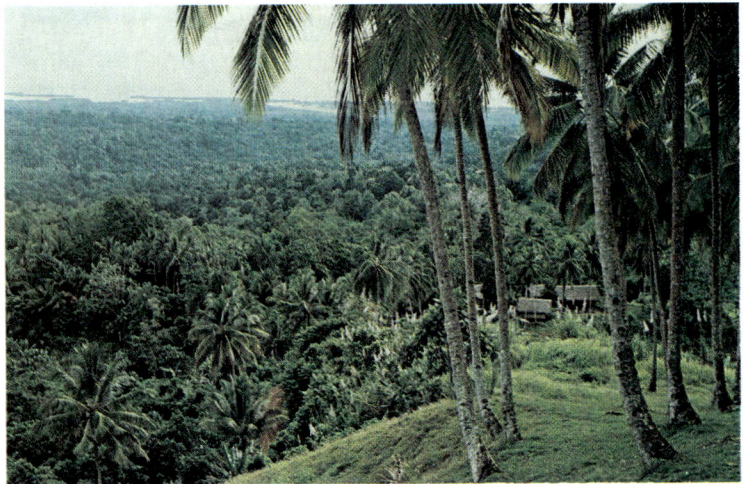

Latitude regions. Geographers often divide the earth into three large temperature regions. They base this division on lines of latitude. A line of latitude marks the angle at which the sun's rays strike the earth at that place at the time of an equinox. The angle indicates whether the heat is spread over a very small area or a larger area. The higher the degree of latitude, the larger is the area over which the heat is spread.

You can see these three regions on the globe map on this page. Between what lines of latitude is the tropical region? The warm and cold regions? The polar regions? In which region would you expect the temperatures to be hot all year long? In which region would you expect temperatures to be cold all year? In which region would you expect great seasonal changes?

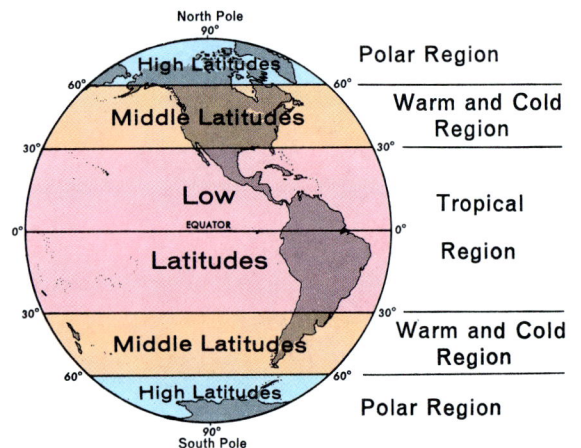

North Pole
90°
High Latitudes
60° — 60°
Middle Latitudes
30° — 30°
Low
EQUATOR
Latitudes
0° — 0°
30° — 30°
Middle Latitudes
60° — 60°
High Latitudes
90°
South Pole

Polar Region

Warm and Cold Region

Tropical Region

Warm and Cold Region

Polar Region

Spring (left), summer (top right), fall (center right), and winter scenes like these are found in middle latitudes north of the equator.

Keeping Facts in Focus

1. Do periods of daylight alternate with periods of darkness because of rotation or because of revolution?
2. What is an equinox? A solstice?
3. On what two dates do the equinoxes fall?
4. Why do geographers use lines of latitude to divide the earth into temperature regions?
5. What answers would you give to the key questions on page 64?

Working with Ideas

1. What would the climate of your city, town, or village be like if the earth were not tilted on its axis?
2. What changes might take place if the earth stopped rotating on its axis?

Unit 1 REVIEW WORKSHOP

Test Your Geographic Knowledge

A. Choose the letter of the item that correctly completes each statement.
1. The prime meridian is a (a) line of longitude, (b) line of latitude, (c) line of magnitude.
2. A place at 40° N latitude is in (a) the low latitudes, (b) the high latitudes, (c) the middle latitudes.
3. September 22 in the Southern Hemisphere is the start of (a) summer, (b) spring, (c) fall.
4. For a map that reads 1:200, one unit of measurement stands for (a) 200 miles, (b) 200 kilometers, (c) either 200 miles or 200 kilometers.
5. A line of latitude that cuts the earth into two hemispheres is (a) the tropic of cancer, (b) the prime meridian, (c) the equator.

B. Tell whether each statement is true or false.
1. A globe is a scale model of the earth.
2. Every map always shows both shape and size accurately.
3. A line graph always has two scales, a vertical one and a horizontal one.
4. A band of sunlight that strikes the earth at a slant spreads heat over a greater distance than a band that strikes the earth directly.
5. Longitude is measured east or west of the equator.

Apply Your Reading Skills

A. Outlining is one way of organizing information. Study the outline given below. Compare it with the headings for the lesson *Using Globes.*

   I. Using Globes
     A. What Globes Show
       1. Scale and distance
       2. Size and shape
       3. Direction
     B. The Global Grid
       1. Parallels of latitude
       2. Meridians of longitude
       3. Using latitude and longitude

     C. Hemispheres
       1. Northern and Southern
       2. Eastern and Western
       3. Land and Water

Now make an outline for one of the other lessons in Unit 1. Follow the form above.

B. Use encyclopedias, books, and other reference tools to do research. Then prepare a report on ONE of the following topics.
1. What myths did Egyptians or other early peoples have that were explanations about daylight following darkness, the changing seasons, or other workings of nature?
2. What have we learned about the mysterious sun?
3. What do we know about the moon?
4. Do scientists think there is life in outer space?

Apply Your Geographic Skills

A. Use a classroom globe or a map in the Atlas to estimate the latitude and longitude of each of the cities listed below.

Calcutta, India	Berlin, Germany
Sydney, Australia	New Orleans, La., U.S.A.
Moscow, U.S.S.R.	Bogotá, Colombia
Peking, China	Ottawa, Canada
Nairobi, Kenya	Tokyo, Japan

B. Tell what kind of graph you would make to show best each of the following sets of facts:
1. The populations of the five largest cities in your state.
2. The hourly temperatures of any day.
3. The percentage of the class that has green eyes, blue eyes, brown eyes, etc.

C. Do research on one of the ideas listed above. Then make a graph based on the information you obtain.

D. Answer the questions below by reading the graph on page 73.
1. Which country is slightly larger than the other?
2. Which country has more people?

Country A	Country B	Legend
■■■■■■■■■▮	■■■■■■■■■▮	■ = 1 million square kilometers
(3 symbols)	(9 symbols)	= 1 million hectares used for farming and grazing
(many persons)	(fewer persons)	= 10 million persons
(9 symbols)	(2 symbols)	= 1 million tons of minerals mined
(5 symbols)	(8 symbols)	= 1 million tons of food produced
(1 symbol)	(4 symbols)	= 1 million tons of food exported
(7 symbols)	(2 symbols)	= 100 million kilowatt-hours of hydroelectric power
(10 symbols)		= 1,000 new housing units
(7 symbols)	(3 symbols)	= $10 billion invested in capital industry
(11 symbols)	(4 symbols)	= $1 billion worth of manufactured goods for export and home use
(6 symbols)	(2 symbols)	= $10 million spent on transportation facilities
(8 symbols)	(1 symbol)	= $10 billion spent on military equipment

3. Which country has more land used for farming and grazing?
4. Which country exports more food?
5. Which country produces more minerals?
6. Which country produces more hydro-electric power?
7. Which country produces fewer manufactured goods for home use and export?
8. Which country spends more money on transportation?
9. Which country has the greater amount of money invested in industry?
10. Which country spends less money on military equipment?

Apply Your Thinking Skills

Answer each question, using the facts in the graph on page 73. Explain your reasoning for each answer.

1. Which country probably employs more workers in industry than in agriculture?
2. Which country is less likely to be thought of as a world military power?
3. Which country is likely to have more people living in farm areas than in city areas?
4. In which country are up-to-date transportation facilities likely to be lacking?
5. In which country are the people likely to earn more per person each year?
6. If one of these countries were the United States, which one do you think it would be?

Discuss These Points

Learning how to study is the main goal of many of the activities in this book. Studying is a skill you need to succeed as a student. It is also a skill you will need as an adult. Suppose you want to buy something that costs a lot of money, such as a car. So you read about cars in magazines and newspapers to compare different makes and models. You test-drive different cars. Once you decide on the kind of car you want, you read the fine print in guarantees and in contracts. After you buy the car, you read the owner's manual to make sure you take good care of the car. What are some other ways and other times you will use study skills as an adult?

Expand Your Geographic Sights

Asimov, Isaac. *How Did We Find Out the Earth is Round?* (Walker, 1972).

Brandreth, Gyles. *Amazing Facts about our Earth* (Doubleday, 1981).

Carey, Helen. *How to Use Maps and Globes* (Watt, 1983).

Hey, Nigel S. *The Mysterious Sun* (Putnam, 1971).

Kraske, Robert. *Is There Life in Outer Space?* (Harcourt, 1976).

Lowenstein, Dyno. *The First Book of Graphs* (Watts, 1976).

Schwartz, Julius. *Earthwatch: Time-Space Investigations with a Globe* (McGraw-Hill, 1977).

Unit 2 Land

The step-like ledges, called benches, of the Bingham Canyon open-pit mine in Utah are silent reminders of the more than 8 million tons of copper ore that have been scooped from the mine since it began operations in 1906.

Lesson 1 The Earth and Its Structure

Reading Focus

1. Good readers, like good drivers, know when to change speeds. Skimming, an important reading skill, involves changes in speed. Most skimming is done very quickly. But sometimes you need to slow down or even come to a stop. In studying a lesson, skim it more than once. The first time, skim it quickly just to get a "feel" for the ideas in a lesson. The second time, skim it to find answers to the questions listed below and at the end of the lesson. Read the questions slowly and carefully, picking out key words. Then skim, looking for these key words. Remember that your skill in skimming is measured by how well you find answers to questions, not by how fast you turn the pages. After you have skimmed a lesson, you will want to give it a slower, more careful reading.

2. As you read this lesson, look for answers to these key questions:
   a. How do some scientists think the earth may have begun?
   b. What are the core, the mantle, and the crust?
   c. What are the three theories scientists use to explain changes in the earth's surface?

Vocabulary Focus

theory	*core*	*crust*	*plate*
geologist	*mantle*	*ridge*	*trench*

Nothing about the earth's beginnings is known with any certainty. Scientists have only theories, or scientific guesses, about how the earth came to be.

According to one theory, a hot, spinning cloud of dust and gas formed in space. Parts of this dust and gas cloud separated, forming the sun, the earth, and the other planets.

The part of the dust and gas cloud that became the earth slowly cooled. As it cooled, it gradually became a solid mass. There are scientists who believe the earth became a solid mass about $4^1/_2$ billion years ago.

The Earth—From Inside Out

Geologists are scientists who study the materials of the earth and how they are put together. It is not difficult for geologists to discover what the earth is like. They can study rocks and rock formations on the surface. They can dig below the surface to study rocks and rock formations hidden from view. They

The Earth's Structure

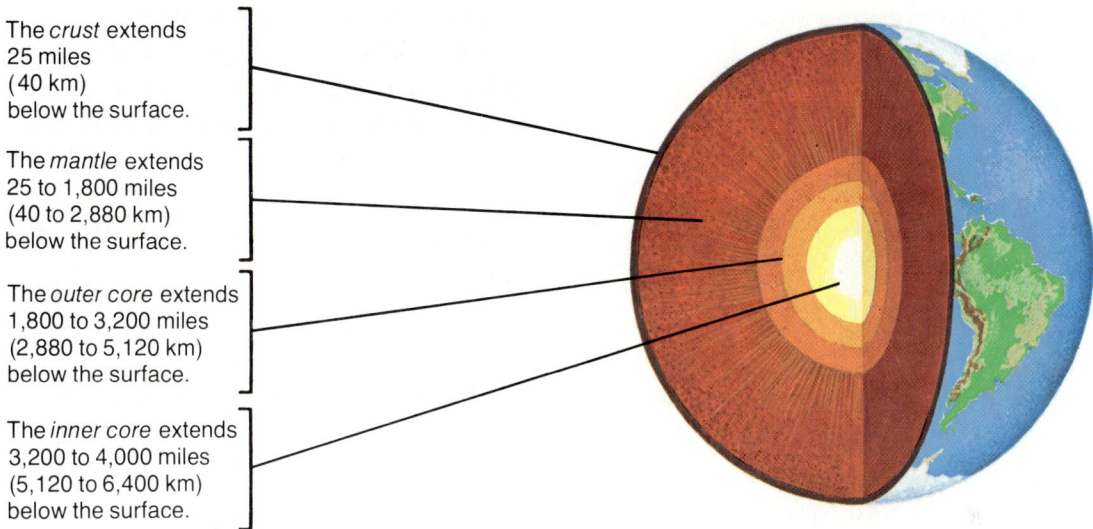

The *crust* extends 25 miles (40 km) below the surface.

The *mantle* extends 25 to 1,800 miles (40 to 2,880 km) below the surface.

The *outer core* extends 1,800 to 3,200 miles (2,880 to 5,120 km) below the surface.

The *inner core* extends 3,200 to 4,000 miles (5,120 to 6,400 km) below the surface.

can even drill more than about 6 miles (9.6 km) below the surface, bring up samples, and study them. Below this level, geologists must rely on their study of earthquake waves to tell them what the earth's inner structure is like.

Like sound waves, earthquake waves carry vibrations. These vibrations travel through the solid materials of the earth. They change speeds and directions as they move through different kinds of rock. From studying these waves, which can be recorded on machines, geologists believe there are three parts to the earth—the core, the mantle, and the crust.

The earth's core. At the center of the earth, in its deepest part, is the core. The core has two parts—an inner part and an outer part. The inner core is solid. It is mostly iron mixed with some nickel. The outer core is also made up of iron and nickel. But geologists think it is a thick liquid rather than solid metal. Temperatures at the edges of the outer

core are at least 4,000° F (2,200° C). At the center of the inner core, temperatures may be as high as 9,000° F (5,000° C).

The earth's mantle. This middle part of the earth is a thick layer of hot rock, most of which is very heavy and solid. Where the mantle meets the core, temperatures may be as high as 4,000° F (2,200° C). At the outer edges of the mantle, close to the earth's crust, temperatures are about 1,600° F (870° C).

The earth's crust. The crust is a thin outer shell of lighter rock covering the earth. Look at the diagram on this page. How far below the earth's surface is the inner core? The outer core? About how thick is the earth's mantle? Compare the thickness of the earth's crust with the thickness of the core and the mantle. Do you see why geologists call the earth's crust a "thin skin"?

Geologists divide the earth's crust into two layers. The bottom layer completely circles the earth. In some places,

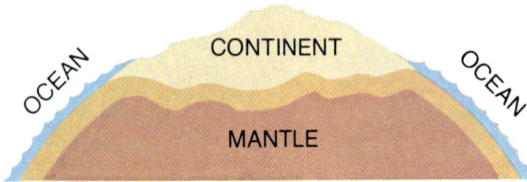

Lighter, upper layer of earth's crust

Heavier, lower layer of earth's crust

mostly on the seafloor under the oceans, it is as little as 5 miles (8 km) thick. At other places, it is as much as 10 miles (16 km) thick. The rock in this layer is lighter than the mantle rock. But it is heavier than the rock in the upper layer of the crust.

The upper, lighter layer of the earth's crust "floats" on the heavier, bottom layer. In most places, the upper layer extends above the surface of the ocean waters to make up the earth's large landmasses—the continents and large islands. The continental landmasses make up less than 30 percent of the earth's surface. It is on this part that most of the earth's 4 billion people live.

Theories about the Earth's Crust

Scientists are not sure how or when the lighter layer of the earth's crust first began to build up. But they are sure that it is much older than the seafloor. Most scientists are also convinced that the earth's continents are slowly moving, or drifting.

Continental drift. Alfred Wegener, in 1912, was the first scientist to propose the theory of continental drift. Wegener said that all continents were once part of one huge landmass surrounded by one huge ocean. He called the landmass Pangaea (Pan·jē′ə) and the ocean Pan-thalassa (Pan′thə·las′ə).

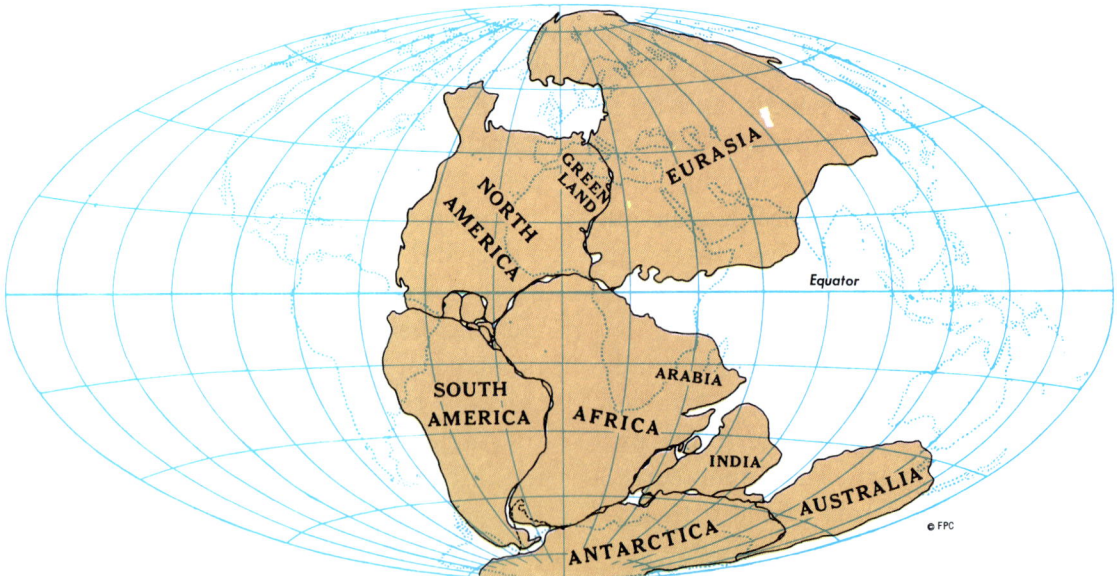

Over 200 million years ago, when today's continents might have been a part of Pangaea, they probably were in the positions shown here.

According to his theory, Pangaea split into two landmasses about 200 million years ago. (Some scientists today believe that the continents may have joined and split several times even before that.) Wegener went on to say that the northern part of Pangaea split into the continents of North America, Europe, most of Asia, and the island of Greenland. The southern part split into the continents of South America, Africa, Antarctica, Australia, and the subcontinent of India. Through the years, these landmasses "drifted" to their present locations. Wegener also said that the continents are still drifting.

A few scientists of Wegener's time supported his idea. For evidence, these scientists noted the way that the eastern bulge of South America seems to fit into the western part of Africa. They also discovered that there were places on the two continents with the same kinds of rocks and plant and animal life. This match seemed to support the idea that the two continents had once been joined.

As the years passed, more scientists began to take Wegener's theory seriously. They began to look for new evidence

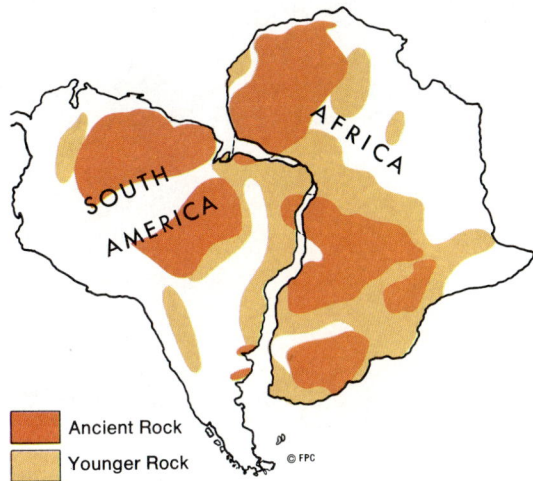

Ancient Rock
Younger Rock

of continental drift. They found their new evidence on the seafloor during the 1950s and 1960s.

Seafloor spreading. During this time, the scientists found a worldwide system of mountainlike ridges in all the oceans. They believe these ridges are formed by new crust rising from deep within the hot mantle of the earth. Where these ridges appear, the seafloor seems to be "cracked" and spreading apart. The new crust becomes part of the seafloor and so renews it. The discovery of seafloor spreading excited scientists all over the

Along cracks on the seafloor, molten lava wells up, quickly cools, and hardens. Over the centuries, lava forms, such as the one shown here, have built up the ridges that are thought to be evidence of seafloor spreading.

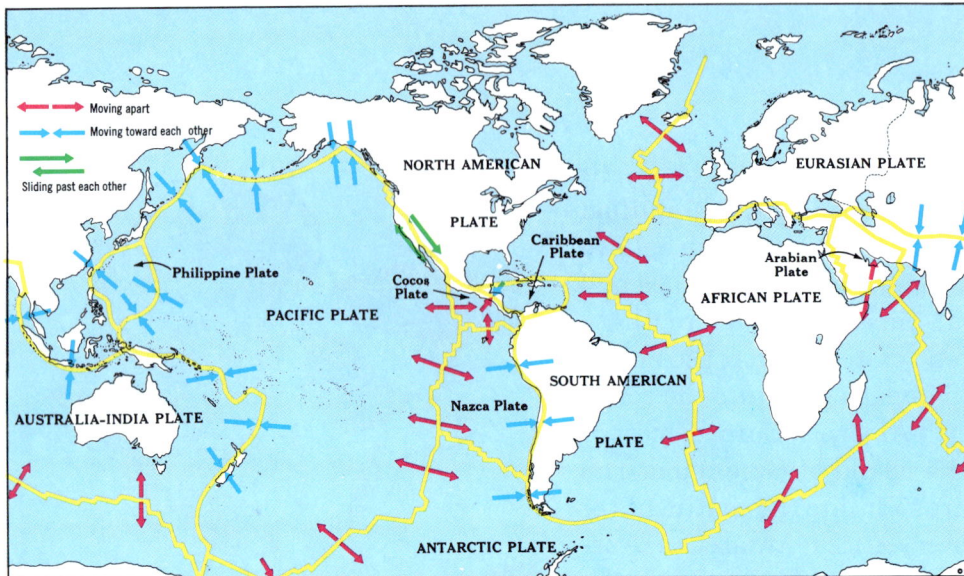

These are the plates into which geologists think the earth's crust is divided. Arrows along plate edges show the directions of movement.

world. It led to a whole new theory—the theory of plates and plate movement.

Plate movement. From studies of the ridges on the seafloor, geologists now think the earth's crust is divided into a number of rigid plates—7 larger ones and several smaller ones. According to this theory, each plate moves very slowly—no more than $1/2$ inch to 4 inches (1.3 to 10 cm) a year—as it "floats" on the heavier rock of the earth's mantle. This movement is very slow on the human time scale, but it is very fast on the geologic time scale.

Changes are always taking place along the edges of these plates. At those edges where plates move away from each other, seafloor spreading takes place. Here, new material from the mantle pushes to the surface. At those edges where plates move toward each other, material is being pushed down into the earth's mantle, forming deep ocean trenches. At these trenches, material from one plate slides underneath the plate next to it. The heat from the

mantle melts this material. Along other edges, the plates may simply grind against each other or even crash into each other. Scientists hope the study of plates will lead them to a better understanding of the earth's structure.

Keeping Facts in Focus

1. How long ago do some scientists think the earth became a solid mass?
2. How does the earth's inner core differ from its outer core?
3. How does the top layer of the earth's crust differ from its bottom layer?
4. What answers would you give to the key questions on page 76?

Working with Ideas

1. For centuries people have used the phrase *the solid earth.* Explain why you would or would not consider this phrase a good description of the earth.
2. Why is it important for people living on earth to have as much information as possible about the structure of the earth?
3. How does the work of geologists benefit you?

Lesson 2 Inner Forces
Change the Earth's Surface

Reading Focus

1. Once you have previewed a lesson by skimming it, you should be able to summarize its main ideas. You already know that the size of a heading in a lesson says something about the importance of the idea it states. The largest heading is used for the title of the lesson. The title states the lesson's main topic. What is the title of this lesson? The next size heading points to main ideas that are related to the lesson's main topic. Find the three main ideas stated in this next size heading. Now look at any smaller headings under each of these main ideas. What are some mountain-building actions and how have they helped to form mountains today? What are some of the different kinds of volcanoes? What does the lesson tell you about earthquakes? Now put all these ideas together. Write a paragraph in your own words that summarizes the main ideas of this lesson.

2. As you read this lesson, look for answers to these key questions:
   a. How many kinds of mountains are there?
   b. What different kinds of volcanoes are there?
   c. What are seismic waves? What are tsunamis?

Vocabulary Focus

fault	*composite*	*intermittent*	*seismic*
molten	*dormant*	*extinct*	*tsunami*
crater			

To people living on the earth, the surface of the earth usually appears firm and unchanging. But geologists know that this is not true. The surface of the earth is always changing. Land-building changes come from deep within the earth. These changes are directly related to seafloor spreading and the movements of the earth's plates, both of which cause continents to drift.

All of these actions produce stresses and strains within the earth as hotter materials from the mantle rise at some plate edges while colder materials nearer the surface sink at other plate edges. When these stresses and strains become very great, the earth's crust gives way. On land and on the seafloor, mountains rise up, molten materials spill out, and earthquakes jolt the earth.

Mountain Ranges and Ages		
Range	Continent	Age (in years)
Cascades	North America	1,000,000
Himalayas	Asia	25,000,000
Alps	Europe	40,000,000
Andes	South America	70,000,000
Rockies	North America	70,000,000
Coast Ranges/ Sierra Nevadas	North America	135,000,000
Juras	Europe	135,000,000
Caucasus	Eurasia	225,000,000
Urals	Eurasia	225,000,000
Appalachians	North America	225,000,000
Green Mountains (Vermont)	North America	500,000,000
Adirondack Mountains	North America	2,500,000,000

Mountain Building

According to most scientists, the great mountain ranges of the world were not all formed at the same time. Nor were they raised in a single uplifting of the earth's crust. Each one took millions of years and many movements to be raised to its greatest height. Look at the chart on this page. What is the youngest mountain range on the earth's surface? What is the oldest mountain range?

Mountain-building actions. Sometimes, sections of the earth's crust break up into blocks. One block, with its layers of rock, moves up or sinks down. Or one block moves sideways along the break.

Such breaks are called faults. You can see a block-faulted mountain in the diagram on page 83.

Pressures sometimes build up on both sides of a section of the earth's crust, squeezing but not breaking apart the layers. As they are squeezed, the layers of rock rise up and sink down in rounded waves, or folds. The result is folded mountains.

At other times, molten materials from deep within the earth push up into the earth's crust without breaking through. As the molten materials push upward, they force the rock layers to rise, forming a rounded, or dome, mountain.

Mountains also form when molten material breaks through the earth's crust. There are times when molten material bubbles up from openings in the crust and only flows along the ground. But there are other times when the material breaks through the crust with great force. Then it forms volcanic peaks or mountains.

Mountains today. All of these actions have helped form the mountain ranges of the world at one time or another. It is not always known which action took place first or how many times it took place. Faulting, we know, helped give the Sierra Nevada Range in California its block shape. Pressure from within the earth lifted the Black Hills of South Dakota, giving them a dome shape. Parts of the Alps, the Himalayas, and the Northern Rockies were formed by folding. Folding also helped to shape the Appalachian Mountains. The Southern Rockies and the Andes were formed

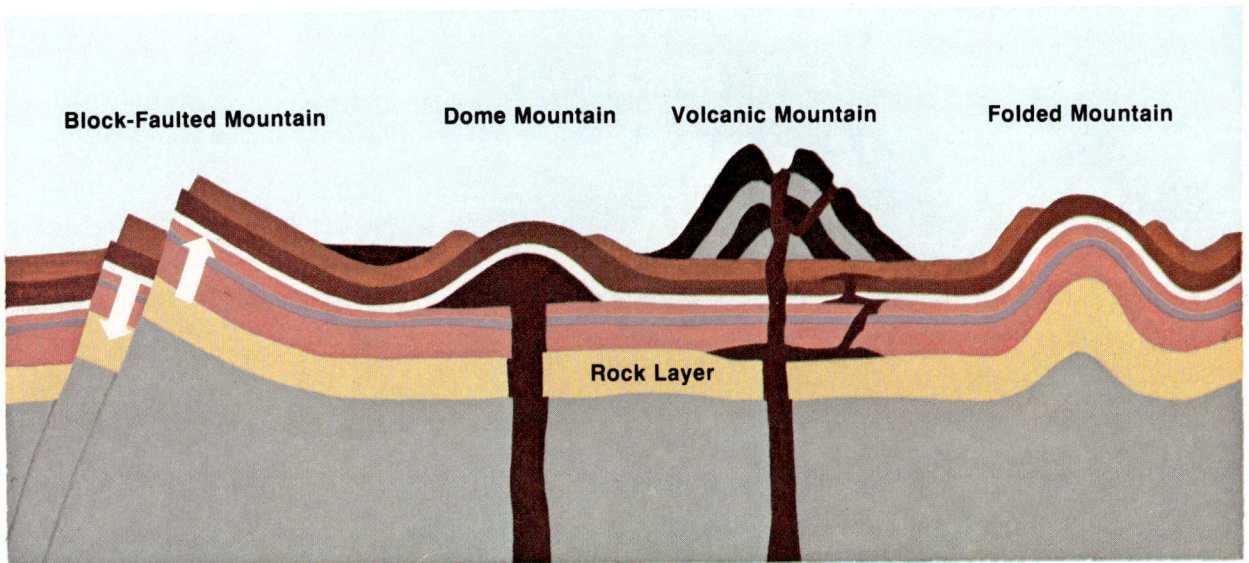

Block-Faulted Mountain   Dome Mountain   Volcanic Mountain   Folded Mountain

Rock Layer

partly by faulting, partly by folding, and partly by molten materials from within the earth.

Volcanoes

Volcanoes, openings in the earth's crust through which lava, hot gases, or rock fragments escape, are found both on the land and on the seafloor. Volcanoes differ in the ways they erupt, in the kinds of materials that escape from them, and in the kinds of landforms they build up. A look at some of the world's famous volcanoes will point out the different kinds of volcanoes.

A shield volcano. The Hawaiian Islands, like many other islands, are the tops of volcanoes that have been built up on the seafloor by many eruptions in the past. There are still some active volcanoes on these islands. An active volcano is one that always shows signs of volcanic action.

In an active volcano, steam and hot gases may rise from cracks or from a crater. Lava may bubble within a crater or flow down its sides. Or lava may be thrown into the air, where it hardens and comes back to earth as ashes, cinders, or pieces of volcanic rock.

One kind of active volcano found in the Hawaiian Islands is Mauna Loa, a shield volcano on the island of Hawaii. A shield volcano seldom erupts with explosive force, because it usually has more than one crater. Steam rises into the air and lava bubbles within its craters. At times, when lava flows over, it flows swiftly and covers a wide area. As a result, a shield volcano is usually wider than it is high. It seldom builds up the cone-shaped mountain that is common to either cinder-cone or composite volcanoes.

A cinder-cone volcano. Parícutin, a cinder-cone volcano in western Mexico, erupted suddenly in 1943 and stopped

Hot molten lava streams from cracks on Mauna Loa, a shield volcano.

erupting just as suddenly in 1952. The first sign was a crack in the ground of a cornfield, followed by rising hot gas and steam. Thirty minutes later, explosions began. Clouds of gas and ashes shot as high as 4 miles (6 km) into the air. For the rest of its active life, the explosions continued. The ashes that fell to the ground formed a cone-shaped mountain that now rises 1,345 feet (410 m) above the level of the cornfield. Because Parícutin has shown no further signs of volcanic actions since 1952, it is said to be dormant, or sleeping. At some future time, it may once again wake up to send more ashes to fall on the earth.

A composite volcano. Some volcanoes are built up in two ways. They have the kind of lava flows that are common to shield volcanoes. They also have the kinds of explosions common to cinder-cone volcanoes. Mount Fuji in Japan is one of the most famous of this type. The drawing on page 85 shows how a layer of

Parícutin is a cinder-cone volcano.

ash and cinder is followed by a layer of lava. The layers build up an almost perfectly formed cone with steep sides.

Like Parícutin, Mount Fuji is a dormant volcano. Some volcanoes erupt from time to time without showing much volcanic action between eruptions. Volcanoes like this are said to be intermittent. Mount Etna in Sicily is an intermittent volcano.

Other volcanic mountains have shown no action for hundreds of years. These are said to be extinct. Mount Aconcagua in Argentina and Mount Kilimanjaro in Tanzania are extinct volcanoes.

What is molten rock called when it is below the earth's surface? What is it called when it reaches the earth's surface? To build up a cone-shaped mountain, composite volcanoes erupt many times, usually with explosive force.

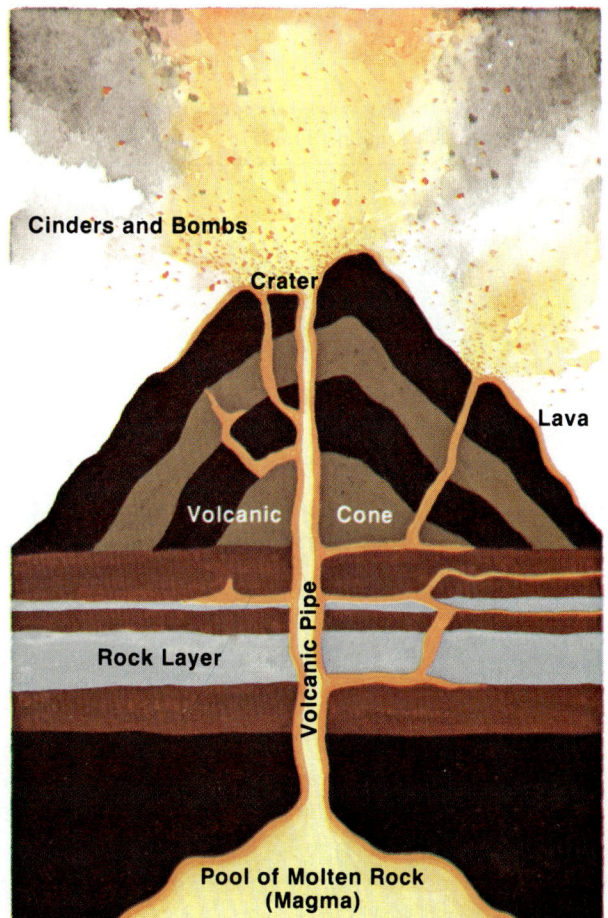

Cinders and Bombs

Crater

Lava

Volcanic Cone

Rock Layer

Volcanic Pipe

Pool of Molten Rock (Magma)

The snow-capped peak of Mt. Fuji towers over the countryside west of Tokyo, Japan.

Most earthquakes and volcanic eruptions take place along the edges of the earth's crustal plates.

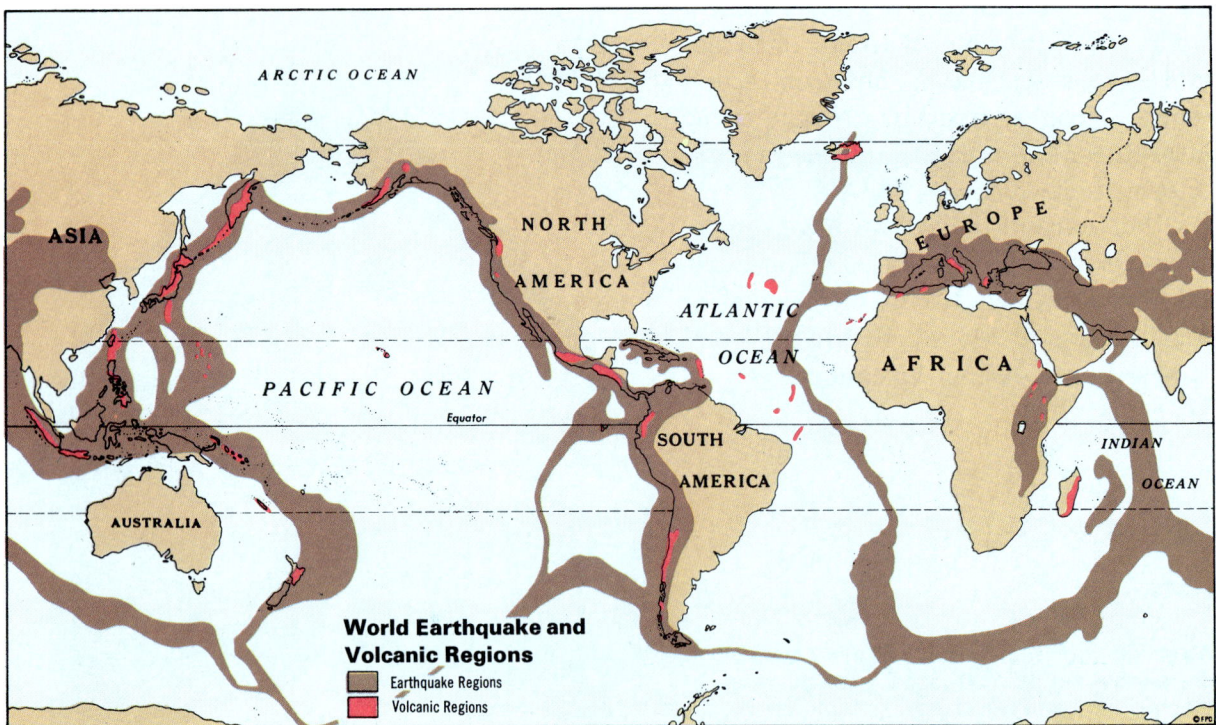

World Earthquake and Volcanic Regions

■ Earthquake Regions
■ Volcanic Regions

Earthquakes

Earthquakes and mountain building go together. Any shift in the earth's crust, no matter how small, causes the earth to tremble. About 800,000 earthquakes are recorded each year. But only about 50,000 of them are strong enough to be felt by people living in the areas where they occur. Earthquakes that take place in heavily populated areas can cause much damage and great loss of lives. These are the earthquakes that make the headlines and make people wonder, "Can an earthquake happen where I live?"

Where earthquakes take place. Earthquakes usually take place along faults or at the edges of the earth's plates. Earthquakes, however, usually have their centers below the earth's surface. A few take place at depths as great as 400 miles (640 km) below the earth's surface. But most have their centers about 25 miles (40 km) below the earth's surface.

Earthquake waves. An earthquake sends seismic, or shock, waves through the earth. These waves carry great amounts of energy in a very short time. The waves cause the rocks through which they move to vibrate and to change shape. Damage is greatest at the center, becoming less severe as the waves move away from the center. This is because energy in the waves decreases with distance traveled. How long the shocks last and how many occur depends on the strength of the earthquake.

Mount St. Helens, located in southwestern Washington, had been dormant for over 100 years when it erupted on May 18, 1980. The eruption blew off 1,300 feet (390 m) of the mountain's top, sent smoke and ash as high as 60,000 feet (18,000 m), leveled over 200 square miles (520 sq km) of the countryside, and took 61 lives. It was the first volcanic eruption to take place in the United States proper since 1917.

Predicting Earthquakes

Picture this scene. It is the middle of the night. Thousands of people are sitting in open spaces far from buildings, bridges, and other structures. They have left their homes, even though the temperature is below zero. Fear shows on their faces as they crowd in small groups and try to keep warm.

This actually happened in Haicheng, China, in 1975. The people in this town in northeastern Manchuria were told that an earthquake would strike within hours. And it did. Because they left their homes when they were warned, over 10,000 people saved their lives.

Only in the last half of this century has progress been made in predicting earthquakes. Seismologists (people who study earthquakes) now have many of the instruments they need to measure changes in the earth's crust.

Recording shock waves. Seismometers record shock waves—strong ones, weak ones, even the ones people do not feel. In using these instruments, seismologists look for changes in the number and speed of shock waves. Such changes often signal that an earthquake is likely to happen soon.

Checking on earth shifts. Creepmeters are solid rods about 50 feet (15 m) long that are placed across a fault. Anchored firmly in the ground only at one end, a rod is free to move when the earth moves. At the free end of the rod, a meter measures any shift or "creep" that takes place between the two ends of the rod. Creeping is another signal that pressures are building up along a fault.

Measuring tilt. Tiltmeters measure changes in the tilt or slope of land. Any change in the slope of the ground is another sign that some kind of pressure is building up. A tiltmeter has a bubble filled with a special liquid. As the ground tilts, the liquid covers or uncovers small rods. The movement of the liquid over the rods sets up an electric current that is fed to a meter.

Reading the meters. In some cases, people read these meters daily or hourly. In other cases, the meters are equipped with radio or other electronic transmitters that send the data back to special centers. One such center has been set up in California. At the centers, seismologists study the data. Many seismologists in the United

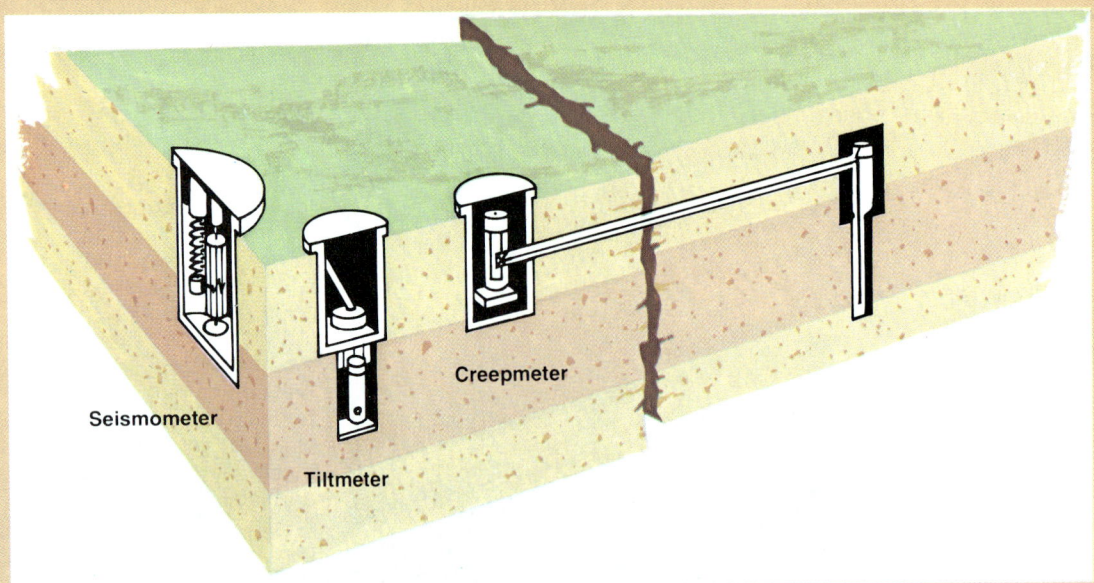

Seismometer

Tiltmeter

Creepmeter

States rely only on data from these instruments. Some seismologists, especially the Chinese, also watch for certain signs in nature.

Studying nature's signs. In warning about the Haicheng earthquake, the Chinese used reports from farmers and other workers they had trained to read nature's signs.

The Chinese believe that animals behave differently just before a quake strikes. The Chinese workers reported strange behavior in their barnyard animals. They also reported that water in their deep wells turned very muddy. The seismologists knew that earth vibrations could muddy water. Adding these reports to what they had learned from their instruments, the Chinese were able to tell the place and the time of the earthquake.

Problems in predicting earthquakes. Accuracy is one problem, because earthquake predicting is in an early stage. Many instruments need to be set up in areas where earthquakes are likely to occur. Many sets of data need to be recorded and studied. Much more needs to be discovered about what triggers earthquakes and what signs coming earthquakes give.

Time is another problem. As yet, very accurate predictions for short time periods cannot be made. Seismologists can say that an earthquake is likely to strike a certain place every 50 to 100 years. They can say one is likely to strike within a year. But they often cannot tell the exact time or even the exact place.

Accurate earthquake predicting on more than a hit-or-miss basis is years away. An earthquake-watch system would take millions and millions of dollars to set up. Do you think such a system is necessary? Would you be willing to have your tax dollars spent on earthquake research and prediction? Give reasons for your answer.

Look closely at this photograph showing tsunami damage in Hilo, Hawaii, in 1960. The tsunami followed an earthquake that struck Chile. What does the photograph tell you about the power of a tsunami?

Tsunamis. An earthquake often sends a fast-moving wave, called a tsunami, racing across ocean waters. If a tsunami reaches a coastline, it causes great damage. In 1883, a tsunami swept 36,000 people to their deaths along the coasts of Java and Sumatra. The wave was so powerful it carried a large ship more than a mile (1.6 km) inland, stranding it 30 feet (9 m) above sea level. It carried 50-ton rocks even farther.

Since the late 1940s, a tsunami warning system has saved many lives. When a warning is sounded, people in low-lying places leave before the tsunami strikes. The warning system cannot stop the damage a tsunami causes, but it does save lives.

Measuring earthquakes. How powerful are earthquakes? Which ones cause the most damage? The Richter Scale rates earthquake power on a scale that ranges from 1 to 8 or more. Earthquakes with a rating of 1 on the scale are barely felt even by the most sensitive of recording instruments. For each step of 1 on the scale, the vibrations in rocks are 10 times greater, and the amount of energy being released is about 100 times greater. Damage to populated areas begins when an earthquake reaches 4.5 on the Richter Scale. The greatest earthquake recorded reached 8.7 on the scale. An earthquake of 9 would carry a force nearly 10,000 times greater than the force carried by the atomic bomb dropped on Hiroshima in 1945. When an earthquake with a rating of 8 or higher occurs in a heavily populated area, thousands of lives are usually lost.

Keeping Facts in Focus

1. What happens to rock layers that form a block mountain? A folded mountain? A dome mountain?
2. How does a shield volcano differ from a cinder-cone volcano? How does a cinder-cone volcano differ from a composite volcano?
3. Why do earthquakes cause great damage?
4. What is the Richter Scale?
5. What answers would you give to the key questions on page 81?

Working with Ideas

An earthquake has just taken place at 30° north latitude and 120° east longitude that measured more than 8 on the Richter Scale. Use maps in the Atlas to identify the city involved. Also name at least five ways the lives of the people in the city will be disrupted by the earthquake.

Earthquake Measurement on the Richter Scale	
Rating	**Effects**
1 & 2	Tremors detected by instruments only.
3	Slight tremors shake the earth. Dishes, glasses, etc., vibrate.
4	Moderate tremors jiggle small objects and tip larger ones back and forth.
5	Strong tremors overturn and upset most objects within buildings.
6	Very strong tremors are felt that crack and split walls and collapse chimneys.
7	The ground shakes violently. It cracks open in places, destroying roads and toppling buildings and bridges.
8 or more	Ground cracks open, land shifts, floods occur when dams burst, and fires break out. Area over earthquake's center is totally destroyed.

Lesson 3 Surface Forces Wear Down the Earth

Reading Focus

1. Every paragraph in a lesson has a main idea. Each sentence in the paragraph points in some way to its main idea. Sometimes one of the sentences in a paragraph states the main idea directly. It may be the first sentence, the last sentence, or one of the other sentences. But there are paragraphs where the main idea is not stated directly. To find the main idea ask yourself these questions: *How are all the sentences related? What idea do all the sentences point to or help explain?*

    To be a good reader, you must learn to recognize every paragraph's main idea. Practice doing this by finding the main idea of each of the first two paragraphs under the heading *Weathering* (page 91). Do the same for each of the two paragraphs under the heading *Erosion* (page 93).

2. As you read this lesson, look for answers to these key questions:
    a. What are the forces that wear down the earth's surface?
    b. What two kinds of weathering wear down the earth's surface?
    c. What are the agents of erosion?

Vocabulary Focus

gully	*erosion*	*soil*
weathering	*chemical*	*loess*

At the same time one set of forces is building up the rock layers of the earth's surface, another set is wearing them down. The second set of forces breaks up rocks and moves the broken pieces— some as small as grains of sand, some as large as boulders—from one place to another. In the process, these forces smooth off mountaintops, fill up valleys, carve deep canyons and gullies in the land, and fill up river bottoms. These forces are weathering and erosion.

Weathering

Weathering, the breakdown of rocks, can take place in either of two ways. In one way, rocks actually break apart into smaller and smaller pieces. This is called mechanical, or physical, weathering. In the other way, the chemicals that make up rocks change by mixing with other chemicals, especially those in water. This kind of rock breakdown is called chemical weathering.

In ages past, weathering and erosion produced this rock formation in Arches Park, Utah. Weathering of more recent times has produced the pile of rock debris at its base.

The physical and chemical changing of rocks go on at the same time, each aiding the other. As rocks break up into smaller and smaller pieces, their chemicals are more easily dissolved in water. As large masses become weakened through chemical changes, they are more likely to break up.

Mechanical weathering. Mechanical weathering is probably as old as the earth. It began as soon as molten materials hardened into the rock layers of the earth's crust. The hardening of these materials produced many cracks in the large rock masses that covered the earth's surface. After the cooling and hardening, other mechanical changes took place.

Water began to collect in these cracks. At times, it froze, melted, and froze again. When it freezes, water expands. The expanding action of freezing water pushed the cracks farther apart until pieces of the rock masses broke off, forming smaller rocks and boulders.

Seeds, too, collected in rock cracks and took root. The growing roots wedged themselves deeper into the cracks and pushed against their sides. This wedging action of roots is another way that rocks split into smaller pieces.

Water fell from the skies and ran over the land. Winds began to blow over the earth's surface. Running water and winds drove rocks and pieces of rock against each other. At times, masses of moving ice called glaciers covered the land, carrying rocks and boulders along. The rocks scraped against each other and scoured others. Grinding, striking, scraping, and scouring actions also break rocks into smaller pieces.

The cooling and hardening of liquid materials, the freezing and thawing of water, the wedging action of plants, the splashing force of falling water, the rubbing of blowing winds, and the actions of moving glaciers—all these are the agents of mechanical weathering. These actions have been repeated over and over and are still at work today. They are most often at work in dry climates. But they also take place where climates are wetter.

Chemical weathering. Chemicals are the simple substances into which all matter can be broken down. Rocks are made up of chemicals. So are air and water. The chemicals in air find their way into water. In water, they react with the water's chemicals, producing new combinations of chemicals. The chemicals in rocks dissolve when they come into contact with water. They react with the chemicals in water, forming still more chemical combinations. Chemical reactions are always breaking down rocks, changing their chemicals in some way into new substances.

Have you ever seen a rock with a reddish stain covering it? This is a widely occurring example of chemical weathering. Tiny grains of iron in the rock, having been dissolved in water, unite with oxygen (a chemical in water). Iron oxide, the result, is the thin reddish stain on the rock. Reddish-colored soil is also produced by iron and oxygen reacting with each other.

Soil formation. Both mechanical and chemical weathering begin the work of forming soil. It takes a long time for soil to form, perhaps thousands of years in the drier areas of the world.

As the rock pieces produced by weathering become smaller and smaller, they form a thin covering over the rocks of the earth's crust. Seeds dropped by the wind or by birds and other animals take root. The plants spring up and eventually die and decay. Animals in the area make the soil rich with their body wastes. The animals also die and decay. The chemicals in the decayed matter go into the soil, helped by the dissolving action of water. Earthworms and other animals burrow in the ground. They mix rock particles and decayed matter, helping rock particles to stick together and giving the soil body. Soil needs body if it is not to be moved too easily by erosion.

Erosion

Erosion moves soil and weathered products from one place and deposits them in another. The agents of erosion are wind, water, and ice.

Erosion and weathering are partners in shaping the earth's surface. Erosion needs weathering to produce small enough pieces to move. Weathering

The eroding force of water running down hillsides without plant cover carves deep gullies and ridges in land. The eroded land here looks like a miniature version of South Dakota's Badlands.

93

Stained black by rain forest vegetation, the Rio Negro keeps its own identity before it merges with the Amazon River. Its water is colored by the tons of brown silt washed down from the Andes.

needs erosion to keep exposing more and more rock to break up, to produce soil.

Wind. Wind acts as a sorter of weathered products. As a sorter, it picks up fine particles and carries them great distances, sometimes forming a very fertile soil called loess. Great deposits of loess are found in northern China and in parts of the Central Plains of the United States. But the wind's sorting actions are not always helpful. Wind also strips fertile soil from fields left unprotected by any plant cover.

Running water. Running water from heavy rains washes soil away. On hillsides without plant cover, it forms deep ruts or gullies. In desert areas, heavy rains produce flash floods. The water rushes down dry streambeds, taking rocks, pebbles, and soil with it. It even cuts through rock, gouging deeper and deeper into the earth to form canyons.

Water running off fields and hills in other areas carries good soil away into streams and rivers. At times of floods, streams and rivers deposit this soil along their banks, restoring fertility to the land. But more often than not, rivers deposit the soil along the bottoms of their riverbeds, building up sandbars that make the rivers more shallow. Or they deposit the soil at their mouths,

Even though efforts have been made to stop shoreline erosion caused by pounding waves, this large home may soon have Lake Michigan at its doorstep. What other forces may be helping to erode the cliff?

The Tasman glacier threads its way through a valley of the Southern Alps, a mountain range on New Zealand's South Island.

sometimes building up land formations called deltas.

Waves and currents. In oceans and lakes, waves and currents eat away at shorelines. Waves form sandy beaches when they pound against rocks. They cut into the land lining the shores, causing parts of cliffs to drop into the water. Currents take pieces of wave-eroded materials and deposit them elsewhere along the shore, forming sandbars and curving points of land called sandspits.

Currents in rivers also erode the land, especially where rivers bend. This is because a current moves faster at the outside of a bend than it does at the inside. The faster moving current along the outside of the bend cuts away at the land. As the material is carried away, it is deposited by the slower moving current at another inside bend farther down the river.

Glaciers. These huge, moving sheets of ice gouge out rocks and soil, carry the materials with them, and deposit them in other places. Glaciers form in cold climates or in high altitudes. A layer of snow remains from one year to the next, increasing in thickness and length. The increasing weight of the snow changes the bottom layers into ice. Then gravity takes over and the ice slowly begins to move from high to low places.

Glaciers that form in high altitudes are called valley glaciers. They are found in the Alps, on the Alaskan coast, and in the mountains in the western United States. Scraping and scouring the land, these glaciers carve out deep U-shaped valleys. Valley glaciers have helped to give the high mountains of the world their rugged look.

Continental glaciers are sheets of ice thousands of square miles wide and

During dry years, winds sweeping across plains pick up tons of soil left exposed by overgrazing or poor farming practices. In the 1930s, dust storms were so common in the Great Plains that people called the area the Dust Bowl.

thousands of feet thick. Continental glaciers once covered most of Canada and the northern parts of the United States, Europe, and Asia. As cold and warm periods followed each other for thousands of years at a time, these glaciers moved forward and retreated several times. The glaciers left the areas they covered with round hills, stony soil, and many lakes and swamps.

The only continental glaciers that remain today cover most of Greenland and Antarctica. Sometimes, these continental glaciers are called ice caps.

People and erosion. Plants are important in saving land from erosion. The roots of living plants help hold soil together, making it more difficult for wind and water to carry soil away. But people often rob the land of its natural plant covers. They cut down trees and plow grassland under. In this century, people and governments have come to realize that they must control erosion.

Keeping Facts in Focus

1. What is mechanical weathering? What are four different ways in which mechanical weathering takes place?
2. Explain how weathering, plants, and animals help form soil.
3. What is erosion? How does wind wear down the earth's surface? Running water? Waves and currents? Glaciers?
4. How do people aid the work of erosion?
5. What answers would you give to the key questions on page 91?

Working with Ideas

1. In what ways is water a weathering agent? In what ways is it an agent of erosion?
2. Two regions of the United States that were affected by glaciers are the Midwest and New England. Describe the effects of glaciation in one of these regions.

Lesson 4 The Earth's Landforms

Reading Focus

1. Some things you read ought to be remembered. Notetaking will help you remember important facts and organize these facts under main ideas. Notes you have taken can also be used to review important points in a lesson. To be a good notetaker, you must do several things. First skim and preview the material in a lesson. Then read the lesson carefully, paying attention to the lesson's headings. Use the headings to form the beginnings of an outline. Under the headings, write down the main ideas and important facts in the paragraphs that follow. Always put notes in your own words, using short phrases rather than sentences. Now follow these steps in taking notes on this lesson. Be sure to use your notes later on to review the lesson.

2. As you read this lesson, look for answers to these key questions:
   a. What four major landforms are found on the earth?
   b. What two factors do geographers use to distinguish one landform from another?
   c. How do these landforms differ from one another?

Vocabulary Focus

elevation   *relief*   *subsistence*   *hydroelectric*

The inner forces of the earth and the forces of weathering and erosion go hand in hand. Together, they have shaped the earth's crust, forming mountains, hills, plateaus, and plains. To distinguish one landform from another, geographers often use two factors—elevation and relief.

Elevation. Geographers measure the height of land on earth from the point where the land meets the sea. This point is called sea level. Its measure is 0 feet or meters. Elevation, then, is the height of land above sea level. Another word for elevation is *altitude.*

Relief. Geographers use the term *relief* to indicate the changes that take place in elevation from one point to another within a given stretch of land. In areas of high relief, land rises or drops sharply—usually within a short distance. In areas of low relief, land rises or falls gently—usually over a greater distance. In some areas of low relief, changes in elevation are barely noticed by people moving from one point to another within the area.

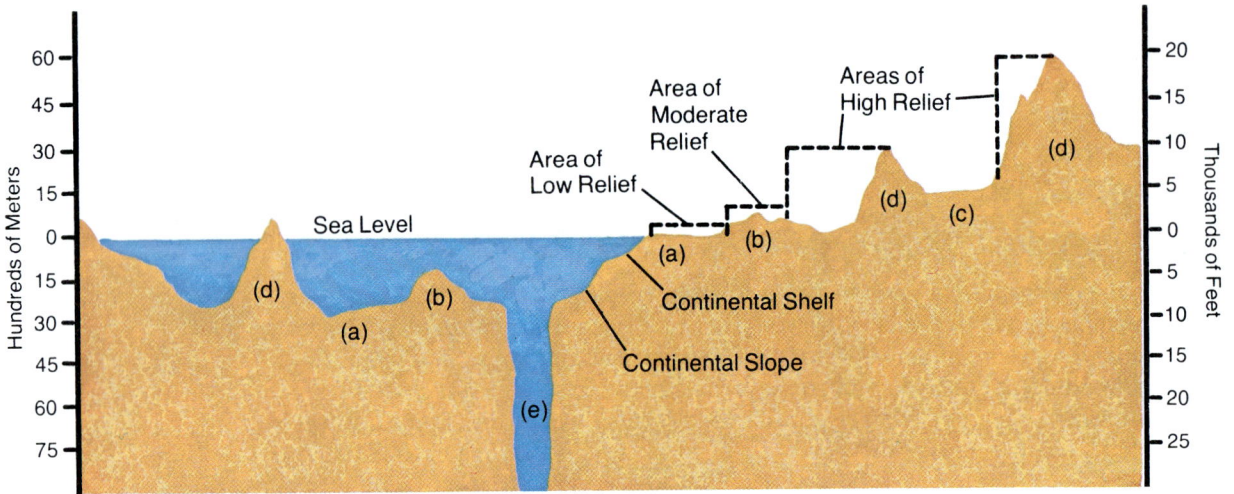

Classified by elevation and relief, the earth's landforms are (a) *plains* on land and on the seafloor, (b) *hills* on land and ridges rising from the seafloor, (c) *plateaus* on land and on the continental shelf, (d) *mountains* on land and mountains rising from the seafloor, and (e) deep seafloor *trenches*.

Mountains

Mountains are landforms with a high elevation—6,600 feet (2,000 m) or more. They also have a very high relief. Within a very short distance, the elevation from one point to another may rise or drop 1,000 feet (300 m) or more.

Mountain living. The people who live in mountain areas often live in valleys or on valley slopes. One valley is usually separated from other valleys by high peaks and impassable ridges. Because of this separation, there is a lack of communication among the people in different valleys. So the people in one valley often develop different customs and language patterns from the people living in nearby valleys. This makes it difficult for people living in mountain areas to

Study the profile of the Pacific Ocean on this page and that of the Atlantic Ocean on the opposite page. At one time, people thought the seafloors under both oceans were flat and featureless, except for scattered mountains that rose above the oceans to form islands. Today, we have instruments to measure ocean depths. We know that seafloors have flat basins (plains), mountain-like ridges, and deep trenches. The Marianas Trench in the Pacific Ocean is the deepest spot on earth. It is 36,198 feet (11,033 m) below the level of the sea.

PROFILE OF THE PACIFIC OCEAN

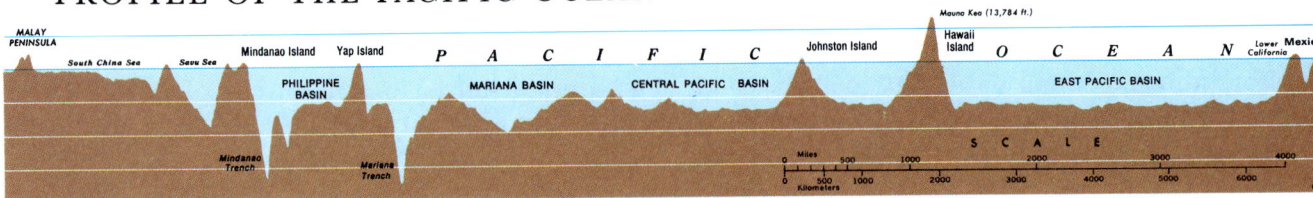

What can you tell about the advantages and disadvantages of mountain living from this photograph? It shows a town nestled against the Pyrenees, a mountain range along the border between France and Spain.

develop a sense of belonging to a nation, a feeling that is important today.

High relief makes all forms of travel difficult. Roads in mountains can seldom follow straight lines. Instead, they must follow a winding line up one side of a mountain and down the other from one valley to the next. Very often the roads are nothing more than dirt tracks. Building modern roads means blasting tunnels and moving tons of rock. Few countries in mountain areas can pay the high cost such road-building demands.

The people who live in mountain areas must be very hardy. At high elevations, air is thinner. It has less oxygen than air at lower elevations. Visitors to mountain areas must give their bodies time to adjust to the strain the thinner air puts on their hearts and lungs.

Mountain resources. People living in mountain areas earn their livings by making use of the resources found there. One of these resources is soil.

Most mountain areas have only thin coverings of soil. But there are places where soil is thicker and where farming can take place. Most farms are only small plots of land on level stretches between rising peaks or on valley bottoms and gentle slopes. Many of the farmers are subsistence farmers. They grow crops to feed themselves and their families rather than to make a profit.

PROFILE OF THE ATLANTIC OCEAN

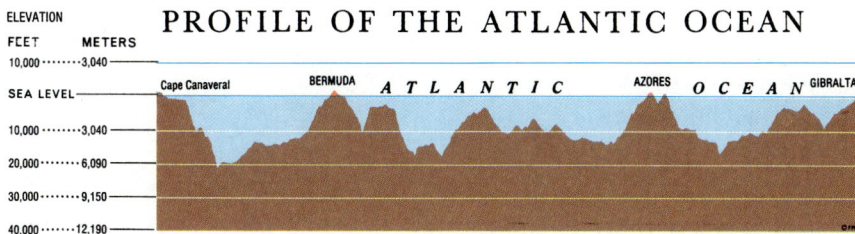

ELEVATION	
FEET	METERS
10,000	3,040
SEA LEVEL	
10,000	3,040
20,000	6,090
30,000	9,150
40,000	12,190

Cape Canaveral BERMUDA *ATLANTIC* AZORES *OCEAN* GIBRALTAR

Using the Land—To What End?

In *America the Beautiful*, we sing of purple mountain majesties, the fruited plain, and alabaster cities. Nowhere do we sing of cans and bottles and paper wrappers littering the land, or of glaring and flashing neon lights and other signs lining streets and roads. Nowhere do we sing the praises of billboards blocking the view of green fields, groves of trees, lakes, and rivers. Nor do we mention the "graveyards" piled high with rusting, wrecked, or worn-out vehicles. And with good reason. All of these things mar the beauty of the land.

Many of our "alabaster" cities are now tarnished, dreary, and run-down. Almost every major city has older sections with burnt-out, abandoned buildings. Suburbs sprawl in all directions from huge manufacturing and trading centers. More and more seashores, farmlands, forests, and other "green spaces" give way to housing developments, roads, factories, and shopping centers.

Are all these forms of land pollution? Are such forms necessary for progress? Or can progress take place without spoiling the land, its beauty, and its usefulness for future generations of living things? If it can, what must be done by public agencies, private groups, and individuals to keep America beautiful? What is already being done to protect our country's beauty?

They use few modern tools or machines because of the high relief, the small sizes of their farms, and the cost of such machinery.

The people also make use of the land by grazing sheep, goats, and other animals on grassy meadows. In the Himalayas people graze herds of yaks. In the Andes they graze llamas and alpacas. The wool from these animals is especially valuable.

Mining is done in those places where the mountains hold deposits of valuable materials. Lumbering takes place where forests grow on the mountainsides. Mountain regions are valuable for another important resource—waterpower. The waters that rush down the mountainsides in swiftly flowing streams have the power to generate electricity. But only a small portion of the total waterpower in mountain areas has been harnessed. The cost of building dams and hydroelectric power plants in rugged, out-of-the-way places is very high. The source of power is often too far away from the industrial cities and population centers of the lowlands where the electricity is most needed.

Mountains have still another kind of resource. They are places of recreation where people can relax and have fun in every season. In the summer and early spring, they offer cooling breezes, trails for hiking and backpacking, and sparkling streams for fishing. When snowfall allows it, their slopes attract skiers of all ages.

Hills

Many hills are the eroded remains of mountains. This is why it is hard to

In highland areas, temperatures are related more to altitude than to latitude. Most of the highland areas of Central and South America are in the low latitudes where temperatures at sea level are very hot. But, in highland areas, the temperature drops about 3.6° F (2° C) for each 1,000 feet (300 m) of altitude. The diagram shows mountain region levels and their Spanish names for hot, mild, and cold. What crops can be grown at each level?

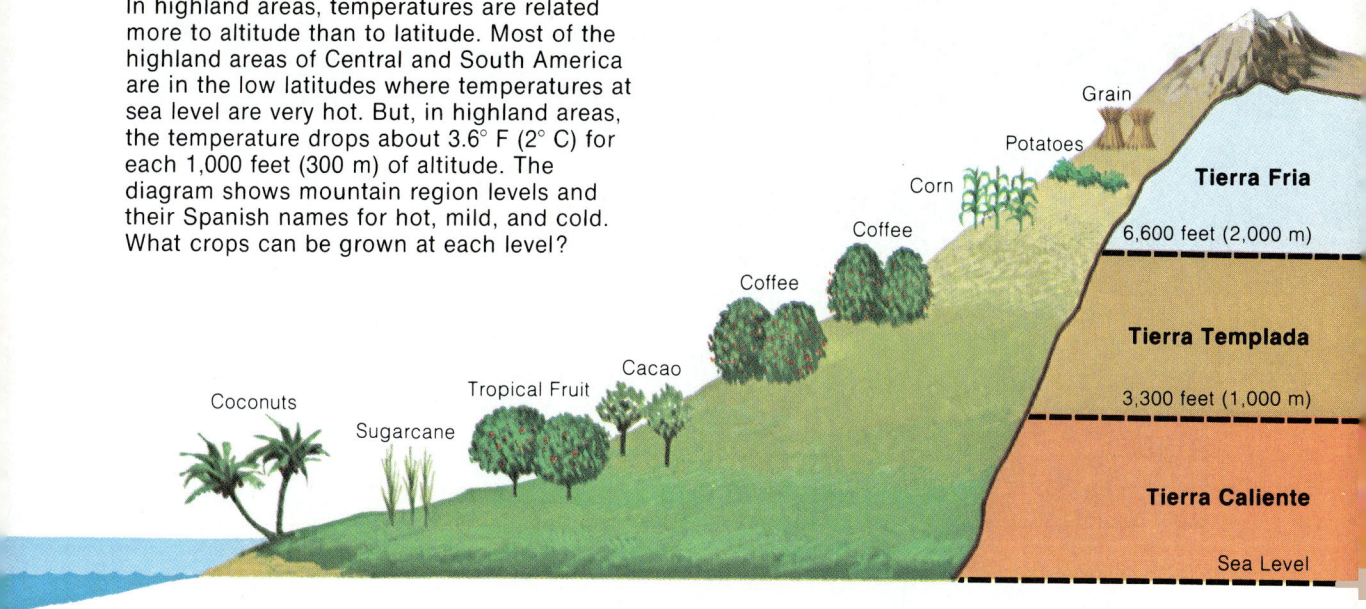

Grain

Potatoes

Corn

Tierra Fria

6,600 feet (2,000 m)

Coffee

Coffee

Tierra Templada

Coffee

3,300 feet (1,000 m)

Cacao

Tropical Fruit

Tierra Caliente

Coconuts

Sugarcane

Sea Level

distinguish a hill from a mountain in some places. The Appalachian Mountains are a good example.

Once, the Appalachian Mountains were probably as high and as rugged as the Rocky Mountains are today. But because of their elevation and relief, geographers say the Appalachian Mountains are really hills. Hills have an elevation of 1,650 to 6,600 feet (500 to 2,000 m). They have a moderate relief. A drop or rise in elevation ranges from 500 to 1,000 feet (150 to 300 m). The change in elevation from one point to another may also be spread over a greater distance than in areas that are mountainous.

Living in hills has some of the same advantages and disadvantages as living in mountain areas. Hills and mountains have many of the same resources. Hill areas, however, are generally more pop-

ulated. Road building is not as difficult, so there are more modern roads. Farms may be larger in the wider valleys, making it possible for some farmers to grow crops for a profit.

Plateaus

A plateau has a generally flat surface. But it is higher than the surrounding countryside, except where a plateau is located between mountain ranges. A plateau has an elevation of 1,650 feet (500 m) or more. Because most of it is flat, it has very low relief, usually less than 500 feet (150 m). On some plateaus, rivers have cut deep canyons through the flat, level land. The Grand Canyon in Arizona, cut through a plateau west of the Rocky Mountains, is one of the most spectacular sights in the world. It is 18 miles (29 km) wide and a mile (1.6 km) deep.

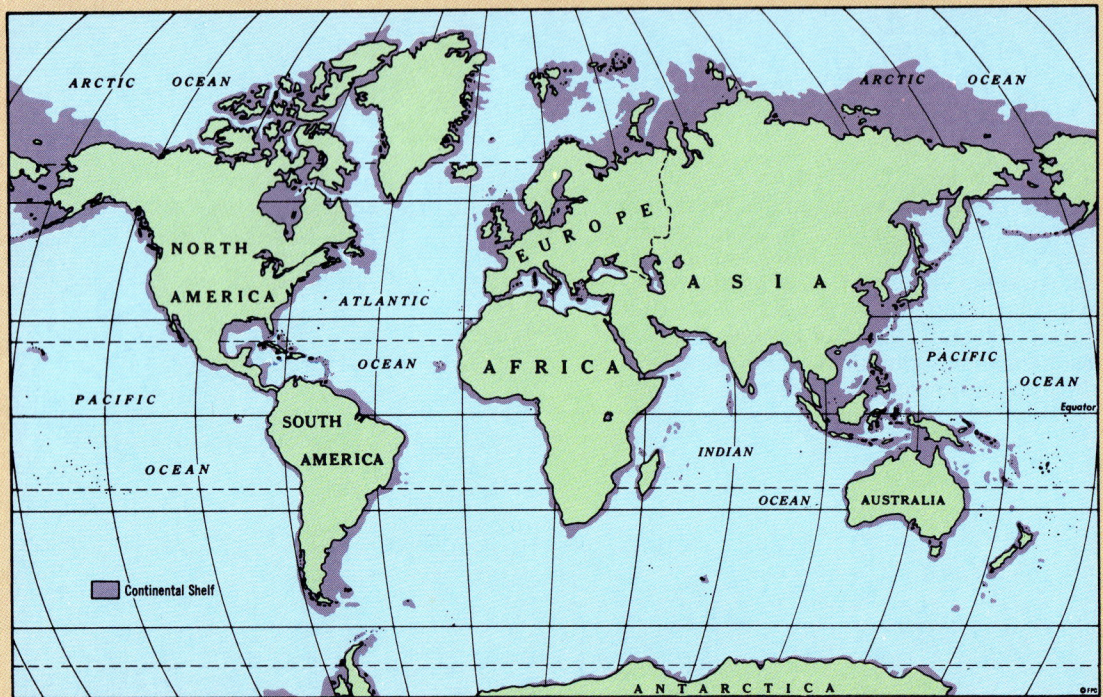

Continental Shelf

The Continental Shelves

The edges of continents slant under the surface waters of oceans and seas. These slanting edges are called continental shelves. From the shoreline, a continental shelf extends outward along most continents for an average of about 100 miles (160 km). The average underwater depth is less than 600 feet (180 m). The continental shelf hardly exists at all off the coast of Peru and Chile. There, it extends only a short distance before it drops off into a deep trench. Along the coast of Siberia in the Arctic Ocean, the continental shelf is much wider. There, it extends for almost 500 miles (800 km).

Where the shelf ends, the continental slope begins. The slope makes a steep drop of 6,000 to 9,000 feet (1,800 to 2,700 m). After this steep drop, the continental edges merge with the crust that forms the ocean floor.

For people living on the earth, the continental shelves are very important. Many of the world's great fishing areas are in the shallower waters over the continental shelves. A few of these areas are in the North Sea, in the Atlantic Ocean off the coast of Newfoundland, and in the Bering Sea. Special conditions in these places attract fish in great numbers.

The continental shelves also hold oil and natural gas trapped in pockets between layers of rock. Scientists and engineers now have the tools and equipment to discover these pockets and to bring the oil and gas to the surface. They have set up off-shore drilling and pumping platforms in the Gulf of Mexico, in the Pacific off the coasts of California and Mexico, and in the North Sea off the coast of Norway. As new pockets of oil and gas are discovered under the seafloor, other platforms will be set up.

Many nations recognize the economic importance of the continental shelves and the waters over them. So they claim them as part of their national territories. The United States and Ecuador are two such nations. They have claimed control over the waters and the land under them for 200 miles (320 km) from the shores of their countries. They say the boats of no other nation may fish in these waters without permission. They say no other nation may drill for oil or gas within the 200-mile (320-km) limit. Other nations claim only a 12-mile (19-km) limit. These nations say the waters over the continental shelves and any oil or gas outside the 12-mile limit belong to all nations. Which side do you think is right? Why? How do you think nations should settle disputes that come up over fishing or drilling rights in waters over the continental shelves?

Plateau areas in the middle and high latitudes are usually windswept, dry, and sparsely populated. Herding is the main occupation. On some plateaus in the middle latitudes, farming is possible with irrigation.

Plateaus in the low latitudes, however, are better suited to supporting large populations. Most of the people of Mexico live on the Central Plateau, which is located between two mountain ranges. The climate is milder than that in the hot steamy lowlands along the Atlantic coast. Much of eastern Brazil is a plateau that rises sharply off the Atlantic coast. Most of Brazil's people live on this plateau.

Plains

Plains are the flat or gently rolling lands of the world. Both their elevation and their relief are less than 500 feet (150 m). A look at the graphic-relief map of the world on pages 12–13 shows that large parts of North America, South America, Europe, Australia, and Asia are plains.

More than three fourths of the world's people live on plains. Most of the world's food is grown or raised on plains. And almost all of the world's largest cities are located on plains.

Farming. Farming on plains differs from farming in highland areas in several ways. Farms are much larger in size. In the United States and Canada, a farm may cover a thousand acres (400 ha) or more. The flat, rolling land allows machinery to be used to plow, seed, and cultivate the land as well as to harvest crops. The yield for crops is many times greater than on the subsistence farms of highland areas.

Not all flat land, however, is good farmland. The treeless plains of the Arctic are frozen most of the year and are of little value for growing many crops. Some plains in the middle latitudes are too dry for the growing of most crops. Plains in the tropics are often too wet to do large-scale farming, although rice growing is important in some parts of tropical Asia.

People and the plains. The forces of nature have changed mountains, hills, and plateaus. But it is people, much more than the forces of nature, who have changed the plains. What are some ways you think that people have changed the plains?

Keeping Facts in Focus

1. What is elevation? What does relief indicate?
2. Why is living in mountain areas difficult?
3. What are some resources of mountain areas?
4. What is the general elevation and relief of hills?
5. What is a plateau? How is it defined according to elevation and relief?
6. What is a plain? What advantages do plains have that mountains do not have as places for people to live?
7. What answers would you give to the key questions on page 97?

Working with Ideas

1. How does the elevation of a place influence its agriculture?
2. Why does high relief make it difficult for people in an area to earn their living through farming?
3. What else besides elevation and relief influence a region's agriculture?

Lesson 5 Riches of the Earth's Crust

Reading Focus

1. A good reader learns to use pronunciation guides and a pronunciation key to pronounce new words correctly. Both of these pronunciation aids are a part of the Glossary of this book. Using the Glossary, look up *titanium,* a word used in this lesson. Immediately following the word listing, find the guide to its pronunciation. The guide divides the word into syllables, tells what syllable gets the most stress, and uses certain symbols to tell what sound a letter has. These symbols are explained in the pronunciation key. Using both the guide and the key, combine the sounds of the vowels and consonants in each syllable. Say each syllable slowly at first. Then repeat the syllables more rapidly, giving the accented syllable the most stress. Follow these same steps to pronounce other new words. Some words you can practice on (besides those in the list below) are *potash, synthetic,* and *silicon.*

2. As you read this lesson, look for answers to these key questions:
   a. Into what two groups can elements and minerals be divided?
   b. What are four ways minerals are mined?
   c. What makes soil fertile?

Vocabulary Focus

element	*sedimentary*	*humus*	*arable*
mineral	*metamorphic*	*fertile*	*leaching*
igneous	*alloy*		

A French farmer drives a tractor through a field, planting wheat. In Zaire, a mill worker processes a big sheet of copper that will be drawn into wire. A truck driver in Germany hauls a load of sand from a quarry to a glass factory. From another quarry—one in Vermont—a crane operator lifts out a block of marble that may become part of a building. These are just a few of the ways that people use the riches of the earth's crust—soil, rock, and minerals.

Minerals, Rocks, and Ores

The earth's crust is made up of elements, or chemicals. An element is the simplest part into which a substance can be broken down. While over 100 chemical elements have been discovered, only 92 of them have been found in nature. Some of these elements exist as solids, some as liquids, and some as gases.

Minerals. A mineral is an element or a combination of elements found in nature, usually as a solid. Gold, copper,

The life story of rocks is ages-old. The processes that produce them, melt them, cool them, and weather them are shown in this diagram. Use this diagram to tell the story of how igneous rock and sedimentary rock are changed into metamorphic rock.

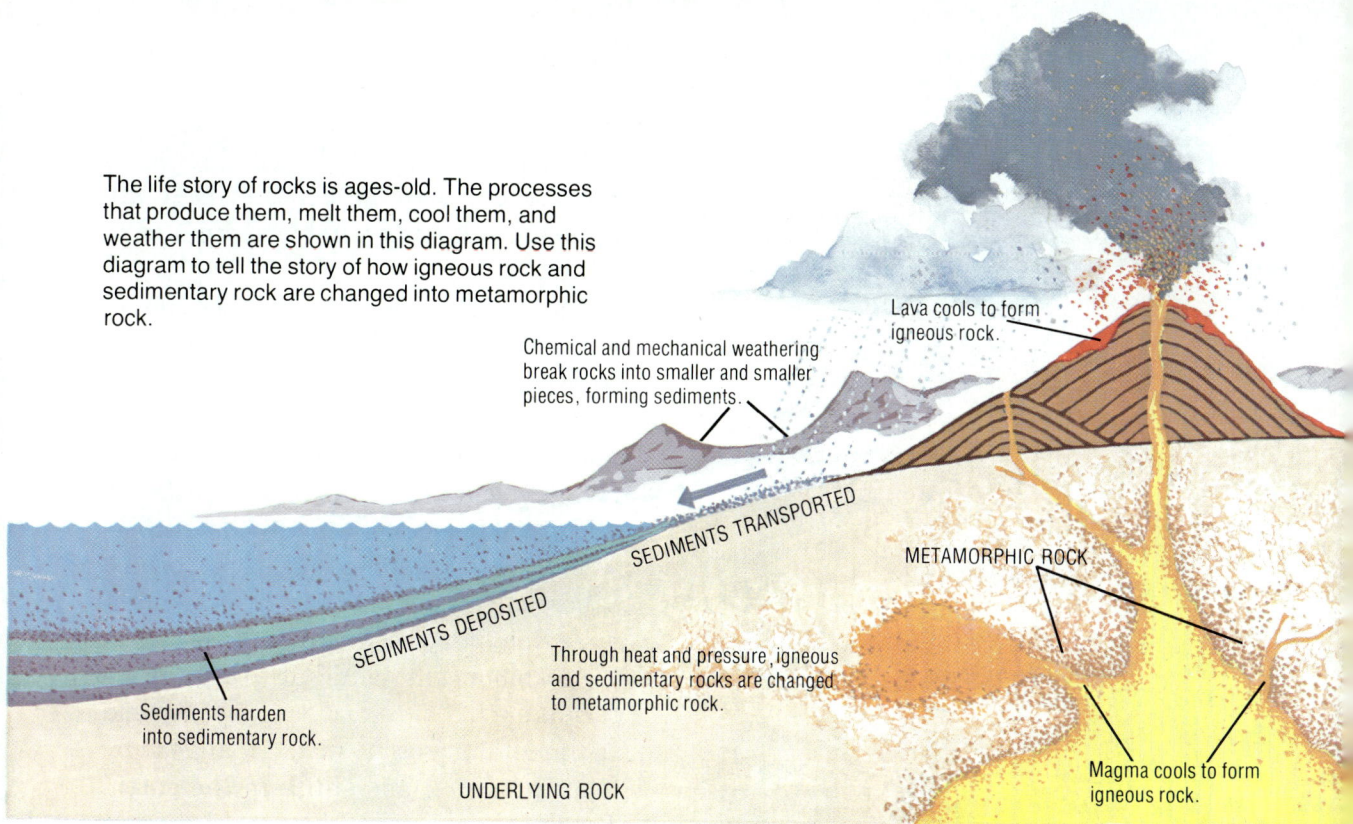

Lava cools to form igneous rock.

Chemical and mechanical weathering break rocks into smaller and smaller pieces, forming sediments.

SEDIMENTS TRANSPORTED

METAMORPHIC ROCK

SEDIMENTS DEPOSITED

Through heat and pressure, igneous and sedimentary rocks are changed to metamorphic rock.

Sediments harden into sedimentary rock.

Magma cools to form igneous rock.

UNDERLYING ROCK

and silver are single elements that are minerals. Most other minerals, however, are combinations of two or more elements. There are about 2,000 minerals known to exist in nature.

Metals and nonmetals. Minerals, like the elements that form them, can be divided into two large groups—metals and nonmetals. One of the chief things that makes an element (or a mineral) a metal is its ability to carry an electrical charge. Metals can be pounded and shaped into objects. They also can be drawn into wire.

Carbon, oxygen, and silicon are examples of nonmetallic elements. These elements combine very readily with other elements to form many of the minerals that make up the different kinds of rocks found in the earth's crust. Nonmetals have many uses. Marble, lime-

stone, slate, granite, sand, and gravel are some of the mineral materials used in the construction industry. The jewelry industry uses many gemstones. Salt is a nonmetallic mineral. Potash is one. So is sulfur, a mineral used in the chemical industry.

Coal, petroleum, and natural gas are often classified as nonmetallic minerals. Because these minerals are fuels—sources of energy—their use will be discussed more fully in a later unit. Other minerals—nonmetallic and metallic—and their uses are listed in the charts on pages 106 and 107.

Rocks. Minerals or combinations of minerals make up rocks. The minerals in many rocks are not very valuable. Or the minerals are present in such small amounts that they are not worth mining. Only a few minerals—about 100—

ROCKS	MAJOR DEPOSIT LOCATIONS	SOME USES
Granite	U.S.A., Canada, Sweden, Finland	Building materials, monuments
Slate	Wales, France, U.S.A.	Roofing materials, flagstone
Marble	Italy, U.S.A.	Ornamental building materials, statuary
Limestone	U.S.A.	Building materials
GEMSTONES		
Garnet	South Africa, U.S.A.	Jewelry; cutting, grinding, and polishing tools; watches
Jade	New Zealand, Burma, Japan, Mexico	Jewelry, carvings
Emeralds	Colombia, Zimbabwe, India, U.S.S.R., U.S.A.	Jewelry; cutting, grinding, and polishing tools
Diamonds	South Africa, Zaire, U.S.S.R.	Jewelry; cutting grinding, and polishing tools; phonograph needles

NONMETALLIC MINERALS	MAJOR DEPOSIT LOCATIONS	SOME USES
Phosphorite (source of phosphorus)	U.S.S.R., Morocco, U.S.A.	Fertilizer, detergent, medicines, chinaware
Kaolinite (clay)	U.S.A., Great Britain, France, East Germany	Pottery, textiles, papermaking
Halite (rock salt)	U.S.A., China, U.S.S.R., West Germany, India, Canada	Food processing, tanning, metal refining
Sulfur (an element in many minerals and chemicals)	U.S.A., Spain, Mexico, Japan, Italy	Gunpowder, matches, fertilizer, pesticide, papermaking, photo-developing
Graphite	U.S.A., Madagascar, Mexico, Norway, Sri Lanka, Austria, China	Pencil lead, batteries, lubricant, synthetic diamonds
Anhydrite (Gypsum)	U.S.A., Canada, France, Great Britain, Spain, Italy, U.S.S.R.	Plaster of Paris, wallboard, paints, cement, insulation
Chrysolite (Asbestos)	Canada, U.S.S.R., South Africa	Fireproofing materials

are concentrated in deposits that are large enough and valuable enough to be mined.

Rocks, and the minerals in them, formed in different ways. Some formed from molten materials that gradually cooled. These are igneous rocks. A common igneous rock is granite. Some rocks formed from sediments—the weathered and eroded pieces of other rocks, such as sand. Sands, cemented together, formed sandstone—a sedimentary rock. Another sedimentary rock is limestone.

As mountain building took place through the ages, some igneous and sedimentary rocks were changed by heat, pressure, and shifts in the earth. These changes produced a new class of rock—metamorphic rocks. Changes in limestone, for example, produced marble. Changes in granite produced gneiss.

Every rock can be put into one of these three classes of rock. By performing tests on them, geologists can tell how they were first formed, what minerals are present in them, and whether or not the rocks have undergone any change since first being formed.

Ores. Any rock material that contains enough metal in it to be mined is known as an ore. There are as many kinds of ores as there are metals. Some ores contain more than one kind of metal. Such ores may be mined for the most abundant metal in them.

Mining and quarrying. How mining is done often depends on the location of a mineral and whether or not it is a solid. Where large deposits of copper or iron ore lie very close to the surface and over a wide area, open-pit mining often takes place.

Coal is a material that is sometimes removed by stripping away the land above it. Where strip-mining takes place, coal usually lies in layers very close to the surface. Strip-mining is done cheaply. But it leaves the land bare of plant life and with piles of waste rock. In some places, laws now require mining companies to level the rock piles and to cover them with soil and plants.

Underground mining is used to remove coal when it lies deep underground. It is also used to mine copper, iron ore, gold, or any other solid minerals located deep within the earth. To reach these minerals, tunnels or shafts must be dug. Supports must be built. Special equipment must be put in to provide for the safety of the miners and to haul out the ores. Underground mining is usually more costly than open-pit mining or strip-mining.

Some minerals, such as petroleum and natural gas, can be reached only by drilling. These minerals are trapped in pockets between layers of hard rock deep within the earth's crust. Once drills bite into the pockets, the minerals are pumped out and piped to refineries.

Quarrying removes valuable rock materials from the earth. Slabs of marble, granite, slate, and limestone are cut from huge pits. Sand and gravel deposits are often scooped from the earth, much like coal that is strip-mined.

Metals and manufacturing. Before metals can be useful for manufacturing purposes, they must be separated from

ORES AND METALS	MAJOR DEPOSIT LOCATIONS	SOME USES
Pyrolusite (Manganese ore)	U.S.S.R., South Africa, Brazil, Gabon, India, Australia, China	Steel alloy, chemicals
Galena (Lead ore)	Australia, Canada, Mexico, Peru, U.S.S.R., U.S.A.	Pipes, storage batteries, electric cables, type, chemicals
Argentite (Silver ore)	Canada, U.S.A., Mexico, Chile, U.S.S.R.	Tableware, jewelry, surgical wire, dental fillings, photographic film
Bauxite (Aluminum ore)	Jamaica, Australia, France, Guyana, U.S.S.R., Surinam, Sierra Leone	Airplanes, foil, utensils, wire, heating equipment
Sphalerite (Zinc ore)	Canada, U.S.S.R., Australia, U.S.A., Peru	Coatings on other metals, especially iron; electric batteries; alloy with other metals; chemicals
Carnotite/Pitchblende (Uranium ore)	Canada, Zaire, France, South Africa, Australia, Spain, U.S.A., U.S.S.R.	Nuclear energy
Wolframite (Tungsten ore)	Australia, Bolivia, China, U.S.S.R., U.S.A.	Steel alloy, electronic equipment, industrial tools
Cassiterite (Tin ore)	Malaysia, Bolivia, U.S.S.R., Thailand, Indonesia	Alloyed materials, coatings, munitions industry
Chalcopyrite (Copper ore)	U.S.A., Canada, Chile, Zambia, U.S.S.R., Zaire, Philippines, Australia, Peru	Electrical wire, piping, ornamental fixtures and hardware
Chromite (Chromium ore)	Cuba, New Caledonia, Philippines, Zimbabwe, U.S.S.R., South Africa, Turkey	Steel alloy, kitchen equipment and utensils, automobiles
Magnetite (Iron ore)	U.S.A., U.S.S.R., Australia, France, China, Canada, Brazil	Machinery, razor blades, toys, automobiles, piping, ships, building materials

the ores that contain them. This often requires many steps and many different kinds of processes before a pure metal is obtained. Much depends on the quality

Mining gold (right) or any other metal that lies in veins far underground is much more difficult and dangerous than stripping coal (below) from the earth's surface.

of the ore and the kind or kinds of metals in the ore.

Some ores are smelted. In smelting, the ore is mixed with other materials in a furnace and heated to a temperature high enough to melt everything in the mixture. Iron, for instance, is mixed with coke and limestone. The coke, limestone, and waste materials form one mixture after heating. The liquid iron forms another. The lighter waste materials are drawn off at the top, forming slag. The liquid iron, called pig iron, is drawn off at the bottom. The pig iron is then formed into ingots, or slabs. In a further process, the ingots are reheated and rolled into sheets or molded into different shapes.

Other ores are separated using different methods. One method uses chemical solutions to dissolve the metal. Then electric currents collect the metal at certain points where it can be drawn off.

Some metals, once they are separated from their ores, are mixed with other metals or nonmetals. These manufactured metals are called alloys. Common alloys are bronze (copper and tin), brass (copper and zinc), and steel (iron and carbon). Steel is often mixed with other metals, such as manganese and titanium. Steel mixed with titanium takes on a great resistance to heat. This alloy is

In this quarry in Italy, workers take out huge chunks of marble. The chief mineral in marble is calcite. Only the purest white marble is used for statues. Other minerals mixed with the calcite give marble its different colorings.

most useful in the manufacture of jet engines.

Minerals and the future. It took nature millions of years to form the mineral materials mined today. Once removed from the earth, minerals cannot be replaced. In today's industrial societies, the use of minerals is increasing rapidly. The reserves of minerals in the world were once thought to be almost endless. Now people are not so sure of their continued abundance. There is concern that mineral resources may soon be used up if new deposits of valuable minerals are not found.

The search for new deposits of metals and other minerals both on land and under the oceans and for the techniques to mine and process them never stops. Neither does the search for manufactured, or synthetic, materials that can be used instead of naturally-occurring materials. Efforts are being made to recycle used mineral materials. There are studies being made to discover ways of removing metals from seawater. The search also goes on to discover new ways

to extract metals from ores. How available many important minerals will be in the future depends, to a great extent, on all of these efforts.

Soil

Hundreds of different kinds of soil cover the rock layers of the earth's crust. What makes one kind of soil different from another kind?

Soil differences. First of all, soil varies in thickness. Weathering, you remember, helps to start the soil-forming process. Erosion then carries soil from high places to low places. How deep soil is depends not only on weathering and erosion. It also depends on location and climate. Soil is only inches deep on most mountain slopes. It is no more than a few feet deep on most plains. Yet, in the tropics, soil may be as much as 100 feet (30 m) deep.

Color is another difference. Soil may be black, brown, gray, red, or yellow. Its color depends on what chemicals it has in it. Its color also depends on whether it is rich in humus—the decayed material

109

Geomorphologists—Searchers into Earth's History

Geomorphologists are people who study the earth's structure, its land materials, and the forces that give these features form and shape. Geomorphologists are interested in rocks, minerals, and soils wherever they exist—on the earth's surface, below its surface, or beneath rivers, lakes, and oceans. They search out rocks, minerals, and soils to find out how they were formed and how long they took to form. Geomorphologists are also interested in changes they see taking place now. Each change today gives them a clue to how changes took place in the past.

By studying these clues, by sorting them out, and by finding patterns of change repeated over and over—geomorphologists have given us a fairly good outline of the earth's history. We know something of the earth's geologic ages and the life forms that appeared or disappeared in each. We know how some landforms build up and how other landforms wear down. We know some of the things that happen inside the earth's mantle when the land shakes, shifts, and splits open. We know some of the things that happen when volcanoes erupt. But there is so much more to know, so much more of the outline to fill in. The work of geomorphologists will continue, either in the field or in the laboratory, until all parts of the outline are as complete as it is possible to make them.

of plants and animals. A soil rich in iron is red. One rich in humus is black or dark brown.

Particles in soil also differ in size and in how they stick together. A sandy soil

Forest Covering

Humus-rich Layer

Leached Layer

Clay-like Layer

Rocky Layer

Crustal Rock

has large particles held together loosely. A clay soil has very fine particles that stick close together. Particle size and makeup determine how much air and water can flow through soil. A sandy soil holds too much air and not enough water. A clay soil holds too much water and not enough air.

Fertile soil. All of these things determine how fertile a soil is. Fertile soil—the kind needed to grow crops—is one of the world's most valuable resources. Where fertile soil is arable (able to be plowed and planted), it grows the crops that feed the world's 4 billion people. A fertile soil has the right mix of chemicals in it. It holds water in the right amounts. It allows enough air to circulate. And it has humus mixed well throughout the top layer. Some of the world's most fertile soil has developed under grass and broad-leaf forest coverings. This kind of plant life forms thick, rich humus, especially where rainfall measures between 20 and 30 inches (50 and 75 cm) a year.

Some soils have the right chemical makeup to start with. But they receive too much or not enough rainfall. Too much rainfall causes leaching. Leaching takes place when water dissolves some chemicals in the soil and washes them away. It leaves other chemicals that make the soil acidic, or sour. Such soils often develop under needleleaf forests or in the tropics. When there is too little rainfall, as in desert areas, the soil becomes too salty to be fertile.

People and soil. The work of weathering and humus-building is always going on. Erosion is always at work moving soil from high places to low places. But certain human practices harm the soil and speed up the rate at which erosion takes place.

The first harmful practice is overuse of the soil by farming. Many crops rob fertile soil of chemicals. Other plants put needed chemicals in the soil. Many farmers today know that they can restore needed chemicals to the soil by using fertilizers and by planting different crops in the same field from year to year. But this has not always been done. Often, farmers have simply moved away when the soil stopped producing good crops. Before nature could restore the soil, erosion turned the area into a wasteland. Parts of North Africa, for example, once grew grains that fed the people of the Roman Empire. But early farmers overused the soil. Today, the area is a desert.

The second harmful practice is overgrazing. As the human population grows in many countries, so does the livestock population. Often, the amount of land on which these animals graze remains the same or becomes even smaller. Overgrazing destroys the grass cover. Then the topsoil washes or blows away. Overgrazing is a problem today in western India, Pakistan, Nepal, North Africa and the Middle East, and the Andean regions of South America.

The third harmful practice is the cutting of trees to clear land for farms and to provide wood for cooking, heating, and shelter. With the disappearance of trees in Morocco, Tunisia, and Algeria, the desert area has grown larger.

The world's population is growing larger every day. The supply of fertile soil in the world is growing smaller. Can you see a problem here? What is this problem? Can you think of ways to solve the problem?

Keeping Facts in Focus

1. What is a chemical element? In what forms are chemical elements found in nature?
2. What is a mineral? About how many different minerals are there?
3. What makes a mineral a metal?
4. What is an ore? An alloy?
5. What three human practices often increase soil erosion?
6. What answers would you give to the key questions on page 104?

Working with Ideas

Which do you think is more important for a nation today to have—large areas with fertile soil or large deposits of one or more of the world's major minerals? Why?

Unit 2 REVIEW WORKSHOP

Test Your Geographic Knowledge

Complete each sentence by filling in the blank with a word from the list below.

erosion	elevation	dormant
lava	weathering	minerals
relief	faulting	seismic
magma	landmass	

1. Many geologists believe that the earth's continents were once joined together in one _____.
2. On high mountains, moving ice is one of the chief agents of _____.
3. When molten material is present beneath the earth's surface, it is called _____.
4. When molten material reaches the earth's surface, it is called _____.
5. Pressures in the mantle often cause the earth's crust to move by folding or _____.
6. Sulfur and salt are _____.
7. A _____ volcano is one that has shown no volcanic activity for several years.
8. The changing of chemicals in rocks by air and water is called chemical _____.
9. The _____, or shock, waves that accompany earthquakes carry great amounts of energy.
10. _____ is the height of land above sea level.
11. Plains are areas of low _____ because there is little change of elevation between the high and low points of the land.

Apply Your Reading Skills

A. Tell whether each sentence below states a fact (F) or an opinion (O).
   1. The earth may have begun about 4½ billion years ago.
   2. Wegener believed that all continents were once part of one huge landmass surrounded by one huge ocean.
   3. Places in both Africa and South America have the same kinds of rocks, plants, and animals.
   4. Scientists think the continents ride on plates and move with them.
   5. Many earthquakes take place where faults are located on the earth.
   6. Seafloor spreading may cause continental drift.
   7. Not all soils are fertile.
   8. Most of the world's food is grown on plains.

B. Use encyclopedias, books, and other reference tools to do research. Then prepare a report on ONE of the following topics.
   1. When did glaciers cover the United States?
   2. When and where did some of the largest volcanic eruptions take place?
   3. How have scientists discovered what the seafloor is like?
   4. What are some of the world's most famous mines, and where are they located?

Apply Your Geographic Skills

Look at the map on page 113. It shows several things about earthquakes in the United States. Using the map, tell whether each statement below is true or false.
1. All earthquakes that can cause major damage will occur only on the West Coast.
2. Only one small part of southwestern Texas is likely to suffer a damaging earthquake.
3. A major earthquake has never occurred in the central part of the United States.
4. A major earthquake took place in Seattle in 1965.
5. Parts of only four states are unlikely to suffer any kind of earthquake damage.
6. More earthquakes with major damage have taken place in Alaska than in California.
7. You live in an area where an earthquake causing major damage can take place.
8. The earliest recorded earthquake causing major damage in the United States took place in Boston.

Apply Your Thinking Skills

A. Each of the statements below refers to one of the following: (a) the earth's core, (b) the earth's mantle, (c) the earth's crust. Match each statement with its correct answer.
   1. It is the thickest of the earth's three parts.
   2. It is the thinnest of the earth's three parts.

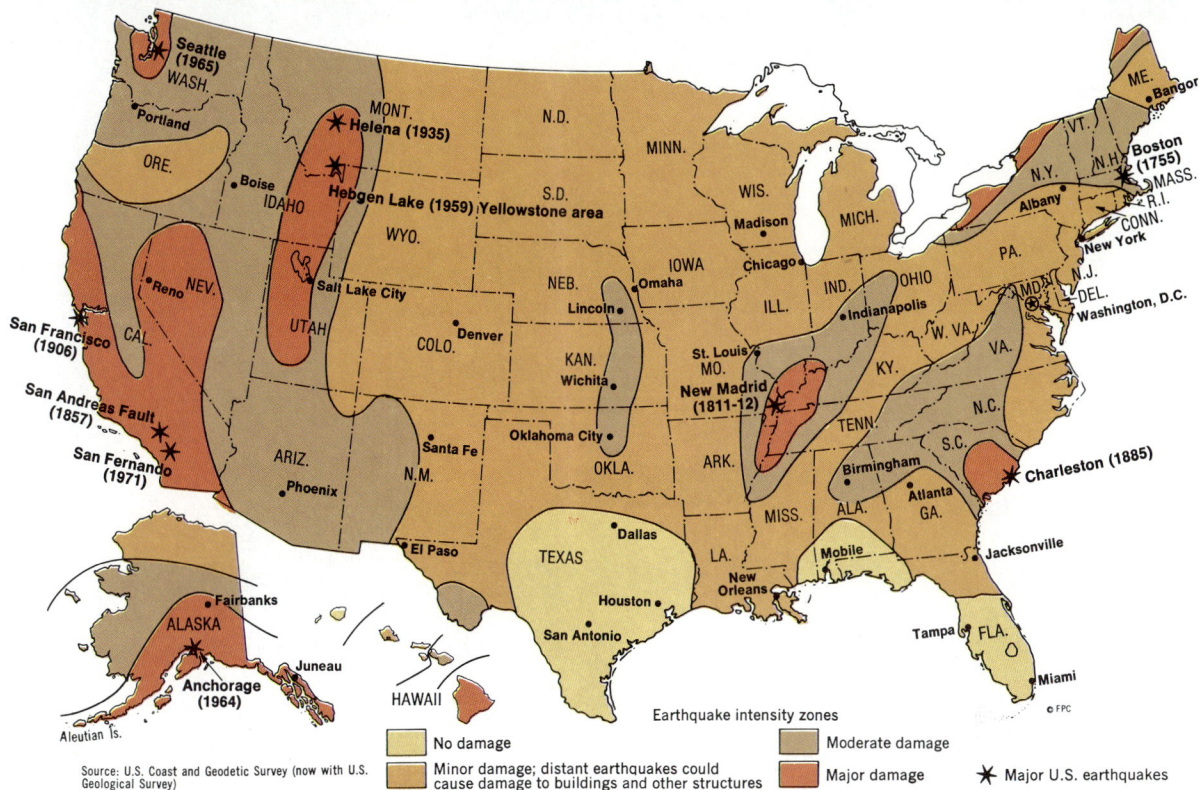

Earthquake intensity zones

| No damage | Moderate damage |

Minor damage; distant earthquakes could cause damage to buildings and other structures

| Major damage | ✱ Major U.S. earthquakes |

Source: U.S. Coast and Geodetic Survey (now with U.S. Geological Survey)

---

3. Its inner part is solid iron and nickel, while its outer part is liquid iron and nickel.
4. It is the hottest of the earth's three parts.
5. Its upper layer "floats" on its bottom layer.

B. Choose the letter of the item that does NOT belong with the first item.
1. **mountain-building forces:** (a) faulting, (b) folding, (c) volcanic pressure, (d) wind erosion.
2. **types of volcanoes:** (a) shield, (b) composite, (c) block, (d) cinder-cone.
3. **agents of erosion:** (a) running water, (b) ice, (c) people, (d) earthquakes.
4. **minerals:** (a) sulfur, (b) oxygen, (c) copper, (d) iron.
5. **mechanical weathering:** (a) chemical change, (b) grinding of rocks against each other, (c) expansion of freezing water, (d) wedging of plant roots.
6. **major landforms:** (a) plains, (b) plateaus, (c) mountains, (d) sandspits.

Discuss These Points

Jim and Barbara are having a discussion. Both want to be scientists. Jim says he wants to do something that will bring an immediate improvement in people's everyday lives, such as increase the supply of fertile soil. Barbara says she wants to find out if there is life in outer space. She says that her work will be just as practical as Jim's. What she discovers may not be useful today. But it is likely to be useful in the future. Do you agree with Barbara? Why or why not?

Expand Your Geographic Sights

Berger, Melvin. *The New Earth Book: Our Changing Planet* (Harper and Row, 1980).

Branley, Franklin. *Shakes, Quakes, and Shifts: Earth Tectonics* (Crowell, 1974).

Brans, Rae. *Rocks and Minerals* (Troll Assoc., 1985).

Elliott, Sarah M. *Our Dirty Land* (Messner, 1976).

Jacobs, Lou. *Shapes of Our Land* (Putnam, 1970).

Keen, Martin L. *The World Beneath Our Feet: The Story of Soil* (Messner, 1974).

Kiefer, Irene. *Global Jigsaw Puzzle: The Story of Continental Drift* (Atheneum, 1978).

Marcus, Elizabeth. *All About Mountains and Volcanoes* (Troll Assoc., 1984).

Matthews, William H. *Introducing the Earth: Geology, Environment and Man* (Dodd, 1972).

Matthews, William H. *The Story of Glaciers and the Ice Age* (Harvey House, 1974).

McFall, Christie. *Underwater Continent* (Dodd, 1975).

Shannon, Tracy. *Sea Searchers: Men and Machines at the Bottom of the Sea* (Dodd, 1971).

Unit 3 Water

All forms of water are present in this scene showing part of Lake Superior and its shoreline. In a few places, currents keep water moving and free of ice. Along the rocky shoreline, frozen spray sheathes rocks in ice and creates ice bridges over gaps between rocks. In the distance, clouds give evidence of water vapor condensing in the air.

Lesson 1 The Wonders of Water

Reading Focus

1. Seeing relationships between two parts of a sentence or between two sentences helps you to make sense out of what you read. Many sentences state cause and effect relationships. Something happens. The something that happens makes something else take place. The something that happens is a *cause*. The something else that takes place is an *effect*. But sometimes an effect is mentioned before a cause. To find the cause, you need to ask yourself some questions. Read example *a.*, below. Does water first gain or lose heat and then change form? Or does water first change form and then gain or lose heat? If you said the gain or loss of heat came first, you have found the cause. That water changes form is the effect. Now study the other examples. Find the cause and the effect stated in each one. Then try to find other cause and effect relationships as you read about water.
   a. Because water can gain or lose heat, it changes form.
   b. Water turns to ice when it releases enough heat.
   c. The temperature of water often drops below 32° F (0° C). When it does, the water turns to ice.

2. As you read this lesson, look for answers to these key questions:
   a. What are some of water's special qualities?
   b. What is the water cycle?
   c. What are two problems that affect the world's water supply?

Vocabulary Focus

water vapor	*condense*	*water cycle*
evaporate	*precipitation*	

Water is everywhere. It covers over 70 percent of the earth's surface, lies underground, and is present in the air that surrounds the earth.

Water supports all forms of life—plants, animals, and humankind. Some very simple forms of life can exist without air. But no form of life can exist without water.

Water shapes and reshapes the crust of the earth. It does this whether it falls as rain, flows in rivers, collects in deep and shallow places, or freezes into ice.

Water plays an important role in determining climate, in weathering rock and forming soil, and in making other natural resources usable. Water does all these things because it has special qualities that few other materials on the earth have. These special qualities make water a wonder on the earth and the earth a fit place on which to live.

Deep underground, water is heated to its boiling point. It follows cracks in the earth's crust until it escapes through an opening to form a thermal, or hot, spring. This thermal spring is in Yellowstone National Park in Wyoming. How can you tell the area contains other thermal springs?

Water—A Special Case

Water's special qualities, taken together, set water apart from all other materials on the earth. What are these qualities?

Water stores heat energy. One of water's most important qualities is its ability to soak up and hold huge amounts of heat energy from the sun. Nearly all materials store some energy. But water stores more of it much more effectively than any other material. As a liquid, water uses the stored energy to change to a gas (water vapor) or releases it to change to a solid (ice).

When temperatures in the air are 32° F (0° C) or lower, water gives up its stored heat energy without any further lowering of its own temperature until it freezes into solid ice. Because water releases heat even as it freezes, it helps to keep air temperatures from getting too cold too fast.

Ice takes on heat energy from the air without changing its own temperature until it melts entirely. Melting, then, takes place slowly, giving the earth time to soak up the meltwater or to carry it away. Colder climates always have some flooding when winter snow and ice melt, but the flooding would be much worse if ice melted swiftly.

Water evaporates, or changes to water vapor held in the air, without raising its own temperature any higher. Heat energy from the sun warms the surface of oceans, lakes, streams, and other bodies of water. This warming starts the evaporation process. How fast evaporation proceeds depends on such things as wind, the amount of salt in the water, and the amount of water vapor already in the air. When heated over a fire, water boils at 212° F (100° C). Boiling water always evaporates rapidly.

When water vapor condenses, or cools in the air, it changes back to water droplets. At the same time, it gives out its stored heat energy to the air without affecting the temperature of the water droplets. Condensation warms the air. It

An ice cap, in some places more than a mile thick, covers Antarctica. Pack ice (so called because pieces of ice jam against one another to form packs or piles of ice) floats in the foreground.

creates strong up-and-down air movements in clouds, turning them into storm clouds that send some form of precipitation back to the earth's surface.

This ability of water in all of its forms to store and release heat energy as it changes form influences the heating and cooling of air. It keeps the low latitudes from becoming too hot for life and the high latitudes from becoming too cold for life. It helps in the exchange of heat and cold from low and high latitudes. It gives variety to weather in the middle latitudes, helps to power all storms and winds, and plays an important role in determining world climate patterns.

Water has a wide temperature range. Water follows its own patterns in its freezing and boiling points. If water followed the pattern set by related chemical combinations, it would boil at $-132°$ F ($-91°$ C) and freeze at $-148°$ F ($-100°$ C). At these temperatures, there could be no liquid water on earth—only steam. But water changes form at much higher temperatures, and the range between its freezing and boiling points is much wider (180° F; 100° C). So water is present on the earth not only as a liquid—its most abundant form—but also as a gas and as a solid.

Ice floats. Most liquids that turn into solids decrease in size and become heavier. Water, when it becomes ice, increases in size and becomes lighter. So ice forms at the top of a water body's surface, floating on the unfrozen water beneath it. If water followed the pattern set by most solids, ice would first form on the bottoms of lakes, rivers, oceans, etc. It would gradually build up layers until the whole water body was frozen. Little melting would take place from year to year. Much of the world's water supply would be locked in ice, making climates much different than they are.

Water dissolves materials. Many minerals and other materials that come in contact with water dissolve in it. Plants and animals need these materials to grow and to build healthy bodies. So do humans. All living things take in dissolved materials when they soak up or drink water. And the water that humans and animals drink helps them to digest the foods they eat.

118

The Water Cycle

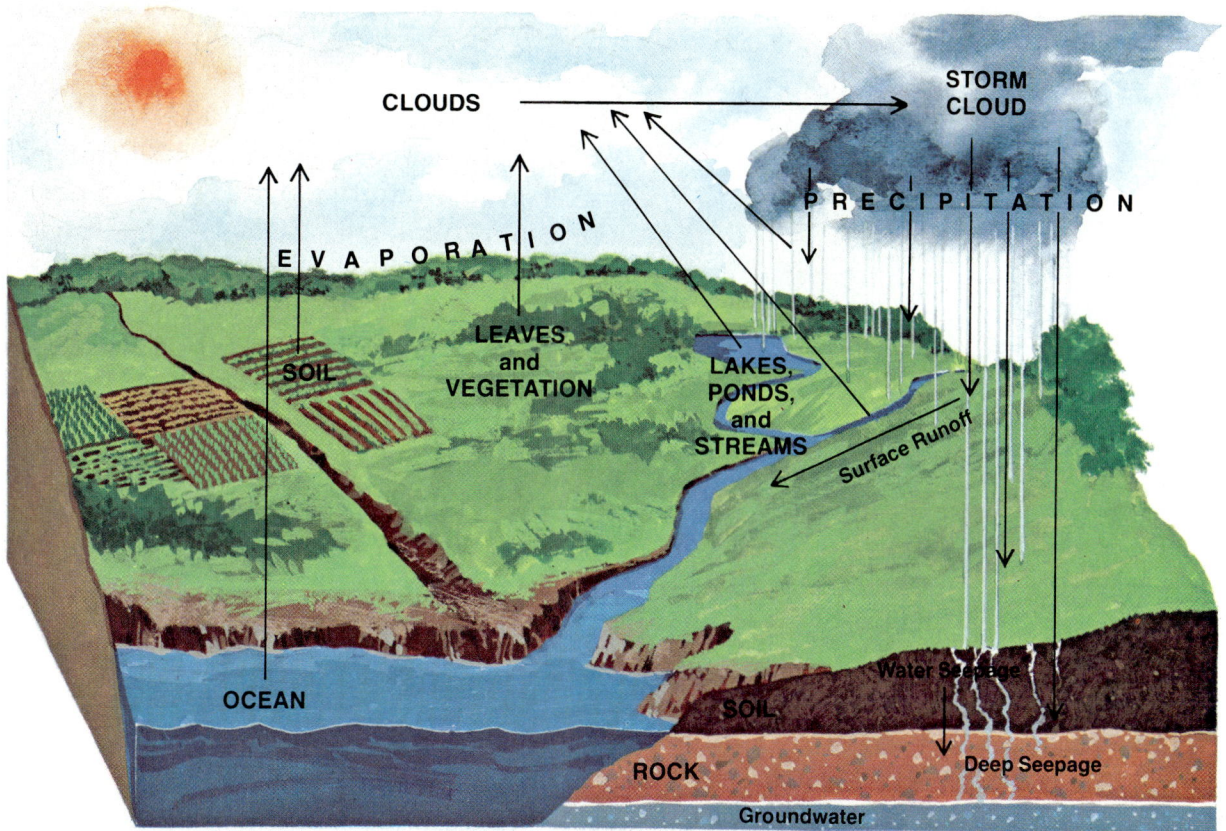

Use this diagram to explain how the water cycle works. What part does the sun play in the water cycle's operations? What happens to water vapor when it rises in the air? What happens to precipitation that falls on land?

Water has still other special qualities. Surface tension, the ability of water to stick to itself very tightly, lets water support some objects that are heavier than itself. It lets water stick to other things, giving water its "wetness" and its cleaning power.

Water can also climb up a surface against the pull of gravity. Because of this quality, water moves up roots and stems of plants to their leaves.

The Earth's Water Supply

Picture a square box. Each side—the height, the width, and the depth—of the box equals a mile (1.6 km). Picture 326 million of these boxes filled with water.

This is how much water there is supplying the earth's needs. It is the same amount that has been on the earth for about 3 million years—thanks to the water cycle.

The water cycle. All the heat energy water soaks up from the sun and all the changes water goes through are what make the water cycle work. Water turns into water vapor and escapes into the air. The water vapor rises to higher altitudes, where it cools and condenses into water droplets. The water droplets join to form clouds. Eventually the water droplets fall back to the earth's surface as a form of precipitation (snow, sleet, hail, or rain). Some of it falls on

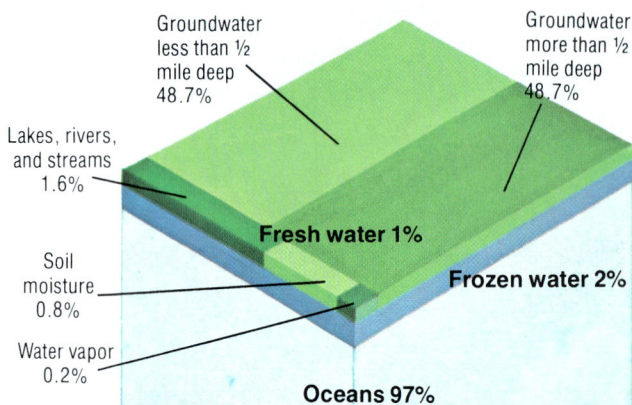

Sources of Earth's Water

Groundwater less than ½ mile deep 48.7%

Groundwater more than ½ mile deep 48.7%

Lakes, rivers, and streams 1.6%

Fresh water 1%

Frozen water 2%

Soil moisture 0.8%

Water vapor 0.2%

Oceans 97%

This diagram shows the sources of earth's water and its distribution.

Salt water. Almost all of the earth's water supply—about 97 percent of it—is in the oceans and seas. Ocean water and the water in some inland lakes and seas is salt water. Salt water cannot be used for drinking, irrigation, or manufacturing. But it does have its usefulness. Salt water supplies most of the precipitation that falls on the earth. When salt water evaporates, the vapor rises in the air, but the salts remain behind. Through evaporation, nature turns salt water into fresh water.

Fresh water Less than 3 percent of the earth's water supply is fresh water. Two percent of the earth's total water supply is locked in ice caps and glaciers as frozen water. The remaining 1 percent— that labeled fresh water—is not frozen. It is ready for use wherever it can be tapped.

water bodies. Some of it falls on land, seeping into the ground or running off into rivers and streams. Most of it finds its way back to the oceans. Day after day, the movement of water from earth to the air and back again to earth takes place. The water cycle never stops renewing the earth's water supply.

Water-supply problems. As you can see from the diagram on page 120, 2 percent of earth's total water supply is locked in ice caps and glaciers. This leaves a useful freshwater supply of just 1 percent for people to draw from. This 1 percent is enough fresh water to supply each person on earth with a million gallons of water each year. Then why do people talk about water shortages?

One problem is distribution. Fresh water is not distributed evenly over the earth. Lands along the equator usually receive more than enough rain. But lands north and south of the equator often receive less than they need. There are some places where a dry season and a rainy season follow each other every year. There are other places that are dry all year.

Parts of India receive over 80 inches (200 cm) of rainfall each year. But they receive most of it from May to October. From November to April, these parts receive less than 10 inches (25 cm) of rainfall a year. In one part of South America, hardly a drop of rain has fallen in over 20 years.

Another problem is pollution. Where water is available, it often has been polluted by the disposal of wastes from homes and factories. One city dweller out of five does not have safe water to

120

How Salts Are Added to Seawater

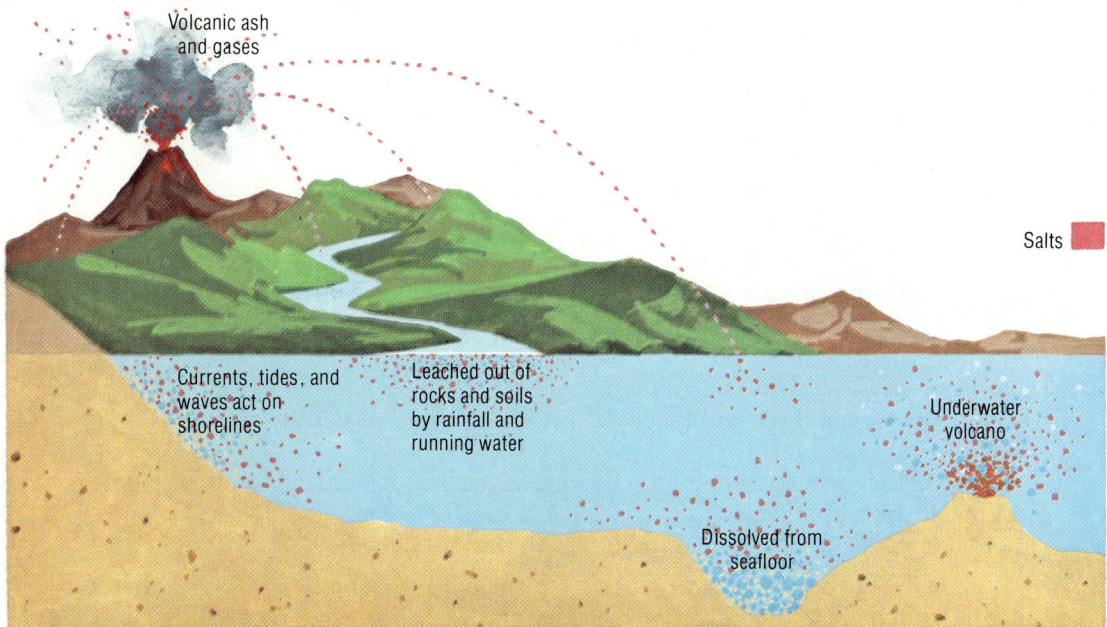

Volcanic ash
and gases

Salts

Currents, tides, and
waves act on
shorelines

Leached out of
rocks and soils
by rainfall and
running water

Underwater
volcano

Dissolved from
seafloor

On the average, ocean water is 3.5 percent salts. Much of the salt is the
kind used to season food. The other kinds of salts found in the ocean are
dissolved minerals. The diagram above shows five ways in which dissolved
salts are added to seawater.

drink. In rural areas, three persons out of four do not have safe drinking water. Every day about 15,000 persons around the world die of illnesses caused by polluted water. To support life, water needs to be kept free from pollution in any form.

Keeping Facts in Focus

1. What do we call the change of water from a liquid to a gas? From a gas to a liquid?
2. Why does water sometimes support objects heavier than itself?
3. What answers would you give to the key questions on page 116?

Working with Ideas

1. How does the changing of water from one form to another influence the temperature of air?
2. How might icebergs benefit a country that suffers from a scarcity of fresh water? What difficulties are involved in a plan to use icebergs to relieve a freshwater shortage?
3. What might the earth be like if ice sank instead of floated?

Lesson 2 The Ocean World

Reading Focus

1. Are you remembering to examine the maps, diagrams, photographs, and other kinds of illustrations that are a part of every lesson? Are you remembering to read carefully any captions that go with these graphics? Check how well you are reading graphics and captions, using the diagram on page 130. After you have studied the diagram, its labels, and its caption, tell which of the following statements are true.
   a. The locks in the Panama Canal allow two-way traffic.
   b. The Panama Canal is 38 miles long.
   c. The Panama Canal has sets of locks at two other locations besides the set shown in the diagram.
   d. If a ship is to be lowered to the next level, water must be drained from a lock until the water levels in the two locks are the same.
   e. Each lock raises or lowers a ship 85 feet to the next level.
2. As you read this lesson, look for answers to these key questions:
   a. In what way is it true to say there is only one world ocean?
   b. What are some of the ways in which oceans move?
   c. What are some of the resources of the ocean world?

Vocabulary Focus

wave	*current*	*fishing bank*
tide	*plankton*	*nodule*

Change the order of the last two words in the title of this lesson and it reads *the world ocean*. This is what really exists on the earth's surface—one world ocean, flowing around and between continents and islands. A look at the map on pages 12–13 shows you that no continent or island completely blocks one part of the ocean from flowing into another part.

But another look at this same map shows you that different parts of this ocean world have different names. These names were given at various times in history, before people realized there was really only one world ocean.

There are four great oceans—the Pacific, the Atlantic, the Indian, and the Arctic oceans. There are other names for smaller parts of these oceans—the Caribbean Sea, the Bering Sea, and the South China Sea. Find the Gulf of Mexico, the Gulf of Alaska, Hudson Bay, and the Bay of Bengal. Find the Mozambique Channel, the Strait of Gibraltar, and the Drake Passage. These are just a few of the many names given to smaller

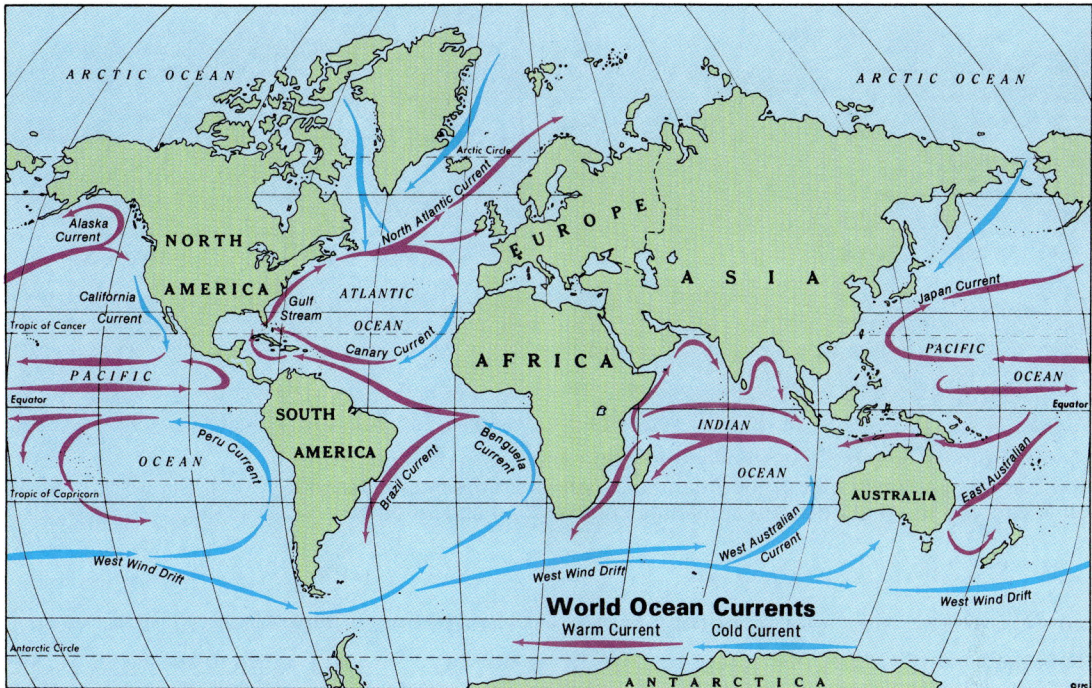

Map labels:
ARCTIC OCEAN — ARCTIC OCEAN
Alaska Current
NORTH AMERICA
California Current
Tropic of Cancer
PACIFIC
Equator
PACIFIC OCEAN
Peru Current
SOUTH AMERICA
OCEAN
Tropic of Capricorn
West Wind Drift
Antarctic Circle
Arctic Circle
North Atlantic Current
Gulf Stream
ATLANTIC OCEAN
Canary Current
EUROPE
AFRICA
ASIA
Japan Current
PACIFIC OCEAN
Equator
INDIAN OCEAN
Brazil Current
Benguela Current
AUSTRALIA
East Australian Current
West Australian Current
West Wind Drift
West Wind Drift

World Ocean Currents
Warm Current   Cold Current
ANTARCTICA

Besides the surface currents shown here, there are ocean currents at deep levels. Both surface and deep currents are an important part of ocean circulation.

parts of the one world ocean. Serving as location guides, these names help people to know exactly what part of the ocean world is being talked about.

The Restless Ocean

Ocean water is never at rest. It moves because different parts of the ocean world have different temperatures. Ocean water moves as waves because winds blow steadily over the face of the earth. It moves as tides because some parts are always pulled by the force of gravity from the moon and the sun.

General ocean circulation. Just as some parts of the earth's land surface are colder than others, so some parts of the world's ocean surface are colder than others. Water in the higher latitudes is colder than water in the lower latitudes. These differences in water temperatures bring about a general mixing, or circulation, of ocean water. Ocean water is always moving from colder latitudes toward warmer latitudes and from warmer latitudes toward colder latitudes.

Colder water is heavier than warmer water. Colder water usually flows at deeper levels and toward lower latitudes. At the same time, warmer, lighter water usually flows at near-surface levels from lower latitudes toward higher latitudes. While they flow at these levels, parts of ocean water are always rising as they become warmer or sinking as they become colder.

Ocean currents. In addition to the general mixing of ocean water, parts of the ocean move as currents, or rivers of water. These currents are warmer or

Compare these photographs showing low tide and high tide along part of the Bay of Fundy, an inlet of the Atlantic Ocean in southeastern Canada. How can you tell that tides are sometimes higher than shown here?

colder than the surrounding water through which they flow. Some things that cause currents are differences in water temperatures and saltiness, winds, the shapes of continents and continental shelves, and the rotation and revolution of the earth.

The map on page 123 shows the world's main ocean currents. Trace the way currents flow north of the equator. Notice that these currents flow in a general clockwise direction. Tracing the way currents flow south of the equator, you notice that they flow in a general counterclockwise direction.

Look again at the ocean currents shown on the map. Most warm water currents flow along the eastern sides of continents, while most cold water currents flow along the western sides. Ocean currents do much to influence the climates of nearby lands. You will learn how they influence these climates in a later unit.

Ocean waves. The movement of ocean waves is very different from that of currents. Currents move forward. Waves move up and down. The only thing that moves forward in a wave is energy. Moving from one wave to the next, energy pushes water up, down, and around in a circle.

There are only two times when waves actually move forward. One time is when they break on hitting a shoreline, creating surf. As the bottom of the wave touches land, it slows down. The top of the wave has nothing to slow it down, so it falls forward. The other time is when the top of a wave is pushed forward by the wind, creating whitecaps at sea.

Winds set most waves in motion. The power of most waves depends on the strength of the wind blowing across the water. During storms at sea, waves may rise to heights of 40 feet (12 m) or more. Some of these waves have the strength to break ships apart.

One very powerful kind of wave is set off by earthquakes and undersea volcanic eruptions. This is the dreaded tsunami. A tsunami is barely felt or seen on the open sea. Only when it touches land, does a tsunami unleash the full force of the energy it carries. Then it becomes a killer wave.

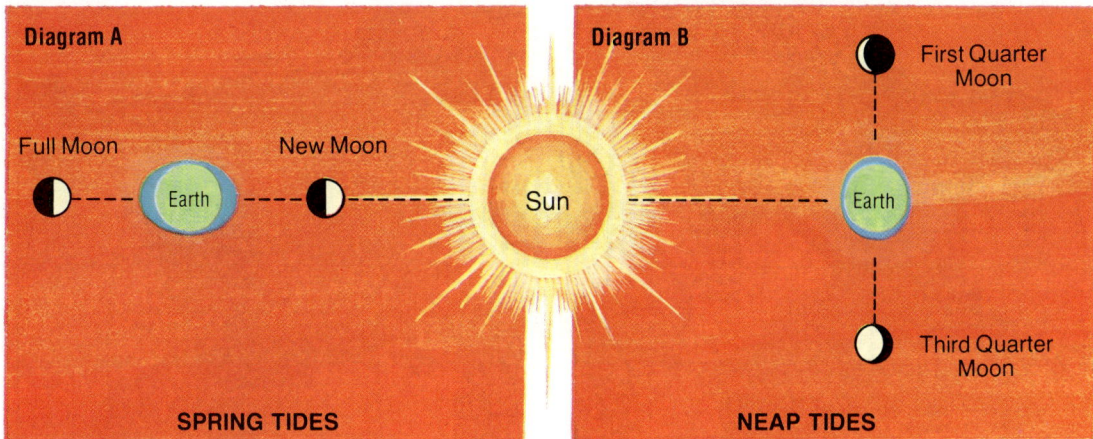

Diagram A

Full Moon · Earth · New Moon · Sun

SPRING TIDES

Diagram B

First Quarter Moon · Earth · Third Quarter Moon · Sun

NEAP TIDES

Are the tides decreasing to their lowest point or increasing to their highest point as the new moon moves to its first quarter position? As the first quarter moon moves to the full moon position?

Ocean tides. Ocean water moves in still another way. It rises and falls with the tides. On most coastlines, the tide moves slowly out from shore for about six hours. This is low tide. For the next six hours, the tide moves back slowly. This is high tide.

Most places along ocean shores have two low tides and two high tides each day. Coasts along the Mediterranean Sea have hardly any tides at all. Coasts along the Gulf of Mexico have only one high tide and one low tide each day.

In some parts of the world, the difference between high tide and low tide is as great as 50 feet (15 m). In others, it is hardly noticeable. The differences between high and low tides in different parts of the world ocean are caused by several things. These are the shape of the coastline; the slope of the continental shelf; and the depth, size, and shape of that part of the ocean.

The tides are produced by the gravity of the moon and the sun pulling on the earth. Most of the pull comes from the moon, because it is closer to the earth. As it orbits the earth, the moon is al-

ways pulling on the side of the earth nearest to it. It also pulls the earth away from the waters on the opposite side.

At two different places in the moon's orbit, the moon is in line with the sun (see diagram A). At these times, the sun's pull joins with the pull of the moon to produce higher tides than usual. These tides are called spring tides.

At two other places in its orbit around the earth, the moon is at right angles to the sun (see diagram B). The sun's pull of gravity, at these two times, works against the moon's pull. The tug-of-war between the pull from the sun and the pull from the moon produces tides that are lower than usual. These tides are called neap tides. At all other times, the tides are either decreasing to their lowest point or increasing to their highest point.

The times when the tide moves in and out from shore are figured out from the movements of the sun, the earth, and the moon and are printed in tide tables. Sailors always check these tables for the times of high tide. High tides, when ship channels are deeper, are the best times

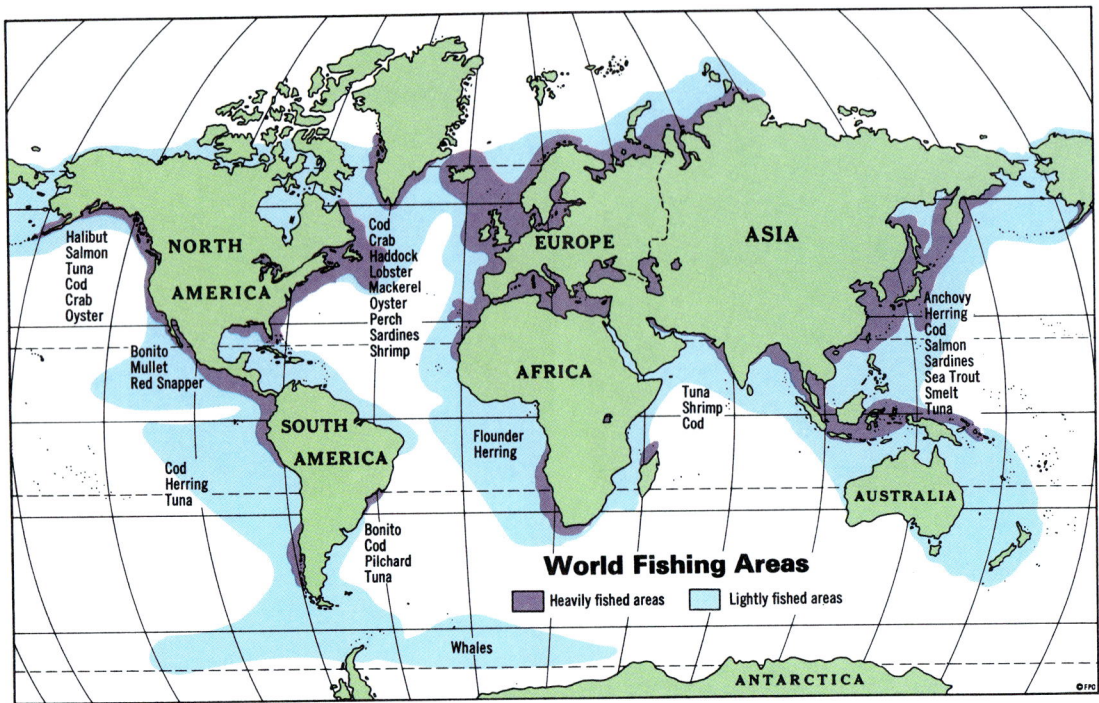

World Fishing Areas

Halibut
Salmon
Tuna
Cod
Crab
Oyster

Bonito
Mullet
Red Snapper

Cod
Herring
Tuna

Bonito
Cod
Pilchard
Tuna

NORTH
AMERICA

SOUTH
AMERICA

Cod
Crab
Haddock
Lobster
Mackerel
Oyster
Perch
Sardines
Shrimp

EUROPE

AFRICA

Flounder
Herring

ASIA

Tuna
Shrimp
Cod

Anchovy
Herring
Cod
Salmon
Sardines
Sea Trout
Smelt
Tuna

AUSTRALIA

Whales

ANTARCTICA

World Fishing Areas

Heavily fished areas   Lightly fished areas

The six largest fishing nations in the world are Japan, the Soviet Union, the People's Republic of China, Peru, Norway, and the United States. Many special groups regulate fishing in different areas of the world. One group, the International Whaling Commission, sets quotas on whales for whaling nations.

for oceangoing vessels to enter and leave harbors.

The in-and-out action of tidal waters keeps harbor waters clean and keeps silt from building up on harbor bottoms. Where tides rush up narrow inlets or bays, they have great power. This power can be used to produce electricity. One tidal power plant has already been built in France. Other places where tidal power plants could be built are being studied. A possible place in North America is the Bay of Fundy off the Atlantic coast of Canada. The tides here are among the greatest in the world.

Using the Resources of the Ocean World

The ocean world does more for people than just make the water cycle work and influence weather and climate. The ocean world teems with life and with many mineral riches. Passenger ships, freighters, and supertankers travel on it, carrying people and goods from one to another of the world's many ports.

Fishing the sea. The map on this page shows where most ocean fishing takes place. Closer to the shoreline in shallower water, fishing for clams, oysters, lobsters, shrimp, and crabs takes place. Deep-sea fishing for herring, haddock, mackerel, cod, tuna, and other fish takes place farther from shore, usually within 200 miles (320 km) of the coastline.

The best fishing grounds for many kinds of fish are those rich in plankton, the small marine plants and animals that fish feed on. These fishing grounds, often called fishing banks, attract plankton in great numbers. This is be-

World's Oceans

Ocean	Area	Greatest Width	Deepest Known Spot	Average Depth	It's a Fact
Pacific	63,800,000 sq mi 165,242,000 sq km	11,000 mi 17,700 km	36,198 ft 11,033 m	14,050 ft 4,280 m	Its area is greater than that of the whole land surface of the earth.
Atlantic	31,800,000 sq mi 82,362,000 sq km	4,150 mi 6,680 km	28,374 ft 8,648 m	14,000 ft 4,270 m	It is the most heavily traveled seaway in the world.
Indian	28,400,000 sq mi 73,556,000 sq km	6,200 mi 9,980 km	25,344 ft 7,725 m	12,780 ft 3,900 m	Most tides along the coasts of the Indian Ocean average 1.5 to 4.5 feet (0.45 to 1.35 m).
Arctic	5,400,000 sq mi 13,986,000 sq km	2,630 mi 4,230 km	17,880 ft 5,450 m	4,360 ft 1,330 m	The shortest air routes between North America and Europe lie over it.

cause they have cooler water flowing through them or coming up from deeper levels. They are also shallower places in the sea, where sunlight can penetrate below the surface. The richest fishing banks are found off the coast of Newfoundland in North America, along the west coast of South America, in the North Sea, and off the coast of Asia in the Pacific Ocean.

Over 5 million people make their livings by fishing. And many more people process, package, and sell fish. The total fish catch each year is enough to supply about 40 pounds (18 kg) of fish for each person in the world. About half of this fish catch comes from the Pacific Ocean. The Atlantic Ocean provides about one third of the total. Only one twentieth comes from the Indian Ocean. The Arctic Ocean supplies the rest.

Almost every method ever devised to catch fish is still in use today. There are people who fish using bows and arrows, spears, or traps. There are people who use small boats, throwing out nets or lines and hauling them in by hand. Many small-boat users never travel very far from their home ports or stay out for more than a day at a time. Most fishing today is done on a much larger scale, using fleets of ships equipped with machines that do much of the work.

Some countries, Japan and the Soviet Union, for example, send out fishing fleets that often stay at sea for two hundred days at a time. The boats in these fleets are equipped with machinery that lowers and raises nets or sends out and hauls in long lines with baited hooks. Refrigerating equipment freezes the tons of fish caught, keeping them fresh for processing. Some fleets even have ships that do part of the processing at sea. Helicopters and electronic gear are often used to locate schools of fish.

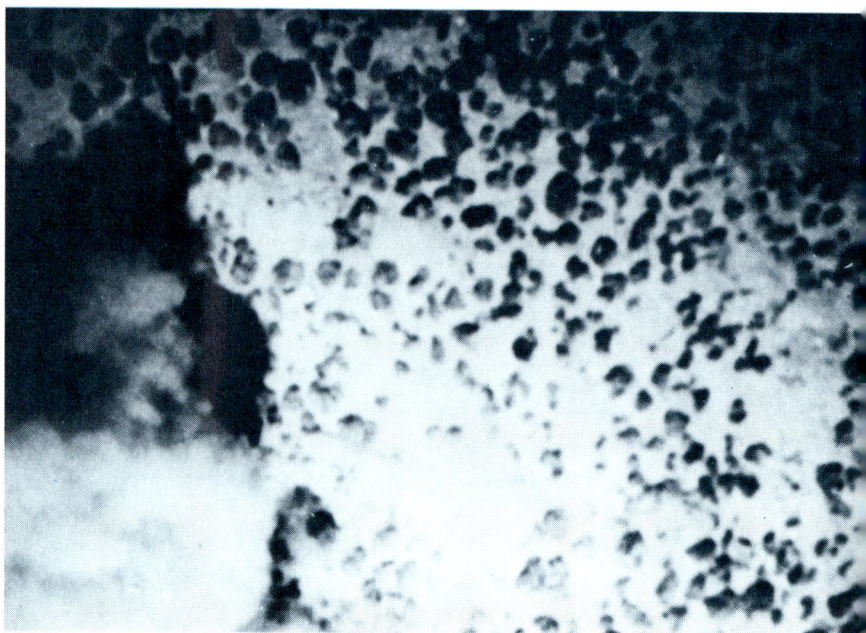

It is estimated that the seafloor of the Pacific Ocean alone contains 1.6 trillion tons of mineral-rich nodules, such as those shown here.

Using this modern equipment, most fishing nations increase their fish catches each year. In one way, this is good. More food is needed to keep up with growing populations. But in another way, it is bad. The ocean world is in danger of being overfished. Fish are being taken from the oceans faster than they can reproduce themselves. What steps do you think nations can take to reduce the problems of overfishing?

Mining the sea. Only in this century have nations begun to mine the mineral riches of the sea. Many nations are now drilling into pockets of petroleum and natural gas trapped beneath the continental shelves off their coasts. With the need for fuels growing greater each day, the search for more pockets continues.

Great areas of the ocean floor are covered with small nodules, or lumps, that contain such minerals as manganese, cobalt, and iron. About six million tons of these nodules form on the seafloors every year. Seawater also contains vast quantities of magnesium, bromine, boron, silver, gold, table salt, and other salts.

The problem is how to mine these ocean minerals. Most of them are still found on land, where they can be mined more cheaply. To prepare for the day when these ocean minerals will be needed, scientists are working on the tools and the processes needed to scoop or suck up these minerals from the seafloor or to extract them from seawater.

Ocean shipping. In times past, people have traveled and carried goods over the oceans and seas in rafts, in reed boats, in dugout canoes, and in wooden boats with wind-driven sails. In recent times, diesel motors have replaced steam engines. Steel ships have replaced wooden ones. Length has increased and so has size. Today there are ships for carrying passengers, ships for carrying grains or ores, refrigerated ships for carrying fruit and meat, container ships that carry prepackaged goods for easy loading and unloading, and supertankers for carrying petroleum. Ships have radar

Geographers and Their Work

Oceanographers—Pioneers of the Ocean Frontier

The pioneers of old have nothing on oceanographers—the new pioneers who seek knowledge about oceans and ocean life. Oceanographers study fish and marine plant life. They explore ocean bottoms to discover how they were formed, what they are made of, and what mineral riches they hold. Oceanographers also study many other things. They experiment with living beneath the sea; study ocean waves, currents, and tides; and trace the effects of ocean pollution.

Many oceanographers work from surface vessels at sea. Some dive to great depths in underwater vessels. Others live in special underwater quarters for days at a time. Still others use special wet suits, masks, and air-breathing gear to do scuba diving and snorkeling. Surface vessels can be as large as freighters. Underwater vessels can be so small that they seat only one or two persons.

The Glomar Challenger is a large ship with a computer-controlled system that is used to hold the ship over a drilling site. By studying samples drilled from ocean bottoms all over the world, oceanographers have obtained a record of what happened on the ocean floor millions of years ago. *Deep Quest* is a sharkshaped underwater vessel that holds four. Using this vessel, oceanographers can stay underwater for hours at a time to take photographs, to study ocean life, or even to work on the legs and drills of oil platforms. *Alvin*, a submarine, seats three and can remain underwater for three days. Using *Alvin*, oceanographers dove over 9,000 feet (2,700 m) below the surface of the Atlantic Ocean. On this dive, the crew found evidence of seafloor spreading and continental drift. With other research vessels, oceanographers have discovered plate boundaries, gathered samples from ocean floors, and studied the effects of underwater earthquakes and volcanoes.

Through the work of oceanographers who study marine life, we are learning more about the conditions that affect fish life. We are discovering new sources of food in the sea and ways to "farm" the sea. Someday, we may farm the sea to an even greater extent than we do at present. As more and more oceanographers work together and share their knowledge, we will soon know as much about the oceans as we do the land.

and other kinds of navigational aids. They have machines that do the work of loading and unloading goods.

Even with these improvements, ocean shipping still has problems. And many of them have to do with size. Their larger sizes make ships more difficult to navigate through narrow or dangerous waters, increasing the risk of accidents, such as oil spills. Ships carrying larger and heavier loads ride lower in the water. They need deeper channels to enter and leave harbors and wider berths at dockside. These needs often limit the number of ports at which these ships can dock. They also limit their use of the Suez Canal, the Panama Canal, and the St. Lawrence Seaway.

The Suez Canal links the Mediterranean and Red seas, shortening the journey between European countries and Asia by thousands of miles. Since it opened in 1869, it has been enlarged several times. Egypt began its most recent enlargement in 1976. The first phase, completed in 1980, deepened and widened the canal to allow ships carrying 150,000 tons fully loaded to pass

A ship passes through the Panama Canal (above). The diagram below shows how a set of locks raises or lowers a ship passing through the Panama Canal. The canal has a total of 12 locks, each with a depth of 41 feet (12 m). Built in pairs (for two-way passage) at three different locations, the locks raise and lower ships 85 feet (26 m) to the level of the oceans at either end of the canal. On an average day, 38 ships pass through the canal. It takes about 8 hours for a ship to travel the canal's 51-mile (82-km) length.

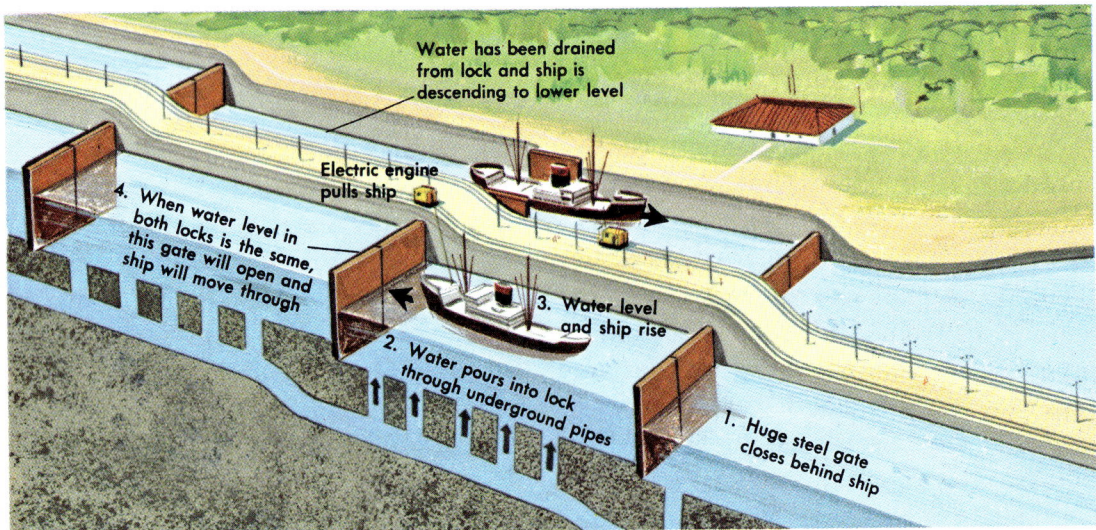

Water has been drained from lock and ship is descending to lower level

Electric engine pulls ship

4. When water level in both locks is the same, this gate will open and ship will move through

3. Water level and ship rise

2. Water pours into lock through underground pipes

1. Huge steel gate closes behind ship

The St. Lawrence Seaway

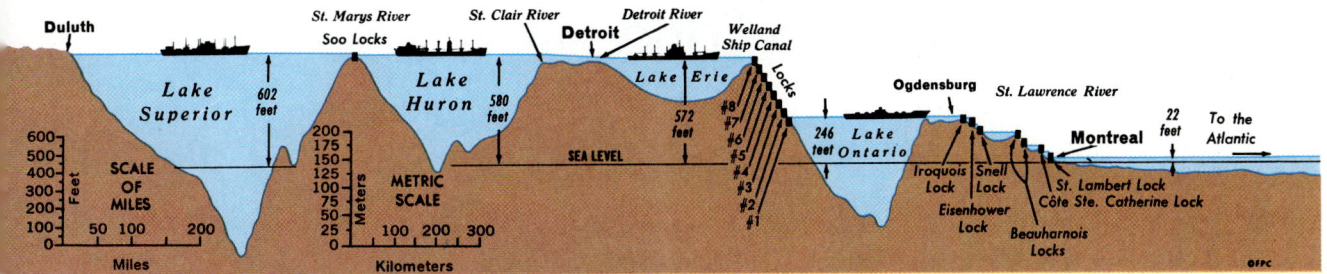

How many locks are there between Montreal and Lake Ontario? How many locks are a part of the Welland Ship Canal? This canal, first opened in 1830, was widened and deepened in 1932. The Soo Canal has five locks—four on the American side and one on the Canadian side. On the map on pages 32–33 of the Atlas, trace the route a ship would take from Montreal to Duluth.

through it. When it is completed, the canal will be 67 feet (20m) deep, 17 feet (5m) deeper than it is now. Even then, supertankers needing channels 100 feet (30m) or more deep will not be able to use it.

The Panama Canal links the Atlantic and Pacific oceans. The task of modernizing the canal, which opened in 1914, is so great and so costly that serious thought has been given to constructing a new canal at another location. Whether this will ever be done depends on shipping needs and on world politics.

The St. Lawrence Seaway opened in 1959. It extends from the mouth of the St. Lawrence River to the western end of the Great Lakes. From about the middle of November to early April, the waters of the seaway are frozen solid. This is probably a greater disadvantage than the size of its locks, none of which can hold the largest vessels.

Keeping Facts in Focus

1. Which of the world's four oceans is the largest? The smallest? The deepest?
2. What is an ocean current? On what side of a continent do most cold water currents flow? Most warm water currents?
3. How does the movement of ocean waves differ from that of currents?
4. What causes the tides? When are tides the highest? The lowest?
5. What are fishing banks? Where are the richest fishing banks found?
6. What are some minerals found on or beneath the seafloor? In seawater?
7. What are some shipping problems caused by the increased sizes of ships today?
8. What answers would you give to the key questions on page 122?

Working with Ideas

1. What causes the ocean water to be constantly moving?
2. What are three laws that you think all nations should accept and obey regarding the use of the world ocean and its resources? How do you think the nations of the world should reach agreement on these laws?

Lesson 3 Stops in the Water Cycle: Water on Land

Reading Focus

1. You have been learning how to use different kinds of reading skills. Are you remembering to use these reading skills, even when you are not told to use them? This is how you become an independent reader. Why not take notes for yourself? Taking notes will help you organize facts. It will help you to connect details with main ideas. With good notes, you can summarize important parts of lessons. Are you remembering to check out the definitions and pronunciations of unfamiliar words by using the Glossary of your book or a dictionary? Practicing these reading skills on your own will make you a better reader.

2. As you read this lesson, look for answers to these key questions:
   a. In what different ways are lakes formed?
   b. What are wetlands? Why are wetlands important?
   c. What is the water table? What part does it play in the water cycle?

Vocabulary Focus

wetland	ecology	porous
pollutant	groundwater	water table

A drop of water travels through the water cycle in a never-ending journey from sea to land and back to the sea. It travels thousands upon thousands of miles as the journey is repeated time and time again. A drop of water may make several stops on land in its journey, staying in one place for a short time—a few hours, a day, or a month. It may stay for a year, even thousands of years. But eventually every drop of water finds its way back to the sea. Consider the stops a drop of water may take once it rises from the sea as water vapor, condenses in the air, and falls back to earth in some form.

Lakes

One possible stop for a drop of water is a lake. Lakes form whenever two things are present. There must be a place lower than the surrounding land. And there must be a source of water to fill it.

The scraping of glaciers over the land has dug out most of the places that form lakes. The eroding action of swiftly running water has dug out others. Still

In Tavira, a town on the southern coast of Portugal, seawater from the Atlantic Ocean is channeled into shallow tanks. As the sun and the wind evaporate the water, the workers scrape the salt into piles along the sides of each tank.

others have been formed when rocks within the earth shifted, leaving gaps in the earth's crust. Or surface rocks have shifted, blocking valleys. People also form lakes by moving earth or building dams to block the flow of water.

Once formed, these low places fill with water from melting glaciers, snow, and ice. They fill with rainwater flowing down from higher places. They also fill with water from underground springs or from layers of water that lie below the earth's surface.

Freshwater lakes. Most of the world's lakes are freshwater lakes. Freshwater lakes have surface rivers or underground streams and springs flowing into them. They also have surface streams and rivers or underground streams flowing out of them.

Saltwater lakes. A few of the world's lakes are saltwater lakes. Saltwater lakes have streams flowing into them. But they have no streams flowing out. Lakes that have no outlet streams cannot renew themselves. As water evaporates, these lakes build up a heavy salt content in their water.

Some inland seas are really saltwater lakes. Because they are surrounded on all sides by land, the Caspian Sea, the Dead Sea, and the Aral Sea are really lakes. They are called seas only because their water is salty. Find the Caspian Sea and the Aral Sea on the map on pages 18–19. Also find the Great Salt

World's Largest Natural Lakes

Lake	Continent	Area	Greatest Depth	It's a Fact
Caspian Sea (salt)	Asia	143,630 sq mi 372,000 sq km	3,264 ft 995 m	It has been shrinking in size because rivers that empty into it bring in less water than it loses by evaporation.
Lake Superior	North America	31,700 sq mi 82,100 sq km	1,333 ft 406 m	About 200 rivers, most of them short, empty into it.
Lake Victoria	Africa	26,830 sq mi 69,490 sq km	270 ft 82 m	The largest lake in Africa, it is the source of the Nile River.
Aral Sea (salt)	Asia	25,660 sq mi 66,460 sq km	223 ft 68 m	This saltwater lake contains many islands.
Lake Huron	North America	23,050 sq mi 59,700 sq km	750 ft 229 m	It drains a 51,700 sq mi (133,900 sq km) area.
Lake Michigan	North America	22,300 sq mi 57,760 sq km	923 ft 281 m	This third largest of the Great Lakes is the largest body of fresh water lying wholly in the United States.
Lake Tanganyika	Africa	12,700 sq mi 32,890 sq km	4,708 ft 1,435 m	It is the longest freshwater lake in the world—420 mi (676 km).
Great Bear Lake	North America	12,280 sq mi 31,800 sq km	1,350 ft 411 m	The Arctic Circle crosses the northern part of it.
Lake Baikal	Asia	12,160 sq mi 31,500 sq km	5,712 ft 1,741 m	It is the deepest freshwater lake in the world.
Lake Nyasa	Africa	11,100 sq mi 28,750 sq km	2,300 ft 701 m	It is part of a chain of lakes in Africa's Great Rift Valley.

Lake in the western part of the United States on the map on pages 34–35. About 200,000 short tons (180,000 metric tons) of salt are mined from this lake each year.

The world's largest lakes. Look at the chart on this page. It lists some of the world's largest lakes. You can find most of these lakes on the map on pages 12–13. Lakes Superior, Michigan, and Huron are part of the Great Lakes. Find these three lakes on the map on pages 34–35. What are the other Great Lakes?

Lakes are found in every part of the world. Lake Titicaca in South America is the most widely used of the world's high-altitude lakes. Its altitude is over 12,000 feet (3,600 m) above sea level.

Thunder Bay is an important port on the Canadian side of Lake Superior. Wheat is a major product shipped from this port.

The Dead Sea, the lowest surface lake in the world, is nearly 1,300 feet (390 m) below sea level. What is the world's largest natural lake? Where is it found?

Lakes and climate. Lakes, especially large ones, modify the climates of their surrounding lands. They do it in much the same way that oceans modify climate. Large bodies of water heat and cool more slowly than land. In summer, the water is cooler than the land. Breezes blowing off a lake cool the land. In winter, the water is warmer than the land. Breezes blowing off a lake warm the land.

The citrus-growing industry in Florida depends on lakes to lessen the effects of cold spells that sweep in from the north. On the eastern shore of Lake Michigan, another fruit-growing belt, winds off the lake keep killing frosts from striking too soon in the fall. In the spring, they keep the trees from budding until winter's killing frosts have passed.

People and lakes. Lakes serve many of the same purposes that oceans do. People use lakes for swimming and boating. They draw fish from lakes for sport and food. They build summer homes around them. Lakes also serve as transportation routes. Along with the St. Lawrence River, the Great Lakes form an important inland waterway in the United States.

The freshwater lakes of the world have other uses that oceans do not have. Lakes provide water for drinking and home use, for irrigation, and for industrial purposes. Lakes backed up behind dams generate electricity.

Wetlands

Wetlands are shallow places in the land where water collects, or stands, either all year long or for only part of a year. Wetlands are often found along the edges of other water bodies. Where wetlands lie near the edges of oceans and seas, their water is salty.

Importance in nature. Both saltwater and freshwater wetlands play an important role in nature. Wetlands "clean" the waters that flow into them by removing pollutants, such as nitrates,

135

Wetlands may be swamps, marshes, or bogs. On the left is a cypress swamp in Florida. The top right photograph shows a bog in New Jersey. The bottom photograph shows a marshy area of Lake Titicaca in South America.

from fertilizers. Wetlands hold the overflow from other water bodies. They reduce flooding from heavy storms or from the melting of snow and ice in spring.

Certain plants—mangrove and cypress trees, cattails, and some kinds of reeds and mosses—grow in wetlands and nowhere else. Some animals, such as the muskrat, are most at home in wetlands. Many insects and some kinds of marine animals begin their life cycles in wetlands before moving to other places. And many wetlands provide nesting places for different kinds of waterfowl, such as ducks, herons, and egrets. Without wetlands, the plants and animals

that live in them or depend on them in some way would disappear.

People and wetlands. Many people do not realize how important wetlands are. They think of them as places with land that could be put to better use. So they drain them and fill them up with trash and land from other places. Then they build homes, offices, factories, or shopping centers on them.

No one knows how much of the United States was originally wetland. But those who know how important wetlands are think that over half of the original wetlands in the United States have disappeared. Still more wetlands disappear

In areas where rock layers below the earth's surface are very porous, groundwater seeps downward. It dissolves minerals, carries them along, and deposits the dissolved minerals elsewhere. Through its dissolving action, groundwater has carved out underground caverns. Where it deposits mineral solutions that later harden, it creates strange formations. This underground cavern is in Virginia.

every year. Protecting wetlands has become a special concern of many people interested in ecology. Are there any wetlands in your area? What can you do to help protect these valuable parts of nature?

Groundwater

There is another place where water collects. This is below ground. Such water is called groundwater. Groundwater helps to keep a balance in the water supply. Plants send their roots in search of it. People tap it for drinking water and for many other uses.

Groundwater is distributed everywhere under the earth's surface. It is found even under the driest deserts of the world. Some of the groundwater lies thousands of feet below the surface, trapped between layers of hard rock. It may have been formed as magma tried to force its way through weak spots in the earth's crust. Most groundwater, however, lies fairly close to the earth's surface. This is the water that seeps through soil and some kinds of rock whenever water flows over the earth.

The water table. Water, like other things on the earth, follows the downward pull of the earth's gravity. As water runs off the land, some of it seeps downward through soil and porous rock, such as sandstone and limestone, until it is blocked by rock that is not porous, such as granite. When it can move downward no farther, water moves along sideways until it fills all the spaces it can, forming a layer of water-filled soil. The top of this layer is called the water table.

The water table follows the general slope of the land above it. At certain places, however, the land dips sharply, exposing the water table. Lakes, ponds, swamps, marshes, and even some rivers are really exposed parts of the earth's water table. So is a spring. Spring-fed lakes and other water bodies are really fed by groundwater.

Groundwater and the water cycle. Water that finds its way beneath the earth's surface does not remain trapped there forever. People tap groundwater by drilling wells. Much of this water returns to the water cycle after it is used. The roots of many plants reach into the water table, drawing off water that eventually returns to the air and the water cycle through the leaves of plants. Some groundwater feeds into lakes, streams, and rivers. Some feeds into the oceans. It becomes a part of the water cycle once again as soon as it is open to the air. Groundwater plays its part in renewing earth's water supply.

Groundwater problems. Although most groundwater is fresh water, not all of it is. Groundwater near ocean coasts is salty. Water trapped deep underground for many years may become heavy with salts dissolved from the rock layers above and below it. Salty groundwater needs special treatment before it can be used for drinking, farming, and other purposes.

Groundwater deep below the earth's surface is likely to have a high mineral content. Such water is often called hard water. Water that is too hard often leaves a coating on water pipes, boilers, coffeepots, and teakettles. Water with a great deal of iron in it is often reddish in color. Such water stains sinks. It also stains clothes washed in it.

Pollution is a problem when the water table lies close to the surface. The same materials that pollute lakes and rivers can also pollute groundwater. Waste materials seep into the ground along with water. The wastes reach the water table and become part of the groundwater. When this happens, wells in the area can become polluted.

Still another problem relates to the overuse of groundwater. Each year the world's population grows larger. Each year there is a greater demand for fresh water. Most of the world's supply of fresh water, you recall, comes from groundwater. Once groundwater is brought to the surface for whatever use, it eventually becomes a part of the water cycle. The water cycle works to keep the world's supply of water (whether salt or fresh) about the same. But the water cycle does not always replace as much water as people have taken from the ground.

Look at the diagram on this page. It tells you what happens to the water that falls on land. Most of the water either evaporates before it can sink into the ground or it runs into lakes, rivers, and streams. Only 10 percent of the water seeps into the ground to become groundwater once again.

What Happens to Water That Falls on Land

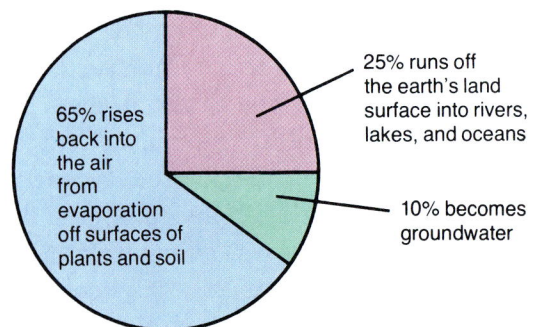

65% rises back into the air from evaporation off surfaces of plants and soil

25% runs off the earth's land surface into rivers, lakes, and oceans

10% becomes groundwater

Safe Water—A Goal Worth Working For

Every body of water—even ones not touched by any human activity—receives some wastes. These wastes come from things that live in water and from once-living things. Wastes from nature usually cause no big pollution build-up, because water has a self-cleaning action. Bacteria, tiny plants, that live in water play an important part in this self-cleaning work. These bacteria use oxygen in the water to break down wastes and to turn them into nutrients (food) for other marine plants and animals.

It is the wastes that wash from the land and that people dump into streams, rivers, lakes, and oceans that cause pollution in water to build up. Years of dumping chemicals, traces of metals, untreated sewage, and heated water into water bodies have weakened water's ability to clean itself and to support life.

The problem is greatest where heavy wastes from industries, farms, and cities pollute water bodies. To break down these heavy wastes, the bacteria must use up too much oxygen from the water. With less oxygen, marine plants and animals die. The result is more waste, more pollution, and a slowly-dying body of water.

In water bodies with heavy pollution, another form of bacteria takes over—one not needing oxygen. These bacteria also break down wastes. But as they do, they give off bad smells, discolor the water, and generally make water unfit for living things.

Most water pollution stems from advances in ways of living and from the growth of population. Industry keeps trying to turn out more and better products in the least amount of time and at the lowest possible cost. Agriculture does the same thing to keep pace with the world's growing need for food. People of all nations keep trying to better their ways of living. All these are goals that are worth seeking.

But equally worth seeking is the goal of clean, safe water. More and more governments are setting up rules for keeping water clean and safe. Industries, cities, and other groups and individuals must follow these rules. In the United States, Congress set this goal: *The streams, lakes, and rivers of our nation must be pure enough to allow safe swimming and to support fish and wildlife by 1983.* While we have not met the deadline, the goal remains. What can you do?

In heavily populated countries and in highly industrialized ones, supplies of groundwater are being used up faster than the processes of nature can replace them. In Israel and on the Great Plains of the United States, for example, people are using most of the available groundwater, mainly for irrigation. The overuse of groundwater in an area has a number of consequences. Because the water table sinks deeper into the earth, drilling becomes more costly. Less land may be irrigated, and fewer crops may be grown.

Keeping Facts in Focus

1. Why are some lakes saltwater lakes?
2. How do lakes modify climates?
3. What is groundwater? What are some groundwater problems?
4. What answers would you give to the key questions on page 132?

Working with Ideas

1. How might heavy pollution in lakes, wetlands, rivers, and other water bodies affect the agriculture of a nation?
2. What are some of the things industries can do to prevent factories from polluting lakes, rivers, streams, other water bodies, and groundwater?

Lesson 4 Further Stops in the Water Cycle: Rivers and Streams

Reading Focus

1. Reference tools such as the Atlas, Index, or Glossary of this book are helpful to you only if you use them. One reference tool you should use often is the Dictionary of Geographical Terms on pages 8–9 of the Atlas. Notice that it has three parts. The first part illustrates many of the terms defined in the second part. Take time to match each illustrated term with its definition. The third part lists many symbols used on maps. Refer to this last part whenever you need to interpret symbols on a map or when you draw a map. Refer to the other two parts whenever you need to "picture" something in your mind or to jog your memory.

2. As you read this lesson, look for answers to these key questions:
   a. What is a river system? A drainage basin? A divide?
   b. How do erosion, transportation, and deposition help rivers do their work of shaping the land?
   c. What makes a river great?

Vocabulary Focus

river system	*sediment*	*floodplain*	*reservoir*
continental divide	*alluvial*	*levee*	*distributary*

Rain falls on the earth, snow and ice melt, and springs gush out of the ground. The water from these sources flows down mountainsides and hillsides, forming tiny streams that run into bigger ones. These streams join small rivers that flow into still larger rivers. Eventually they join a main river, one that empties into the sea.

River Systems

A main river and all its tributaries, or branches, form a river system. The land a river or river system drains is a drainage basin. Each river or stream has its own divide—the hill or ridge that determines the direction a river or stream first takes. All of the rivers and streams flowing into a main river form a larger drainage basin, one usually determined by the higher peaks and ridges of a mountain range. Mountain ranges form continental divides.

Look at the map on page 141. It shows that the high ridges of the Rocky Mountains and the Appalachian Mountains form divides on either side of the conti-

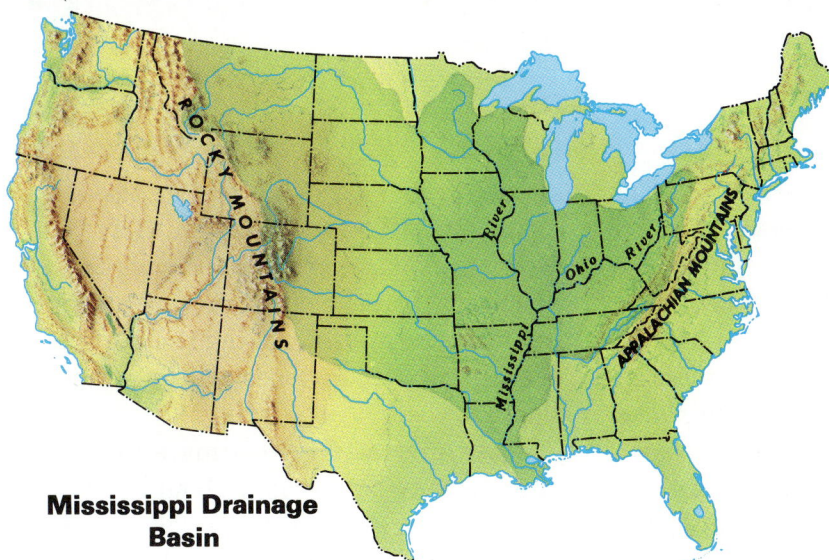

Mississippi Drainage Basin

The Mississippi River system drains almost two thirds of the contiguous United States. Where the Ohio River joins the Mississippi, the river doubles the amount of water it carries. Other major tributaries of the Mississippi are the Illinois, the Missouri, the Arkansas, and the Red rivers.

nental United States. The rivers on the landward sides of these two continental divides follow downhill courses until they flow into the Mississippi River. Other, smaller river systems and drainage basins drain the seaward sides of these mountain ranges. Use the map on pages 34–35 to find the names of some of these rivers.

What wearing-down processes have helped to carve out this mini-canyon?

A river's source. Every river and stream has a source—a place where it begins. The source may be a lake, a spring, or a place where meltwater from a glacier or a snow-covered area collects to begin its journey to lower ground.

A river's mouth. Every river and stream has a mouth—a place where it empties into another body of water. The mouth may be a pond, a swamp, a marsh, another river or stream, a lake, or an ocean.

A river's flow. The path a river takes is called its channel. The bottom of the channel is a river's bed. The movement of water in a river's channel is its current. The term *downstream* indicates a river's flow toward its mouth. *Upstream* is the direction toward a river's source. Moving upstream means moving against a river's current.

The Work of Rivers and Streams

Rivers and streams are always at work on the land, gouging out bits and pieces of rock and soil (erosion), washing them away (transportation), and putting them down some place else (deposition). The first two processes—erosion and transportation—wear down the land, changing highlands into lowlands. The last process—deposition—builds up the land. Together these three processes keep a balance between the high places and the low places of the earth.

Wearing down the land. Water, you recall, is one of the agents of weathering and erosion. As a standing liquid, water changes rocks and soil chemically through its dissolving action. As ice, it breaks up rocks mechanically. But when it moves as ice or flows as water, it does its work of erosion.

Every splash of a raindrop dislodges tiny bits and pieces of rock and soil. In heavy rains, sheets of water flow over parts of the land, carrying the dislodged bits and pieces away. The water gathers into gullies and rivulets, streams and rivers, all the while cutting into the land and increasing its load of eroded materials.

How much erosion takes place depends on the slope of the land, the hardness or softness of the underlying rock, the amount of the water supplied by the source, and the amount of plant cover on the land. Fine soils are more easily eroded than coarse soils. Softer rock wears away faster than harder rock. Hillsides that are bare lose their soil faster than hillsides with plant cover. Water moving from great heights flows faster than water moving over a gradual slope. Fast-moving water has more power to erode the land.

The eroded materials carried by flowing water move at different levels in a stream or river. The largest and heaviest pieces settle to the bottom. Here the force of the water pushes and rolls them along. As these pieces move, they dig into the streambed, cutting the channel deeper and deeper, especially on steeper slopes.

Other pieces of eroded material, not quite so heavy or so coarse, bounce along above the streambed, hitting and striking one another. Sometimes, they settle on the bottom of the streambed, only to move along again after being struck by other bits and pieces.

The finest and lightest pieces, called sediments, float in the water and are carried along with the current. The more sediments water holds, the muddier it looks.

Building up the land. The work of deposition, of building up the land, is done where rivers and streams slow down in their courses.

Wherever rivers move more slowly, they begin to drop the eroded materials they carry. Boulders and other large pieces are dropped first, then smaller rocks, and finally the sand, silt, and clay particles that make up sediments. Eroded materials laid down by rivers are called alluvial deposits. Landforms created by these deposits are also described

A

B

C

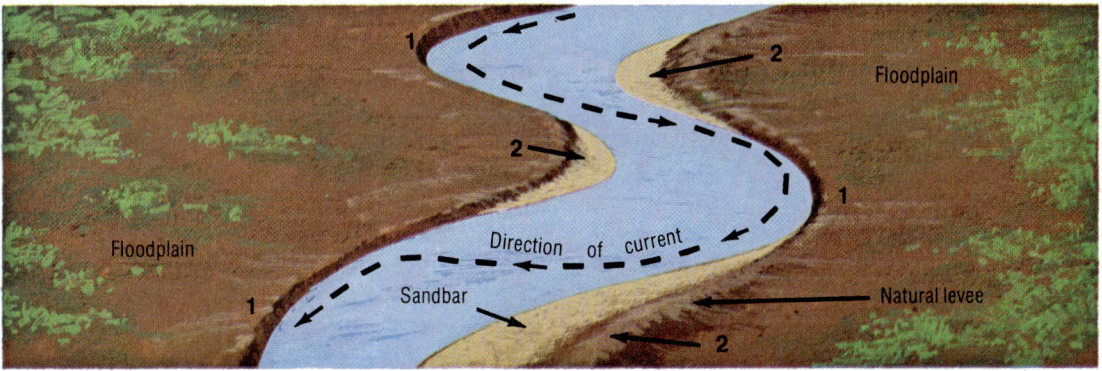

Floodplain

Floodplain

Direction of current

Sandbar

Natural levee

D

as alluvial. Alluvial deposits usually make very fertile soils.

Two of the chief landforms shaped by rivers are alluvial fans and alluvial plains. Alluvial fans develop in a fan shape where rivers and streams flow out of canyons onto a flat or nearly flat plain (see diagram A). Repeated erosion, transportation, and deposition widen and lengthen the alluvial fan (diagram B). Where several canyon openings are nearby, as at the foot of a mountain range, the fans from each canyon may meet and form a piedmont (foot of the mountain) plain (diagram C).

Most alluvial plains are formed where rivers swing back and forth (meander) in their courses (diagram D). At each bend, a river erodes the land at the outside of the bend (point 1), where the current is faster moving. At the inside bend (point 2), where the current is moving more slowly, rivers deposit some of their sediments. At times of floods,

From the Uinta Mountains in northeastern Utah, a river meanders through the floodplain it has built up through the centuries. Is this a youthful river or one in its old age?

when a river carries a full load of water and sediments, it may change its course, cutting off old bends and creating new ones. It is in this way that the floodplains of rivers have developed.

With each flooding, a layer of sediments is laid down. Thick rich soil develops. Because the heavier materials are laid down first, levees (banks higher than the surrounding land) are created. These levees hold back water within the river's channel for a little longer the next time a river floods.

Where the danger of flooding is great and where a river passes through a well-populated area, people often reinforce these levees with sandbags or other materials. Sometimes, they build artificial levees of concrete. Dams are also built at certain places along some rivers to create reservoirs where water can be stored and released slowly to control flooding.

A delta is another landform created by rivers that empty into a sea, bay, or gulf. A delta at the mouth of a river forms in much the same way that an alluvial fan does on land. At a river's mouth, the slope of the river's course approaches sea level, slowing the river's current. Great amounts of sediments gradually build up land on which swamps and marshes develop. Streams branch out from the main channel. Each branch is called a distributary. (A distributary is any stream that leaves a main river but does not enter it again.)

A number of the world's longest rivers have built up deltas at their mouths. The most widely-known examples are the deltas of the Mississippi and Nile rivers.

From youth to old age. Streams and rivers, like mountains, can be described as young or old. Streams flowing down steep slopes are considered young. They have fairly straight courses. Young streams are still cutting into their streambeds and forming their valleys, which are narrow and V-shaped. The land between stream valleys is usually high and broad.

As rivers become older, their currents slow down. Less erosion takes place, and more materials are deposited along their banks. Streambeds gradually

144

Starting in the upper left of this Landsat "false-color" image, follow the main channel of the Mississippi River. It skirts Lake Pontchartrain and the city of New Orleans. Then it flows through its delta toward the Gulf of Mexico. The sediment-filled waters at its mouth show up as light blue.

widen and flatten out. Then floodplains build up. As the rivers advance into old age, they begin to change course. Instead of flowing straight through the land, they swing from side to side, forming wide bends. At times of flash floods, old bends may be cut off. As new bends form, the old bends may remain as lakes.

Once rivers reach old age, they have done just about all the cutting they can, except for times of flooding. Sometimes, earth shifts uplift parts of river valleys. Then old-age streams suddenly become youthful again.

Almost every part of the world has streams that are youthful and streams that are old. The Colorado, coursing down from the high slopes of the Rockies, is still a youthful stream. The Mississippi is one that is in its old age. Can you tell why?

Great Rivers of the World

River	Continent	Length	It's a Fact
Nile	Africa	4,190 mi 6,700 km	Once the Nile deposited silt along its floodplain, renewing Egypt's soil yearly. Today, the Aswan Dam prevents this silt from flowing downstream. Farmers below the dam must now use fertilizers.
Amazon	South America	4,000 mi 6,440 km	The Amazon falls over 16,000 feet (4,800 m) during the first 600 miles (960 km) of its course and only 800 feet (240 m) during the rest of its journey to the Atlantic Ocean.
Yangtze	Asia	3,430 mi 5,520 km	It is China's most important river, even though it is navigable for only 680 miles (1,090 km) of its length.
Hwang	Asia	3,000 mi 4,830 km	The Hwang Ho is often called the "Yellow River" from the color of the earth carried by its water. It is also called "China's Sorrow" because of the suffering and death brought by its many floods.
Irtysh	Asia	2,760 mi 4,440 km	This river is the chief tributary of the Ob, a river that is 2,287 miles (3,681 km) long.
Zaire	Africa	2,720 mi 4,380 km	The Zaire flows in both the Northern and Southern Hemispheres, crossing the equator twice in its course.
Amur	Asia	2,700 mi 4,350 km	Like many other rivers located outside the low latitudes, this river is frozen during the winter months. At other times, it is navigable its entire length.
Lena	Asia	2,650 mi 4,270 km	The Lena has formed a delta that is now about 250 miles (402 km) wide.
Mackenzie	North America	2,640 mi 4,250 km	The longest river in Canada, it is more than a mile wide along much of its length.
Niger	Africa	2,600 mi 4,180 km	Where the Niger empties into the Gulf of Guinea, it has formed a delta larger than that formed by the Nile.
Yenisei	Asia	2,570 mi 4,140 km	Ocean-going ships travel 400 miles (640 km) to a port upriver to pick up timber, gold, coal, and graphite taken from land along its banks.
Mississippi	North America	2,350 mi 3,780 km	Its channel first formed more than 2 million years ago, when great glaciers covered much of the United States.
Missouri	North America	2,320 mi 3,730 km	Often called "Big Muddy" because of the many tons of silt it carries, it drains about half the area of the Mississippi drainage basin.
Murray	Australia	2,310 mi 3,720 km	Along with the Darling River, its chief tributary, the Murray drains an area as large as Texas and California. The waters of this river system are used chiefly for irrigation and generating electric power.
Volga	Europe	2,290 mi 3,690 km	A canal connecting it to the Don River gives shipping on the Volga an outlet to the Black Sea.

Great Rivers of the World *(Continued)*			
River	**Continent**	**Length**	**It's a Fact**
St. Lawrence	North America	1,900 mi 3,060 km	The St. Lawrence and the Great Lakes form the St. Lawrence Seaway, an inland navigable waterway for ocean-going vessels.
Paraná	South America	1,830 mi 2,950 km	The mouth of this river is an estuary called the Río de la Plata.
Danube	Europe	1,780 mi 2,870 km	The second longest river in Europe, it is navigable its entire length.
Rhine	Europe	820 mi 1,320 km	Canals connect the Rhine to the Danube, Rhône, Marne, Elms, Weser, Elbe, and Oder rivers, making the Rhine one of Europe's most important inland waterways.

The World's Great Rivers

The chart on pages 146–147 shows some of the great rivers of the world. There are many different things that make a river great. One thing is length. Even though the Amazon flows through an area where few people live, it is a great river. Not only is it about 4,000 miles (6,440 km) long, but it also carries more water in its system than the Mississippi River, the Nile River, and the Yangtze River put together.

On the other hand, the Rhine River in Europe is only about 820 miles (1,320 km) long. But it, too, is a great river. It flows through an area where great numbers of people live. Its waters are used to manufacture many industrial products, to generate power, to transport goods and people, and to provide water for home use, for sport and recreation, and for agriculture. Use is another measure of a river's greatness.

Kenyans fish in the clear, shallow waters along the eastern shore of Lake Victoria. The lake's only outlet forms one of the headwaters of the Nile River system.

People and Rivers

Whether long or short, every river has its usefulness. Rivers are a part of the water cycle, which is always working to keep earth's water supply constant. Rivers renew the land by depositing fertile soil. Rivers supply water for home use, for irrigation, and for industry. As important transportation routes, they connect inland cities with seacoasts and with other inland cities. Riverboats and barges carry heavy and bulky materials, such as grains, liquids, and ores, more cheaply than some forms of land transportation. River waters are home to different kinds of fish—many of which provide food for people. Where waters flow swiftly or where dams can be built, people often use rivers to generate electricity. Rivers also provide places for recreation.

While rivers aid people in many ways, people do not always use rivers wisely. Nor do they always realize how closely connected rivers are to the rest of the natural environment. Sometimes people make one improvement only to find it has caused problems somewhere else.

Chemicals sprayed on crops rid plants of diseases. Those sprayed on the land fertilize the soil or kill weeds and insects. But chemicals from the land get washed into rivers, polluting their waters and all living things in them. Other actions that may bring harmful changes to the environment include leveling land, draining swamps, and altering the courses of rivers. Sometimes the bad effects do not show up until years later.

Keeping Facts in Focus

1. Does a river's current flow upstream or downstream?
2. What is a stream or river called if it flows into another stream or river? What is it called if it flows out of a stream or river?
3. What are some of the ways people use rivers?
4. What answers would you give to the key questions on page 140?

Working with Ideas

1. How does the speed of a river's current affect the buildup of land along its bank and at its mouth?
2. What are some things that might cause the Grand Canyon to look different a million years from now? How might the area now covered by the Grand Canyon look one million years from now?

The Water People Use

Where do people get the water they use? An Indian in South America's rain forest may fill a large pot with water from a small tributary of the Amazon. A Navajo in New Mexico may fill a barrel with water from a nearby spring. A nomad in the Sahara may draw water from a well at an oasis. People living in towns and cities usually have a water supply as close as the nearest faucet. Regardless of where people live, water for personal and industrial use comes from one of three sources: groundwater, rivers, or lakes.

Tapping groundwater by digging or drilling wells is one way people have an available water supply. Wells are drilled not only on farms or in isolated villages of the world but also in small towns and cities. Water from wells is drawn out by hand using pulleys and buckets or is pumped out with machines of one kind or another. In the United States, most cities and towns with fewer than 5,000 people get their water from wells. Even cities with larger populations may pump water from wells. In these cities and towns, water towers are familiar landmarks.

Rivers are not always a dependable source of water. In rainy seasons, the rivers are full. But in times of drought, they often slow to a trickle. To make sure water is always available, many government groups build dams and reservoirs behind the dams. Then water can be stored to be used as needed. Often the dams and reservoirs are built in highland areas. A system of aqueducts and pipes carries the water to other reservoirs near cities. Pumping stations then feed the water into city water systems.

Lakes are natural reservoirs that store the water fed into them by rivers, streams, and springs. Many cities, towns, and villages tap the water in these natural reservoirs and pump it into homes, stores, and other buildings.

But water from rivers, lakes, and some wells in populated and industrial areas must be treated before it is safe to drink and to use in other ways.

Treating water. The treatment of water to make it fit for use generally takes place in several stages. First, the water is mixed with a chemical, such as aluminum sulfate. Bacteria, mud, and other wastes in the water collect around the chemical. After mixing, the water is passed into a large tank or basin. There the wastes and the chemical clot together and sink to the bottom, where they settle.

Next, the purer water at higher levels in the settling basin is usually passed through a filter at a filtration plant. The filter, often a layer of sand over a layer of coarser gravel, strains out other impurities. This water is then passed along to another reservoir, tank, or building. At this stage, other chemicals, such as chlorine, are added to kill disease-carrying bacteria. Other things may also be done to water to improve its taste, to remove odors, or to soften it.

Distributing water. Once water is fit to drink, it is distributed to users through a system of underground water mains and smaller pipes. Once used, waste water, or sewage, must be carried away by a separate system of sewers. Many cities and towns treat this water once again before returning it to rivers or lakes. Cities that dump untreated sewage into rivers and lakes create many problems.

Water rights. Cities, towns, and private users of water from rivers and lakes do not have the right to use all the water they want. Other cities and towns and other private users also draw water from these sources. In every case where rivers, lakes, or underground sources are shared by more than one city or town, state or country, agreements need to be worked out. These agreements determine how much water each user may draw from the source so the source does not dry up or become used up.

The Great Lakes is one area where water use is carefully regulated. Cities in both the United States and Canada draw water from the lakes. Should too much water be drawn off, the level of the lakes could drop. This would affect shipping. It would also affect the rivers that empty into them and all people living along their tributaries. The area drained by the Colorado River is another one that is carefully regulated. There are demands for irrigation water, demands for water to supply cities, demands for water to generate electric power. Finally, there are demands to guard the area's scenic beauty. Not all of these demands can be met to each user's satisfaction, but some of them can be met through cooperation and compromise.

Unit 3 REVIEW WORKSHOP

Test Your Geographic Knowledge

A. The term in italics makes each statement below false. Replace the italicized term with another one that will make the statement correct.

1. The land a river or river system drains is called a *continental divide*.
2. Water increases in size when it changes from a liquid to a *gas*.
3. *Deltas* are built up where rivers flow out of canyons onto a flat or nearly flat plain.
4. About *97 percent* of the earth's water is fresh water.
5. The *Atlantic Ocean* is the world's deepest and largest ocean.
6. *Tides* are rivers of warmer or colder water that flow through an ocean.
7. The place where a river begins is called its *mouth*.
8. The Amazon is a major river of *Asia*.
9. Almost two thirds of the United States is drained by the *Colorado River* and its tributaries.

B. Fill in the missing words in this paragraph about water.

As heat from the sun warms water on the earth's surface, some of the water turns into (1) _____, a gas that rises into the air. At higher altitudes, the water vapor (2) _____, or cools into water droplets. The water droplets join together to form (3) _____ in the air. Eventually, the water droplets become heavy enough to fall back to earth in some form of (4) _____. Some of the water seeps into the ground to become part of the (5) _____. Most of it, after a few stops on land, finds its way back to one of the four great (6) _____ of the world. Because of the work of the (7) _____, the earth's water supply has stayed the same for millions of years.

C. A good title for the above paragraph is
1. How Clouds Form
2. The Unending Water Cycle
3. Why Water Changes Form

Apply Your Reading Skills

A. Tell whether the italicized part in each item below states a cause or an effect.

1. *Water travels up the roots and stems of plants* because it can move against gravity.
2. The earth's water supply stays the same year after year *because of the water cycle.*
3. *Polluted water carries germs.* People who drink polluted water often get sick and may even die.
4. Ocean waters are always moving. *They have differences in temperature and saltiness. Winds blow and the earth turns.*
5. *The top of a wave hitting a shoreline falls forward* because it has nothing to slow it down.

B. Use encyclopedias, books, magazines, or other reference tools to do research. Then give a report on ONE of the following topics.

1. What have scientists discovered about the possibility of people living under the sea?
2. What is the ocean food chain and how does pollution in oceans affect it?
3. What things are known about icebergs?
4. What are some of the methods used to turn salt water into fresh water?
5. Why can productive wells be drilled in some places and not in others?
6. What are some interesting facts about _____? (Select one of the world's oceans or major rivers or lakes to report on.)

Apply Your Geographic Skills

Some graphs present a great deal of information. These graphs need careful study. Tell which of the following things you can learn from the graphs at the top of page 151.

1. The average precipitation for each month for Ankara, Baghdad, and Beirut
2. The average water needs for each month of the year for these cities
3. How many inches of snow fall in the winter months

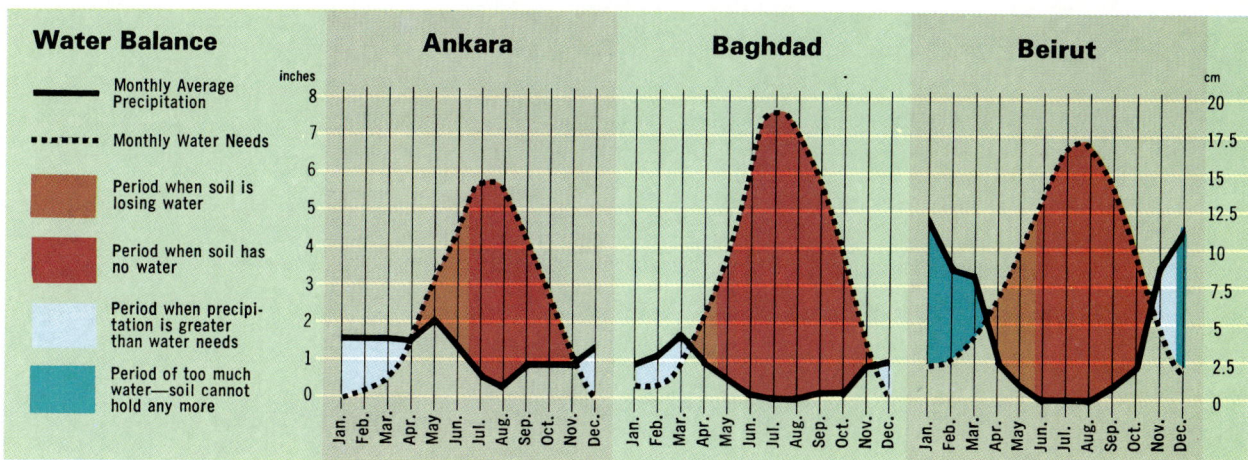

Water Balance

— Monthly Average Precipitation
······ Monthly Water Needs

Period when soil is losing water

Period when soil has no water

Period when precipitation is greater than water needs

Period of too much water—soil cannot hold any more

Ankara   **Baghdad**   **Beirut**

4. Months when the soil has too much water
5. Months when the soil is completely dry
6. The number of days it rains in each city
7. Differences between water needs and the average monthly precipitation

Apply Your Thinking Skills

Using the graphs, tell whether each statement is True or False.

1. Only Beirut has months when the soil holds too much water.
2. All three cities have periods when precipitation is greater than water needs.
3. Only Baghdad has a period when the soil is losing water.
4. In January, Beirut receives more precipitation on the average than Baghdad or Ankara.
5. Ankara receives less precipitation in May than Baghdad or Beirut.
6. In July, Baghdad needs almost 8 more inches (20 cm) of precipitation than it receives.
7. December is a very dry month in all three cities.
8. Both Beirut and Baghdad have months when little, or no, precipitation falls.
9. Baghdad has a longer period of dryness than Ankara or Beirut.
10. All three cities receive almost the same amount of average precipitation in April.

Discuss These Points

Hardly a day goes by without someone, some place in the world, mentioning a problem about water. The problem may be one of water supply. It may be one of water quality. Find out what problems there are with water in your state or city. Choose one of the problems to investigate. Find out how many different points of view there are about the problem. Discover some of the possible solutions to the problem. Take a vote of your classmates to see which solution they think will best solve the problem.

Expand Your Geographic Sights

Archer, Sellers. *Rain, Rivers, and Reservoirs: The Challenge of Running Water* (Coward, 1973).

Bergaust, Erik. *Colonizing the Sea* (Putnam, 1976).

Boyer, Robert. *The Story of Oceanography* (Harvey, 1975).

Brindze, Ruth. *Charting the Oceans* (Vanguard, 1972).

Mulherin, Jenny. *Rivers and Lakes* (Watts, 1984).

Naden, Corinne J. *The Mississippi: America's Great River System* (Watts, 1974).

Polling, Kirk. *Oceans of the World: Our Essential Resource* (Putnam, 1983).

Pringle, Lawrence. *Estuaries: Where Rivers Meet the Sea* (Macmillan, 1973).

Sabin, Francene. *Swamps and Marshes* (Troll Assoc., 1985).

Schultz, Given. *Icebergs and Their Voyages* (Morrow, 1976).

Woodbury, David O. *Fresh Water from Salty Seas* (Dodd, Mead, 1967).

Zim, Herbert S. *Waves* (Morrow, 1967).

Unit 4 Air and Weather

As sunlight passes through the raindrops of a morning storm over
Boulder, Colorado, it creates the unusual sight of twin rainbows.

Lesson 1 The Air That Surrounds Us

Reading Focus

1. Every word used in this book is not listed in the Glossary. Only the words listed in each *Vocabulary Focus* and other words in each lesson necessary to the understanding of geographical ideas are listed. Some words used in this lesson, but not listed in the Glossary, are *pollen, microscope,* and *merge.* If you do not know the meaning or are not sure of the pronunciation of one or more of these words, you should use a dictionary. A dictionary does the same things the Glossary of your text does. It tells how to pronounce words. It tells the meaning or meanings of words. Look up the words listed above in your classroom dictionary. Find their meanings and how they are pronounced. Remember, when your Glossary does not list a word you need to understand, you can find help in the nearest dictionary.

2. As you read this lesson, look for answers to these key questions:
   a. What does air contain?
   b. Into how many layers is the earth's atmosphere divided?
   c. Why is the troposphere important to people on the earth?

Vocabulary Focus

oxygen	*nitrogen*	*troposphere*	*ozone*
carbon dioxide	*atmosphere*	*stratosphere*	*ionosphere*

Air is everywhere. No one has to pay to use it. No one has to store up air, trade for it, mine it, harvest it, or even package it. Air is the one natural resource that is free to everyone, everywhere, at all times.

The air that surrounds the earth gives our planet an environment that makes life possible. Layers of air wrap the earth in a blanket that keeps the earth from receiving too many of the sun's burning rays. At the same time, this blanket of air keeps the earth's heat from escaping into space. Without air, the earth would be too hot for life during the daytime hours and too cold for life during the nighttime hours.

Air supplies oxygen, the gas people and animals need for life. Air supplies carbon dioxide and nitrogen, two gases that plants need for life. Air is an important part of the water cycle, carrying water vapor from ocean to land and back again to the ocean.

The Earth's Atmosphere

The earth's atmosphere has four layers. As each layer gradually merges into the layer above it, air becomes thinner and thinner until outer space is reached. There is no air in outer space, which is limitless. In the layers colored blue on the diagram, temperatures decrease with altitude. In the layers colored orange and yellow on the diagram, temperatures increase with altitude.

The top three layers contain only about 1 percent of the air in the atmosphere, The troposphere contains the other 99 percent.

Electrically-charged particles in the ionosphere shield the earth by reflecting many ultraviolet and other harmful rays back into outer space. These particles also help to give the sky its blue color to viewers on earth.

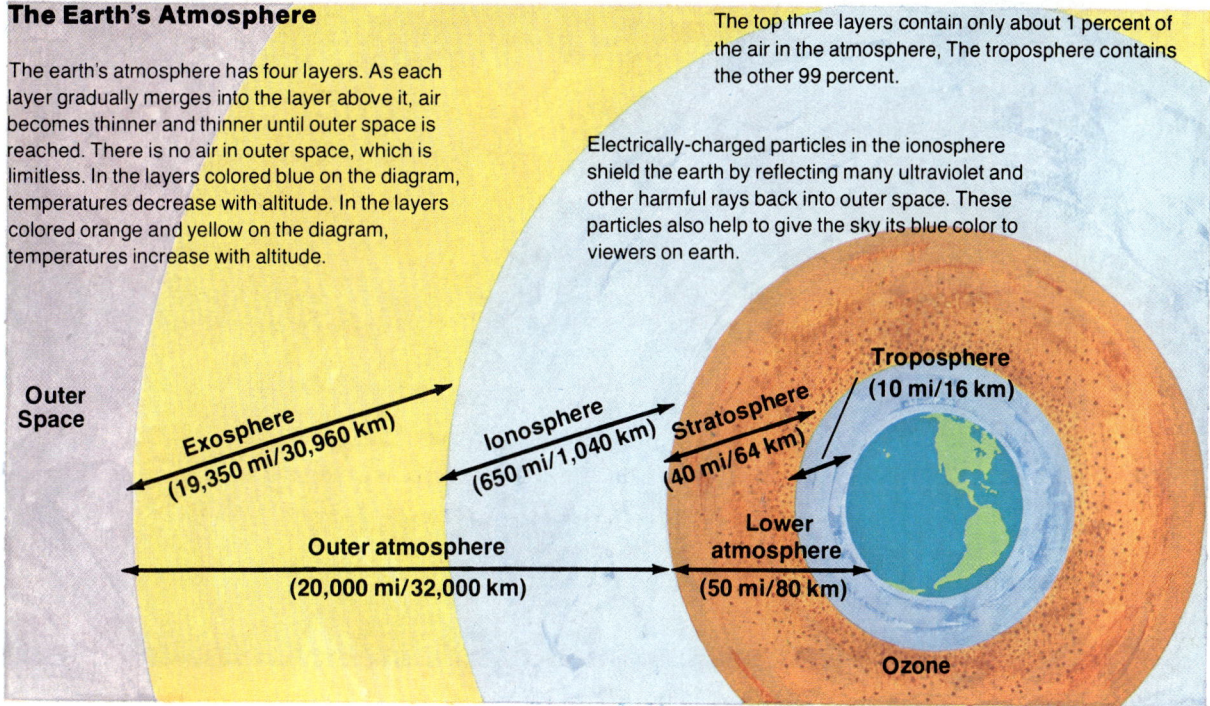

Outer Space

Exosphere
(19,350 mi/30,960 km)

Ionosphere
(650 mi/1,040 km)

Stratosphere
(40 mi/64 km)

Troposphere
(10 mi/16 km)

Outer atmosphere
(20,000 mi/32,000 km)

Lower atmosphere
(50 mi/80 km)

Ozone

What Air Is

Air—pure air—has no color, no taste, and no odor. Pure air is a mixture of gases. Nitrogen makes up about 78 percent of air. Oxygen about 21 percent. Tiny amounts of about twelve other gases, including carbon dioxide, make up the remaining 1 percent of the air.

But air closest to the earth is rarely pure and never completely dry. Besides the gases of pure air, air holds water vapor and water droplets; particles of dust, soot, and ashes; small bits of salt thrown into the air from ocean spray; pollen from plants; and living things so small they can only be seen under a microscope. Together, these things make up the atmosphere, the layers of air that surround the earth.

The Layers of Earth's Atmosphere

The blanket of air that wraps around the earth stretches upwards for hundreds of miles. These layers are shown in the diagram above. Name the layers that make up the earth's atmosphere. Which layer is closest to the earth? Which one is farthest from the earth?

The weather layer. The most important layer of air for people on earth is the one immediately above the earth's surface. This is the troposphere, or weather layer. Like all of the other layers of the atmosphere, it is widest at the equator and narrowest at the poles. Practically all of the atmosphere's oxygen and all of its water vapor are found in the weather layer.

When solar storms are at their height (about every 11 years), the *aurora borealis* (northern lights) and the *aurora australis* (southern lights) give an eerie glow to the night sky in the high latitudes.

Air in the weather layer is denser than air in the other layers. The air is densest near the earth's surface. As altitude increases, the air gradually thins. At altitudes above 10,000 feet (3,000 m), people who are not used to the thinner air of higher altitudes find breathing more difficult.

The conditions of the air in the troposphere change very often. These changes are what make weather. You will learn more about these changes and why they take place further on in this unit. Just remember that all weather changes take place in the troposphere—the earth's weather layer.

The air-travel layer. The upper reaches of the troposphere gradually merge into the stratosphere about 10 miles (16 km) above the earth's surface. The air in this layer is clearer and much thinner than air in the weather layer. Because it is free of water vapor, the stratosphere has no clouds. It is also free of strong up and down movements of air. For these reasons, jet planes use it for long distance air travel.

The stratosphere is important for another reason. It contains ozone, a special form of oxygen. Ozone protects the earth by blocking out many of the most damaging of the sun's rays.

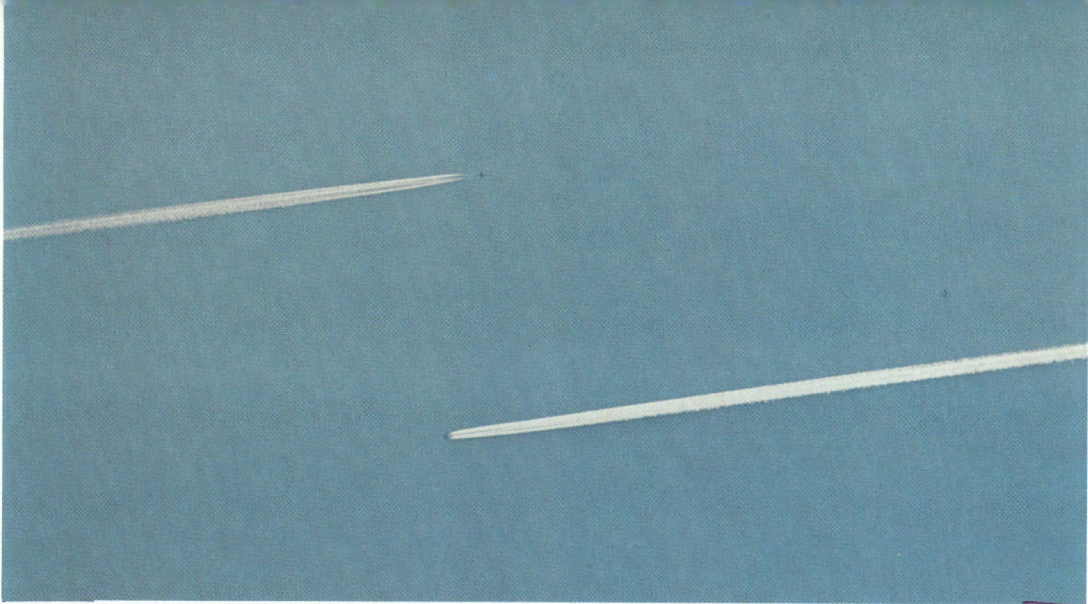

When flying in the usually cloudless stratosphere, jet planes often leave contrails, or vapor trails, behind them. The water vapor from the jets' exhausts condenses. The moisture often freezes into ice crystals, forming cloud-like trails.

The radio layer. Radio waves travel in straight lines. They travel only a short distance before the earth curves away beneath them. Without ions, the charged particles present in the ionosphere, radio waves would travel far into space and never be received on the ground. But the radio waves hit the ions, which reflect the radio waves toward the earth, where they can be received hundreds, even thousands, of miles from their sending points.

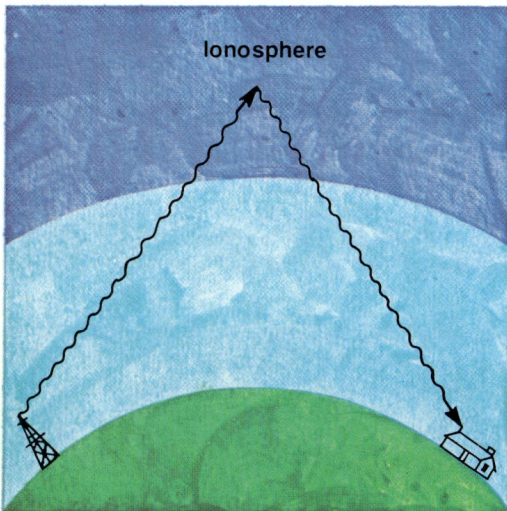

Ionosphere

Keeping Facts in Focus

1. In what two ways does earth's blanket of air affect the earth's temperature?
2. What percent of the air is oxygen? Nitrogen? Carbon dioxide?
3. In what layer of the atmosphere is the air the densest? The thinnest?
4. What is ozone? How does it protect the earth?
5. What are ions? Where are they found in the atmosphere? How do they make long-distance communication possible?
6. What answers would you give to the key questions on page 154?

Working with Ideas

1. Why are scientists working to explore outer space?
2. Science fiction stories have human beings exploring outer space in huge city-like spaceships. Do you think human beings will ever explore or land on one of earth's neighboring planets? Why or why not?
3. How does the air-travel layer of the troposphere differ from its weather layer?

Clean Air—Is It Worth the Price?

You have all heard the saying *Where there is smoke, there is fire.* Obvious, you might say. But perhaps not so obvious is this: *Where there is smoke, there is air pollution.*

Smoke, fire, and air pollution go hand in hand because most air pollution is the result of burning. Whenever anything is burned, some pollutants escape into the air. The pollutants in smoke may be in the form of a gas. Or they may be tiny particles of solid or liquid materials, such as ash, soot, or traces of metals.

Air pollutants harm many living and nonliving things. They eat into concrete, metals, building stones, and the marble of statues and other things. Air pollutants make breathing difficult for some people, cause headaches in others, and generally irritate the eyes. Air pollutants harm animals in the same ways they harm humans. The chemicals in air pollution harm plant leaves—clogging their openings, discoloring them, or causing them to drop off. These chemicals also harm fruits and vegetables.

Air pollution also affects weather. Winds carry pollutants in air far from their sources. Winds also blow pollutants from the land into the air. Rain or snow forms around pollutants in air and falls to the ground. The pollutants harm the things they touch. At other times, a layer of warm air settles over a layer of cooler air near the ground. If pollution is heavy in the area, the warm air holds down the cooler air and prevents the pollution from rising and scattering. What results is smog. Smog hangs over some of the world's great industrial cities for days at a time. Often it causes great discomfort and sometimes even loss of life.

Ozone, a special form of oxygen, is always found in the stratosphere, where it works to protect life. However, ozone sometimes forms in the weather layer. Then it works in a harmful way by making breathing difficult for many people. It often forms on warm humid days when pollution from city traffic fills the air. At such times, ozone alerts are issued to warn people to limit their outdoor activities.

Air pollution even affects the ozone in the stratosphere. This pollution comes from the fluorocarbons released from aerosol spray cans. These fluorocarbons can rise to the stratosphere where they act on the ozone, destroying it. If the ozone layer were to be destroyed, the earth's temperatures would rise sharply. Then the earth would become too hot to support life. The ozone destruction in the stratosphere can be stopped by no longer manufacturing or using aerosol sprays (and this is being done). But other forms of air pollution are not so easily stopped.

The burning of materials produces energy. We use this energy to turn raw materials into finished products and to take raw materials from the earth. We use it to power airplanes, automobiles and other land vehicles, boats, and ships. We use the energy from fuels to heat homes, factories, and office buildings. Devices can be manufactured and installed that "scrub" smoke from factories and "clean" exhaust fumes from vehicles. Because these devices are expensive, they add to the cost of running automobiles and other means of transportation. They add to the cost of manufacturing materials and to the cost of disposing of waste materials.

How willing are you and your families to pay the cost for clean air? Does paying the cost mean only spending money, or does it mean much more—sacrificing some of our convenience and our time to work for clean air?

What comparisons can you make about these two views showing polluted air over Gary, Indiana, a steelmaking, port city on Lake Michigan? In which photograph do the pollutants seem to be forming a dense cloud trapped under a warmer layer of air?

Lesson 2 Weather Makers

Reading Focus

1. In this lesson, you learn that there are five weather makers. You need to understand how each weather maker works. You also need to discover how each weather maker is related to the others. One way of doing this is by asking yourself questions after you have read about each weather maker. Here are some examples:
   a. What is *temperature?* Why do different places on the earth have different temperatures?
   b. What is *air pressure?* What causes air pressure to change?
   c. What is *wind?* How are winds and pressure centers related?
   d. What is *humidity?* How does humidity change with temperature?
   e. What are *clouds?* How are clouds related to humidity?

2. As you read this lesson, look for answers to these key questions:
   a. Why are there differences in temperatures on the earth's surface?
   b. What three things work together to set up wind belts?
   c. What are the effects of cloud cover during the day? During the night?

Vocabulary Focus

temperature	*air pressure*	*relative humidity*
humidity	*prevailing wind*	*dew point*

Air in the earth's weather layer is always changing. Daily, even hourly, air temperatures change. Air moves up and down. By rising or falling, air changes pressure. Air moves across the land, sometimes blowing with great force and sometimes blowing gently. The amount of water vapor, or humidity, in the air changes. Finally, the water vapor held in the air turns into water droplets or ice to form clouds. The story of changes that take place in air is the story of what makes weather.

Temperature: Weather's Starting Point

The story of weather begins with heat, a form of energy that radiates from the sun. Temperature (recorded in degrees Fahrenheit or Celsius) is the measure of this energy. The warmer the air, the more energy it has. The colder the air, the less energy it has.

Heating the earth. Some of the heat energy from radiation remains in the upper layers of the atmosphere. Particles in the lower parts of the atmosphere

What Happens to Solar Heat Energy

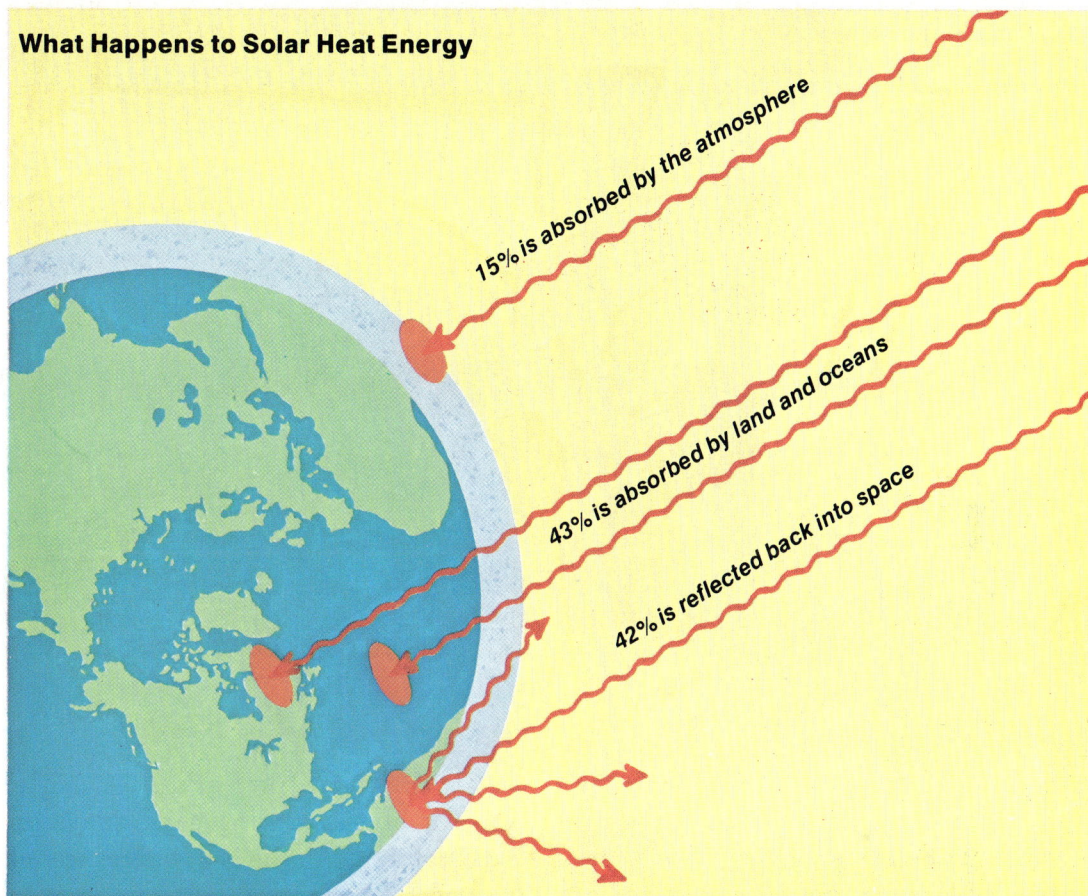

15% is absorbed by the atmosphere

43% is absorbed by land and oceans

42% is reflected back into space

block some radiation by bending it back toward space or by breaking it up. Radiation that is broken up in space loses some of its heat energy in the atmosphere. Some of the sun's radiation passes untouched through the atmosphere's layers, striking the earth with its full amount of heat energy.

Only 43 percent of the sun's radiation actually reaches the earth itself and heats it. This 43 percent, however, is enough to keep the average temperature of the earth's surface at about 59° F (15° C).

Heating the weather layer. About 15 percent of the sun's radiation remains in the weather layer. Water vapor in this layer absorbs, or soaks up, this radiation. This soaked-up radiation helps to heat the weather layer. The rest of the heat in the weather layer comes to it in a second hand way from the radiation that reaches the earth's surface.

Water and land surfaces soak up heat energy from incoming radiation. But they also reflect, or send, heat energy back into the air as outgoing radiation. The outgoing radiation is soaked up by

In Death Valley, summer air temperatures often are greater than 120° F (49° C). How can you tell there is a fairly constant wind blowing over the floor of Death Valley?

water vapor and other particles in the weather layer. Because the outgoing radiation is soaked up much faster and more easily than the incoming radiation, it supplies most of the heat energy in the weather layer.

Surface temperature differences. The sun's rays spread over different parts of the earth in different ways and in different amounts at different times of the year. The shape of the earth, the tilt of its axis, and the ways the earth moves in space account for these differences.

Remember that the earth is a sphere. Because of its rounded shape, the sun's rays do not strike all parts of the earth at the same angle. The earth also moves in space, revolving around the sun once each year and rotating on its axis once every twenty-four hours. As the earth

moves, its axis is tilted at a $23^1/_2°$ angle. The tilt of the moving earth plays an important part in the heat differences of the earth's surface.

If the earth's axis were *not* tilted, the sun's direct rays would always strike the equator. Other rays would always strike the same spot on the earth at the same angle. Rays at the Tropics of Capricorn and Cancer would always strike at a $23^1/_2°$ angle. Those at the Arctic and Antarctic circles would always strike at a $66^1/_2°$ angle. At the poles, rays would always strike at a 90° angle.

But the earth's axis *is* tilted. Because of this tilt, all of the sun's rays move to different places at different times of the year. The direct rays move as far north as the Tropic of Cancer and as far south as the Tropic of Capricorn. The slanting rays move, too. As they move, seasons change. The hours of daylight and darkness in a day also change.

Other surface temperature differences. Some land surfaces soak up more heat energy than others. Dark-colored surfaces soak up more heat than light-colored surfaces. Snow-covered surfaces soak up less heat than snow-free ones. Dark-colored or snow-free surfaces, in turn, give off more outgoing radiation. The air above them is usually warmer than air over light-colored or snow-covered surfaces.

Sun rays striking water do not remain on the surface as they do when they strike land. Instead, the rays travel several feet below the surface of the water. The movement of the water forces the heat energy to spread even deeper. This causes oceans and other large bodies of water to heat more slowly than large land areas. It also causes them to cool more slowly than land areas.

The change from daylight to darkness also affects surface temperatures. During the warmest daylight hours, the earth soaks up more heat energy than it loses to the air. But around four o'clock in the afternoon, the earth begins to lose more heat to the air than it soaks up. The cooling continues at faster rates as darkness sets in. Usually the coolest time of the day is just before sunrise.

Air temperature differences. Surface temperatures affect air temperatures. This is because air takes on the temperature of the surface over which it is passing. Over warm surfaces, air becomes warm. Over cool surfaces, air becomes cool. Where will air be

Why is there only a slight change in the temperature of the air over this Yucatán rain forest as daytime changes to nighttime?

warmer—over dark-colored or over light-colored surfaces?

Because there are differences in temperatures between land and water surfaces, there are also differences in the temperatures of the air over land and over water. The air over water is usually cooler than that over land during the warmer months of the year. During the colder months of the year, the air over water is usually warmer than the air over land.

Keeping Facts in Focus

1. What is heat? What is the source of the earth's heat?
2. How much of the sun's radiation is soaked up by the earth's surface? How much is reflected back into space? What happens to the rest of the radiation that comes from the sun?
3. How is the earth's weather layer heated?
4. Which surfaces soak up more heat energy from the sun—dark-colored surfaces or light-colored surfaces? Which surface—land or water—soaks up heat energy from the sun faster? Why?
5. How is the temperature of the air related to the temperature of the surface below it?

163

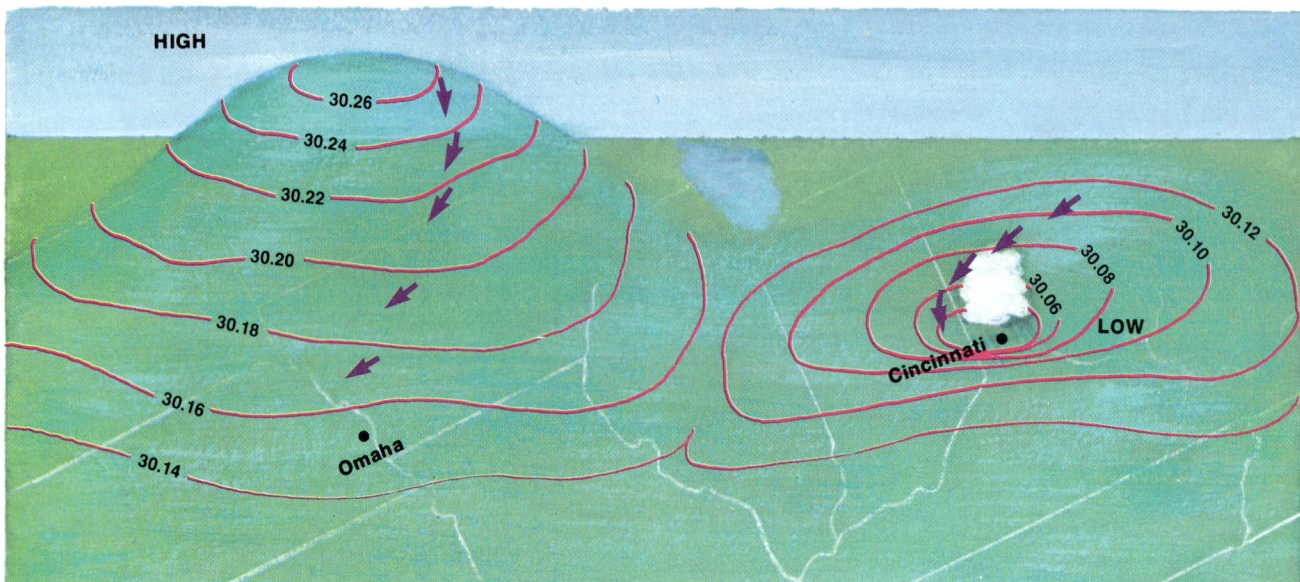

A high pressure center is like a *dome* of air, while a low pressure center is like a *bowl* of air. To discover high and low pressure centers, weather forecasters connect all places in an area reporting the same pressure readings with lines called isobars. What is the highest pressure shown above? What is the lowest? Do winds blow out from the center of a low or out from the center of a high pressure area?

Air Pressure

Like everything else on earth, air has weight, or pressure. Air pressure is another part of the weather story.

The layers of air over the earth weigh about 5 million million tons. The earth and all things on the earth bear this weight because air surrounds everything. At sea level, the full weight of the atmosphere presses on the earth's surface. This averages out to 14.7 pounds for a column of air covering one square inch of surface and reaching from the bottom to the top of the atmosphere.

Pressure differences. At different altitudes, air has different pressures. Pressure is highest near or at the earth's surface. Pressure becomes lower as altitude increases. But at the *same* altitude, air pressure also differs from place to place and from time to time. Differences in surface temperatures cause differences in air temperatures.

And differences in air temperatures cause differences in air pressure. Some areas have high pressure and some areas have low pressure. The air in low pressure areas is thinner than air in high pressure areas.

Low pressure, high pressure. Depending on the location, on the season of the year, and on the time of day, air is always being warmed or cooled. Air that is warmed rises. The rising air expands and gets thinner and lighter. Its pressure becomes lower. Air that is cooled sinks. The sinking air contracts and gets denser and heavier. Its pressure becomes higher.

The low latitudes, those near the equator, receive the most heat energy from the sun. The high latitudes, those near the poles, receive the least heat energy from the sun. Where would you expect air pressure to be higher—near

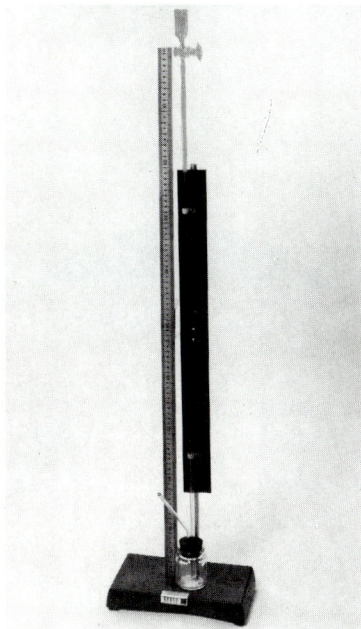

A barometer records air pressure in inches or millibars as mercury rises or falls within a column. The world's average sea level pressure is 29.92 inches (1013.2 millibars). The highest recorded air pressure is 32.04 inches (1084 mb). The lowest is 25.93 inches (877 mb).

How do you know the winds in this mountain valley blow mainly from the same direction?

the equator or near the poles? Over land in winter or over water in winter? Over dark-colored surfaces or over light-colored surfaces?

Winds, Wind Belts, and Pressure Centers

Air moves because of differences in temperature and pressure. Differences in surface temperatures set up air currents, causing air to rise or to sink. Differences in air pressure set up winds and wind belts, causing air to move across the earth's surface.

Air circulation and pressure centers. Due to sinking air over cold surfaces, a permanent high pressure center is set up over the poles. Due to rising air over warm surfaces, a permanent low pressure center is set up over the equator. Air moves (as wind) from an area of

high pressure to an area of low pressure. The sinking cold air over the poles moves toward the equator at low altitudes. The rising warm air over the equator moves toward the poles at high altitudes. This sets up a north-to-south and south-to-north circulation of air over the earth.

But not all of the cold air from the poles reaches the equator. Nor does all of the warm air from the equator reach the poles. Several things break up this movement of air. As the air moves, its temperatures and its pressures begin to change. As warm air rises, it gets cooler and its pressure increases. As cool air sinks, it gets warmer and its pressure decreases. At the same time, the spinning motion of the earth on its axis works on the winds. The winds no longer blow straight south or straight north. They take on a curving motion.

165

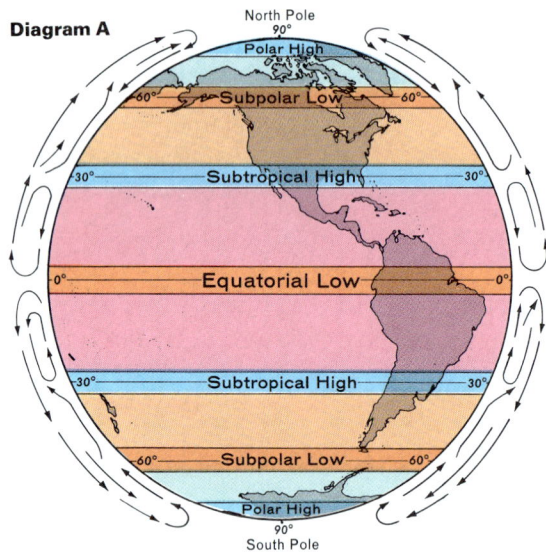

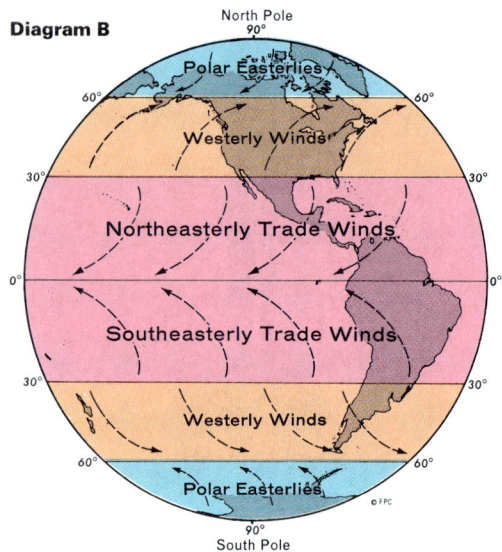

Winds blow from areas of high pressure to areas of low pressure, as shown in diagram A. Diagram B shows prevailing wind belts near the earth's surface. High altitude winds carry highs and lows across the middle latitudes.

Wind belts and pressure centers. By studying the two diagrams on this page, you will gain a better understanding of how temperature differences, pressure differences, and the earth's spinning motion set up wind belts and pressure centers in the weather layer.

Diagram A shows the high and low pressure centers of the earth. By the time air from the poles has reached 60° north and south latitude, it has become warmer. As part of this air begins to rise, it forms a low pressure center in each hemisphere. What is this pressure center called?

By the time air from the equator has reached 30° north and south latitude, it has become cooler. As part of this air begins to sink, it forms a high pressure center in each hemisphere. What is this pressure center called?

Now study the wind belts of the earth in diagram B. Keep in mind the location of the high and low pressure centers from diagram A. Notice how the winds blow from areas of high pressure to areas of low pressure. In the Northern Hemisphere, the winds move out from the high toward the low in a clockwise direction. In the Southern Hemisphere, the winds move out in a counterclockwise direction. From what high pressure center do the trade winds blow in each hemisphere? The westerlies? The polar easterlies?

The winds in each belt come from the same general direction much of the time. For this reason, they are called

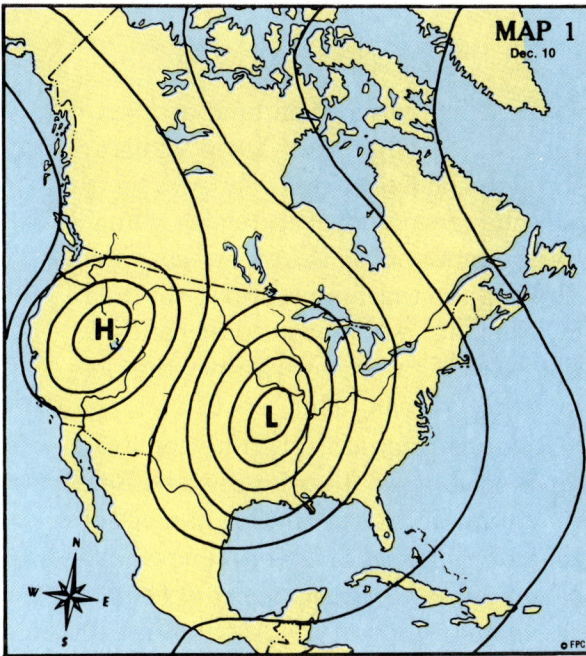

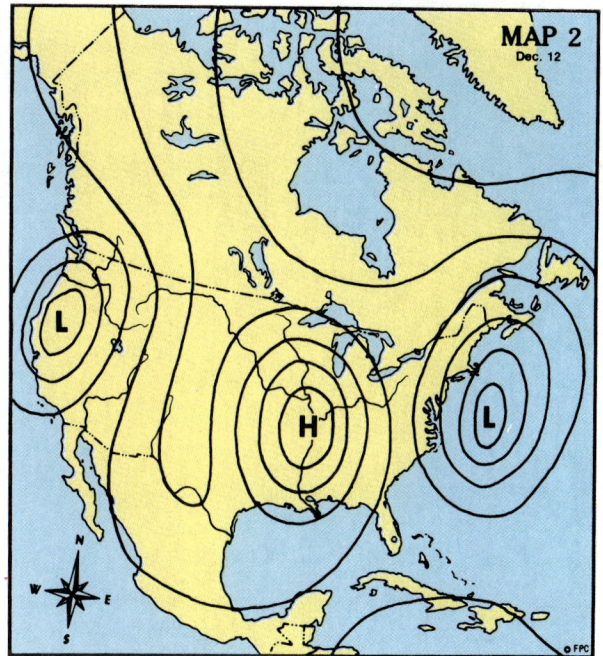

Pressure patterns over the middle latitudes are always changing. For what day does Map 1 show pressure patterns? Map 2? Most stormy weather comes when a low pressure center passes over an area. In Map 2, the low pressure center shown on Map 1 has moved off the East Coast. What other area of the country is probably having stormy weather on Dec. 12?

prevailing winds. Can you tell how winds are named? Do the westerlies blow from the west or toward the west? From what direction do the polar easterlies blow? What is the prevailing direction of the trade winds in the Northern Hemisphere? Of the trade winds in the Southern Hemisphere?

Changing pressure locations. Diagram A is a model. It shows the location of pressure centers as if they were wide bands completely circling the earth. But you know that the earth is divided into continents and oceans. You know that pressure and temperature over land is almost always different than pressure and temperature over water. This leads to changes in the location of pressure centers at different times of the year.

Pressure centers change locations as seasons north and south of the equator change. In general, the high pressure centers form over land in winter and over water in summer. The low pressure centers form mainly over water in winter and over land in summer.

There is another way that pressure centers change locations. Smaller pressure systems develop as air moves over the earth's surface away from the larger, more permanent pressure centers shown in the model. These pressure systems are carried by the prevailing winds to the middle latitudes of the earth. It is these smaller, moving pressure systems with their changes in wind speeds and wind directions that cause weather to change from day to day.

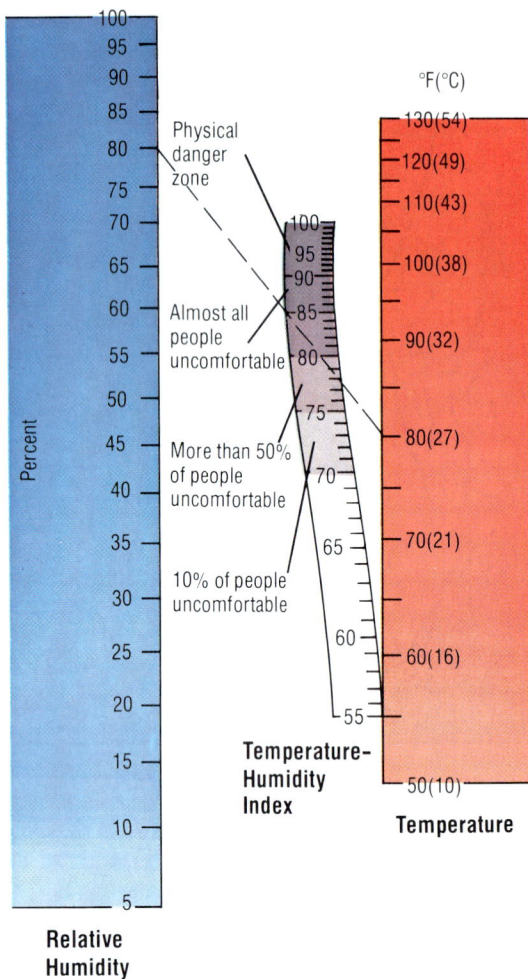

Physical danger zone

Almost all people uncomfortable

More than 50% of people uncomfortable

10% of people uncomfortable

Percent

Relative Humidity

Temperature-Humidity Index

°F(°C)

Temperature

Humid air can be very uncomfortable, especially on a warm summer day. Study the diagram above, which shows the relationship of discomfort to temperature and relative humidity. What is the temperature-humidity index reading if the relative humidity is 80 percent and the temperature is 80° F? What percentage of the people might feel uncomfortable outdoors or in places that are not air-conditioned? If the relative humidity remains at 80 percent, but the temperature increases to 90° F, does discomfort increase or decrease?

Humidity

Another important weather maker is humidity—the amount of water vapor, or moisture, in the air. Humidity changes with temperature, just as air pressure does. Warming air gains moisture by evaporating water from land and water surfaces. Cooling air loses moisture through condensation.

A change in temperature also changes the amount of moisture air can hold. The higher the temperature, the more moisture, or water vapor, air can hold. The lower the temperature, the less moisture air can hold.

Relative humidity is an important term used by weather forecasters. Relative humidity is the amount of moisture the air holds at a given temperature compared to the amount it *could* hold at that same temperature. Air that has 100 percent humidity has all the moisture it can hold. Air that has 80 percent relative humidity could hold still more moisture at the same temperature. If the temperature of the air rises, the air can hold more moisture. If the temperature of the air drops, the air can hold less moisture.

Another important term used by weather forecasters is *dew point.* The dew point is the temperature air must reach for water vapor to turn back into droplets of water or bits of ice. When air cools to its dew point or below, some form of condensation or precipitation usually takes place.

Clouds and Cloud Cover

Clouds in the sky are a sign that condensation has taken place. For condensation to take place, the air must have water vapor in it. Particles of dust, soot, sea salt, and other things must also be present in the air. And the temperature of the air must have reached its

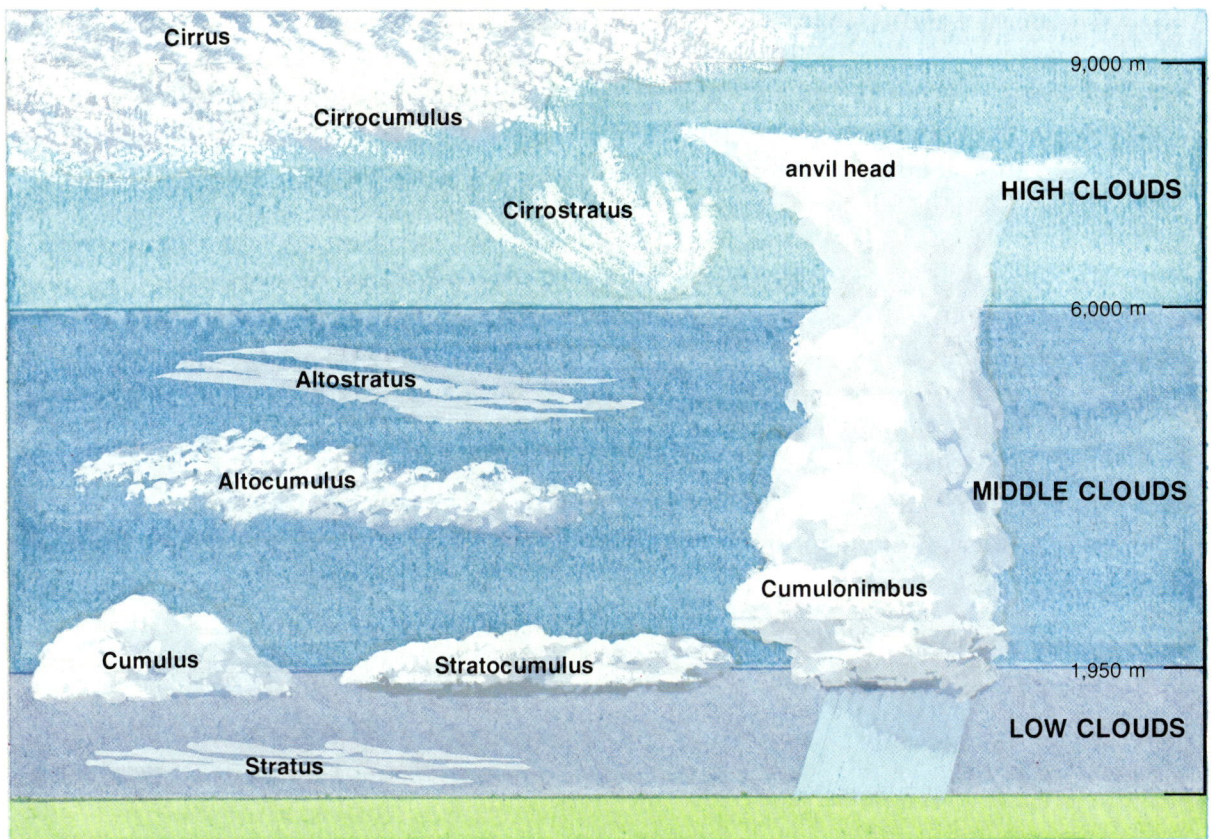

Cirrus

Cirrocumulus

Cirrostratus

anvil head

HIGH CLOUDS

9,000 m

6,000 m

Altostratus

Altocumulus

Cumulonimbus

MIDDLE CLOUDS

Cumulus

Stratocumulus

1,950 m

Stratus

LOW CLOUDS

dew point. On humid days, rising air often reaches its dew point.

Rising air. There are several ways in which air is forced to rise. Air rises when it is heated from below by passing over warm land or water surfaces. High coastlines, mountains, and hills force air to rise as it moves over them. Air is also forced to rise when it meets a mass of colder air.

Cloud terms. There are five key terms weather forecasters use to describe the clouds that appear most often in the sky. Once you know the meaning of each of these terms, you can tell something about a cloud's characteristics. You can

tell whether the water vapor is changing into water droplets or into bits of ice. You can tell whether the air in the cloud is rising slowly or rapidly. You can also tell whether the cloud is forming in the low, middle, or high part of the weather layer. The key terms and their meanings are:

stratus: sheet- or layer-like clouds
cumulus: puffy, heaped-up clouds
cirrus: thin, wispy clouds
alto: middle weather-layer clouds
nimbus: stormy or rain-producing clouds

The first three terms describe the basic forms that clouds take. Each one is

169

a name for a certain kind of cloud. But each term (with a slight change in its ending) is often joined with one of the other terms to name other clouds.

Stratus clouds. Stratus clouds cover large portions of the sky and have a dull, gray appearance. They form in the lower part of the weather layer, mostly below 6,500 feet (1,950 m), when air is slowly rising.

Layer-like clouds can also appear in the middle and high parts of the weather layer. Cirrostratus clouds are found in the high parts of the weather layer, while altostratus clouds are found in the middle part.

Cumulus clouds. Puffy, heaped-up piles of clouds with flat bottoms that seem to change shape minute by minute are cumulus clouds. Their billowing shape comes from strong currents of rising air. Whether they form in the low, middle, or high part of the weather layer depends on how fast and strong the air currents are. It also depends on the temperatures in different parts of the weather layer.

Some cumulus clouds form close to the ground but have very strong air currents. These clouds rise to great heights. The air moves up and down so fast that bits of ice form near the top of the cloud while water droplets form at lower levels. A towering cumulus cloud often develops a top shaped much like the anvil used by a blacksmith. Then it becomes a cumulonimbus cloud, or thunderhead. What kind of weather do cumulonimbus clouds bring?

Cirrus clouds. Cirrus clouds form high in the sky, usually at heights above 25,000 feet (7,500 m). They have a thin, wispy appearance because they are mainly ice crystals. Cirrostratus clouds form a thin sheet or layer of ice crystals across parts of the sky. They seem to form a halo, or ring, around the moon or sun shining above them. Cirrocumulus clouds look like tiny tufts of cotton hanging high in the sky.

Effects of cloud cover. Cloudy days are often cooler than clear days because clouds block out much radiation from the sun. Cloudy nights are usually warmer than clear ones because clouds trap the heat from the earth beneath them.

Keeping Facts in Focus

1. What is air pressure? Does air over the equator generally have high pressure or low pressure? What pressure does air over the poles generally have?
2. In what ways does air move? What happens to air as it moves away from the equator? From the poles?
3. What is relative humidity? What is the dew point?
4. In what ways is air forced to rise?
5. What are the three basic forms of clouds?
6. What answers would you give to the key questions on page 160?

Working with Ideas

1. Why does a place near the South Pole receive less energy from the sun than a place near the equator?
2. Why does a coastal city located along the same parallel as an inland city usually have cooler temperatures in summer and warmer temperatures in winter than the inland city?

Lesson 3 Weather Packages

Reading Focus

1. Because weather is something you experience every day, keep a class "weather diary" for a week or more, using weather reports (sources: newspapers, radio, TV) and your personal observations. Record information for each day in columns, such as the following (the column labels are in italics and the kinds of information to be recorded are in parentheses): *date/time of day, temperature* (high and low), *air pressure* (rising, falling, remaining steady), *wind speed* (15 to 20 miles an hour, etc.), *wind direction* (NW, SE, etc.), *relative humidity* (80%, 50%, etc.), *precipitation* (none, rain, snow, etc.), *cloud cover* (clear, partly cloudy, overcast), and *personal observations* (kinds of clouds, lightning, thunder, etc.). Discuss your findings for each day. Try to discover connections between the weather in your area and the movements of air masses and weather fronts.

2. As you read this lesson, look for answers to these key questions:
   a. What is an air mass? How do air masses differ?
   b. What is a weather front? How do weather fronts differ?
   c. What are jet streams?

Vocabulary Focus

air mass   *source region*   *front*   *jet stream*

Air, the always moving air, is like a big machine that turns out one product—weather. The product comes in two packages—fair weather and stormy weather. Almost every part of the world has a good supply of each.

All packages have the same weather makers—temperature, pressure, winds, humidity, and cloud cover. Each weather maker gives its own flavor to the package. How the weather makers are mixed and how much of each one is present are what make weather packages so different in so many places around the world. As these weather makers change, times of fair weather follow times of stormy weather throughout most parts of the world.

Air Masses: Earth's Fair Weather Packages

Air masses are large packages of air that may cover distances of 1,000 miles (1,600 km) or more. The places where air masses form are called source regions. You can see these source regions on the map on page 172.

Depending on their source regions, air masses are either cold or warm, moist or

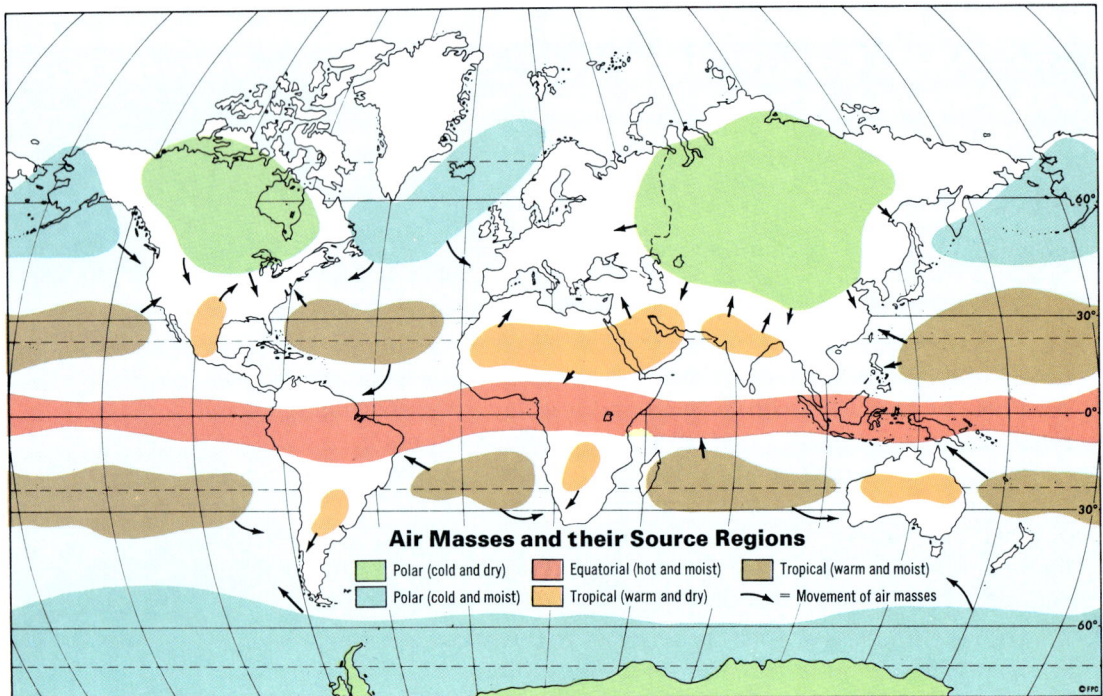

Air Masses and their Source Regions

■ Polar (cold and dry)	■ Equatorial (hot and moist)	■ Tropical (warm and moist)
■ Polar (cold and moist)	■ Tropical (warm and dry)	◄— = Movement of air masses

An air mass formed over land is generally stable. It usually changes very little as it moves away from its source region. An air mass formed over water is generally unstable. Because a moist air mass has more energy, it tends to change more rapidly as it moves away from its source region.

dry. Which air masses are cold to start out with—those that form over tropical regions or those that form over polar regions? Which air masses are dry to start out with—those that form over water or those that form over land?

Air-mass movement. Another look at the map of air masses and their source regions shows that most of the air masses move. Air masses that form over land continue to move over land. The air masses that form over water move first over water and then over land.

As they move away from their source regions, air masses usually carry their original characteristics of temperature and humidity with them. But they can and do change, usually very slowly. The kinds of changes that take place depend on two things.

The first thing is how long an air mass stays over its source region before it moves. An air mass that has stayed over its source region for some time before moving away does not change as readily as an air mass that moves out quickly from its source region.

The second thing that is important in determining air-mass weather is the relation of the surface temperature to the temperature of the air mass. An air mass is cold only if its temperature is cooler than the surface over which it is passing. An air mass is warm only if its temperature is warmer than the surface over which it is passing.

Air-mass weather. If the air mass is *cold* and *dry,* skies will become clear. But as the lower layer of the air mass is heated from the surface below, patches of cumulus clouds usually will begin to dot the once clear sky.

If the air mass is *cold* and *moist,* thick stratus clouds form, from which heavy amounts of precipitation may fall.

If the air mass is *warm* and *dry,* skies become clear. Little precipitation is likely to occur because the lower layer of the air mass is being cooled by the surface below it.

If the air mass is *warm* and *moist,* the cooling from below will produce condensation. Stratus clouds form. Sometimes, when pollution in an area is heavy, haze or smog will develop. At other times, when cooling is rapid, a heavy fog may blanket the area. But again, little precipitation is likely to fall.

Local conditions. For the most part, air-mass weather is fair weather—weather without precipitation. But air is always subject to local conditions—the shape of the land, the presence of a large body of water, and heavy amounts of pollutants in the air. When air is lifted over mountains, precipitation can take place. Winds moving over a body of water pick up additional moisture. Rapid condensation of this moisture can cause precipitation. Pollutants collect drops of moisture around them. If the water droplets become heavy enough, precipitation can fall or a fine mist form.

There is only one thing about weather that is certain—weather, even air-mass weather, is subject to changes that can cause precipitation or condensation.

Fronts: Earth's Stormy Weather Packages

The general flow of winds from centers of high pressure to centers of low pressure carries air masses from their source regions (review the map on page 172). Polar air masses and tropical air masses come close together over the middle latitudes of the world. Where these air masses almost meet, a narrow band forms in the weather layer. This narrow band is called a front.

Fronts may cover a distance as narrow as 30 miles (50 km). Most fronts, however, are larger—60 to 150 miles (100 to 250 km) wide. It is along fronts between air masses that most of the world's stormy weather takes place.

Frontal movement. Along a front, a depression, or low, forms in the weather layer. This depression usually has a lower pressure than the air masses around it. Local winds, blowing from a center of high pressure, move into the low pushing it ahead. And prevailing winds in the middle latitudes carry fronts with their low pressures in an easterly direction about 30 miles (48 km) an hour. Every place in a front's path will experience the stormy weather a front usually brings with it.

Cold-front weather. Wherever a cold front forms, the leading edge of a cold-air mass pushes against the retreating edge of a warm-air mass. The cold front

Meteorologists—Decoders of Weather's Mysteries

Lightning, thunder, a rainbow, or a streak of light across the night sky were all strange and fearful happenings to early peoples. The early Greeks called any strange happening in the air a meteor. Today, we call the men and women who study the atmosphere and explain its mysteries meteorologists.

Because of the work of meteorologists, we know that the buildup of electrical charges in air causes lightning. We know that air expands and contracts to cause thunder. We know that particles in the air break up light rays to form rainbows at certain times. We also know that solid objects falling into earth's atmosphere burn up, leaving trails of light in the sky.

Meteorologists of the past discovered all that we know about weather makers. They explained how each one works and how each reacts with the others. They developed the tools and instruments to record temperatures, amounts of precipitation, pressure, and other things. Then, they began to keep daily records and yearly averages. Plotting this information on maps, they learned about worldwide weather patterns.

Meteorologists of the present continue the work of the past. Some are still trying to discover new things about the earth's atmosphere by using space probes, radar, satellite pictures, and computers. Others work in laboratories trying to duplicate the conditions that cause storms. They want to know why and how storms form. Then they can predict more accurately when storms, such as tornadoes, will take place. Others experiment with weather control, seeding clouds to reduce the power of hurricanes or to bring rain to dry areas. Finally, there are meteorologists who predict the weather on a daily, weekly, or longer-range basis.

The goal of all meteorologists is the accurate forecasting of weather. It is seldom easy to do this. So much about weather varies from day to day. When weather is not predicted accurately for a day, it usually does not mean that meteorologists have made a mistake. Instead, something has happened in the weather layer that could not be forecast from readings taken on the ground or in the air.

At the present time, the most accurate forecasts are those made for periods of two or three days. Forecasts made for periods of ten or thirty days are less accurate. Nevertheless, these longer-range forecasts are useful to many people. Farmers use them. So do builders and other people who work outdoors.

Radar is a great aid to meteorologists and weather forecasters in tracking storms and tracing their outlines.

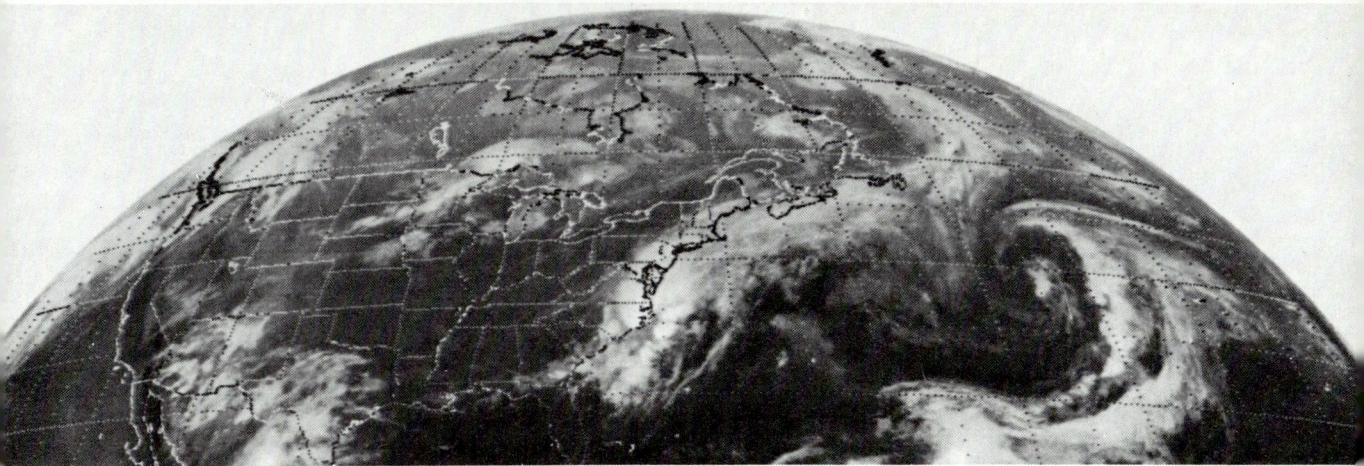

In a satellite view of the United States, lines of clouds mark the locations of three weather fronts. One lies over Minnesota. Another covers the eastern seaboard. A third forms around a low over the Atlantic Ocean.

marks the boundary between the colder air behind the front and the warmer air ahead of it. Where a cold front forms, cold air pushes under warmer air. The cold air forces the warmer air to rise rapidly. Some of the moisture in the warm air begins to condense, usually forming billowing cumulus clouds.

More warm air continues to rise and to condense, forming clouds. The water droplets or bits of ice that make up the clouds become heavier and heavier. Eventually, some form of precipitation falls to earth. The passing of a cold front

often brings very stormy weather to all places in its path.

Warm-front weather. Sometimes tropical air does the pushing against polar air, forming a warm front. The warm front marks the boundary between the warm air behind the front and the cooler air ahead of it. Because the warm air is thinner and lighter, it cannot push under the denser, heavier air of the cold air mass. So the warm air moves up and over the cooler air. The moisture at the leading edge of the warm front forms thin, wispy cirrus clouds at higher alti-

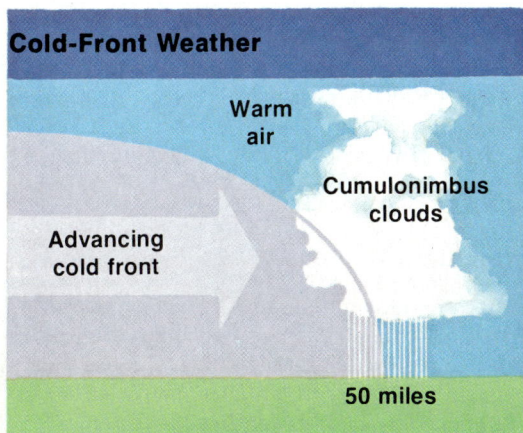

Cold-Front Weather

Warm air

Cumulonimbus clouds

Advancing cold front

50 miles

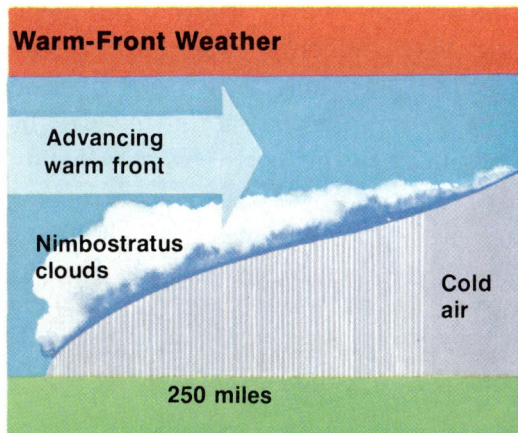

Warm-Front Weather

Advancing warm front

Nimbostratus clouds

Cold air

250 miles

tudes. These clouds give way after about twenty-four hours to stratus clouds.

These stratus clouds usually produce a steady, drizzling rain. Or they may produce snow. How long the rain or snow lasts over an area depends on the width of the front, on how fast the air masses ahead and behind are moving, and on the changes that take place in upper level winds.

Jet streams. Jet streams are high-altitude winds that affect air-mass circulation and the location of high and low pressure centers in the middle latitudes. Mainly westerlies, these winds flow at high speeds—up to 300 miles (480 km) an hour. The jet stream winds usually follow a gentle wave-like path over the United States and other countries in the middle latitudes. But sometimes a jet stream develops deep bends in its movements that upset normal weather patterns. Weather forecasters cannot always predict when these deep bends will develop.

When the jet stream swings far to the south (in the Northern Hemisphere), it carries polar air with it. Places that usually have warm weather all year long may suddenly experience cooler weather. They may even have frost or snow in winter. When the jet stream swings far to the north (in the Northern Hemisphere), it pulls warm tropical air with it. Then places may suddenly experience warmer weather, even in the wintertime.

Weather in between fronts. After a front passes a place, the weather takes on the characteristics of air-mass

The ice crystals in cirrostratus clouds often appear as a halo around the sun or the moon. They are too thin to dim sunlight or moonlight, but they often give the sky a milky appearance.

weather. A region within a cold-air mass is likely to have cooler, clear weather. Within a warm-air mass, the weather is likely to be warm and humid. Condensation may also collect on objects when rapid cooling takes place.

Keeping Facts in Focus

1. Is a dry or a moist air mass likely to form over land? Where is a warm-air mass likely to form? A cold-air mass?
2. What causes an air mass to move out from its source region?
3. What happens when cold air pushes under warm air along a cold front? What happens when warm air rises over cooler air along a warm front?
4. What answers would you give to the key questions on page 171?

Working with Ideas

Use the map on page 172. From what source regions do the air masses that influence weather in your area originate? Which of these source regions is likely to produce hot weather in summer? The coolest weather in winter?

Lesson 4 Stormy Weather

Reading Focus

1. Many people use checklists. Pilots use checklists before taking off to make sure the flight will run smoothly. Shoppers use checklists. A checklist is helpful when there are many steps to be taken in getting something done right. Reading is like this. There are many steps you need to take to understand what you are reading. Review the "Focus" sections you have studied so far. Make a checklist for yourself of the things you need to do to understand what you read. Some headings your checklist might have are: *Previewing, Skimming, Understanding Words, Understanding Ideas, Organizing Information, Summarizing.*

2. As you read this lesson, look for answers to these key questions:
   a. Why do the earth and its people need storms?
   b. Which forms of moisture result from condensation? From precipitation?
   c. What are some of the kinds of storms that take place on the earth?

Vocabulary Focus

tornado	cyclone	dew	glaze
hurricane	fog	frost	monsoon
typhoon			

Each latitude region of the world—the high, the middle, and the low latitudes—has its own brand of stormy weather packages. The high latitudes have their blinding snowstorms. The low latitudes have their daily afternoon thunderstorms. The middle latitudes have a greater number of stormy weather packages than the other two regions.

Stormy weather packages in the middle latitudes come in all shapes and sizes. Some are labeled thunderstorms. Some, tornadoes, or twisters. Others, hurricanes, typhoons, or cyclones.

Some storms only make it inconvenient for people to be outdoors. Others sweep across land and water with great fury, causing loss of life and much property damage. But wherever they occur and whatever their size, shape, and power, stormy weather packages are important to the earth and its people.

Storms evaporate great amounts of water from the oceans, condense it, and return it to earth as precipitation. Without evaporation, condensation, and precipitation, rivers would run dry. Farm lands would become unfit for growing crops. Water tables would sink deeper and deeper below the earth's surface. Storms keep the water cycle working.

Forms of Condensation and Precipitation

Condensation takes place in the air, changing water vapor into water droplets or ice crystals. Condensation forms clouds, fog, dew, and frost.

Precipitation is the falling of some form of moisture through the air to the ground. Rain, snow, sleet, and hail are forms of precipitation.

Fog. Fog is really a cloud that forms near the ground. Ordinarily, air is warmer near the ground than it is at higher levels of the weather layer. But sometimes a layer of cooler air forms between the ground and a warmer layer of air. This happens when the ground loses heat very quickly during the nighttime. The air just above the cooling ground chills to its dew point. Then a thin blanket of fog covers the land. At times, this fog is very patchy. At other times, it covers a large area.

Warm, moist air moving inland from the ocean or some other large water body also forms fog. As the air passes over the cooler land surface, it chills and water vapor condenses, forming a thick, rolling fog. Such a fog may even last into the daylight hours.

Dew and frost. Dew and frost are forms of water vapor that has condensed on the ground or on solid objects on or near the ground. When the temperature of the air is above freezing, dew forms. When the temperature of the air is below freezing, the water vapor turns directly into ice crystals, forming frost on objects.

Rain and snow. Rain and snow are the most common forms of precipitation. Whether rain or snow falls depends on the temperature of the air at every level from the cloud to the ground. Snow falls only when the air temperature is below freezing all the way from the cloud to no more than a few hundred feet above the ground. As a general rule, 10 inches (25 cm) of snow is equal to 1 inch (2.5 cm) of rain in the measurement of a region's total amount of precipitation.

Sleet. Sometimes raindrops fall into a freezing layer of air near the ground. The raindrops freeze and form particles of falling ice, or sleet.

Glaze. Sometimes rain freezes on contact with freezing ground surfaces, forming a glaze, or coating of ice, on everything. Such "ice storms" often cause great damage. Telephone wires and electric power lines break, bushes collapse, and tree branches snap under the heavy weight of the ice coating. Fruit trees sometimes do not recover from ice storm damage.

Hail. In some storm clouds, there is a repeated and a rapid rising and falling of air that produces hail. At the top of the cloud, ice particles form. Strong air currents carry the ice particles downward. With each fall, the ice particles collect more and more water around them. At lower levels, strong air currents carry the particles upward. With each rise, the water freezes, making small balls of ice called hailstones. Some hailstones can become as big as base-

Convectional Storms

Clouds form as moist air cools

Moist hot air rises

Orographic Storms

Clouds form as cooler air condenses

Moist air moves up a mountainside and becomes cooler

Frontal Storms

Thunderhead forms and rain falls

Clouds form and rain falls

Cold air mass meets warm air mass

Warm air rises rapidly

Warm air cools as it is pushed to higher levels

Cold air mass pushes under warm air mass

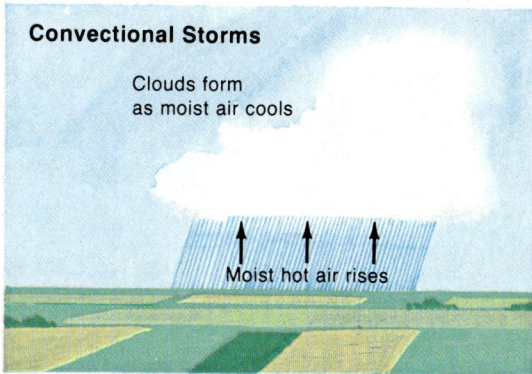

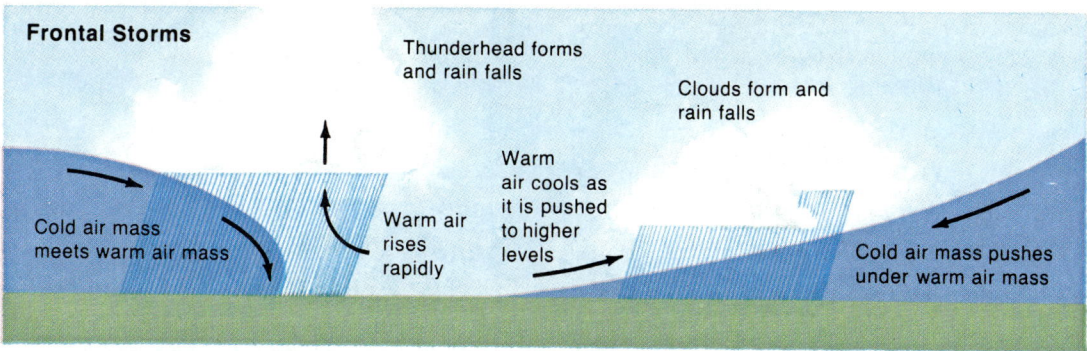

Convectional storms form frequently on hot summer days. How do orographic storms occur? Study the frontal storms diagram. Which part of the diagram shows a storm occurring along a cold front?

balls if the air currents are strong enough to move the hailstones up and down many times. Most hailstones are much smaller. But all are made of layers of ice that show how many times each hailstone rose and fell within the storm cloud.

Storms

Some rising air changes temperature rapidly. It has powerful updrafts and downdrafts. It condenses or evaporates great amounts of moisture, forming dark, billowing cumulonimbus clouds. Thunderstorms, tornadoes, and hurricanes develop under these conditions.

Thunderstorms. More common than any other kind of storm, thunderstorms occur on every continent except Antarctica. About 2,000 thunderstorms are in progress at all times over the earth.

Thunderstorms form when two things are present—rapidly rising air and great amounts of water vapor. Study the illustration on this page. It shows the different conditions that cause thunderstorms. In each case, air is lifted rapidly, causing condensation and precipitation. Why do thunderstorms form over flat land on a hot summer day? Why do thunderstorms often form over mountains? How can the passing of a cold front produce thunderstorms?

Sharp flashes of lightning and loud claps of thunder mark the arrival of a thunderstorm. Electrical charges build

Ice storms occur when warm air overrides cold air and the temperature of the cold air is below freezing.

up in towering clouds. When these electrical charges separate within the cloud, lightning flashes. Lightning flashes usually take place from cloud to cloud or within a cloud. On the average, only about one in every one hundred flashes reaches the earth's surface.

Thunder follows the discharge of lightning. The lightning heats the air around it very rapidly and causes the air to expand. Just as rapidly, the air cools and contracts. It is the rapidly cooling and contracting air crashing into the air around it that makes thunder.

Tornadoes. The same cumulonimbus clouds that give rise to thunderstorms can also give rise to tornadoes, or twisters. The updrafts of warm air and the downdrafts of cool air in parts of the cloud begin to spin when acted on by upper level winds in the middle part of the cloud. A funnel-shaped cloud forms and drops beneath the main cloud! The center of the funnel cloud has extremely low pressure. Winds in these funnel clouds are sometimes as high as 500 miles (800 km) an hour.

Once a funnel cloud forms and drops below the main cloud, it may move aloft, never touching ground. It may hop, skip, and jump, hitting one spot on the ground and missing others. Funnel clouds stay on the ground longest and travel the straightest path over flat, open land. Funnel clouds and the storms that carry them normally move parallel to cold fronts in a generally northeastern direction. But they can change paths when they move into an area with hills or high buildings.

The width of the surface covered by a funnel cloud varies from a few feet to about one mile. The average distance a funnel cloud travels on the ground is about 25 miles (40 km). The cloud itself moves at an average speed of about 25 miles (40 km) an hour.

Most of the damage caused by tornadoes comes from the high winds and the debris the winds pick up. The winds push in walls and windows, rip off roofs, knock down fences, and pick up objects of all shapes and sizes. Sometimes the rapid drop in pressure that comes as the

Tornadoes form over water (left) as well as land (above). Tornadoes that form over water are called waterspouts.

funnel cloud passes over an area causes a building, especially one with few openings, to explode.

Tornadoes occur in most parts of the world. But the area of their most frequent appearance is the United States, where they have struck in all states and in all months of the year.

In the United States, the tornado season usually begins in February in the Southern states bordering the Gulf of Mexico. The season hits its peak in these states in April and May. Then the season moves northward during June, July, and August, hitting Oklahoma, Kansas, or other Midwestern states. In fall, it moves southward again.

Whenever conditions exist that make a tornado possible, weather bureaus issue a series of warnings. A *tornado watch* simply means that violent thunderstorms are approaching, or may soon form, that could give rise to a tornado. A *tornado warning* alerts people in the area that a funnel cloud has been sighted. In either case, the general location and path of the storm (or the funnel cloud) are reported.

Hurricanes, typhoons, and cyclones. These storms form in late summer and early fall over tropical waters. Most of them form on the western sides of oceans between latitudes 5° and 20° north and south of the equator. Those that form in the Atlantic and off the western coast of Mexico in the Pacific are called hurricanes. Those that form in the Pacific between the international date line and the coasts of Asia and Australia are called typhoons. Those that form in the Indian Ocean are most often called cyclones.

High winds, thunderstorms, and tornadoes combine to make these tropical storms the most damaging of all storms. Sailors at sea fear them. So do people

Patrolling the Weather

Keeping a watch on the weather—either by trying to control it or by trying to track its changes and its dangerous moments—takes many people besides meteorologists. These people include airplane pilots, computer and radar operators, weather observers, persons who track icebergs, and workers who make, launch, and track weather satellites.

Controlling weather. Rain or no rain can mean life or death for people, plants, and animals. For centuries, people have tried to bring on rain when it was most needed, using many means. Today, people sometimes seed clouds in an attempt to control weather.

Cloud seeding is done with dry ice or silver iodide dropped from airplanes or shot into the air from rockets. Cloud seeding speeds the forming of water droplets or ice crystals in humid air at upper levels. In drier areas of the world, cloud seeding is done to bring on rain. In areas where hurricanes form, seeding clouds helps to weaken a storm's power.

Experiments with cloud seeding do not always succeed. Sometimes, they only serve to increase a storm's power. Cloud seeding may bring on so much rain that flash floods take place, causing loss of lives, property, and crops. Much more needs to be learned about how weather makers react under different storm conditions before cloud seeding will always work.

Tracking the weather. One group of weather patrollers are the thousands of weather observers who make readings from weather instruments. Nearly every country in the world has weather observers who send reports four times a day to the World Meteorological Organization (WMO). Meteorologists in the WMO put these reports together and send them to national weather centers around the world. In the United States, reports from weather observers are plotted on weather maps that are issued daily by the National Weather Service in Washington, D.C.

Hurricane hunters fly into tropical storms. Their planes are equipped with radar and other instruments that record conditions in a storm. By studying the data recorded on hurricane-hunting flights, meteorologists are able to predict the movement and power of these storms and warn people where they will strike.

Another group helping to save lives is the International Ice Patrol of the United States Coast Guard. This group tracks the courses of icebergs that break off from Greenland's ice cap to float in the shipping lanes of the North Atlantic. Twice each day, the Ice Patrol issues radio warnings that tell ships about these floating giants. Whether tracking icebergs, flying into hurricanes, or taking weather readings, these weather patrollers help to make lives safer for many people.

living along coastal areas where they are likely to strike.

Hurricanes, typhoons, and cyclones are really air-mass storms. They begin in a warm-air mass that has temperatures higher than 80° F (27° C) and uniform air pressure. But something, probably strong updrafts of very humid air, triggers an upset in wind circulation within the air mass. Moist air begins to move rapidly toward the center of the warm-air mass. The rising air causes rapid condensation. This, in turn, causes an increase in the temperature of the air. As the air mass begins to move out from its source region, it builds up more and more heat energy.

Moving slowly at first, the tropical storm gradually gains in strength. As it increases its speed, it may change directions several times. As long as the storm remains over water, it gains in power.

Not all tropical storms move inland, but those that do often cause great damage in coastal areas. At times, the effects of a tropical storm are felt far inland. The longer a storm remains over land, the weaker its winds become. If a storm moves back over cold water, it generally blows itself out. But if it moves back over warm water, it will renew its power. A storm such as this may strike a part of Central America, move back over the warm waters of the Gulf of Mexico, and then strike the Gulf coast of the United States.

To be called a hurricane, typhoon, or cyclone, a tropical storm must have winds with speeds greater than 74 miles (119 km) an hour. The center of the storm is called the eye. Winds here are light and variable, seldom blowing more than 15 miles (24 km) an hour. Just outside the eye, winds may reach speeds greater than 120 miles (193 km) an hour. These winds may cover a band 180 to 300 miles (290 to 483 km) wide. Another band 360 to 480 miles (580 to 773 km) wide carries winds that vary in force from 72 miles (116 km) to 32 miles (51 km) an hour. The air mass that carries a tropical storm moves very slowly—about 9 to 18 miles (14 to 29 km) an hour.

Along with the high winds come great amounts of rainfall. One typhoon in the Philippines dumped over 60 inches (160 cm) of rain on the area over which it passed. Great ocean waves 15 to 20 feet (4 to 6 m) high pound coastal areas, causing flooding. Storms with the most flooding and the highest waves usually hit at times of high tide.

Monsoons

Monsoons are seasonal shifts in wind patterns that affect India, other countries east of it, and the islands of the East Indies.

During the winter months (November through April), winds blow from the land toward the sea. The winter monsoons bring a dry season to countries in southern Asia. During the summer months (May through October), the winds blow from the sea toward the land. The summer monsoon, when it reaches land, usually brings great amounts of rain.

Keeping Facts in Focus

1. What is condensation? Precipitation?
2. If the temperature of the air at ground level is below freezing, will dew or frost form on objects on the ground? If the temperature of the air from cloud to ground is above freezing, will rain or snow fall?
3. How much snow equals one inch of rain?
4. What are the three ways in which air can be lifted to produce thunderstorms?
5. What is a tornado? In what part of the world do tornadoes most frequently appear?
6. Is a hurricane, typhoon, or cyclone a frontal storm or an air-mass storm? Where do most tropical storms form?
7. What answers would you give to the key questions on page 177?

Working with Ideas

1. How does the warm-water ocean current that is the North Atlantic Drift influence the weather patterns of England?
2. Why is snow almost always present on high mountain peaks, even in the tropics?

Unit 4 REVIEW WORKSHOP

Test Your Geographic Knowledge

Choose the letter of the item that correctly completes each statement.

1. All the layers of air that surround the earth are called the (a) atmosphere, (b) troposphere, (c) hemisphere.
2. Most of the world's stormy weather takes place along (a) wind belts, (b) fronts, (c) source regions.
3. The amount of moisture in the air is called (a) humidity, (b) pressure, (c) temperature.
4. When air rises, it becomes (a) heavier, (b) denser, (c) thinner.
5. Fog is a form of (a) condensation, (b) evaporation, (c) precipitation.
6. Clouds that are thin, wispy, and mainly ice crystals are (a) cumulus clouds, (b) cirrus clouds, (c) stratus clouds.
7. Polar air masses that form over land are (a) warm and moist, (b) cool and moist, (c) cool and dry.
8. All weather takes place in the (a) ionosphere, (b) troposphere, (c) stratosphere.
9. When air sinks, it becomes (a) denser, (b) cooler, (c) thinner.

Apply Your Reading Skills

A. Rearrange the sentences, using their numbers, to tell the correct order of events in the development of a thunderstorm over a desert.
   1. Electrical charges build up in the cloud.
   2. Water vapor in the air condenses, and a small cumulus cloud begins to form.
   3. Moist air over heated ground is forced to rise.
   4. Following the first thunderclap, drops of rain begin to splatter the ground.
   5. The cloud grows into a cumulonimbus cloud with an anvil head.
   6. After one more clap of thunder, the cloud "bursts," releasing a downpour that lasts several minutes.
   7. Lightning flashes from cloud to cloud, and thunder follows.
   8. Updrafts and downdrafts in the cumulus cloud make the cloud increase in height.

B. Use encyclopedias, books, and other reference tools to do research. Then give a report on ONE of the following topics.
   1. What are some of the instruments used on or near the ground to collect information for weather forecasting? Describe what each does.
   2. How do satellites help in weather forecasting? How does radar help in weather forecasting?
   3. What are some sources of air pollution? What are some steps that can be taken, or are being taken, to reduce air pollution?
   4. What are some things people are doing to try to control weather?
   5. What kind of training and education does a meteorologist need? What are some of the kinds of work done by meteorologists?

Apply Your Geographic Skills

Newspapers often print weather forecast maps and weather report maps issued daily by the National Weather Service. The map on page 185 is a weather report map. It reported the actual weather for a day in late November in the United States. Study the map and the key. Then tell whether each of the following statements is true or false, according to the information on the map.

1. The map reported the temperature, sky condition, wind speed, and wind direction for selected cities.
2. On this day, only one area of low pressure was influencing the weather.
3. Only Denver reported no winds.
4. Cloudy skies caused by a passing warm front brought rain to Chicago and Detroit.
5. Freezing temperatures were reported as far south as Oklahoma City.
6. The air pressure over Seattle was higher than the air pressure over Sault Ste. Marie.
7. Winds at Boise, Idaho, were blowing faster than winds at Salt Lake City, Utah.
8. Jacksonville, Florida, reported the highest temperature shown on the map.

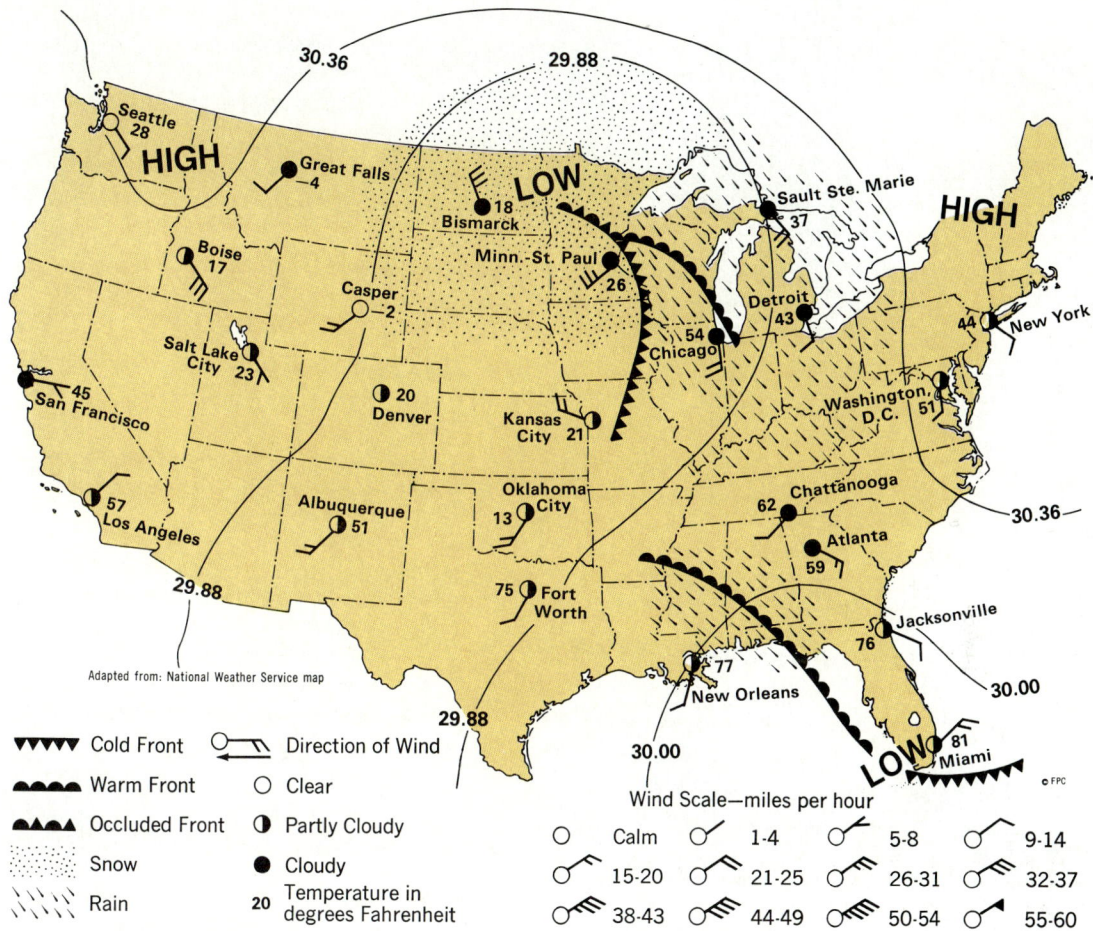

Adapted from: National Weather Service map

Map labels (weather stations, temperatures, pressures):

30.36 — 29.88 — 29.88 — 30.36 — 30.00 — 30.00 — 29.88

HIGH — LOW — HIGH — LOW

Seattle 28
Great Falls −4
Bismarck 18
Sault Ste. Marie 37
Boise 17
Minn.-St. Paul 26
Detroit 43
New York 44
Casper 2
Salt Lake City 23
Chicago 54
Washington D.C. 51
San Francisco 45
Denver 20
Kansas City 21
Los Angeles 57
Albuquerque 51
Oklahoma City 13
Chattanooga 62
Atlanta 59
Fort Worth 75
Jacksonville 76
New Orleans 77
Miami 81

© FPC

Legend:

Symbol	Meaning	Symbol	Meaning
▼▼▼▼	Cold Front	⊙→	Direction of Wind
●●●●	Warm Front	○	Clear
▲▲▲▲	Occluded Front	◐	Partly Cloudy
∴∴∴	Snow	●	Cloudy
⁄⁄⁄	Rain	20	Temperature in degrees Fahrenheit

Wind Scale—miles per hour

Symbol	MPH	Symbol	MPH
○	Calm		
○⁄	1-4		5-8
	9-14	○⁄	15-20
	21-25		26-31
	32-37	○⁄	38-43
	44-49		50-54
	55-60		

Apply Your Thinking Skills

Tell whether each of the following statements refers to weather in the low latitudes (L), in the middle latitudes (M), in the high latitudes (H), or in all three (A).

1. Moisture from the earth's surface evaporates into the air, condenses, and falls to earth in some form of precipitation.
2. In these latitudes, the sun's rays always strike the earth at direct or nearly direct angles.
3. In these latitudes, the sun's rays always strike the earth at very slanted angles.
4. During most of the year, the snow-covered surfaces give off very little heat to the air above.
5. In these latitudes, warm- and cold- air masses meet to produce days of very stormy weather.
6. Prevailing winds flow over the earth from the same general directions.
7. Most hurricanes, typhoons, or cyclones form over water in these latitudes.

Discuss These Points

It is especially important to know how to protect yourself when stormy weather strikes. What safety rules should you follow if an electrical storm takes place when you are outdoors? Are there special rules to follow when you are also indoors? Do tornadoes strike often in your area? How do you protect yourself indoors? Outdoors? For what other kinds of stormy weather in your area should you prepare safety rules?

Expand Your Geographic Sights

Barry, Frederick and Frank, Sidney. *Your Future in Meteorology* (Rosen, 1970).

Donnan, Jack and Donnan, Marcia. *From Rain Dance to Research: The Story of Weather Control* (McKay, 1978).

Ford, Adam. *Weatherwatch* (Lothrop, 1982).

Stambler, Irwin. *Weather Instruments and How They Work* (Putnam, 1967).

Unit 5 Climate and Vegetation

Climate, in one way or another, places limits on the people of a region. What limits do you think climate places on desert dwellers in Morocco's Atlas Mountains?

Lesson 1 The Importance of Climate

Reading Focus

1. By now, you have noticed that the opening pages of every unit list an outline for each lesson of the unit. This outline, however, lists only the major topics presented in each lesson. It is a "bare bones" outline. It is up to you to "flesh out" each lesson's outline by finding subtopics and the details that support them. Review the numbering and lettering system used for outlining that is presented on page 72. Also review the *Reading Focus* on page 91. It takes time and practice to become a good outliner. Practice your outlining skills on this lesson and on the following lessons of the unit. But remember—outlining is not something you do and then file away in a notebook, never to look at again. Use your outlines both for study and for review.

2. As you read this lesson, look for answers to these key questions:
   a. What is climate?
   b. What is climate's role in the environment?
   c. What changes in climate have taken place in the past? What changes in climate might take place in the future?

Vocabulary Focus

climate   *natural environment*   *glacial*   *technology*

How the air behaves at any given place at any given moment is weather. The weather at any given place over a long period of time, at least twenty or thirty years, is *climate*. Climate is the average weather of a place or region. But to say that climate is only average weather is to fall far short of describing the importance of climate to the natural environment.

Climate's Role in the Environment

Climate has relationships to all other parts of the earth—its land, its water, and its atmosphere. These relationships work two ways. Land, water, and the changes that take place in air play their parts in shaping climate. At the same time, climate plays its part in helping to shape landforms and soils. Climate

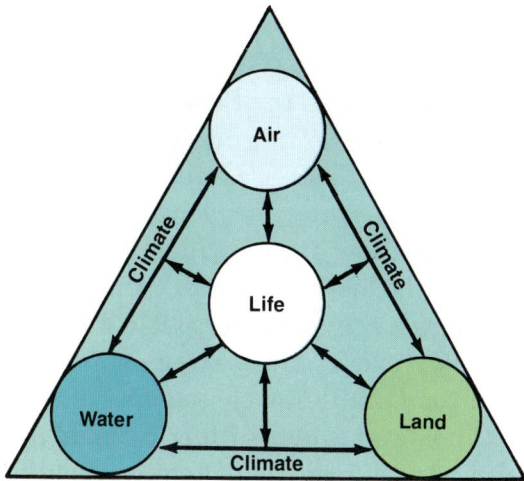

The Natural Environment

helps to keep the water cycle working. Climate, as average weather, also helps to determine what changes take place in the air from month to month and from year to year.

The relationships among land, water, air, and climate go even further. They have a direct bearing on the kinds of plants and animals that live in a region.

Land, water, air, climate, vegetation, animal life—all are important parts of the natural environment. No part acts alone, stands alone, or is more important than any other part. There is an ongoing interaction among all parts. Put together, these interacting relationships give climate a central role in helping to form the natural environment.

In its central role, climate is always there, forming the background, the setting, for life. It is always there placing limits on where plants can grow or animals can live. It is always there placing limits on people's choices about how they will use the land, the water, and the forms of life found in a region.

Climates of the Past

Climate probably began to play its central role as a part of the natural environment as soon as the earth took form and settled in its orbit around the sun. We have no certain way of knowing what all the past climates of the earth were like. Neither do we know for certain how many different climates there were at a given period. But we do know some things about climate changes in the past.

Lands around the equator have probably always been hot. But, there have been periods when climates in the middle and high latitudes were warmer than they are today. There have been other periods when climates outside the

"THE FORECAST FUR TODAY IS ICE AGE, FOLLOWED IN 10,000 YEARS BY FAIR AND WARMER."

tropics were much cooler than they are today. These cooler periods usually resulted in the buildup of ice on the earth. Glaciers spread over large parts of the earth's land surface. Packs of ice covered large parts of the world's oceans and lakes.

But always, the glacial periods were followed by warmer periods. This cycle of warming, cooling, and warming again has repeated itself several times in the past. The earth's last glacial period gave way to a warming period about 11,000 years ago.

Climate and Life

After the last great continental glaciers withdrew to the lands around the poles, the climates we know today emerged. Now we have hot climates around the equator, cold climates around the poles, and—as a general rule—climates with warm and cold seasons in the middle latitudes.

Climate and plants. Where temperatures are hot to warm and where precipitation is heavy to moderate, forests cover the land. Where temperatures are warm enough but where precipitation is lighter or falls only in one season, tall grasses cover the land. In very hot but dry places, special forms of plants grow by storing water in their stems or by taking moisture from the air. In very cold places, tiny mosses and some very short grasses manage to survive.

Climate and animals. Special forms of animal life, too, live in certain climate regions. Animals that swing from branches and climb up and down tree trunks live in forests. Animals that can hide among grasses or can outrun their enemies survive in grasslands. In very dry regions, there are animals that can go without water for long periods of time or that burrow underground to escape the heat. Animals with furs or thick skins live in very cold regions. Finally, every climate region has its own communities of insects and birds.

Climate and human life. Human life also has differed from climate region to climate region. From the hotter regions of the earth, groups of human beings spread to every continent except Antarctica. The earliest people were hunters of wild animals and gatherers of fruits, nuts, and berries from wild plants. These people moved from place to place looking for food. Later groups of people changed from hunters and gatherers to farmers who settled in villages.

Having discovered they could grow crops, tame and raise animals, and settle in one place, people began to make changes in the land and in its vegetation patterns. Having spread to different parts of the earth, people discovered that they could survive in many different climates. But people also discovered that climate did set limits to the kinds of things they could do to meet their needs.

Climate's Limiting Role

People discovered they could not farm the same way or grow the same crops in every climate. So they learned to farm in different ways, to irrigate some lands, and to drain others. They learned to grow crops that were suited to each

Why are camels (top) sometimes called "ships of the desert"? What are
some animals that can be found living on Antarctica (bottom, right)?
Bison, or buffalo, were once plentiful in many areas of the United States.
Today only a few thousand survive as protected animals in such places as
Custer State Park, South Dakota.

Climatologists—Investigators of Climates Past and Present

Climatologists, people who study climates, are also meteorologists. Like meteorologists, they study changes in the air. But unlike the people who study daily weather data, climatologists study weather averages of past years. Instead of making daily forecasts, they look for long-term trends. These trends show whether climates are getting warmer or colder, wetter or drier.

Climatologists do their work in many ways. Some teach in colleges and universities. Some study climates of the past, doing research in the field or in the laboratory. Others study climates of the present. Some of these climatologists work for private industries. Others work for governments or international organizations, such as the United Nations. They give advice on how the climate of a region will affect products. Or they advise farmers, builders, and manufacturers, showing how climates can be affected by changes in land surfaces or by the activities of people on the land.

One climatologist working for an international organization helped rid a town in Yugoslavia of a bad air pollution problem. Located at the end of a narrow valley in a mountainous region, the town's buildings were badly damaged by an earthquake. In deciding to rebuild, the town officials wanted to get rid of the air pollution problem. They called in a climatologist for advice about the problem.

The climatologist worked in a laboratory with a model of the town's land features. From the model and a study of the town's wind patterns, the climatologist told the planners where to put up the buildings. He told them what height and what shape they should be. Today, the city seldom has an air pollution buildup because air now flows freely through the town.

In many other parts of the world, climatologists, such as this one, are working quietly to do their part in keeping all parts of the natural environment interacting smoothly.

climate region. Whether farming, building houses, or making clothing, people have learned to make choices within the limits set by each climate region.

In making their choices, people have developed different patterns of living. Climate is one of the reasons why the Eskimos chose to be fishers and hunters rather than farmers. Climate is one of the reasons why the people of the Amazon Basin have hardly ever chosen to wear fur clothing, even though furry animals live in their environment.

For the same reason, the Plains Indians of the United States developed a pattern of living that was different from the one developed by the Woodland Indians. The Plains Indians moved from place to place, following the bison that lived in the grasslands west of the Mississippi River. They needed shelters that could be easily moved. The tepees the Plains Indians made were much different from the bark-covered lodges of the Woodland Indians. Living in the forests east of the Mississippi River, the Woodland Indians cleared plots of land for farming. They settled in villages. Besides growing corn, squash, and other crops, they gathered fruits, nuts, and berries from the forests. They also hunted forest animals.

People and Climate Today

Over the centuries, there have been many advances made in ways of living. Most of these advances have been brought about by technology. In today's world, climate often seems to play less of a limiting role for many people. But

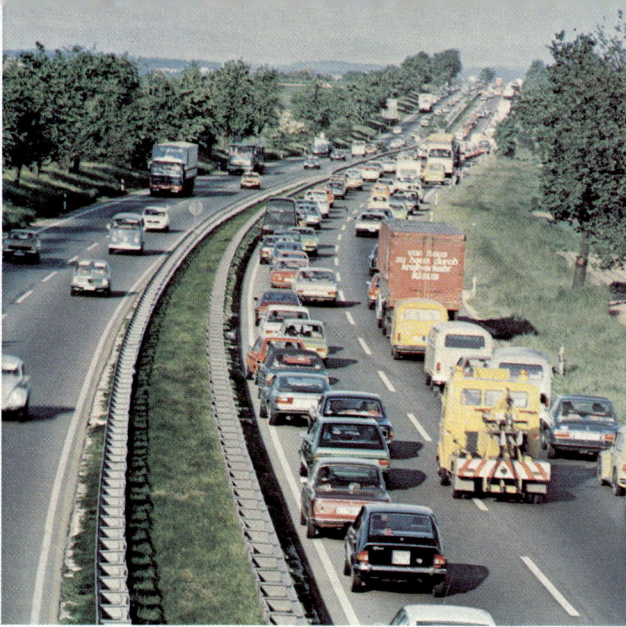

On one side of West Germany's *autobahn* (highway), the idling motors of vehicles in a bumper-to-bumper traffic jam heavily pollute the countryside.

no matter how many advances are made in the use of machines and tools, people must still deal with the relationships between climate and the other parts of the natural environment. Changing one part of the natural environment often triggers changes in other parts that may affect climate.

For example, we pollute the air with automobile exhausts and with smoke from homes, factories, and other buildings. Adding chemicals to the air upsets air's balance. This, in turn, affects the exchange of heat energy between the sun, the air, and the earth's surfaces. The loss or addition of heat energy affects precipitation rates. This can lead to changes in plant growth and, perhaps, in the animal life of a region. Eventually, these changes can lead to changes in the present climate patterns of the world and to unwanted changes in the ways people live.

No one wants to speed the coming of another Ice Age. But a lowering of world temperatures by only a few degrees over a long period of time could speed its coming. Neither does anyone want to raise world temperatures by a few degrees. But if this is done over a long period of time, the polar ice caps would melt, raising ocean levels. The effects of an Ice Age or of rising sea levels are good reasons for people today to continue making choices within the limits set by climate.

Keeping Facts in Focus

1. Where are hot climates generally found today? Cold climates? Climates that have warm and cold seasons?
2. Where are forests usually found growing? Grasslands? What are some ways that plants survive in very hot but dry climates? What are some of the plants that survive in very cold climates?
3. What are some of the ways climate limits people's choices about how they will live and meet their needs?
4. What answers would you give to the key questions on page 188?

Working with Ideas

1. How might the coming of another Ice Age begin to change the patterns of your life?
2. How important is climate to the business and industry of your state?

Lesson 2 Climate Regions of the World

Reading Focus

1. The Index of your book is a helpful tool. Turn to the Index at the back of your book. Read the information at the top of the page. Now skim several listings of terms and topics on the page. In many cases, there is a series of numbers. Sometimes a comma separates two page numbers. At other times, a dash joins two page numbers. When a comma separates two page numbers, read the two numbers as two separate pages. On each of the pages, the term or topic may be mentioned only once. Or it may be mentioned several times. You must skim the page to find how many times it is referred to. When a dash joins two page numbers, read the two numbers together. For example, 124–125 means page 124 and page 125; 256–260 means page 256 through page 260. When using the Index for research or for review, first look up the page numbers joined by a dash. On these pages, the term or topic is usually discussed more fully.

2. As you read this lesson, look for answers to these key questions:
   a. What are the things that shape climates?
   b. What weather averages are used to give an overall view of a region's climate?
   c. How does temperature range give a clearer picture of a region's climate?

Vocabulary Focus

windward   subtropical   temperature range
leeward   subarctic

You already have a very general view of today's climate patterns. These larger climate patterns are based on the temperature differences that exist at different latitudes between the equator and the poles.

But there are smaller climate patterns within these latitude regions that are based on precipitation and temperature differences. These differences enlarge the number of major world climate regions from three to thirteen.

Things That Shape Climates

To understand the things that shape climate patterns, you need only to review some of the things you have learned about air and the things that cause changes in the air.

Latitude. Because of the earth's rounded shape and the ways it moves in space, all places on the earth's surface are neither heated nor cooled equally. Places in the low latitudes, you recall, are hot all year round. They always

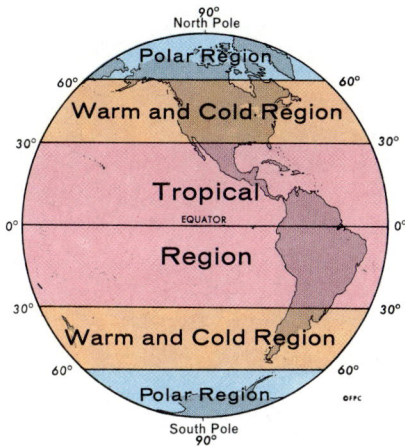

Air moving over water picks up much more moisture than air moving over land. Sometimes the air over water blows toward the land, where its moisture falls as precipitation. This happens where air moves over warm ocean currents. At other times, air moves toward the land over cold ocean currents. Then its moisture falls over the ocean rather than on land. Some of the driest parts of the world are located along the coasts where cold ocean currents flow.

Drier Leeward Side   Wet Windward Side   Prevailing Winds

receive the direct or nearly direct rays of the sun. Places in the high latitudes are cold all year round. They receive only the very slanted rays of the sun. In the middle latitudes, places are warm in summer when the sun's direct rays are closest to them. They are cold in winter when the sun's direct rays are farthest from them. Latitude is one of the main factors determining temperature differences from place to place.

Altitude. Another factor determining temperature differences is altitude. Temperatures, you recall, drop about 3.6° F (2° C) with each rise of 1,000 feet (300 m) above sea level. Because of their higher elevations, highland regions have climates that are cooler than the climates of the surrounding lowlands.

Land and water divisions. The division of the earth into land and water surfaces affects both temperature and precipitation. Land areas, you recall, heat and cool faster than water areas. Air also takes on the characteristics of the surface over which it lies or moves. So the air over land areas is always hotter in summer and cooler in winter than the air over water areas.

The slope of land areas is also important. Air follows the slope of the land. On the windward side of a mountain, the side from which the prevailing wind blows, air rises. As it rises, the moisture in it condenses and falls as precipitation. But on the leeward side of a mountain, air sinks. Sinking air picks up moisture rather than dropping it. The leeward sides of mountain ranges are usually dry places.

Shifting pressure centers. Precipitation is usually heavier where air is rising. Air is usually rising over the equator and around 60° north and south latitudes. In these places, air generally has a low pressure.

Precipitation is usually lighter where air is sinking. Air is usually sinking around 30° north and south latitudes and over the poles. In these places, air generally has a high pressure.

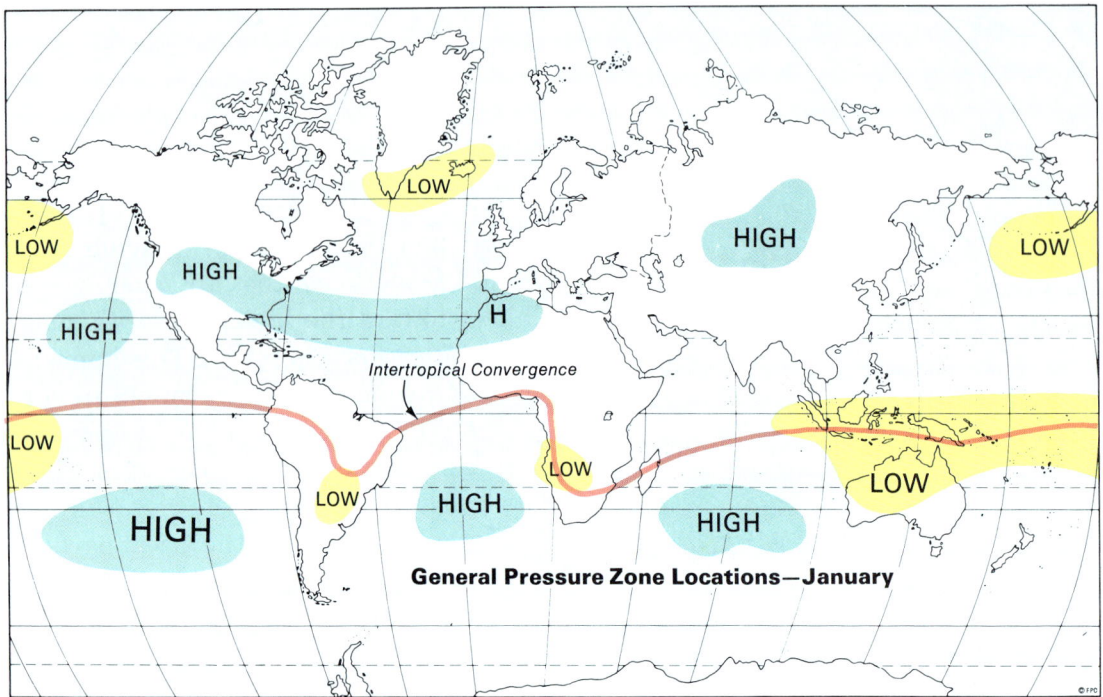

General Pressure Zone Locations—January

Notice the line called the Intertropical Convergence (ITC) on each map. Along the ITC, weak and variable tropical trade winds from the Northern and Southern Hemispheres meet as they flow at or near the earth's surface. The air, heated from below, rises. The ITC, then, is an area of low pressure and thunderstorms. In January (above), the ITC forms farthest south of the equator when it is over a continent. By July (below), the ITC has shifted to its most northerly locations. At the same time the ITC shifts, pressure zones in both hemispheres also shift, forming farthest south in January and farthest north in July. What other changes in pressure zones over land also take place between January and July?

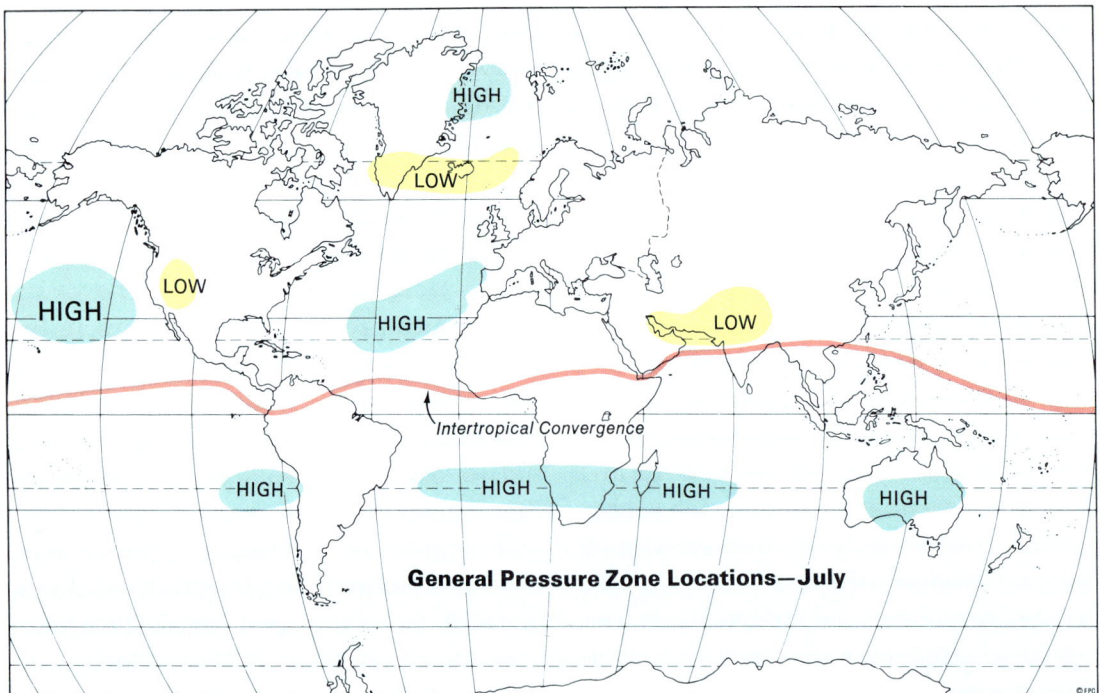

General Pressure Zone Locations—July

Daytime: Sea or Lake Breeze

Cool air

Warm air

Ocean

Land

Nighttime: Land Breeze

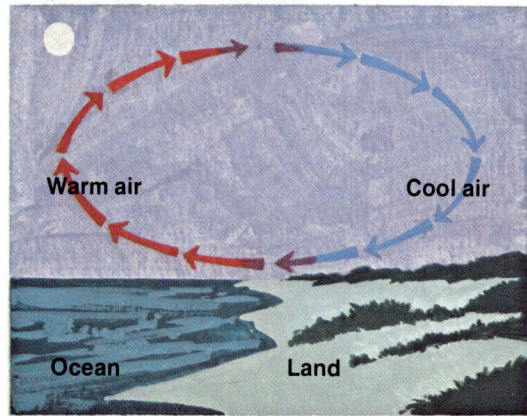

Warm air

Cool air

Ocean

Land

There are surface wind patterns, other than prevailing winds, that often affect climates. One of these wind patterns is found along the coastlines of large water bodies. During the day, the wind often blows strongly from the cooler water toward the land. In the tropics, a breeze blowing from the sea may lower midafternoon temperatures on land by as much as 20°F (7°C). As the sun sets, the breeze dies down. During the night the pattern reverses itself. The cooler air over the land then blows toward the water.

But, remember, there are temperature differences between large land and water bodies. The temperature differences at any season of the year at any one latitude region are slight. But they are enough to cause pressure differences between the air over water and the air over land. Instead of one wide band of high pressure or low pressure at any one of these latitude regions, there are alternating pockets of high pressure and low pressure.

In summer, the high pressure forms over the water and the low pressure forms over the land. In winter, the high pressure centers and the low pressure centers trade places.

This back and forth shifting of pressure centers is one of the reasons why there are seasonal differences in precipitation in some climate regions. When high pressure centers form over the oceans, winds may blow across the oceans toward the land. These winds bring heavier amounts of precipitation to the land. But when high pressure centers form over the land, winds may blow over the land toward the ocean. These winds usually are dry.

Pressure centers shift another way. They follow the north-south movement of the sun's direct rays. When the sun's direct rays are south of the equator, most pressure centers often shift farther to the south. When the sun's direct rays are north of the equator, pressure centers often shift farther north. This north-south shift also causes seasonal differences in precipitation. At one season of the year, a region may be under the influence of a low pressure center. Then it receives much precipitation. At another season, a region may be more under the influence of a high pressure center. Then it usually receives very little precipitation.

Prevailing winds. Much of the work prevailing winds do in shaping climates has already been discussed. How the prevailing winds influence a climate depends on whether the winds are blowing from the land toward the water or from the water toward the land. It depends on whether the winds are blowing toward

197

the land over cool ocean currents or over warm ocean currents. It also depends on whether winds are blowing up a mountainside or down a mountainside. Explain what happens in each case.

Climate Patterns

The map on pages 200–201 shows the different climate regions of the world. There is also a brief description of each climate region at the bottom of these pages. Refer to the appropriate description and to the map to answer any questions in the following paragraphs.

Hot climates. Two climates are hot all year long. They are the *tropical wet* and the *tropical wet-and-dry* climates. The main difference between these two climate regions is precipitation. Which of these two regions gets most of its precipitation in one season of the year? At what time of the year does this region receive most of its precipitation?

Locate these two climate regions on the map. In most places where they are found, a tropical wet climate region merges into a tropical wet-and-dry climate region. Shifting pressure centers and prevailing wind directions are among the main factors shaping these tropical climates.

Semiarid and desert climates. Locate these climate regions on each continent. The first thing to notice about these climate regions is that they are found in the middle latitudes as well as in the low latitudes. In the low latitudes, the temperatures of these climate regions are usually hot all year long. In the

middle latitudes, the temperatures in both *semiarid* and *desert* regions usually vary with the seasons. Desert regions in these latitudes often have sharp differences between daytime and nighttime temperatures.

The next thing to notice is that a semiarid region borders a tropical wet-and-dry region on most continents and usually merges into a desert region. What does the description say about how precipitation differs in these two regions? Semiarid regions usually receive about 10 to 20 inches (25 to 50 cm) of precipitation in a year. Desert regions, on the average, receive less than 10 inches (25 cm) a year.

Several factors contribute to the shaping of these climates. In most of the Western Hemisphere, precipitation is light in these regions because of prevailing winds blowing over high mountain ranges. In Africa and Australia, shifting pressure centers and prevailing winds blowing over cool ocean currents help to shape the climates. In Asia, prevailing winds, mountain barriers, and shifting pressure centers often are the main influences.

Hot summer, cool winter climates. Both the *Mediterranean* and the *humid subtropical* climate regions have hot summers and cool winters. Which of these two climate regions has precipitation the year round? In what season of the year is its precipitation heavier? In what season of the year does a Mediterranean climate usually receive most of its precipitation?

Locate these two climates on the map. Notice that in the Western Hemisphere,

Mediterranean climate regions are located on the west coast of a continent. In the Eastern Hemisphere, this climate region is located mostly at the southern edges of a continent. On what side of a continent is the humid subtropical climate region usually found? What climate regions border each of these hot summer, cool winter climates?

Warm summer climates. Both the *marine* and the *continental* climate regions have warm summers. But their winters differ. How do they differ? Which of these two climate regions has heavier precipitation in the winter?

Locate these two climate regions on the map. Notice that the marine climate regions are found mostly within the middle latitudes on the west side of a continent. Notice that they also border a Mediterranean climate region on the north or on the south. Finally, you should note that continental climate region is found only in the middle latitudes of the Northern Hemisphere. This climate region usually is located in the deep interior of a continent or along a continent's eastern edges.

Cold climates. The *subarctic*, the *tundra*, and the *polar* climates are the high latitude climates of the world. Which one of these climates is cold all year round? What difference is there between winter and summer in the other two climates? Is precipitation in these three climates heavy or light?

Locate these climates on the map. What climates border the subarctic climate regions on each continent where they are found? The tundra climate? The polar climate is found only in

In those parts of southern Florida where winters are likely to be frost-free, oranges and other citrus fruits are grown. In what climate region is Florida located?

Greenland and Antarctica, although Antarctica is not shown on this map.

Highlands climate. The *highlands* climate is really several climates rather than one. Because temperatures become colder as altitude increases, climates change at different levels of a mountain range. Places at the foot of a mountain range have one climate. Places at the middle levels have another. Places at still higher altitudes have yet another climate. Added to this, the climate at each level on the windward side of a mountain range is different from the climate at each level on the much drier, leeward side.

WORLD CLIMATES

- ■ Tropical Wet
- ■ Tropical Wet-and-Dry
- ■ Semiarid (Steppe)
- ■ Desert (Arid)
- ■ Mediterranean (Dry Summer Subtropical)
- ■ Humid Subtropical
- ■ Marine
- ■ Continental
- ■ Subarctic (Boreal)
- ■ Tundra
- ■ Polar (Ice)
- ■ Highlands

World Climate Regions

The following descriptions will help you interpret the climate regions shown on the map.

Tropical Wet. Temperatures: hot all year round. Precipitation: heavy all year round. Vegetation: rain forests.

Tropical Wet-and-Dry. Temperatures: hot all year round. Precipitation: heavy when sun is overhead; dry when not. Vegetation: tall grasses and scattered trees.

Semiarid. Temperatures: variable. Precipitation: light. Vegetation: short grasses.

Desert. Temperatures: variable. Precipitation: very light. Vegetation: scrub, cactus, grasses.

Mediterranean. Temperatures: hot summers; cool winters. Precipitation: dry summers; wet winters. Vegetation: grasses, scrub, some trees.

Humid Subtropical. Temperatures: hot summers; cool winters. Precipitation: year round,

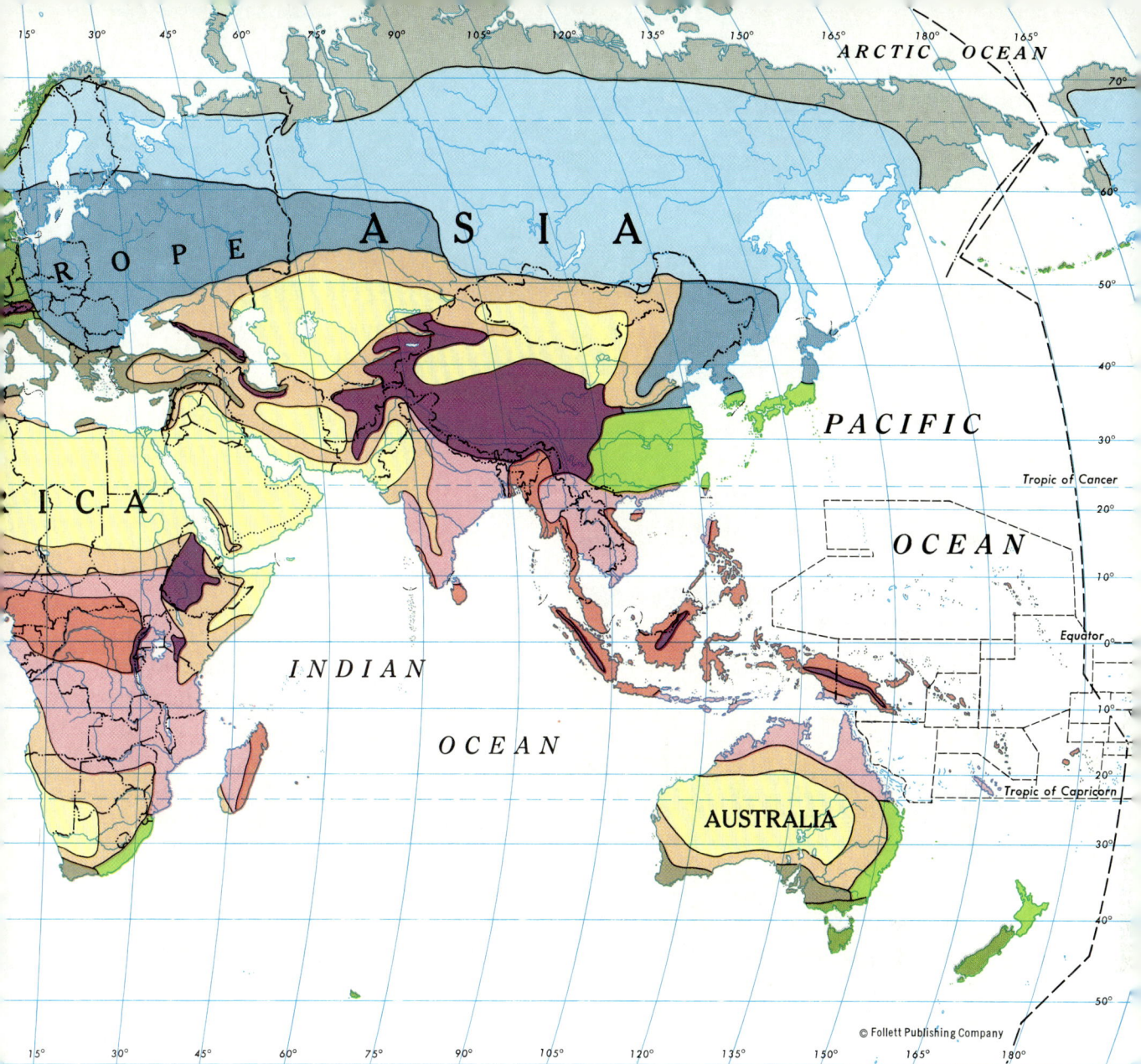

© Follett Publishing Company

but heavier in summer than in winter. Vegetation: forests and grasses.

Marine. Temperature: warm summers; cool winters. Precipitation: year round, but heavier in winter than in summer. Vegetation: forests and grasses.

Continental. Temperature: warm summers; cold winters. Precipitation year round, but heavier in summer than winter. Vegetation: trees and grasses.

Subarctic. Temperature: cool summers; cold winters. Precipitation: light. Vegetation: trees, some grasses and mosses.

Tundra. Temperature: cool summers; cold winters. Precipitation: light. Vegetation: grasses, mosses, lichens.

Polar. Temperature: cold all year round. Precipitation: light. Vegetation: none.

Highlands. Temperature, precipitation, and vegetation all vary greatly in these areas depending on altitude and direction of prevailing winds.

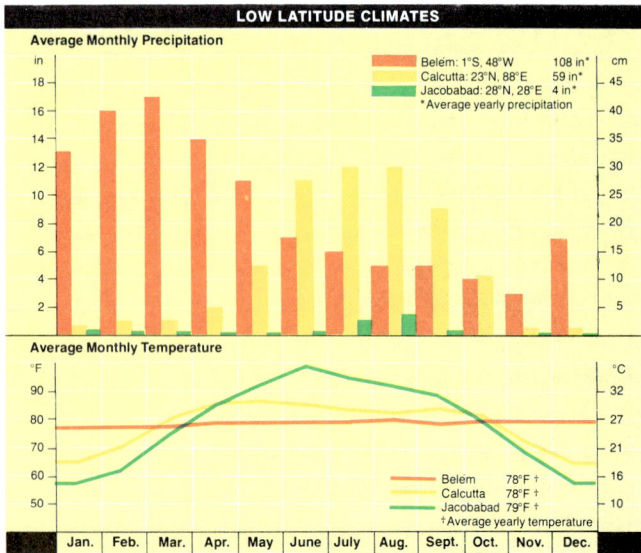

LOW LATITUDE CLIMATES

Average Monthly Precipitation

Belém: 1°S, 48°W 108 in*
Calcutta: 23°N, 88°E 59 in*
Jacobabad: 28°N, 28°E 4 in*
*Average yearly precipitation

Average Monthly Temperature

Belem 78°F †
Calcutta 78°F †
Jacobabad 79°F †
†Average yearly temperature

Jan. Feb. Mar. Apr. May June July Aug. Sept. Oct. Nov. Dec.

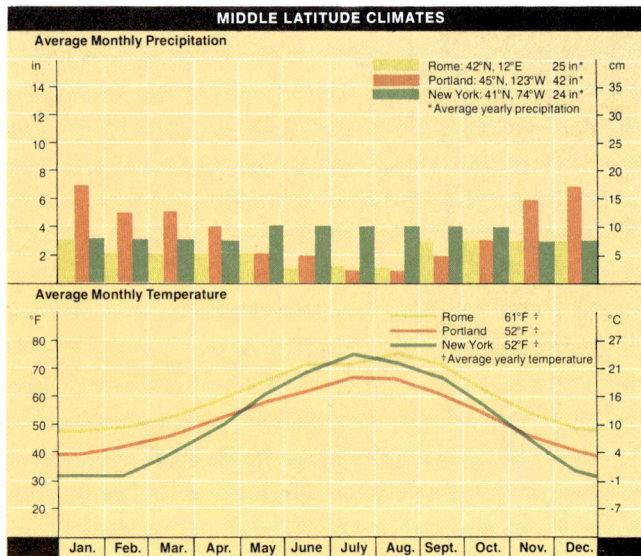

MIDDLE LATITUDE CLIMATES

Average Monthly Precipitation

Rome: 42°N, 12°E 25 in*
Portland: 45°N, 123°W 42 in*
New York: 41°N, 74°W 24 in*
*Average yearly precipitation

Average Monthly Temperature

Rome 61°F †
Portland 52°F †
New York 52°F †
†Average yearly temperature

Jan. Feb. Mar. Apr. May June July Aug. Sept. Oct. Nov. Dec.

Understanding Climate as Average Weather

Climate gives us a general, or average, view of weather. To arrive at this average view, temperature and precipitation figures are evened out for days, months, and years. Daily averages are evened out into monthly averages. Monthly averages are evened out into yearly averages. All the years over a twenty- or thirty-year period are evened out to give the average picture of a climate's temperature and precipitation.

Yearly averages. The *average yearly temperature* figure tells us how heat energy is balanced over the years. It tells us how warm or cold a place usually is. The *average yearly precipitation* figure balances wet years with dry years. It gives us a general idea of how humid or how dry a place is over the years.

Monthly averages. Monthly averages show seasonal patterns. *Average monthly temperatures* show us which months are usually hot and whether or not some months are cool or cold. *Average monthly precipitation* figures indicate how precipitation is spread out over the year. They tell us whether precipitation falls all year long or mainly in one season.

Temperature range. Range—the average difference between the highest and the lowest temperature—can be figured on a daily, monthly, or yearly basis. The *average yearly temperature range* tells us the difference between the highest average monthly temperature and the lowest average monthly temperature. In New York City, for example, the hottest month is July. In this month, temperatures average 73° F (23° C). The coldest month is January. Its temperature average is 26° F (3° C). What is New York City's average yearly temperature range?

Weather averages and graphs. Differences between climate regions are very easily seen when average weather figures are plotted on graphs, such as those on pages 202 and 203.

Look at the top graph, on page 202. It

shows weather averages for three places in the low latitudes. Each place is in a different climate region. Notice that there are two parts to the graph. The lower part of the graph shows temperature averages for each place. The upper part of the graph shows precipitation averages for the same places. The key in the upper part of the graph also shows the latitude and the longitude of each place.

Using the latitude and the longitude given for each place, locate its climate region on the world climate map. In what climate region is Belém? Calcutta? Jacobabad? What is the average yearly precipitation for each place? What is each place's average temperature? Which place has the highest average yearly temperature range? Which has the lowest? What are some other things you can tell about each of these three climate regions from the graph?

Now examine the graph showing three places in the middle latitudes. Locate each place in its climate region, using the world climate map. In what climate region is Rome? Portland? New York? What is Rome's lowest average monthly temperature? Its highest? Which is greater—Rome's average yearly temperature range or New York's?

The graph on page 203 shows two places in the high latitudes. In what climate is Yakutsk found? Sagastyr? Of these two climate regions, which one receives the lower amount of precipitation? Does Yakutsk usually receive most of its precipitation in the winter or in the summer months?

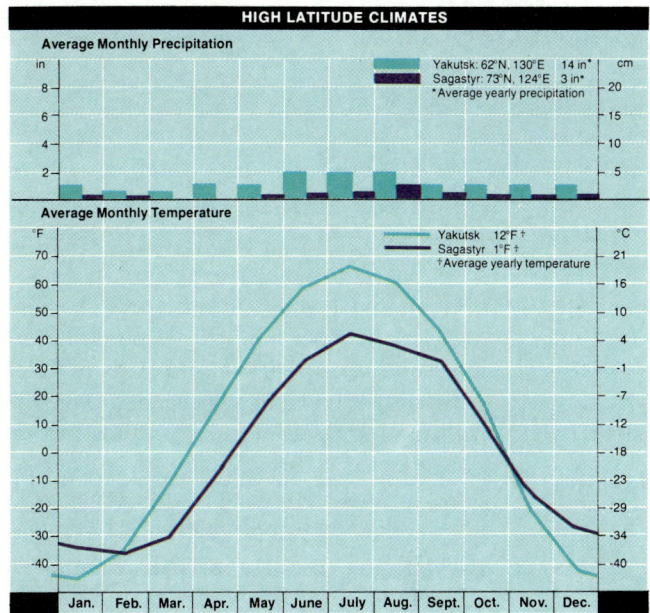

HIGH LATITUDE CLIMATES

Average Monthly Precipitation

Yakutsk: 62°N, 130°E 14 in*
Sagastyr: 73°N, 124°E 3 in*
*Average yearly precipitation

Average Monthly Temperature

Yakutsk 12°F †
Sagastyr 1°F †
†Average yearly temperature

Jan. | Feb. | Mar. | Apr. | May | June | July | Aug. | Sept. | Oct. | Nov. | Dec.

Keeping Facts in Focus

1. In what latitude region would you expect to find climates that are hot all year long? That are cold all year long? That have seasonal differences in temperature?
2. Which usually brings more precipitation to an area—air with high pressure or air with low pressure?
3. Which is usually drier—air that moves over water or air that moves over land?
4. At what season of the year does air over a large landmass usually have a lower pressure than air over a large water body?
5. What answers would you give to the key questions on page 194?

Working with Ideas ·

1. Why is it possible for two cities located along the same line of latitude to have different climates?
2. Why does the windward side of a mountain receive more precipitation than the leeward side?

Lesson 3 Vegetation Regions of the World

Reading Focus

1. Climate maps and vegetation maps are special-purpose maps. Like political maps, they show boundaries. Political boundaries separate different kinds of political divisions exactly. On one side of a political boundary is one division, such as a country or state. On the other side is another political division. The boundary lines on climate and vegetation maps look as if they also separate the earth into exact divisions, or regions. But, like the lines on many other special-purpose maps showing natural regions, they do not. One natural region gradually merges into another. Tropical forests gradually give way to grasses. A wet region gradually becomes a dry region. Boundary lines on most special-purpose maps are really average lines. Keeping this idea in mind will help you interpret special-purpose maps more accurately.

2. As you read this lesson, look for answers to these key questions:
   a. Why is a natural vegetation map useful?
   b. What are the four major kinds of vegetation regions into which the land areas of the earth are divided?
   c. What determines the kind of vegetation that grows at any one place in a highlands area?

Vocabulary Focus

evergreen	*savanna*	*steppe*	*permafrost*
broadleaf	*deciduous*	*prairie*	*lichen*
needleleaf			

Natural vegetation is the kind of plant life that grows wild. It is the kind that grows without the help of people. The natural vegetation regions of the world are shown on the map on pages 206–207. Take a look at this map.

In studying this map, you must remember that it is a model. As a model, it gives us a picture of what vegetation regions would be like if people had not cut down forests or plowed under grasslands. Even though it shows vegetation growing in places where cities and farms exist today, a map such as this is useful. It helps us to see the close relationship between climate and vegetation. By comparing vegetation regions with the climate regions shown on the map on pages 200–201, you can see this relationship.

Look at the vegetation map. Notice where evergreen broadleaf forests grow. Then look at the climate map. Notice that these forests grow in regions that have a tropical wet climate. Now check the maps for other comparisons. Mixed forests grow in regions with a continental, marine, or humid subtropical climate. Needleleaf forests grow in regions with a marine or subarctic climate. Scrub forests grow in a Mediterranean climate region. Savanna vegetation is typical of places with a tropical wet-and-dry climate. Grasses are natural to regions with a semiarid climate. Finally, notice the desert, tundra, polar, and highlands vegetation regions. In what climate do we find each of these regions?

Forestlands

More of the world's land was once covered with forests than with any other kind of vegetation. Today, forests cover only about one fourth of the world's land surfaces.

Evergreen broadleaf forests. Forests are named according to the kinds of trees that grow in them. Trees in tropical wet climate regions have broad, flat leaves. The trees are also green all year round. There is never a time when evergreen trees are without leaves. When a leaf falls off, a new one grows in. Evergreen broadleaf forests are also called rain forests. They are called rain forests mainly because they grow in regions where the air is very hot and filled with moisture.

In a rain forest, the tallest trees grow to over 200 feet (60 m). Their leaves form a canopy, or roof, that almost completely shades everything below it. As the plants beneath the canopy struggle to find sunlight, they grow to different heights, forming layers. Shorter trees form one or two layers immediately below the canopy. Woody bushes form

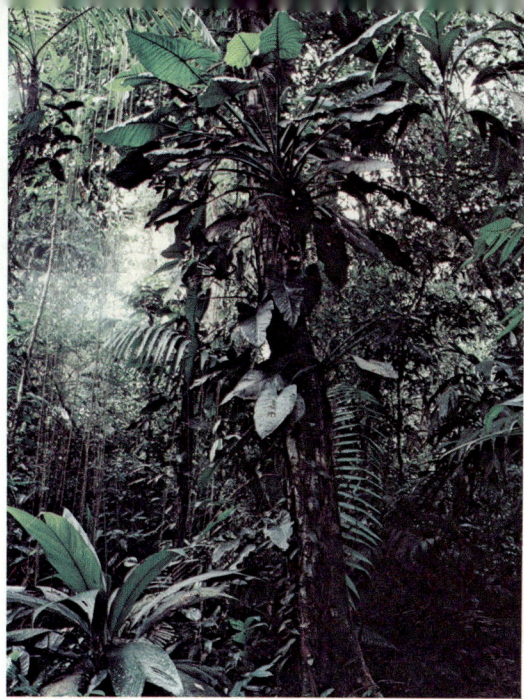

How many layers do you see beneath the canopy in this Costa Rican rain forest?

a still lower layer. Low-lying, shade-loving plants usually form the lowest layer—the forest floor.

Where hardly any sunlight reaches the forest floor, there are open spaces and clear paths between the tree trunks. Where sunlight does reach the forest floor, a tangled growth of bushes, vines, and shorter trees rises to the canopy. This tangled growth is found mainly along riverbanks. It is also found where tall trees have been felled or where clearings have been made and then abandoned.

In one acre of rain forest, there may be as many as sixteen different kinds of trees. Many of the trees are valuable, either for their wood or for the products they yield. One kind of palm tree produces a wax that has many uses. Teak, ebony, and mahogany trees are logged for their fine wood. But the different kinds of rain forest trees grow singly and at some distance from each other. For this reason, gathering products

Each vegetation region takes its name from the type of plant community that is most dominant. Grasses may be found in forests and groves of trees in grasslands. But trees are dominant in forestlands. What vegetation dominates grasslands? Desert lands? The tundra?

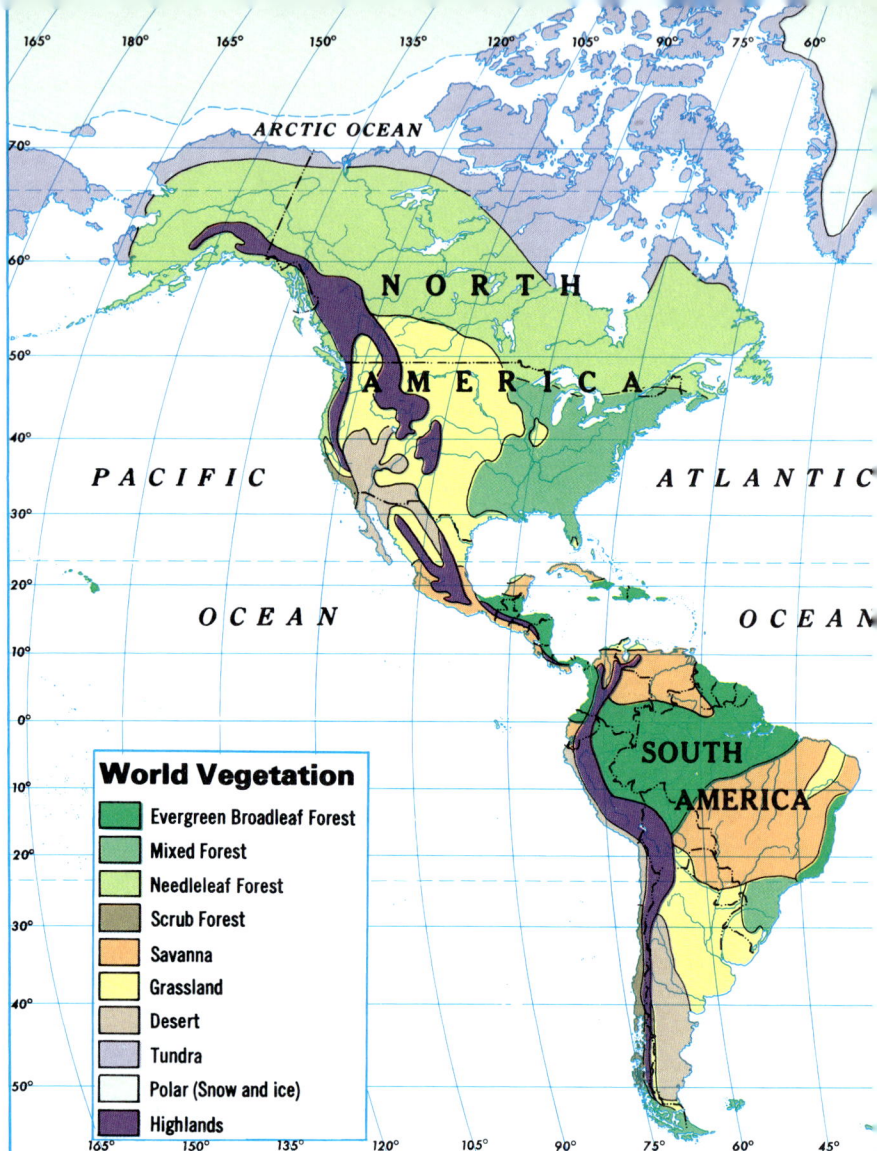

World Vegetation

- Evergreen Broadleaf Forest
- Mixed Forest
- Needleleaf Forest
- Scrub Forest
- Savanna
- Grassland
- Desert
- Tundra
- Polar (Snow and ice)
- Highlands

from rain forests is often difficult, time-consuming, and costly.

Farming, too, is difficult. To make clearings for crops, people must cut down the tall trees. They must burn away any undergrowth. But once they plant crops—such as corn, squash, and beans—the soil begins to lose its fertility. With the canopy gone, the heavy rains soak into the ground. The water leaches, or drains away, the minerals needed to keep the soil fertile. After a number of years, the clearing must be abandoned and a new one begun.

However, there are places in tropical wet climate regions—mostly in Central America and in East Asia—where volcanic ash has made the soil rich in minerals. There, farming is less difficult, and farms are often much larger. Banana, rubber, and other trees are grown on plantations. Tea and rice are grown on both plantations and smaller farms.

Needleleaf forests. Needleleaf trees are the chief form of plant life where climates are cold for much of the year. Most needleleaf trees are evergreens. But unlike the evergreens of the rain forest, needleleaf trees have very narrow or spike-like leaves. The narrower

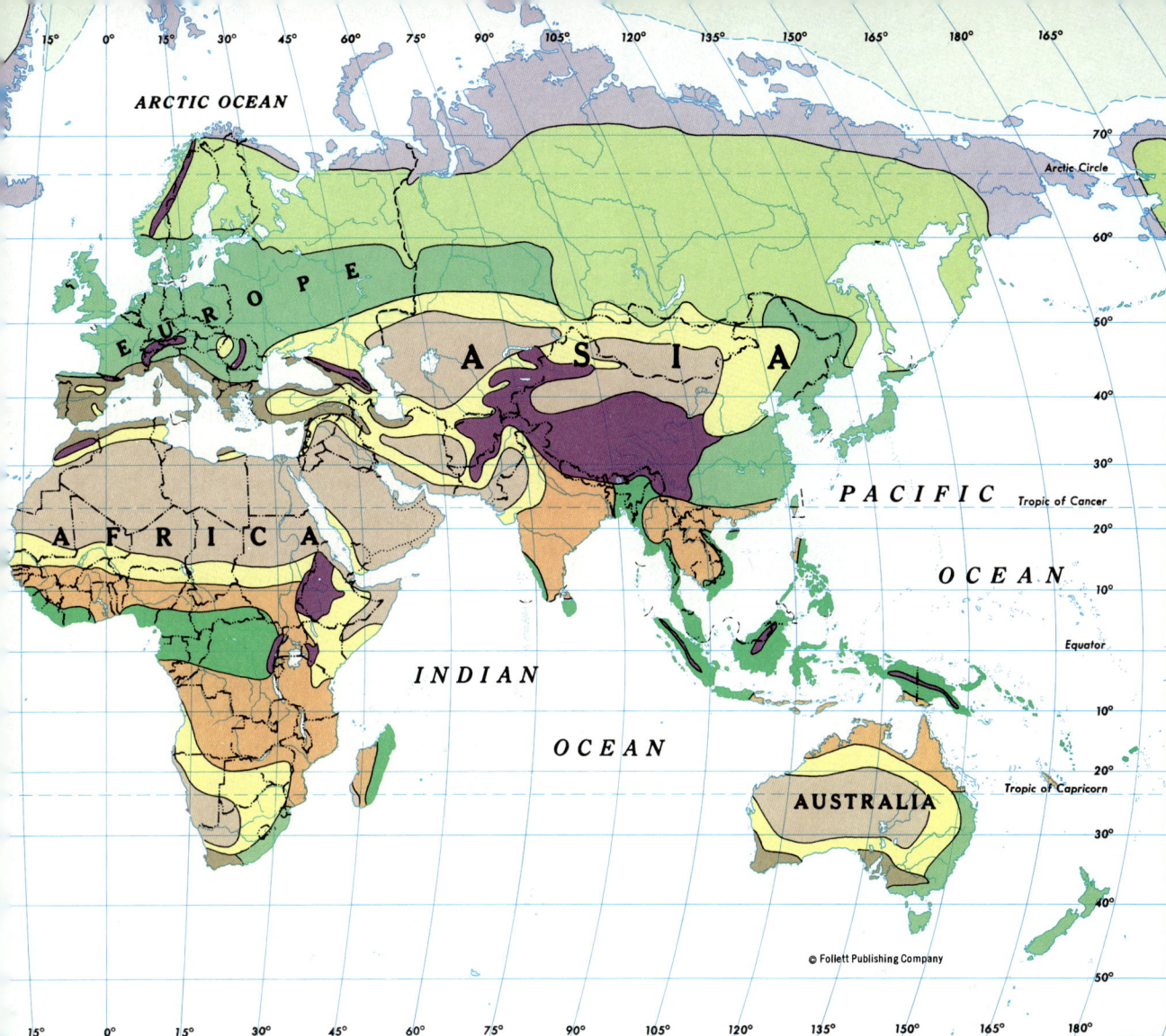

leaves allow the trees to survive in the colder and drier areas of continental climates and in subarctic regions.

In a needleleaf forest, the same kinds of trees grow close together. Logging these trees is easier than logging trees in a rain forest. Often called softwood trees, some common needleleaf trees are pine, spruce, fir, hemlock, and juniper. Softwood trees are often turned into wood pulp and then into paper, plastics, rayon, or other products. People also log softwood trees for lumber to build homes or to make boxes, crates, telephone poles, and railroad ties.

Mixed forests. In mixed forests, stands of broadleaf trees and stands of needleleaf trees grow near one another. The needleleaf trees are usually evergreens. The broadleaf trees, however, are mostly deciduous. This means that the trees lose their leaves either in a cold season or in a dry season. In a mixed forest, deciduous trees usually drop their leaves in the fall of the year and remain bare until spring comes.

Some common broadleaf trees are oak, hickory, maple, beech, tupelo, elm, and chestnut. Often called hardwood trees, broadleaf trees provide fine wood

207

In Sequoia National Park, California, a mature redwood with its deeply furrowed bark and limbless trunk dwarfs other needleleaf varieties.

for furniture and other wood products. Many nuts and fruits also come from hardwood trees. Find the mixed forest regions on the vegetation map. Notice that the largest areas are found in the Northern Hemisphere.

Scrub forests. Found chiefly in Mediterranean climate regions, scrub forests have woody bushes and short trees. Thorns protect many of the trees and bushes from grazing animals. Olive, live oak, cork oak, and eucalyptus are some common trees found in scrub forests. In the United States, the vegetation of scrub forests is often called *chaparral* (shap·ə·ral′). In Western Europe, it is often called *maquis* (mə·kē′).

Grasslands

Most trees need good amounts of precipitation all year long. In places with a dry season or in places where precipita-

tion is less than 20 inches (50 cm) a year, trees find it hard to survive. Grasses, however, are hardy plants. They can grow in almost every climate region. But they thrive best in the tropical wet-and-dry climate regions and in the semiarid climate regions of the world.

Tropical grasslands. Grassland regions in the tropics are called savannas. They have single trees or patches of trees widely scattered throughout the region. The grasses in the savanna are 8 to 10 feet (2 to 3 m) tall. They are also tough and very coarse. In the wet season, the heavy rains keep the grasses green and tender. But in the dry season, the rainfall drops off to less than 2 inches (5 cm) a month. Then the grasses wilt, turn brown, and become brittle.

Fires, set by lightning or by people, are common sights in grassland areas, especially toward the end of the dry season. The fires burn off the dead stems and make room for the new shoots that spring up when the wet season arrives.

Middle latitude grasslands. On the map on pages 206–207, find the regions labeled *grasslands*. In the middle latitudes, wild grasses grow in many different sizes. Where yearly precipitation is under 20 inches (50 cm), grasses grow only a few inches high. These grassland regions are called steppes. Where precipitation is greater than 20 inches, the grasses grow anywhere from 4 to 10 feet (1.2 to 3 m) tall. Areas with such taller grasses are called prairies.

In steppe regions, grazing is usually the chief occupation of the people. Even though the soils are fertile, precipita-

A baobab tree with its thick, stubby trunk stands amid the trampled grasses, woody shrubs, and scrawny trees of a savanna in southeastern Kenya. What tells you the tree is a deciduous one?

tion is light and varies in amount from year to year. So crops can be grown only where irrigation is possible. In prairie regions, precipitation is heavier and more reliable from year to year. So crops are grown on a much larger scale. The fertile soils of the prairies now grow much of the world's corn, wheat, and other cereal grains.

Desert Lands

Find the desert vegetation regions on the map on pages 206–207. Deserts are very dry and windy places. In some places in a desert, there is no soil—only rocks, pebbles, or sand. In other places, there is soil. But it is usually too thin to support much plant life. Or, it is too heavy with chemicals that are harmful to most plant life. Even so, there are plants that can grow in deserts.

Some plants that grow in deserts are deciduous. These shrubs shed their leaves during long dry periods and grow them quickly again when cloudbursts take place.

Evergreen plants also grow in deserts. Some have thick and leathery leaves. These plants survive because their leaves can withstand the sharp changes in temperature that are common to most desert regions. Others have leaves with shiny or waxy surfaces that limit evaporation. Still others, such as the cactus, do not grow leaves. Instead, they grow thorny stems in which they store water.

There are also many flowering plants that grow in desert areas. Some of these plants lie dormant in dry seasons and spring to life again when rain comes. There are others that die in a dry season. But they drop seeds that lie on or in the ground. When rain comes, the seeds take root. The plants sprout quickly and bloom. They die just as quickly when the moisture they need evaporates under the glaring sun or in the ever-blowing winds. But before they die, they drop more seeds.

In a southern Arizona desert, prickly pear cactus plants and desert shrubs surround a giant saguaro cactus. Why do you think the sky over the desert is so cloudless?

What does this summertime photograph tell you about life in the highlands climate of the Swiss Alps below the timber line? Above the snow line?

Tundra Lands

Find the tundra vegetation regions on the map on pages 206–207. The tundra's vegetation comes to life in a short period lasting about three months or less. During this short growing season, the top layer above the tundra's permafrost (soil permanently frozen) thaws to a depth of a few feet. When it thaws, the soil remains very marshy and is not very fertile. Only dwarf shrubs, marsh grasses, mosses, lichens, and tiny flowering plants dot the land. Tundra land is largely unoccupied except for scattered groups of people, such as Eskimos who fish in Arctic waters and Lapps who herd reindeer.

During the long days of summer, the sun melts the snow cover of the tundra. Because the subsoil remains frozen, the surface in many places is wet and marshy. What kind of vegetation do you see in this part of Alaska's tundra?

Highlands

Vegetation in the highlands, like climate, changes with elevation and with location on the windward or the leeward side of a mountain range.

On the lower slopes of highlands, vegetation usually resembles that of the nearby lowland. As elevation increases, the vegetation changes, much as vegetation at sea level changes with latitude, or distance from the equator.

In the low latitudes, for example, vegetation may begin as rain forest on a windward slope, change to needleleaf forest at higher levels, and to grassy meadows above the forests. On the drier, leeward side, vegetation is likely to begin as scrub forest or savanna and then change to shorter grasses or desert-like vegetation.

Each highlands area, depending on its latitude, has an elevation above which trees cannot grow. This is the timber line. In the low latitudes, the timber line often lies above 10,000 feet (3,000 m). At high latitudes, it usually lies below this altitude. Above the timber line and below the snow line (the elevation above which snow lies all the time), only tundra-like plants may grow in scattered places.

Keeping Facts in Focus

1. What is the main difference between an evergreen and a deciduous tree? Between prairie grasses and steppe grasses?
2. What is the timber line? The snow line?
3. What answers would you give to the key questions on page 204?

Working with Ideas

1. How do animal and human activities change the native vegetation in an area?
2. National, state, and local governments sometimes put aside sections of the natural environment as parks or wildlife preserves. Why do governments find it necessary to do this?

Unit 5   REVIEW WORKSHOP

Test Your Geographic Knowledge

Choose the THREE endings that complete each statement correctly.
1. A climate region found in the low latitudes is (a) desert, (b) highlands, (c) continental, (d) tropical wet.
2. One of the things that shapes climate is (a) prevailing winds, (b) altitude, (c) ocean currents, (d) tornadoes.
3. A low pressure center forms (a) where air is sinking, (b) over land in summer, (c) over water in winter, (d) over the equator.
4. A rain forest usually has (a) a long dry season, (b) a canopy, (c) plants growing in layers, (d) hot, humid air.
5. A desert is sometimes located (a) on the leeward side of a mountain range, (b) where rainfall is greater than 10 inches (25 cm), (c) where cool currents flow offshore, (d) around 30° north or south latitude.
6. A region that is located on the west coast of a continent may have a climate that is (a) subtropical, (b) marine, (c) desert, (d) Mediterranean.

Apply Your Reading Skills

A. Study the sample index listed below. Use its listings to tell whether the statements that follow it are true or false.

> **Mediterranean lands,** 82–
>   121, 126, 130, 137, 170,
>   327, 373, 401, 407, 411
>   agriculture, 149–152, 157
>   climate, 83
>   vegetation, 85
> **meridians,** 23–24, 26
> **Middle East,** 397–415
>   geography, 397–400
>   nations of, 401–415
>   physical environment,
>   397, 398
>   problems of, 399
> **monsoons,** 366, 367, 384,
>   385

1. References to monsoons can be found on four pages in the book.
2. You will find information about meridians by reading pages 23, 24, and page 26.
3. On every one of the pages from 149 through 157, you will find information about the kinds of agriculture practiced in Mediterranean lands.
4. Pages 397 through 415 contain only information about the geography of the Middle East.
5. To determine what nations are a part of the Middle East, you can skim pages 401 through 415.

B. Use encyclopedias, books, and other reference tools to do research. Then prepare a report on ONE of the following topics.
1. What are some of the things people are discovering by living on Antarctica?
2. What are tree farms and how will they help to preserve forest resources?
3. What are some of the animals that live in tropical forests?
4. What was the earth like when ice covered large parts of it?
5. How do Eskimos survive in the tundra?

Apply Your Geographic Skills

Use the information in the chart on page 213 to correctly answer the following questions.
1. What is the world's average yearly temperature?
2. What is the world's average yearly precipitation?
3. What is the highest temperature ever recorded in the world? The lowest?
4. What is the world's greatest one-year rainfall total? Where was it recorded? When?
5. What continent holds the record for the lowest recorded temperature and the world's lowest average yearly temperature?

Apply Your Thinking Skills

You can often draw conclusions from evidence, or information, given in a paragraph. Using the information given in the paragraph below, tell whether each of the conclusions that follow it is true, false, or doubtful (a conclusion is doubtful if evidence is lacking).

In the northeastern part of Bolivia, there are tropical lowlands where rainfall is high

WORLD CLIMATE AVERAGES AND WEATHER EXTREMES

Climate Averages

Yearly Average Temperature	about 59° F (15° C)	
Highest Yearly Average Temperature	94° F (34° C)	Dallol, Ethiopia
Lowest Yearly Average Temperature	−70° F (−57° C)	Plateau Station, Antarctica
Yearly Average Precipitation	40 in (102 cm)	
Highest Yearly Average Precipitation	472 in (1,199 cm)	Mt. Waialeale, Hawaii
Lowest Yearly Average Precipitation	0.03 in (0.08 cm)	Arica, Chile

Weather Extremes

Highest Recorded Temperature	136° F (58° C)	El Azizia, Libya, 9/1922
Lowest Recorded Temperature	−127° F (−88° C)	Vostock, Antarctica, 8/1960
Highest One-Year Total Precipitation	1,042 in (2,647 cm)	Cherapunjii, India, 1860-1861
Highest 24-Hour Total Precipitation	74 in (188 cm)	Réunion Island, 3/1952

throughout the year. There are other tropical lowlands in the southeastern part of the country where the rainfall is seasonal. Moving from north to south in these lowland areas, the vegetation changes from rain forest to scrub forest to savanna. To the west of the lowlands stretch a range of the Andes. Rich tropical valleys—where cacao, sugar, and other tropical crops are grown—nestle between high peaks. On the treeless, windswept Altiplano—a high plateau rising 13,000 feet (4,000 m) above sea level—there is little farming.

1. There are three climate regions in Bolivia—tropical wet, tropical wet-and-dry, and highlands.
2. Many of Bolivia's large cities lie in the fertile valleys.
3. Mining is an important industry in the Andes.
4. The tropical wet climate region is found in the northeastern part of the country.
5. The Altiplano is part of the highlands climate region.
6. The lowlands of Bolivia all have the same climate.

Discuss These Points

1. If you were planning a two-week trip to a place in Canada, what average weather figures would you look up to help you plan the kinds of clothing you would take along on your trip?

2. How does the keeping of weather records for many places around the world help us to understand how climates differ and how they change?
3. How does the climate of the area where you live place limits on the kinds of choices you can make in your everyday activities?

Expand Your Geographic Sights

Ames, Gerald and Wyler, Rose. *Story of the Ice Age* (Harper-Row, 1956).

Harrison, C. William. *Forests* (Messner, 1969).

Hicks, Sam. *Desert Plants and People* (Naylor, 1966).

Jenners, Aylette. *Dwellers of the Tundra: Life in an Alaskan Eskimo Village* (Macmillan, 1970).

Johnson, Sylvia A. *Animals of the Tropical Forests* (Lerner, 1976).

Lyle, Keith. *Weather and Climate* (Silver, 1984).

Naden, Corinne J. *First Book of the Grasslands Around the World* (Watts, 1970).

Shannon, Terry and Payzant, Charles. *Antarctic Challenge: Probing the Mysteries of the White Continent* (Children's Press, 1973).

Silverberg, Robert, *The World of the Rain Forest* (Hawthorn, 1967).

Updegraffe, Imelda and Updegraffe, Robert. *Continents and Climates* (Penguin, 1983).

Unit 6 Food and Farming

In the clear air of India's dry monsoon period, a worker winnows grains of wheat. Using the wind to separate the grain kernels from their outer coverings probably dates back to the first wheat growers.

Meeting Food Needs

Reading Focus

1. Newspapers and magazines often print articles about food, nutrition, or dieting. Many of these articles give facts and opinions on all sides of a question, expressing many viewpoints. But other articles express only one point of view. The authors of these articles may use only certain facts and omit any that do not support the expressed viewpoint. Bring in an article on food, nutrition, or dieting. Discover whether the article has expressed one or several points of view. Check the facts used to back up any viewpoint. Try to discover if any important facts have been omitted. Check to see what conclusions have been drawn. See if there is enough evidence to back up a viewpoint. Then decide which, if any, viewpoint you agree with.

2. As you read this lesson, look for answers to these key questions:
   a. What are the five essential nutrients each person needs?
   b. What causes malnutrition?
   c. What kinds of conditions may influence the ways people meet their food needs?

Vocabulary Focus

carbohydrate	*protein*	*mineral*	*malnutrition*
fat	*vitamin*	*calorie*	

Food is one of the basic needs shared by all people. Unfortunately, more than half of the world's people regularly fail to meet their food needs.

In this lesson you will learn what our food needs are. These needs are not the same for everyone, and many factors determine each person's needs. You will also learn about the serious consequences people must face when they fail to satisfy their food needs.

How Food Needs Are Met

To meet food needs people must (1) eat enough food and (2) eat the right kinds of foods. Food must do more than satisfy hunger pains. It must supply the body with five essential nutrients. These nutrients are carbohydrates, fats, proteins, vitamins, and minerals.

Calories. Food gives us energy. A calorie is the unit used to measure the energy in a given amount of food. Ener-

AVERAGE DAILY CALORIE NEEDS		
Age	**Boys or Girls**	
1–3	1,300	
4–6	1,800	
7–10	2,400	
	Men	**Women**
11–14	2,800	2,400
15–22	3,000	2,100
23–50	2,700	2,000
50+	2,400	1,800

gy needs differ from person to person. So do calorie needs. Calorie needs depend on a person's sex, age, height, weight, and degree of physical activity. A man generally needs more calories than a woman, a boy more than a girl. A growing child needs more calories than an old person, a tall person more than a short person, and a manual laborer more than an office worker.

Calorie needs also differ according to the temperature of the region in which a person lives. People in tropical regions generally need fewer calories than those in polar regions. In the middle latitudes, a person needs fewer calories in the summer than in the winter. The chart above shows the average number of calories needed each day for fairly active persons of either sex and of different ages living in a mild climate of the middle latitudes.

Essential nutrients. Study the chart on page 218 that shows the roles each of the five essential nutrients plays in our bodies. Notice the kinds of food that contain each of these nutrients. Which two nutrients are our chief sources of energy? Why do we need to eat foods that are high in protein? What are some animal foods that are high in protein? Some plant foods? What are two minerals and four vitamins needed by our bodies? What roles do these minerals and vitamins play in the proper functioning of our bodies?

What is needed for good nutrition is a variety of nutrients consumed over a period of several days or several meals. Over a period of time, about 45 percent of the food we eat should be carbohydrates, 40 percent fats, and 15 percent protein. People who balance their diets in this way will also receive most of the 45 minerals and vitamins that are now known to be needed for good health.

Hunger and Malnutrition

Of all the people in the world, less than 30 percent may meet their food needs. About 500,000 people a year, 10,000 a week, or 60 an hour die from the effects of hunger—the failure to eat enough food—or from malnutrition—the lack of one or more of the essential nutrients.

Starvation is the most severe form of hunger. Starvation usually takes place during times of famine—periods when droughts, plagues of insects, earthquakes and other natural disasters, or the upheavals of war destroy food crops or farmland. Hunger saps the body of its strength, weakening its ability to ward off diseases. Some people die of actual starvation, but many more die from diseases that take hold in their hunger-weakened bodies.

ESSENTIAL NUTRIENTS—THEIR ROLES AND SOURCES

Essential Nutrients	Roles in the Body	Food Sources
Carbohydrates	The sugars and starches that are carbohydrates give the body energy to operate muscles and nerves.	Sugarcane, sugar beets, honey, milk, and most fruits are sources of sugar. Starches are found in grains, potatoes, dried beans, and peanuts
Fats	Fats furnish more than twice as much energy as the same amount of carbohydrates. The body uses fats to form protective tissues around vital organs.	Meat, cream, butter, margarine, shortening, salad oils or dressings, cheese, olives, nuts
Proteins	The body uses the amino acids in proteins to build body and muscle tissue, to repair damaged cells, and to keep all body processes working. Protein also helps the body to fight off disease and infection by producing antibodies in the bloodstream.	Meat, poultry, fish, milk, cheese, eggs; also soybeans, dried beans, oatmeal, potatoes, and peanuts
Minerals	The body needs small amounts of a number of different minerals. These minerals help the body to use carbohydrates, fats, and proteins efficiently. Two examples of needed minerals are:	
	Calcium builds strong bones and teeth and aids blood circulation and the workings of the nervous system.	Milk, cheese, green leafy vegetables—especially collard, mustard, or turnip greens; also salmon and sardine bones
	Iron combines with protein to make red blood cells.	Liver and shellfish; also enriched breads and cereals and dark green leafy vegetables
Vitamins	The body cannot manufacture vitamins. So it must get them from food or from vitamin preparations. Some examples are:	
	Vitamin A helps to keep vision normal and skin healthy.	Green and yellow vegetables, peaches, cantaloupe, tomatoes; also milk and cod-liver oil
	Vitamin B_1 aids the appetite and digestion and enables the body to use carbohydrates.	Meat; also whole grain cereals or breads
	Vitamin C holds body cells together, keeps gums healthy, and aids in the fight against infection.	Citrus fruits, potatoes, tomatoes, peppers, broccoli, raw cabbage
	Vitamin D enables the body to use calcium.	Eggs, cod-liver oil, liver; also exposure to sunlight

Fruits and vegetables are an important part of the diets of many people in most areas of the world. What fruits or vegetables sold in this open-air market in Bombay, India, can you identify?

About 60 percent of the world's people probably suffer from some form of malnutrition. Where diets consist mostly of cereal foods high in carbohydrates, people are likely to suffer from protein malnutrition. Many plant foods, as well as animal foods, contain protein. However, the kind of protein a food contains is more important than the amount. No one plant food contains all of the amino acids that are the building-block materials that make up protein. To get all of the building-block materials needed to

PERCENTAGE OF CALORIES IN DIETS FROM PLANT AND ANIMAL PROTEIN FOODS

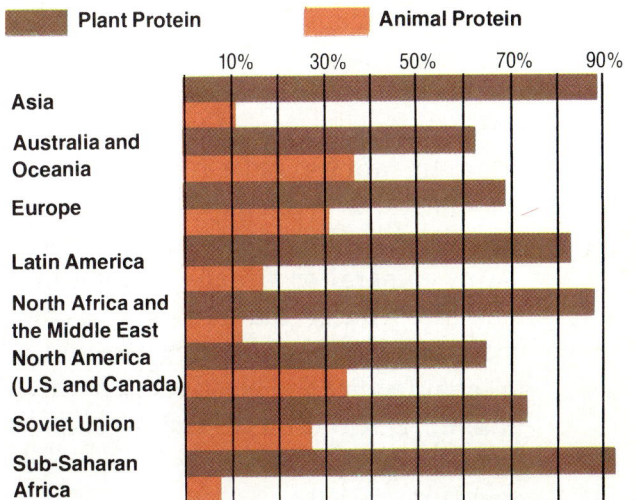

Plant Protein   Animal Protein

10%   30%   50%   70%   90%

- Asia
- Australia and Oceania
- Europe
- Latin America
- North Africa and the Middle East
- North America (U.S. and Canada)
- Soviet Union
- Sub-Saharan Africa

219

In the near future, more high-rise buildings may stand on the land occupied by a rural family of Jamshedpur, India. In what ways is city expansion an advantage? A disadvantage?

build or repair body tissues, people must eat a variety of plant foods or include such animal foods as meat, fish, milk, cheese, and eggs in their diets. Protein requirements are greater for people who eat mainly plant foods.

Protein malnutrition bloats parts of the body and damages internal organs. It also turns hair and skin a reddish color. In infants and children, lack of protein stunts both mental and physical growth and often causes death.

Any mineral or vitamin lacking in a diet also causes malnutrition. A diet without iron robs the body of its ability to manufacture red blood cells. Without vitamin D, children develop rickets—a disease that deforms or weakens bone structure.

Why Food Needs Are Not Always Met

It should not be surprising that there are places in the world where people do not meet their food needs. To produce food, people depend on the natural resources of their environments. Land level enough and fertile enough to farm profitably, water supplies, climates favorable for farming—all these resources are distributed unevenly over the earth. Subtract the mountainous land, the land that is too wet or too dry for farming, the polar regions, and the lands used for cities and towns. Subtract the land used for other things. What is left for herding and farming is only about a third of the world's land area.

Only with irrigation can corn be grown on this land in the province of Alberta, Canada.

What effects did a hard frost have on an orange grove in Florida?

Many things influence the choices people make about what crops to grow and what animals to raise on the world's farmland. Natural conditions influence some choices. Cultural, economic, or political conditions influence others.

Natural conditions. Each crop grown in the world has a certain set of natural conditions under which it grows best. Two important natural conditions are soil makeup and water supply. Soil makeup, you recall, includes particle size, air circulation, moisture content, and mineral content. Water supply depends on the depth of the water table, seasonal differences in precipitation, drainage characteristics of the land, and the development and use of irrigation technology.

A third natural condition is the length of the growing season—the period between the last killing frost of spring and the first killing frost of fall. Most crops must be planted, cultivated, and harvested within this period. How long a growing season is depends mostly on latitude and the differences in altitude of places within a latitude region. Growing seasons vary from year long within the low latitudes to less than three months for those farming areas that stretch into the high latitudes. (For maps showing growing seasons, see pages 21 and 31 of the Atlas.)

Still other natural conditions influence farming and herding. Important among these are natural enemies—the diseases and pests that damage growing

Many farmers in India grow sugarcane in small patches and process it themselves for home use.

crops or livestock. The tsetse (tet′ sē) fly afflicts cattle in Africa's savanna regions. In corn-growing regions of the United States, the corn borer is one of 350 kinds of insects that feed on the roots, stalks, leaves, or kernels of corn. Fungus diseases, such as rot, spoil wheat and other grain crops.

Cultural and other conditions. The farming and food practices of groups of people are influenced by cultural, economic, and political conditions.

Cultural conditions include all those things that are a part of a group's pattern of living. Values, attitudes, beliefs, and tastes are some of these things. Depending on what they value, believe, or the attitudes they form, some groups of people refuse to accept changes in farming practices that will increase food supplies. Or they refuse to change eating habits to improve nutrition and health.

Economic conditions include all those things that influence how people meet their needs and wants. Some farm products are in great demand the world over. These products usually have a high market price. Where people produce such products, they often make a profit. The profits can be used to increase standards of living and to improve farming practices.

With better farming practices, people produce more food with less work. Countries that produce surplus food can afford to develop other sources of income—to develop industry. In countries where farmers barely grow enough food to feed themselves, let alone other people, no money is available to buy such things as fertilizer, machinery, or improved seeds. The majority of the people become locked into a life of poverty from which they see little chance to escape.

Land ownership patterns also influence a group's ability to furnish food. In some areas of the world, such as parts of Latin America and Asia, fewer than 10 percent of the people own most of the land. In these regions, the majority of people who farm the land are sharecroppers—people who work the land for a share of the crop. Or they are tenant farmers who rent the land. In either

The Food and Agricultural Organization

An organization formed within the United Nations, the Food and Agricultural Organization (FAO), has the job of wiping out hunger in the world. Through its activities, it promotes good nutrition and offers aid to farmers throughout the world.

In Sri Lanka, agriculturalists of the FAO have conducted experiments with cover crops to preserve and increase soil fertility during dry periods. In Thailand, experts have shown the people how they might raise fish in flooded rice fields, canals, and swampy land. In Latin America and other parts of the world, they have helped stem outbreaks of hoof-and-mouth disease in cattle, taught farm management courses, set up breeding stations to improve herds of livestock, and helped farmers to expand the numbers and kinds of crops they can grow. In Africa and other places, they have conducted programs to wipe out malnutrition diseases.

The news media seldom headline the work of the FAO. Yet, its work in all phases of food production is making progress in helping food supply keep pace with population growth.

In an FAO-sponsored program in Botswana, a young woman learns how to operate a tractor.

case, the land these farmers work is often poor and very small in size.

Wars, political strife within a country, or long periods of control by another country have kept some nations from making any economic progress. Only since the 1960s have a number of countries in Africa and Asia received independence. Their governments often pour money into building up industry and do little to improve agriculture. Yet it is in Africa and Asia that food is shortest in supply and that great numbers of people are unable to meet their food needs.

Keeping Facts in Focus

1. What two basic things must people do to meet their food needs?
2. What is a calorie? What characteristics determine a person's calorie needs?
3. On what two continents is food shortest in supply?
4. What answers would you give to the key questions on page 216?

Working with Ideas

1. How does hunger differ from malnutrition?
2. What are some of the natural, economic, and cultural conditions that have combined to make the United States a great food-producing nation?

Lesson 2 The World's Food Supply

Reading Focus

1. *Supply, demand, one-crop economy,* and *income* are economic geography terms—terms that say something about how people earn their livings and use resources to satisfy needs and wants. Geographers look at *supply* as the amount of a resource available for use and *demand* as the need for a resource. Other factors include the ability and the willingness of people to produce and consume these resources. Many nations in Africa, Asia, and Latin America have a one-crop economy. In a one-crop economy, one major export crop provides most of a nation's income, or money received. When world market demand for a product is greater than its supply, prices paid for it are high. At such times, a nation with a one-crop economy experiences a period when income grows. When world market demand for a product is less than its supply, prices and income decrease sharply. Pay close attention to the meanings of these and other economic terms used in geography, such as those listed in the *Vocabulary Focus* below. Knowing the meanings of these economic terms will help you get a clearer picture of the problems related to producing the world's food supply.

2. As you read this lesson, look for answers to these key questions:
   a. About what percentage of the population is usually involved in farming in industrialized nations? In developing nations?
   b. What kind of farming is found mainly in industrialized nations? In developing nations?
   c. Why does the world's food supply fall far short of world food needs?

Vocabulary Focus

industrialized	*profit*	*developing nation*	*commercial*
cash	*market*	*production*	*cost*
invest			

Feeding more than four billion people is no easy task. That, however, is the job facing the world's farmers. The task is easier for farmers in nations that have a high level of industrial development. It is more difficult for farmers in those nations that are just beginning to develop modern industries.

A highly industrialized nation has a great many sources of income. More of its income is from manufacturing rather than from farming. As few as 10 percent of the people may be directly involved in farming. Most of the farmers sell crops and livestock for cash, investing their profits in fertilizer, machinery, and

Of these three farms—one in France's grape-growing region (top), one in Iowa (lower left), and one in Ecuador—which is an example of a mixed farm? A specialized farm?

better varieties of seeds or breeds of livestock. They make use of modern methods of transportation to get their products to market. The United States, Canada, Australia, New Zealand, many of the countries of Europe, the Soviet Union, and Japan are the major industrialized nations of the world.

A developing nation has some industry. Yet, it still depends on agriculture as the main source of income for its people. In the least developed nations, as much as 70 percent of the population is directly involved in farming. Most of the labor on farms is done by hand or with the help of animals rather than with machines. Only a few crops are sold for cash. Developing nations usually have little money to invest in ways to increase crop production. Most develop-

ing nations are found in Africa, Asia, and Latin America.

Farming in Industrialized Nations

Most of the farmers in industrialized nations practice commercial farming. They grow crops or raise livestock for cash. Commercial farming is either specialized or mixed farming. About 5 percent of all farming in the United States is mixed farming.

Specialized farming. In specialized farming, one crop or one kind of livestock brings in more than half of a farm's income. The crop may be wheat or some other grain, tobacco, cotton, sugarcane, sugar beets, grapes, some other kind of fruit, or vegetables. Specialized livestock farms include large cattle and sheep ranches and smaller poultry, dairy, and fur farms.

Mixed farming. In mixed farming, farmers raise a variety of crops and one or more kinds of livestock. Most of the crops are sold for cash or fed to livestock raised on the farm. On a typical mixed farm in the United States, a farmer may grow corn, alfalfa, soybeans, or other fodder crops.

Farming in Developing Nations

Most of the farming practiced by people in developing nations is subsistence farming. Some subsistence farmers, especially those living in regions with a tropical-wet climate, practice shifting agriculture. In this type of subsistence farming, people abandon farmed land that has lost its fertility and make new clearings in another area. Another kind of subsistence agriculture is nomadic herding. Nomadic herders move about with their herds of cattle, sheep, goats, camels, or other animals in search of water and good grazing land. Nomadic herders often live in arid or semiarid climate regions.

In many developing nations, commercial farming takes place on plantations that often cover large areas. Bananas in Central America, coffee in Brazil or Colombia, tea in Sri Lanka, or rubber in Malaysia are some of the cash crops produced on plantations.

Food Supply and Population Growth

The world's farmers produce billions of tons of food every year. Yet this huge supply of food falls far short of world needs. Each day, the world's population increases by more than 236,000 people. The rate at which population is growing is greater than the rate at which farmers are able to increase food production. Moreover, population is growing fastest in those regions where the food supply is lowest—in the developing nations. Ways must be found to make food supply and population growth keep pace with each other.

Encouraging people in developing nations to limit the sizes of their families is one effort that is being made. However, parents living at the subsistence level often resist this effort. Such parents usually regard children as a means of support. To have someone take care of them in their old age, these parents believe that they must have many children. Experience has taught them that few of their children survive the effects of hunger and disease. When parents in developing nations see more children surviving into adulthood, we may see a slowing of population growth in Asia, Africa, and Latin America.

Another effort that seems attractive to leaders in many developing nations is industrialization. The populations of nations that are highly industrialized grow more slowly than those of developing nations. Highly industrialized nations usually provide enough food for their people. What food their farmers are unable to produce, their people are able to buy from other nations. By concentrating their efforts on building up industry, leaders in developing nations hope to bring about the series of changes

Agriculturalists—Problem Solvers for Farmers

Name a problem relating to food production—an agriculturalist probably had something to do with solving it. Agriculturalists come from many fields of science. They may be geographers who study land and vegetation patterns and their relationships to other parts of the natural environment. They may be biologists or specialists in certain kinds of plant and animal life. They may be soil experts, chemists, or ecologists. Whether working alone or as a member of a team, an agriculturalist has one aim—to increase crop and livestock production.

Agriculturalists in many parts of the world have given us new varieties of grains and better breeds of livestock. They have found ways to control certain insect pests and plant and animal diseases, to increase soil fertility, and to improve farming techniques. More efforts along these lines are in progress. Efforts are also under way to find new crops and new products that can be made from crops. There are even efforts to modify the structures of plants.

In one such effort, agriculturalists are working with kinds of bacteria in the soil that take nitrogen from the air and use it to manufacture a natural fertilizer. Such nitrogen-fixing bacteria are present in nodules attached to the roots of plants in the legume family (alfalfa, beans, peas, and soybeans). Agriculturalists are trying to find out why these kinds of bacteria are present in legumes and not in other kinds of plants. They are experimenting with other plants, trying to modify them so that nitrogen-fixing bacteria will form on their roots. Should these experiments prove successful—and there is promise that they will—grasses with nitrogen-fixing bacteria may someday increase the fertility of soil in the tropics and other places.

that took place in Western Europe, the United States, and other nations during the period from the 1700s to the present.

The growth of factories, mechanization, the development of energy resources, improvements in transportation and food production—all these things and others brought about an increase in the production of goods and services. As standards of living rose, people earned more, spent more, and even saved more money. People ate more nutritious food, had better medical and sanitation services, lived longer, and began to limit the sizes of their families.

In many developing nations, these and other changes are taking place. But they are not happening fast enough or in enough developing nations to keep population growth and food supply in balance. Efforts that attack more directly the problem of increasing food supply where it is most needed must also be undertaken.

Increasing the Food Supply

One direct way of making more food available is to increase the amount of arable land in the world. There is little arable land left that is not already in use. So nations must turn to wasteland—land that is too steep, too cold, too dry, or too wet to grow crops or that is too lacking in vegetation to support livestock. Some wetlands may be reclaimed by draining them. Some dry lands may be reclaimed by irrigating them. Some soils may be reclaimed by fertilizing them. There are places in the world, such as in Israel's dry land, where wasteland has been or is being

Plants for the Future

Scientists have discovered a number of wild plants that can be cultivated profitably.
The five shown here could be used to turn wasteland into farming or grazing land.

Water Hyacinth

Tamarugo Tree

Guayule

Jojoba

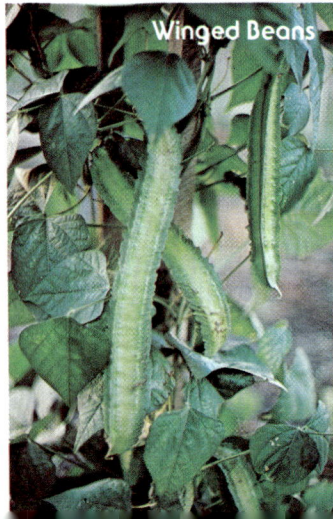
Winged Beans

Long considered a weed with no value, the *water hyacinth* clogs waterways of the wet tropics. New studies, however, have shown it has value. In waterways, it "cleans" water of pollutants. Harvested and placed on farmland, it makes a good fertilizer and a cheap one.

Because of the *tamarugo tree,* sheep now graze in the Atacama Desert and goats in the Canary Islands, two very dry and salty places. Planted in other salt desert regions, tamarugo trees may help develop livestock industries in places where none existed before.

The *jojoba* (hō•hō′bə) and the *guayule* (gwī•ōō′lē) plants grow wild in desert areas of North America. From the jojoba plant, we can obtain an oil that is remarkably like that obtained from the sperm whale. The guayule plant contains up to 26 percent rubber. Its use would reduce the world's dependence on petroleum—a product we use to produce synthetic rubber.

The *winged bean* is 37 percent protein and 20 percent oil. Above ground, the plant has pods that can be eaten as a vegetable or that can be processed for oil. The pods can also be dried and fed to animals. Below ground, the plant grows a large root that contains more protein than the cassava—the chief root crop of the tropics. The winged bean plant also enriches the soil in which it grows. If it were widely planted outside of southeast Asia, its native home, it could reduce protein malnutrition. It could also be rotated with other crops to increase soil fertility.

When farmers in Niger store millet in these granaries made of sun-baked clay, they cannot be sure that fungus, rot, insects, or other pests will not destroy much of their surplus.

reclaimed. But in many other places, the cost of reclaiming land would be greater than any return from the products raised. So people are trying to find other ways to increase food supply.

Increasing yield. Many people believe that increasing the productivity of the arable land already in use is the most profitable way of increasing the food supply, especially in developing nations.

Developing new and better varieties of grains and breeds of livestock is one way to increase yield. There has been a continuing search to find varieties of wheat, corn, rice, and other grains that produce more kernels per plant, that have shorter growing seasons, that have stronger stalks, or that resist diseases. Many efforts have also been made to improve livestock by selective breeding or by wiping out pests and diseases that kill them. In one such effort, teams of scientists in Africa's grasslands are trying to wipe out the tsetse fly.

Fertilizing farmland will also increase crop yields. Enough fertilizer to supply worldwide demand can be made from chemicals extracted from natural gas. But new methods of fertilizing land must be found because natural gas is in short supply. In the future, fertilizer may come from the sea or from plants. It may come from materials not yet discovered or be produced through processes that have not yet been devised.

Other methods. The food supply can be increased by teaching farmers in developing nations contour plowing, crop rotation, and ways to improve storage. What are some other efforts you think people can make to increase the world's food supply?

Keeping Facts in Focus

1. How does farming in developing nations differ from farming in industrialized nations?
2. Name at least five things people are doing to increase the amount of food production.
3. What answers would you give to the key questions on page 224?

Working with Ideas

1. Why does the percentage of the agricultural population usually decline as a developing nation becomes more industrialized?
2. A rich and powerful nation helps a developing nation meet its food needs. Discuss the advantages and disadvantages of this action from the viewpoint of giver and receiver.

Lesson 3 Agriculture and Its Products

Reading Focus

1. To get the most you can from each lesson, remember that photographs must be "read" and studied just as closely as the words of the text. As you study this lesson, see what information you can obtain from each photograph, even before you read its caption. See what the photographs tell you about people, their environments and resources, the kinds of agricultural products they produce, and the methods and tools they use to produce them.

2. As you read this lesson, look for answers to these key questions:
   a. What makes agriculture an essential part of each nation's economy?
   b. What are the world's important grains? What are seven important food products other than grains with a worldwide market demand?
   c. How does agriculture help people to meet many of their nonfood needs?

Vocabulary Focus

agricultural	*import*	*winnowing*
export	*threshing*	*cellulose*

People engaged in agriculture produce the world's food supply, feeding themselves and all the people who work in nonagricultural occupations. In addition, agriculture provides raw materials for such industries as food processing, papermaking, lumbering, clothmaking, leather making, and medical supply. Agriculture is an essential part of each nation's economy, whether it is an industrialized or a developing nation.

Agricultural Exports and Imports

An agricultural product is anything usable that comes from a plant or an animal. It need not be a food product.

Many agricultural products enter the world market as exports or imports. Nations that produce a surplus, or oversupply, usually export products. Those that do not produce enough to meet their needs or that lack some resources usually import products. However, almost every nation imports some agricultural products and exports others.

By far, the greater number of agricultural products come from plants. At one time or another, people have cultivated, or domesticated, 4,000 of the more than 300,000 known plants in the world. While many of these plants are still cultivated today, less than a hundred enter the trade markets of the world.

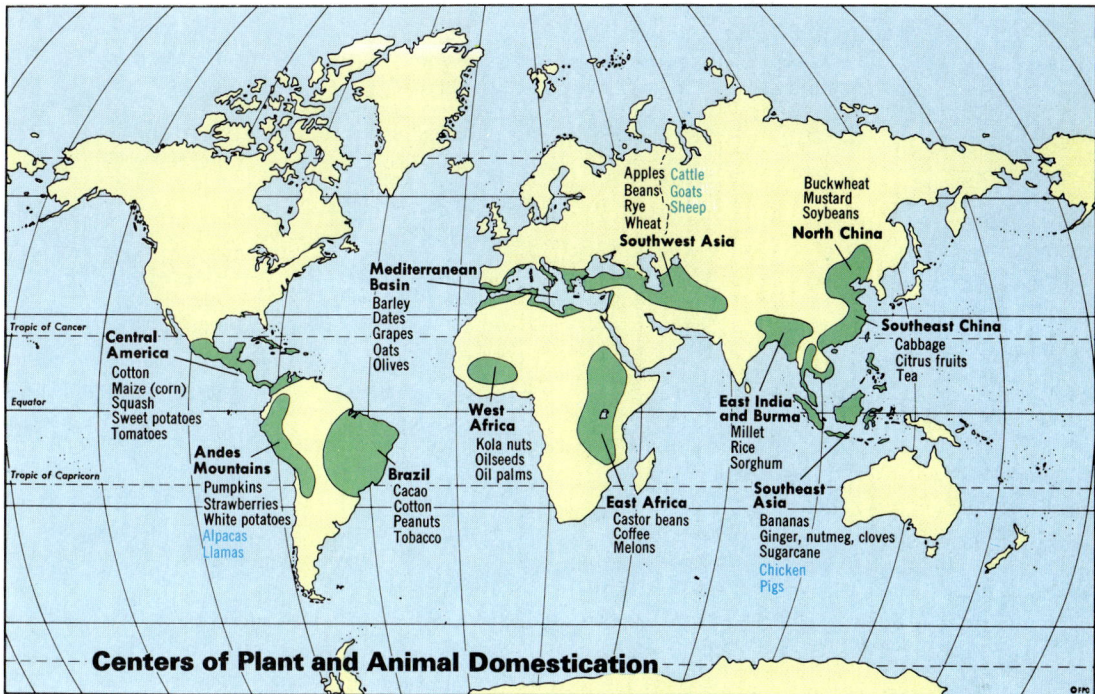

Centers of Plant and Animal Domestication

Southwest Asia
Apples
Beans
Rye
Wheat
Cattle
Goats
Sheep

North China
Buckwheat
Mustard
Soybeans

Mediterranean Basin
Barley
Dates
Grapes
Oats
Olives

Southeast China
Cabbage
Citrus fruits
Tea

Central America
Cotton
Maize (corn)
Squash
Sweet potatoes
Tomatoes

West Africa
Kola nuts
Oilseeds
Oil palms

East India and Burma
Millet
Rice
Sorghum

Andes Mountains
Pumpkins
Strawberries
White potatoes
Alpacas
Llamas

Brazil
Cacao
Cotton
Peanuts
Tobacco

East Africa
Castor beans
Coffee
Melons

Southeast Asia
Bananas
Ginger, nutmeg, cloves
Sugarcane
Chicken
Pigs

Tropic of Cancer

Equator

Tropic of Capricorn

Some of these plants are food crops, such as grains. Others are nonfood crops, such as rubber, tobacco, and cotton.

Grains

Grains, the seeds of grass or cereal crops, are the main source of food for many people. In order of total amount produced worldwide, the most important grains are wheat, rice, and corn. Other important grains are barley, oats, rye, sorghum, and millet.

Wheat. There are more acres of land planted with wheat than with any other grain. At any month in the year, wheat is planted or harvested someplace in the world. Wheat can grow in climates that are hot or cold, wet or dry. It can grow at altitudes up to 10,000 feet (3,000 m) in some places. It can grow in so many different climate conditions because over 30,000 varieties of wheat have been developed. In the United States, farmers plant about 200 of these varieties each year. Only the Soviet Union produces more wheat than the United States (see the chart on page 232 for other wheat-growing countries).

Wheat varieties may be either winter wheat or spring wheat. Winter wheat varieties are planted in the fall and

Explain why you think farmers in this farming community in Saskatchewan, Canada, are harvesting spring wheat rather than winter wheat.

Major Grains	World's Yearly Production in metric tons*	Chief Producing Nations	Food Uses	Nonfood Uses
Wheat	481,050,000	U.S.S.R., U.S.A., India, Canada, Argentina	Bread, spaghetti, noodles, other pastas, cereal and baked goods, livestock feed	Synthetic rubber, munitions, straw
Rice (wet)	411,897,000	China, India, Indonesia, Bangladesh, Japan	Cereal, flour, candy, vinegar	Cosmetics, glue, laundry starch, paste, vinegar, alcohol, straw
Corn (maize)	455,351,000	U.S.A., China, Brazil, Rumania, U.S.S.R.	Puddings, syrup, margarine, cornmeal, cereal products, candy, chewing gum, livestock feed	Paints, linoleum, adhesives, felts, cleaning compounds, plywood
Barley	160,288,000	U.S.S.R., China, Canada, France, United Kingdom	Cereal, malt, beer, livestock feed	Straw
Oats	45,278,000	U.S.S.R., U.S.A., China, Canada, Poland	Oatmeal, bread, other cereal products, livestock feed	Solvents
Millet	29,166,000	China, India, U.S.A., U.S.S.R., Zimbabwe, Nigeria	Flour, cereal, livestock feed	Animal bedding
Sorghum	69,113,000	U.S.A., India, Argentina, Nigeria, Mexico	Flour, cereal, livestock feed	Brooms, brushes
Rye	30,126,000	U.S.S.R., Poland, West Germany, East Germany	Flour, alcoholic beverages	Thatching, mattress stuffing, paper, hats

*Figures are rounded

harvested in early summer. They are usually planted in regions where precipitation is greater in the winter months, but where temperatures are not so cold as to damage the crop. In the United States, for example, winter wheat is planted in the southern part of the Great Plains. The snows of winter keep the dormant plants from freezing and also supply moisture for the crop's renewed growth in the spring. In regions with a Mediterranean climate, winter wheat does not have a dormant period. It grows throughout the period of winter precipitation and is usually harvested in early spring.

Spring wheat varieties are planted where winters are more severe, but where the growing season is at least three months. Planted in the spring after killing frosts have passed, spring wheat ripens quickly. It is usually harvested by the end of summer.

Rice. Over 90 percent of the world's yearly rice crop is produced in China, Japan, India, and other countries in Southeast Asia. For people in these countries, rice—rather than wheat—supplies most calorie needs.

Where land is hilly and somewhat dry, people grow upland or dry varieties of rice. Dry rice grows in dry, rather than flooded, fields in much the same manner as wheat and other dry grains. Only a very small amount of the world's rice crop is dry rice. Most of the world's rice crop is lowland rice, or wet rice.

To grow wet rice, farmers need flat surfaces and standing water. Very few lowlands—whether deltas or flood-plains—are as flat as they need to be. So to keep water standing in the fields, which are usually called paddies, wet rice farmers build low-lying dikes, or levees. The dikes keep the water in a paddy at a desirable level while the rice is growing. By opening and closing gates in the dikes, farmers direct the flow of water through the paddies. The dikes also separate one farmer's paddy from another.

In tropical regions, farmers harvest at least two crops of wet rice each year. Each crop begins in specially prepared beds, or nurseries. After the seeds take root and become seedlings, the young shoots are transplanted to the flooded paddies by hand. As this crop grows, a new crop of seedlings takes root in the nurseries. It will be ready for trans-planting when the first crop has been harvested and the fields are made ready.

In many of the countries where rice is grown, it is often harvested by hand, using knives and sickles. Threshing—separating the heads from the stalks—takes place once the rice is dried. Some farmers hit the stalks against a screen with openings large enough for the rice heads to fall through.

Winnowing separates the kernels from the heads. In some places, farmers toss the heads of rice into the air. The wind blows away the lighter material while the heavier kernels fall to the ground ready to be hulled.

Winnowed rice is still covered with a rough outer covering called the hull. The hull must be removed either by hand or by machine. Hand-hulling by grinding or pounding usually produces a brown rice that is much more nutritious than the white (or polished) rice pro-duced by machine-hulling. Hand-hulling does not remove all of the bran coat—an inner covering rich in minerals and vitamins.

Some wet rice is grown outside of Asia in regions with a humid subtropical climate. In the United States machinery does much of the work of planting, culti-vating, and harvesting. Low-flying air-planes may drop sprouted seeds onto flooded fields. Or machines called grain drills may plant the seeds. After the grain ripens and the fields are drained, combines harvest and thresh the grain in one operation.

Corn. It is as grain, rather than as a vegetable, that corn has worldwide eco-nomic importance. Most grain corn vari-eties are fed to livestock or are processed for the starch, sugar, and oil they con-tain. The United States, as the leading corn-growing nation, produces almost half of the world's total corn crop.

Other grains. Barley and oats are grown in the middle latitudes, mostly in climates that are too cool for corn or too moist for wheat. In the chief barley growing areas, barley replaces corn as an animal feed. Oats are fed to horses, are ground into flour, or are processed into oatmeal.

A field of millet (top left) surrounds a village in Chad. In Ghana, a cacao pod swells to full growth (bottom), while in Costa Rica, coffee berries ripen on a highland slope. Which of these crops is unlikely to be an agricultural export?

In hot, dry areas where no other grain can thrive, sorghum and millet are the chief grains. In the United States, sorghum is used as livestock feed or is processed into molasses. In Africa and parts of Asia, the very poorest people grind the starchy seeds of sorghum and millet into flour to make flatbreads.

Over 90 percent of the world's rye crop is harvested in northern and eastern European countries. Grown generally for human consumption, rye ranks second to wheat as a bread flour.

Other Agricultural Products

Besides grains, there are other major plant and animal products that are used either as food or as raw materials for industries.

Sugarcane and sugar beets. Both sugarcane and sugar beets contain a sweet juice that is processed into white sugar, powdered sugar, or brown sugar. Sugarcane grows best in the hot, humid climates of the low latitudes. Sugar beets grow best where temperatures are mild and cooler, but where the growing season is at least five months.

Coffee, tea, and cacao. The farmers growing grains, vegetables, sugarcane, or sugar beets get a return from their crops the same year they are planted. But the growers of coffee, tea, and cacao crops—like most fruit growers—must tend their plants for a number of years before harvesting a marketable crop.

Coffee trees are first planted as seeds in nurseries. As seedlings, they are transplanted to fields or hillsides. After about five years, the trees bear a full

Other Major Crops	World's Yearly Production in metric tons*	Chief Producing Nations	Food Uses	Nonfood Uses
Sugarcane	888,735,000	India, Brazil, Cuba, China, Mexico	White sugar, brown sugar, powdered sugar, molasses	Nylon, plastics, carbon paper, phonograph records, paper, wallboard, cosmetics, synthetic rubber
Sugar beets	271,002,000	U.S.S.R., U.S.A., France, West Germany, Poland	Same as above	Same as above
Coffee	5,537,000	Colombia, Brazil, Ivory Coast, Mexico, Uganda	Beverage, flavoring agent	Medicines with caffeine
Tea	2,020,000	India, China, Sri Lanka, Japan, U.S.S.R.	Beverage	Tannin
Cacao	1,557,000	Ghana, Ivory Coast, Brazil, Nigeria	Cocoa powder, cocoa butter, chocolate	Medicines
Cotton	67,600,000	U.S.S.R., China, U.S.A., India, Egypt	Salad oil, margarine, shortening, livestock feed	Textiles, soap, linoleum, phonograph records, paper, plastics, paints
Tobacco	6,000,000	China, U.S.A., India, U.S.S.R., Brazil	Manufactured niacin, a vitamin used as a food supplement	Cigars, cigarettes, etc., insecticides, pesticides
Rubber (natural)	3,866,000	Malaysia, Indonesia, Thailand, Sri Lanka, India	None	Tires, tubing, waterproof material

*Figures are rounded

crop of berries. Hand-picked when it is bright red, each berry contains two coffee beans. After picking, the beans are dried, sorted, roasted, and then shipped to markets around the world.

The tea plant also takes about five years to mature. Then it produces new tea leaf shoots about once every two weeks. A worker picks about 40 pounds (18 kg) of these young shoots a day. By processing tea leaves in different ways, manufacturers produce black, green, and oolong teas.

The cacao tree, once it is transplanted as a seedling, takes from three to eight years to bear pods, depending on the variety. But once it begins to bear pods, the tree produces pods twice a year. Each pod contains between 20 and 40 beans, about the size of large lima beans. First, the beans are separated from the pulp surrounding them. Then they are processed. Processing turns the beans into cocoa butter (the natural fat of the cacao bean), cocoa powder, or chocolate. West Africa produces almost 70 percent of the world's cacao crop, mostly on small farms.

Meat and dairy products. Cattle supply beef and veal. Sheep give us lamb and mutton. From hogs, we get pork and bacon. Poultry give us meat and eggs.

Chemicals in Our Food—Two Sides of an Argument

Much of the food that leaves a farmer's field undergoes several processing steps before it reaches our tables. Grain is ground into flour or meal, often enriched with vitamins and minerals lost in processing, and then turned into cereal or bakery products. Meat is smoked, pickled in brine, or otherwise treated to keep it from spoiling. Salt, pepper, garlic, other herbs and spices, and flavorings improve the taste of food. Dyes injected into fruit or added to other foods make them more eye-appealing. Vitamins, minerals, smoke, dyes, thickeners, salt, and other flavor enhancers, substances with chemical-sounding names—all these things are food additives.

In the past, most—if not all—food additives came from nature. Today, technology has made it possible for many natural additives to be replaced with manufactured ones—additives made by putting chemicals together.

Some people say we should be just as concerned with the effects these manufactured additives have on our bodies as we are with the effects of chemicals on the land, in the water, and in the air. These people fear we are polluting our bodies with too many chemicals. They point to laboratory studies showing that certain manufactured additives, when given in large doses over a short period of time, cause diseases such as cancer in animals. They think these chemicals will do the same thing in humans.

Food processors try to justify their use of manufactured additives. They point out that every living and nonliving thing on earth is made up of chemicals or combinations of them. Even when eating "natural" foods, people are eating chemicals. Processors argue that manufactured additives do all the things that natural additives do, but they do them with less cost to the consumer. They also point out that manufactured additives do some things that natural ones cannot do. For example, they enrich milk, flour, and other foods with vitamins and minerals lost in processing. Or they make foods tastier for people who cannot eat salt or sugar.

Food companies admit that large doses of some manufactured additives can cause diseases. They say, however, that it is impossible for a human being to receive a dose large enough to cause disease by eating processed foods—at one meal or even in a lifetime of meals.

At present, our government is carefully listening to both sides of the argument. It has banned the use of some manufactured additives. It has added words of caution on the packaging of some products. And it continues to test all foods for the safety of our health.

Cows and goats provide milk, butter, and cheese. Almost every part of the flesh of animals supplies the makings for some kind of food rich in protein.

Fish. In many parts of the world, fish—rather than meats—are a main source of protein. Most fish are not strictly agricultural products. But there are fish that are raised on special kinds of farms. Salmon, trout, and catfish are some of the fish raised in artificial lakes and ponds. Oysters are raised in specially prepared beds in the shallow waters along some seacoasts. If overfishing continues in some parts of our oceans and lakes, we will have to step up our efforts to raise other kinds of fish.

Nonfood agricultural products. Farmers help us meet other needs and wants besides those of food by turning out raw materials for the manufacture of nonfood products. These products provide clothing and shelter and supply people with touches of beauty and moments of pleasure. They also make it possible for developing nations to break away from one-crop economies.

Even though the use of tobacco is dangerous to health, many people smoke or use tobacco in other ways. As long as there is a demand for it, farmers will continue to grow tobacco, an important cash crop in over sixty countries. Important tobacco-growing areas exist

A duck seller in a market in China displays his live wares for prospective buyers.

where there are warm climates and where soil is well drained and fertilized.

Cotton is the most important fiber used to make clothing in many parts of the world. All parts of the cotton boll can be used. The lint is used in making textiles. The seeds provide an oil that has both food and nonfood uses. The meal that remains after the oil has been removed is fed to livestock.

Fiber from the flax plant is woven into products that range from coarse rope to fine linen and lace. From its seed, we get linseed oil, a product that has many industrial uses. We use flax straw to make special grades of paper, such as the kind used for diplomas or other kinds of certificates. Farmers in many European countries, China, the Soviet Union, and Turkey raise most of the world's flax crop.

Coarser kinds of cloth and kinds of twine or rope are made from hemp, henequen, and sisal. Eastern Africa is the leading producer of sisal. Most of the world's henequen is produced on the Yucatán Peninsula in Central America. Hemp is grown in parts of Europe, Asia, and North America.

As a renewable resource, trees have almost limitless value. They can be burned as fuel and used to produce energy. They supply lumber and raw materials for papermaking. Cellulose, the fiber in wood, can be used to make a number of things, such as clothing or plastics.

The coats of many animals also provide fibers for clothing, carpeting, and other products. By tanning the skins of many animals, we make leather. The horns and hooves of animals provide materials for making such things as buttons and glue. Chemical compounds extracted from animal parts are used to manufacture medicines. The parts of some animals are even used to repair vital organs of the human body.

Keeping Facts in Focus

1. Describe how wet rice is grown, threshed, and winnowed by hand in some countries.
2. What countries lead the world in the production of wheat? Sugarcane? Cotton? Rubber?
3. What answers would you give to the key questions on page 230?

Working with Ideas

1. Some developing nations are dependent on one-crop economies. Why is it important for a developing nation to break away from a one-crop economy?
2. Why are more acres of land in the United States planted with wheat than with rice?

UNIT 6 REVIEW WORKSHOP

Test Your Geographic Knowledge

A. Choose the ending that best completes each of the following statements:
1. To keep vision normal and skin healthy, people should eat some foods that contain (a) vitamin A, (b) vitamin C, (c) carbohydrates.
2. Foods that are mostly sugars and starches are (a) minerals, (b) proteins, (c) carbohydrates.
3. The main reason why there are people who are hungry or malnourished is (a) industrialization, (b) the uneven distribution of natural resources, (c) too few subsistence farmers.
4. A developing country (a) depends on farming for most of its income, (b) does most of its farming with machinery, (c) has no industry whatsoever.
5. Over 90 percent of the world's yearly rice crop is harvested in (a) Asia, (b) Latin America, (c) the United States.

B. Answer these questions:
1. What are the five essential nutrients needed for healthful living?
2. What is a balanced diet?
3. In terms of amount produced, what are the world's three major grains?
4. What causes malnutrition?

Apply Your Reading Skills

A. Read this summary of a report from the Department of Agriculture. Then use the facts in the summary to tell which of the statements following the summary are true and which are false.

Bumper crops of corn and soybeans are pouring into American markets. American farmers have harvested more corn and soybeans than in any previous year.

These grains make up most of the feed used to produce meat, poultry, and dairy products. These products account for more than one third of what the average American consumer spends on groceries.

In spite of bumper grain crops, food prices continue to climb. One reason for the food-price climb has been a reduced supply of beef caused by farmers and ranchers sharply decreasing herds after four years of not making profits. Another reason is weather damage to fruits and vegetables.

1. Much livestock feed comes from corn and soybeans.
2. Americans, on the average, spend at least one third of their grocery budgets on meat, poultry, and dairy products.
3. A bumper crop sometimes breaks the record for the total amount produced in a previous year.
4. Food prices have climbed four years in a row only because of weather damage.
5. Food prices are high because there is a shortage of livestock feed.
6. Lately, farmers and ranchers have been sending fewer head of beef cattle to market.

B. Use encyclopedias, books, and other reference tools to do research. Then prepare a report on ONE of the following topics.
1. What kinds of jobs are agribusiness jobs?
2. What kinds of foods will people be eating in the future?
3. Can plants be grown without soil?
4. How do food customs differ around the world?
5. How were vitamins discovered?

Apply Your Geographic Skills

The paragraph below tells some of the important early events in the history of humankind. But the sentences are not in proper order. Use the time line on page 239 to renumber the sentences in correct order. (Note that the time line has two parts. The bottom line shows the time span that is covered. The top line is an enlarged view of a smaller period within the time span.)

(1) Several thousands of years later, people had begun to use the wheel, a great labor-saving device. (2) The first important step took place when hunters and gatherers learned to use fire.

About 9,000 B.C. Beginnings of village life in Middle East

About 7,000 B.C. Domestication of sheep, goats and cattle

About 7,000–6,000 B.C. People begin to practice agriculture

About 6,500 B.C. People begin to make pottery

About 3,500 B.C. People begin to use wheel

About 2,500 B.C. People begin to make things of iron

500,000 400,000 300,000 200,000 100,000 B.C. A.D.

About 500,000 B.C. People begin to use fire

About 500,000–3,500 B.C. People live mainly by hunting animals and gathering plants

About 60,000 B.C. People begin to use bows and arrows

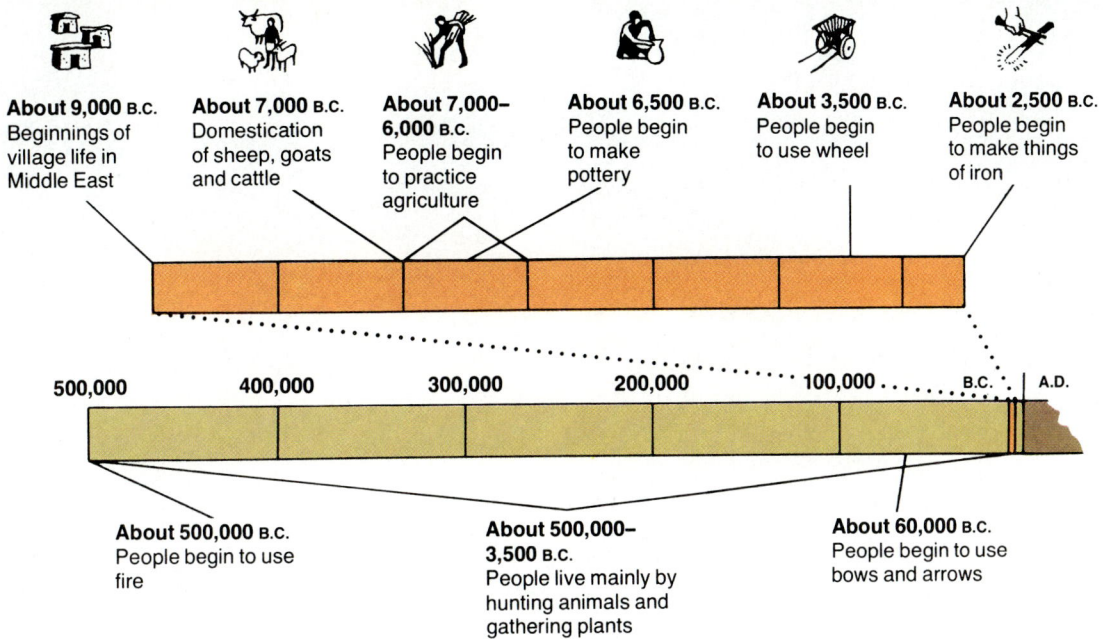

(3) Another important step took place when groups of people began to live in villages. (4) Some people living in villages learned to make pottery in which they could cook and store foods. (5) Finally, when people began to make tools and other things from iron, they took another huge step forward in learning to use the resources of their environments. (6) The next step took place when the hunting of wild animals for food and clothing was made easier with the invention of the bow and arrow. (7) Within two thousand years of the beginning of village living, people had learned to tame, or domesticate, certain animals—first the dog, then sheep, cattle, and goats. (8) By the end of the next one-thousand-year period, people living in villages had begun to cultivate crops.

Apply Your Thinking Skills

Decide whether each of the following statements describes an industrialized country (I), a developing country (D), or both (B).

1. People meet most of their calorie requirements by eating foods made from grains.
2. More people are likely to be overnourished than undernourished.
3. Less than 10 percent of the people make their livings by farming or herding.
4. Most farmers are likely to be subsistence farmers.
5. Some farmers practice commercial farming.
6. The people need a diet that includes all five essential nutrients.

Discuss These Points

1. How do customs or tastes in your family influence the kinds of foods you usually eat? Do you think you can get a balanced meal at a fast-food restaurant?
2. Developing nations are sometimes called "have-not" nations. What are some of the things they need to become "have," or industrialized, nations?
3. A viewpoint sometimes expressed by persons in an industrialized country is this: "I know there are needy people in the world. But I don't want my tax money going to help needy people outside my country. Let people in other countries take care of their own problems." Do you agree or disagree with this viewpoint? Why?

Expand Your Geographic Sights

Asimov, Isaac. *How Did We Find Out About Vitamins?* (Warne, 1976).

Haggarty, James J. and Sebrell, William H., Jr. *Food and Nutrition* (Silver, 1967).

Hawkes, Nigel. *Food and Farming* (Watts, 1982).

Houlehen, Robert. *Jobs in Agribusiness* (Lothrop, 1974).

Perl, Lila. *The Global Food Shortage* (Morrow, 1976).

Sullivan, George. *Understanding Hydroponics* (Warne, 1976).

Wolfe, Louis. *Aquaculture: Farming in Water* (Putnam, 1972).

Unit 7 Energy

Anchored firmly in the floor of the continental shelf under the Gulf of Mexico, an oil platform drill probes for the oil needed to power the world's many industrial activities.

Lesson 1 Making the World Work

Reading Focus

1. In this unit, you will study the many sides of energy. The first lesson will tell you where energy comes from. As you read through this lesson, note how each energy source is related to another. Be aware that all energy is caused by some action. And, for every cause, there is some effect. For example, the sun causes the wind to blow as its radiation creates differences in air temperature and pressure on the earth. This action causes many effects. The effects, in turn, cause other things to happen. By understanding this very basic rule, you will discover some very interesting things about energy in general. As you read this lesson, try putting together a cause-and-effects list. Start with the sun, and note the types of energy that were or are now being caused by its energy.

2. As you read this lesson, look for the answers to these key questions:
   a. What is energy?
   b. What are the three major fossil fuels?
   c. What happens to energy once it is used to complete a specific job?

Vocabulary Focus

nuclear	*fusion*	*radiation*	*thermal*
hydrogen	*helium*	*fossil fuel*	*turbine*

Energy is the ability or capacity to do work. Your body possesses energy, and so has the ability to perform work. Every time you work, you also spend energy. Yet, do you know how and from where you originally received the energy? Do you know what happens to energy once it is spent? For the answers, let's take a 93-million-mile journey to the sun.

The Sun

If you could stick a thermometer into the sun, you would find extremely hot temperatures. Scientists estimate the outer surface is thousands of degrees hot. The temperatures in the core are in the millions. The core is the center of the sun's activity. It is also the source of most of the earth's energy.

The sun's core. The sun's center is a giant nuclear power plant. There, the pace is frantic and the action never stops. Hydrogen, the chief element of the sun, is the main worker. With the aid of the intense heat, hydrogen atoms are able to go through a process known as nuclear fusion. During this process, countless hydrogen atoms collide with one another. The collision causes them to fuse, or join together. The fusion causes the hydrogen to change its chemical makeup. The atoms are no longer hydrogen but the heavier element, helium. It is this change that causes huge amounts of energy to be released. This energy, mostly in the form of heat and light, is then scattered throughout our solar system. A small but very important part of this energy reaches the earth.

The sun's many roles. The heat energy in the sun's radiation warms the layers of earth's atmosphere and its land and water bodies.

Plants take the light energy in the sun's radiation, combine it with chemicals in air and water, and make it into food for themselves. Animals eat plants or plant-eating animals. People eat both plant and animal products. This energy exchange is called the food cycle. Without the sun's radiation, the food cycle would not exist.

Besides providing light and heat and setting the food cycle in motion, the sun's radiation greatly affects the weather. To be more specific, it actually creates the weather. Air pressure differences, winds, clouds, and rains are all caused by the exchanges of energy that take place as radiation from the sun moves through the atmosphere to the earth's land and water surfaces.

Indestructible energy. An important law of nature applies to energy of all types. This law says that energy cannot be destroyed—it is simply changed. A good example of this law is the earth's supply of fossil fuels.

Fossil fuels—petroleum, natural gas, coal—are thought to be the remains of prehistoric plants and animals. The fossils collected in thick layers over various parts of the world. With time, water and/or layers of earth covered the fossils. The covering layers prevented air from reaching the matter. Thus, the dead materials were not able to decay completely. Instead, the fossils were pressed against rock beneath them by the tremendous weight of the overlying soil and water. This great pressure caused heat that then changed the matter into fossil fuels. The complete process, from living matter to fossil fuels, took many years.

The formation of these fuels clearly shows that energy is not destroyed. Rather, it is changed into other forms. Energy is: (1) Formed in the core of the sun as *nuclear energy*. (2) Passed down to earth in the form of *solar energy*. (3) Stored as *chemical energy* in the leaves of plants. (4) Changed into *food energy* for animals. (5) Changed into a more concentrated form of chemical energy (fossil fuels).

People Enter the Energy Picture

When humans appeared on the earth, their chief concern was to feed and clothe themselves. To do this, they turned the stored chemical energy in their bodies into muscle power. Using their muscle power, they were able to hunt and forage for plants and animals. Their relationship with nature's energy system was direct and basic.

Human use of energy increases. As years went by, people began to gain more control over the environment. They learned to use the thermal energy (heat) given off by burning wood. The heat provided them with warmth. It also allowed for cooking, which made their food easier to chew. Without knowing it, people were converting one form of energy into another. This enabled them to perform more work and other jobs that they previously were not able to do.

In the years that followed, people learned to use other forms of energy. They found the wind was useful for sailing ships. They discovered certain animals could pull plows. They used water to turn mills. All these changes lessened the amount of body energy people had to use. The energy people saved could be used for other jobs.

Within recent history, people discovered many more ways to use energy. They learned that heated water produces steam. They discovered steam could be used to drive machinery. This discovery enabled the United States and European nations to participate in industrial revolutions. In essence, the more ways people found to use energy, the more their lives changed.

Still later, other scientists discovered many more beneficial ways to use energy. Thomas Alva Edison showed the world some of the many jobs electricity could do. Other inventors learned how to produce power from a gasoline-burning engine. As a result, the automobile came onto the scene, and the world was never to be the same again.

Today's energy situation. Energy abounds all around us. Its potential to do work is enormous. Yet, most of it today is not being used. This is true even though there is now a shortage of fossil fuels. The many uses we have found for fossil fuels and the increased demand for them have caused the shortage. If the use of fossil fuels continues at a rapid rate, the world may run out of petroleum and natural gas within the next forty or fifty years.

Cars of the future may use electric energy stored in batteries (right) rather than gasoline.

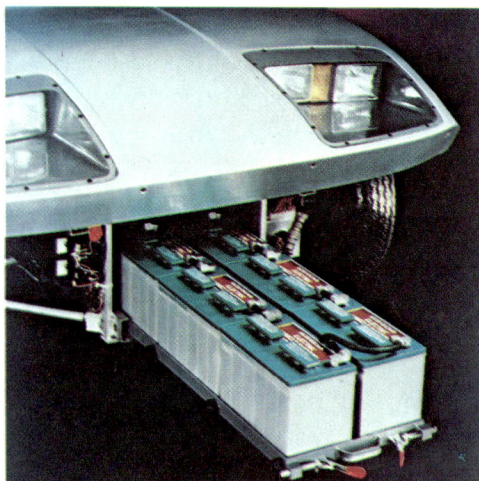

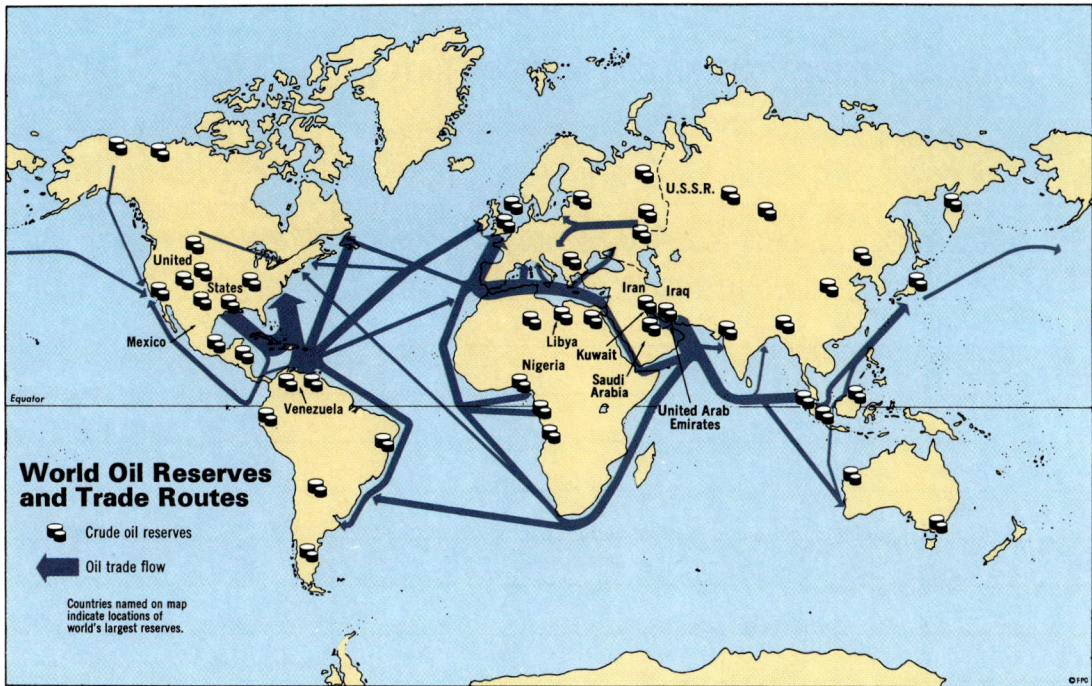

World Oil Reserves and Trade Routes

🛢 Crude oil reserves

➜ Oil trade flow

Countries named on map
indicate locations of
world's largest reserves.

U.S.S.R.

United States

Mexico

Venezuela

Equator

Iran
Iraq
Libya
Kuwait
Nigeria
Saudi Arabia
United Arab Emirates

Fossil Fuels

Today, fossil fuels make the world run. About 95 percent of all the energy generated in the world comes from fossil fuels. Most essential for the world, at least now, is petroleum, or crude oil.

Petroleum. Besides providing fuel for homes and factories, petroleum is made into many other valuable products. Gasoline is undoubtedly the most important. This fuel powers most of the world's automobiles. Kerosene, matches, perfumes, soaps, and clothing are but a few of the many other useful petroleum products. Crude oil now supplies the world with about 45 percent of *all* its energy.

Because petroleum has so many uses, the demand for it becomes greater all the time. As population rises, so does the demand for crude oil products. This demand has severely cut into the earth's supply of this precious fuel. So much, in

fact, that some experts say our known reserves will last only twenty-five or thirty years. Even if new wells are found, oil supplies may not last much longer than fifty years at the very most.

To add to the supply problem, the distribution of oil reserves around the world is uneven. Close to 70 percent of the world's reserves lie in the Middle East and North Africa. Countries there export most of their oil to industrial nations such as the United States, Japan, and the countries of Europe. Because of demand, oil-exporting nations have much control over what price they can charge.

Natural gas. This is the world's third most used fuel. Natural gas provides about 25 percent of the world's energy. In the United States, it ranks second behind oil. Natural gas provides about one third of all U.S. energy. It is often known as crude oil's twin because it generally is found near oil.

NATURAL GAS PRODUCTION AND CONSUMPTION RATES

(in billions of cubic feet)   ■ Production   ■ Consumption

Bar chart comparing production and consumption of natural gas (in billions of cubic feet, scale from 0 to 20) for the following countries:

- United States
- U.S.S.R
- The Netherlands
- Canada
- Rumania
- United Kingdom
- Mexico
- Iran
- West Germany
- Italy

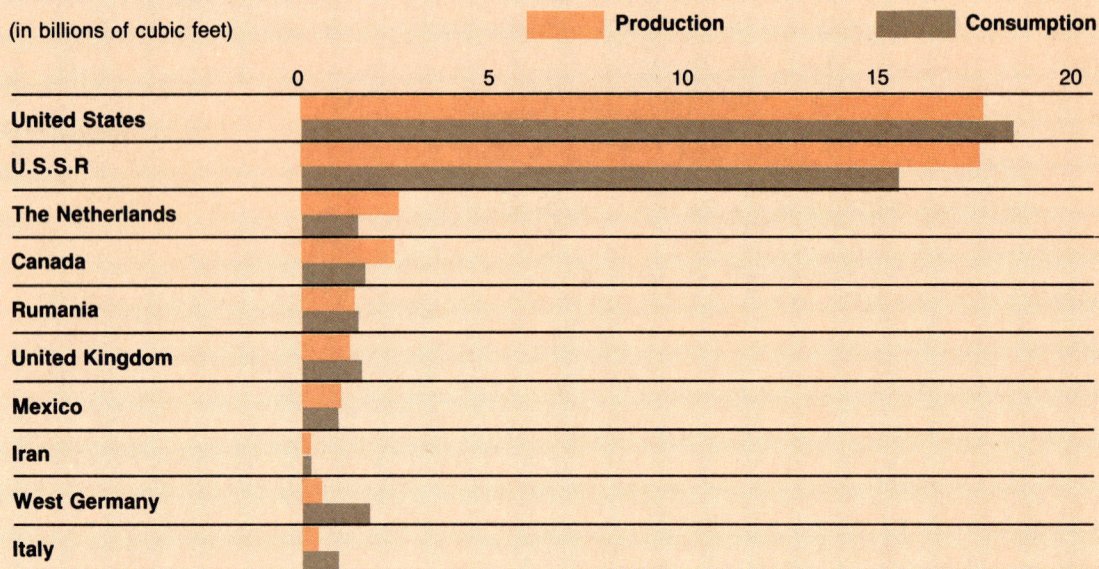

Up until the 1920s, natural gas was not used commercially. Oil drillers who found it near the oil simply let it escape into the air. Sometimes it was burned. New technology in the late 1920s made it possible to use natural gas. People soon became aware of its advantages. They found it burned much cleaner than petroleum or coal. They also found it was much easier to ship via pipeline. Back in those days, natural gas was also a very inexpensive fuel.

Today, natural gas is used to heat many homes and industries. It is used to run water heaters, clothes dryers, and cooking stoves. Utility companies use it to generate electricity. Plastics, fertilizers, and many other products are also made from natural gas.

Like petroleum, natural gas is in short supply today. It is also very expensive. Most of the world's reserves are in the Soviet Union, the United States, and

Canada, and nations of the Middle East. More of the fuel is being used each year. Of the three major fossil fuels, natural gas will probably run out first.

Coal. This resource is undoubtedly the most abundant fossil fuel. It is in a slightly different class from gas or oil. For one thing, it is a solid fuel. For another, there is plenty left. Experts say that at the present rate of use, our known coal reserves could supply the world with all its energy for almost 1,000 years. Coal now provides about one third of the world's energy. About 20 percent of the energy in the United States comes from the burning of coal. It is used primarily to generate electricity. The United States and the U.S.S.R. have the largest reserves of coal.

There are different types of coal. Anthracite, or "hard coal," is the most valuable of all coals. It gives off great heat and burns with a smokeless flame.

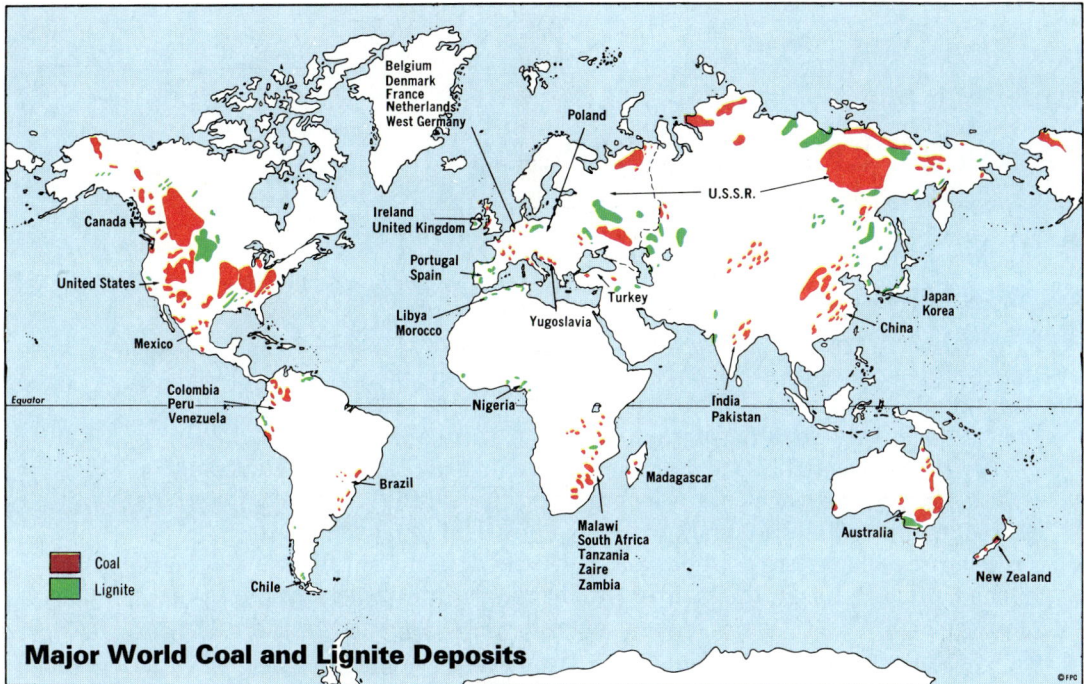

Major World Coal and Lignite Deposits

Legend: ■ Coal ■ Lignite

Unfortunately, anthracite is in very short supply.

Poorer quality coals are called "soft coals." In this category are bituminous and lignite coals. Most of the world's coals are of the soft variety. When burned, soft coal gives off a thick cloud of black smoke. Because most coal is not very clean burning, its smoke must be "scrubbed" during the burning process. Scrubbing is expensive and does not filter out all dangerous pollutants. Because of these facts, many people would rather not use coal.

Fuels and the future. The demand for fossil fuels, still the world's major source of energy, increases every day. But the days of cheap and abundant fossil fuels are over. Supplies, once thought endless, are now limited. Costs, no longer inexpensive, continue to rise. The supplies left must be used wisely.

Once all the fossil fuels are gone, what will people use to cook their food and heat their homes? What will drive the turbines that now produce electricity? What choices do we have? You will find some answers to these questions in the next lesson.

Keeping Facts in Focus

1. Why is the sun's core described as a "giant nuclear power plant"?
2. Which forms of energy depend upon the sun?
3. How were fossil fuels formed?
4. What answers would you give to the key questions on page 242?

Working with Ideas

1. How has the use of motor vehicles led to the increased use of fossil fuels?
2. How has industrialization led to the increased use of fossil fuels?
3. Why should people do as much as possible to conserve energy and use fossil fuels wisely?

Lesson 2 Energy Options

Reading Focus

1. In this lesson, you will explore some energy options, or choices. These are the energy sources that are and will be taking over the jobs of fossil fuels. Some of these sources will be more important than others. Some will be important in only certain geographic areas. Can you now think of a type of energy that is located in one place and not in another? As you read through this lesson, weigh each of the energy options carefully. Watch for key sentences that will tell you how to discover levels of importance. Afterward, make a list of the energy options you believe will perform the greatest jobs in the United States in the future. Number them from one to five, with the most important first. Compare your answers with those of your classmates.

2. As you read this lesson, look for the answers to these key questions:
   a. What are ten energy options that could replace the use of fossil fuels?
   b. What are two drawbacks and two pluses connected with the use of nuclear energy?
   c. How can bioconversion, gasification, and the use of oil shale help to relieve energy shortages?

Vocabulary Focus

hydropower	uranium	bioconversion
renewable resource	radioactivity	gasification
kinetic energy	geothermal	oil shale
fission		

There is great energy potential in many rivers and streams. Unlike the energy found in fossil fuels, the energy of rushing water, hydropower, is a renewable resource. A renewable energy resource can be easily replaced by more of the same energy. In this case, hydropower is considered renewable because of the sun's never ending job. The sun continually evaporates water from the earth and then drops it in the form of precipitation. This is how rivers maintain their water supply.

Hydropower

Although hydropower has been used since ancient times, modern methods used to harness it are more advanced. The principles are basically the same,

Compare these inside views of two electrical generating plants. On the left are the generators that are a part of the Hoover Dam power plant. At the right is a core of uranium rods. What is the source of the energy produced in each power plant?

however. Today, the potential energy of water is turned into kinetic (motion) energy with the help of huge dams. The dams hold water and then release it when it is needed. Water pouring through a dam is very powerful. Its force turns giant turbines. The turbines, in turn, generate electricity.

The world today gets about 2 percent of all its energy from hydropower. The United States gets about 4 percent. The percentage of electricity produced, however, is usually much higher. In Brazil, for example, more than 80 percent of all electricity comes from hydropower. In Japan, 40 percent is developed from hydropower. In the United States, about 14 percent of all electricity comes from this source. Why do you think some nations produce more electricity from hydropower than others?

The further development of hydropower could help ease the energy squeeze in many areas of the world. Some nations, however, already have harnessed most of their large rivers.

Others, because of a lack of rivers, obviously cannot depend upon hydropower. Other sources that might provide more versatile and reliable energy will have to be developed. Nuclear energy is one such source.

Nuclear Energy

By far, nuclear energy is the world's most controversial energy option.

Fission. The nuclear energy used to perform work in the world today is derived from fission. Fission occurs in a nuclear reactor when atoms of a certain heavy element split. Uranium atoms are used for this purpose. When the first atoms split, they cause others to split. Soon, many atoms are splitting. This is called a chain reaction. The reaction causes tremendous heat (thermal) energy to be released.

Why is the storage of liquid nuclear wastes, as in these tanks under construction, a possible danger for people living near a storage site?

Nuclear Wastes

One of the biggest problems with nuclear power is storing the wastes it produces. These wastes are highly radioactive and very dangerous. They are capable of poisoning the soil and the plant life it supports. Water sources can also be poisoned. Some of the matter can remain dangerous for thousands of years.

In the United States, power companies and the military store their own nuclear wastes in locations of their own choice. The most dangerous materials are stored in liquid form in huge underground tanks. Many people feel this storage system is not good enough. They fear wastes could leak out and enter water supplies or escape into the air.

The government also does not like the present storage system. It hopes one day to be able to store all nuclear wastes at one or two large sites. The government believes the most ideal places to store the materials would be in abandoned salt mines. Salt mines are usually located in areas where earthquakes are unlikely to occur.

Most states would not welcome a storage site within their own borders. Everyone wants to store the nuclear wastes safely, but no one wants a site nearby.

Confusing laws and opposition to the location of any disposal site will probably hold things up for quite a while. Meanwhile, more and more wastes are piling up. In the long run, this could eventually mean the slowdown of all nuclear power programs. Some plants might even have to be shut down until proper disposal sites can be found.

Today, the world uses nuclear energy primarily to generate electricity. Heat from the nuclear reaction is used to boil water. The resulting steam is then used to drive a turbine. The turbine then changes the steam to electricity.

Nuclear power plants are becoming a common sight around the world. They now produce about 2 percent of all the world's energy. If certain plans are realized, the energy of the atom might produce 25 percent of the energy in the United States by the year 2000. This would certainly fill a large hole in the gap the dwindling supplies of fossil fuels will leave behind by that year.

Nuclear drawbacks. Many people are concerned about leakage of radioactive

The huge mirrors of a solar plant built in the Pyrenees Mountains in France reflect the sun's rays into a receiver. The concentrated heat changes water to steam that produces electricity for a wide area. In many other places around the world, simpler, rooftop solar collectors heat water for other uses.

materials from nuclear power plants. Radioactivity is caused by the destruction of parts of certain atoms in elements such as uranium. You cannot see radioactivity. It can only be detected with machines called Geiger counters. Experts agree that even hairline cracks in reactor walls could let out harmful amounts of radioactivity. Because of this danger, most nations have strict rules concerning the construction of nuclear reactors. Other people also fear the possibility of accidental explosions at power plants. The planned use of nuclear materials to cause or threaten an explosion is another story. A nuclear bomb in the hands of an enemy nation could obviously cause serious problems.

Nuclear pluses. Beyond the drawbacks surrounding all types of nuclear power, there are several clear-cut benefits. For one, nuclear power is a relatively clean power. It does not produce the types of pollution fossil fuels do when they are burned. (In some instances, reactors can produce thermal pollution. So far, scientists do not know to what

extent this type of pollution will harm the environment.) For another, a very small amount of uranium will produce a large amount of energy. A six-pound (2.7 kg) piece of uranium will yield as much power as 18,000 pounds (8,100 kg) of coal.

Solar Energy

As we know, solar energy is the source of most of the world's energy. There is an endless supply of this energy. In less than an hour, the sun delivers enough energy to earth to supply all human needs for a full year. Concentrating it into a usable form, however, has always been a problem. Methods used to catch the rays of the sun and put them to work were too expensive. Inexpensive fossil fuels were always there. So, who needed to spend a lot of time, money, and effort trying to put the sun to use? Now, of course, the story is different. Scientists from many fields are working on new solar-catching devices. The progress has been so good that many now believe

In a nineteen-mile-per-hour wind, the blades of this experimental windmill in Ohio will generate enough electricity to supply thirty homes in the area.

also used in western parts of the United States during the nation's early development. The main job of the early windmills was to grind corn and other grains. They were also used to pump water. Although many of these windmills still stand in many parts of the world, very few do the jobs they once performed.

Today, the shortage of fossil fuels has sparked the building of different types of windmills. Their job is to generate electricity. Electricity produced by a windmill costs about the same as that produced from coal. The expensive part is the windmill itself.

Wind power is not free of problems. Since winds vary from place to place, windmills are more practical in some areas of the world than in others. In some places, they do not work at all. Some good locations for using windmills include West Africa, the west coast of South America, and the eastern Great Plains of the United States.

Geothermal Energy

Geothermal energy is another option that is getting attention these days. This energy comes from the intense heat that is stored within the earth.

Like wind energy, geothermal energy is usable in only some parts of the world. A good example of geothermal power are the geysers found in Yellowstone National Park in Wyoming. The geysers are the result of groundwater seeping down to hot layers of rock. When the water hits the rock, steam is produced. As we know, steam can be used to generate electricity.

large amounts of energy can be obtained. The prices of some of the devices are already starting to come down. In some instances, energy from the sun is already competing with other fuels.

Wind Energy

People have long used the energy of the winds. Perhaps the most familiar form of wind power is that derived from windmills. Old-style windmills were widely used in Europe and the Middle East for hundreds of years. Many were

HOW GEOTHERMAL ENERGY IS TAPPED

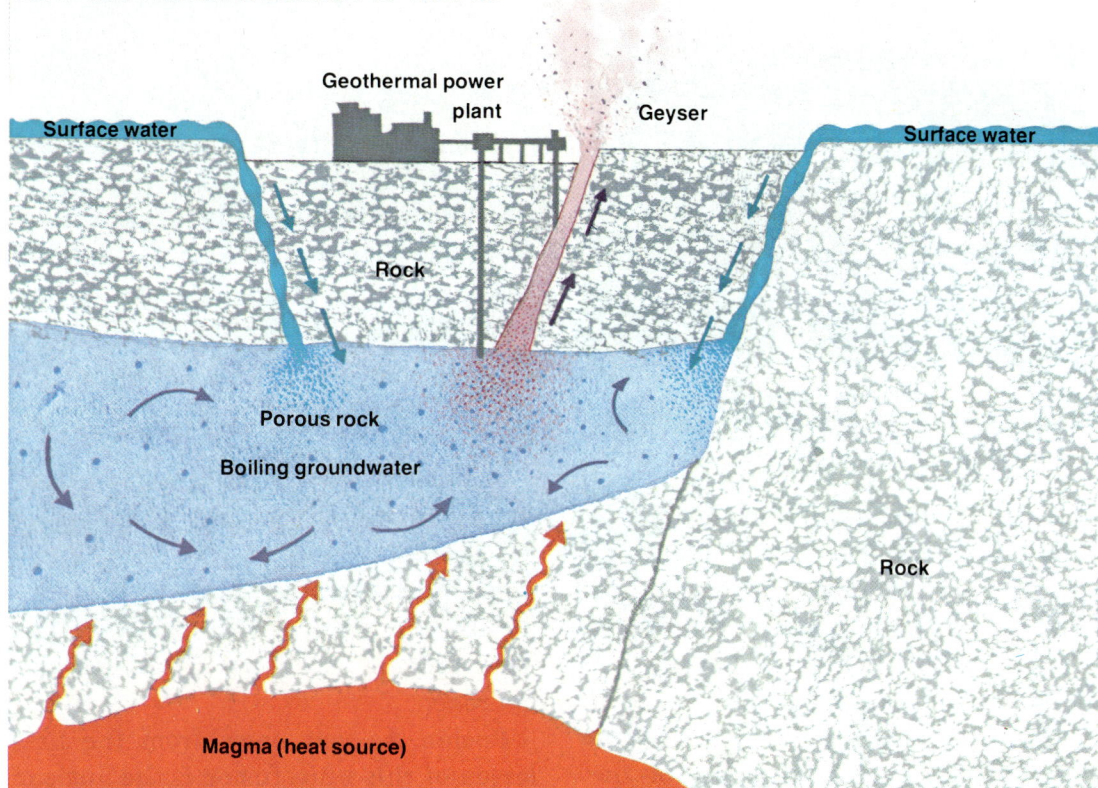

Geothermal power plant

Geyser

Surface water

Surface water

Rock

Porous rock

Boiling groundwater

Rock

Magma (heat source)

The diagram above shows how a geothermal energy source can be tapped to generate electricity. In New Zealand, a geothermal plant releases unused steam into the air. Other geothermal plants are located in Italy, Mexico, Japan, Iceland, the U.S.S.R., and the U.S.A.

What changes might take place at the tidal plant on the Rance River as the tide comes in?

Currently, the total work output of energy from *all* geothermal plants in the world is equal to that of *one* nuclear power plant. The future does not look all that bright, either. There just are not enough good drilling locations. At best, experts say geothermal power could one day supply about one percent of the electric power for the United States. The same holds true for most other countries with sites that can produce geothermal energy.

Tidal Energy

Tidal energy is another option that is limited by location. Tides vary in strength from place to place, and only the most powerful ones are suited for generating electricity. As with hydroelectric plants, dams are used to control the flow of water. The force of the moving water, as the tides move in and out, turns turbines that generate electricity.

There are only two tidal power plants in the world today. One is situated on France's Rance River. The other is in the Soviet Union. With methods presently used, the electricity from a tidal plant costs more than that from most other types of plants.

Scientists say there are only two ideal locations for tidal energy plants in North America. One is in Canada's Bay of Fundy. The other is in Cook Inlet, Alaska. Locate these two sites on the map on pages 34–35.

Other Energy Possibilities

Besides the options already mentioned, there are a number of other possibilities. Some of these may have great impacts on the world's energy supply in the coming years. Two of the most promising options are fusion and hydrogen.

Fusion. As you know from the first lesson of this unit, fusion is the nuclear process that keeps the sun burning. Scientists know how fusion takes place. They have even used the process to explode powerful hydrogen bombs. But they have yet to learn how to control this energy and put it to more useful work. This will not be easy to do. Extremely high temperatures are needed to start a fusion reaction. With methods now used, it is very difficult to achieve such high temperatures. There is also a problem with containing the reaction. No known substance can stand up to the intense heat without melting. If these problems were solved, fusion could one day supply almost endless amounts of energy. Some experts say fusion could supply *all* the world's energy needs.

Hydrogen. This element, which is one of the earth's most plentiful elements, might also become a good worker. By removing it from seawater and then turning it into a liquid, hydrogen may

Research on coal gasification and its costs is taking place in this plant in Chicago, Illinois.

become an excellent fuel for cars. Pure hydrogen is a very clean fuel. But, like many of the other energy options, it has problems of its own. In its pure form, hydrogen can explode very easily. Further studies may make it safer to work with. Do you remember what role hydrogen plays in the sun's fusion process?

Bioconversion. Still other energy options are coming up all the time. Recently, the term *bioconversion* has been in the news quite a bit. An example of this is the use of garbage as a fuel. Several cities have already produced sizable energy amounts by burning garbage. Another example of bioconversion is the growing of certain plants specifically for use as a fuel. These plants could be grown on special plantations, or "energy farms." Some plants high in sugar content are well suited for this purpose. The sugar can be turned to alcohol. Alcohol, in turn, can be used as a fuel. It is too early yet to know if the

cost of producing fuels through bioconversion will be able to compete with that of other fuels.

Coal gasification. Our remaining fossil fuels may also be made to work more efficiently in the future. One process, called coal gasification, might make coal more useful. It would convert coal into a clean gas or synthetic petroleum. Presently, more research on methods of coal gasification is needed.

Oil shale. This type of rock, common to arid regions, is so called because it contains crude oil. Scientists estimate shales in Wyoming, Colorado, and Utah have more oil than the known reserves of the entire Middle East.

Most oil shales are found in dry regions, far from the huge amounts of water needed to separate the oil from the rock. Bringing water and oil shale together makes mining oil shale very costly. Nevertheless, a number of oil companies have begun the work of mining oil shale reserves.

Keeping Facts in Focus

1. What is a renewable resource? Name some renewable energy resources.
2. How does fission differ from fusion?
3. If solar energy is so abundant, why can we not harness more of it?
4. What answers would you give to the key questions on page 248?

Working with Ideas

1. Why do some people support the generation of nuclear energy?
2. Why do some people oppose the generation of nuclear energy?
3. How might farmers help increase the world's energy supply?

Lesson 3 Transition into the Future

Reading Focus

1. This lesson will give you an idea of the difficulties involved when switching from one energy source to another. It also briefly sketches several differing energy plans and gives a futurist's view of energy. While going through this lesson, you might find yourself wanting to learn more about a particular subject. Instead of just wondering about it, why not do some energy research of your own? Encyclopedias would be a good place to start. They will provide useful information on a variety of energy-related topics. Then, search the contents pages of magazines for recent articles on your subject. You might also ask your teacher or librarian how to use guides to periodical literature. These handy references tell you where to find magazine articles on particular subjects. To get the latest information, watch your daily newspaper closely. Clip and save articles dealing with your subject of interest.

2. As you read this lesson, look for the answers to these key questions:
   a. How do the United States and Europe differ in regard to the breeder reactor?
   b. What might be some possible consequences of plutonium production?

Vocabulary Focus

transition   *conservation*   *futurist*

There are obviously a great number of energy options available today. With more research, many other energy doors could also open. But developing new energy sources takes time and money. Many factors must be considered. Is the source capable of damaging the environment? Will the source be renewable? How easily can it be transported?

Because of these factors, many of the energy options mentioned in the last lesson may not help us very much in the immediate future. For our purposes, the immediate future is from now until the early part of the next century. This space of time can be called the transition period. It is a time that will see the world shift from its dependence on fossil fuels to one or more other sources of energy.

How will we get along during this transition? We already know it will be difficult for oil and gas to meet increasing needs. What can be done now? How can we protect ourselves from the crisis that will certainly come if the earth's fossil fuels dry up before new sources are ready?

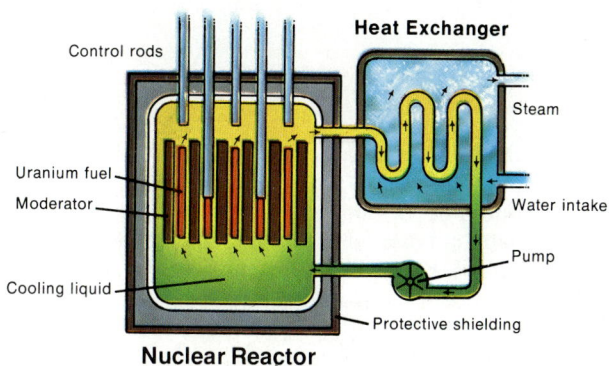

Nuclear Reactor

Control rods
Uranium fuel
Moderator
Cooling liquid

Heat Exchanger
Steam
Water intake
Pump
Protective shielding

In a standard reactor (above), nuclear fuel produces the heat that changes water into the steam used to generate electricity. What advantage does a breeder reactor have over a standard reactor? How many countries in the world have working nuclear power plants?

MAJOR NUCLEAR POWER PRODUCERS

Country	Working plants	Plants planned or under contruction	Capacity of working plants*
United States	81	100	65,420
France	32	32	27,633
U.S.S.R.	43	43	38,001
Japan	25	12	17,342
West Germany	12	17	10,316
United Kingdom	32	10	8,979
Sweden	10	4	7,715
Canada	12	13	7,597
Switzerland	4	3	2,034
East Germany	5	2	1,830
Belgium	5	4	3,621
Italy	3	5	1,330
Spain	4	15	2,047

*In thousands of kilowatts, as of December 30, 1982

The Breeder Plan

Some of the nations of the world believe the answer lies with the nuclear breeder reactor. Many European nations belong to this school of thought. Great Britain and France already have working breeder reactors. West Germany, Belgium, and the Netherlands are jointly building one. Farther east, Japan is also developing one. Some of these nations plan to build breeders for other nations as well.

The breeder converts energy faster than standard reactors. Experts say breeders can get up to 70 times more energy out of an equal amount of uranium than can standard reactors.

The United States, at least for now, is shying away from the breeder. Strong public sentiment has succeeded in slowing down its development. People against the breeder fear the bomb-making material (plutonium) it produces could easily fall into the wrong hands. Only a small amount of plutonium, about the size of an apple, is needed to make a powerful bomb. In addition, many U.S. citizens are concerned about the storing of radioactive wastes.

Why do the European nations insist on going ahead with breeders while the United States is stalling? The simplest answer is that the United States has a far greater variety of other fuels available. Many European countries are now almost solely dependent on Middle East oil for their energy. To become less dependent, these nations have decided to turn to the breeder.

The Coal Plan

Other nations believe the transition period should be one that calls for the increased use of coal. The United States

257

A Futurist's View

As a scientist, the geographer must be equipped to perform many functions. One of these is visualizing the future—what it *might* be like or *could* be like. Geographers or other experts who offer opinions about the coming years are called futurists. Sometimes their ideas may seem pretty "way out" or impossible. Yet often, with the passage of time, many of these visions become reality. Here is one futurist's idea of what might or could happen in the United States sometime in the future.

Dr. Theodore Taylor, a leading scientist, has envisioned the future United States in terms of his idea of "micropolis." In his plan, the nation's population would be divided into groups of about 3,000. Each group would live within its own micropolis, a community with an area of about one square mile.

Unlike the towns and cities of today, each micropolis would be self-sufficient—each community would be responsible for taking care of its own needs. All food for each community would be grown within its own borders. And all workers would be employed within their own micropolis. This would eliminate the need for cars and other fuel-consuming vehicles. People would get around on bicycles or on foot. Some electric vehicles might also be used.

Each community would produce its own energy within its borders. The sun would be called upon to do most of the work. Solar cells would convert sunlight directly into electricity. Organic wastes would be converted to hydrogen. The hydrogen could generate added electricity. Solar-heated water and air would keep homes warm or cool as needed.

The sources of energy would vary from community to community. Where there is plenty of rushing water, hydropower could be used to generate electricity. Where winds are steady and strong, windmills could help supply power. Geothermal sources and the tides might even be utilized in some locations.

Although communications would be open with other communities, each could very well get along on its own. By using renewable resources, the possibility of another energy crisis would be eliminated. By using clean energy sources, pollution would be reduced.

Perhaps this plan may seem more like fantasy than a possible reality, but other people recognize the possibility of such communities. In fact, one such place is now being planned in Arizona. It will be modeled after many of Dr. Taylor's ideas. Who knows? Perhaps someday, you will live in such an energy-efficient place.

is heading in this direction. It believes coal can supply many of its energy needs until more permanent and cleaner sources are put to work.

Currently, the U.S. government is urging industries and utilities to switch to coal-burning furnaces and boilers. It is also urging mining companies to produce more coal. At the same time, the government is making these changes difficult. Pressure from certain citizen groups has forced the government to place many limits on both the burning and mining of coal.

Methods to clean coal are available. However, they are very costly. Therefore, industries are hesitant about converting to coal.

WORLD ENERGY SOURCES

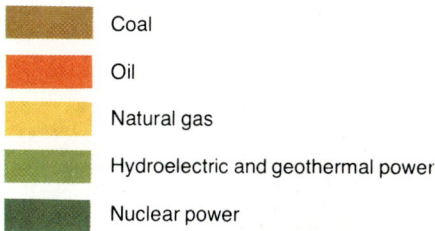

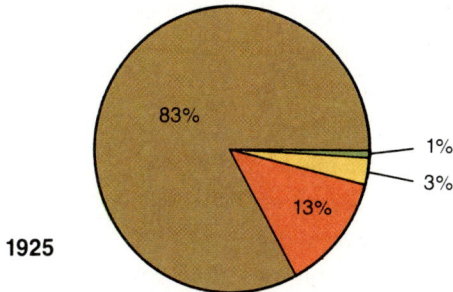

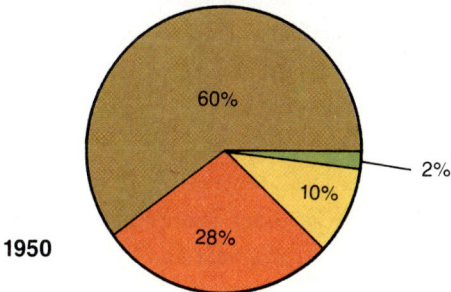

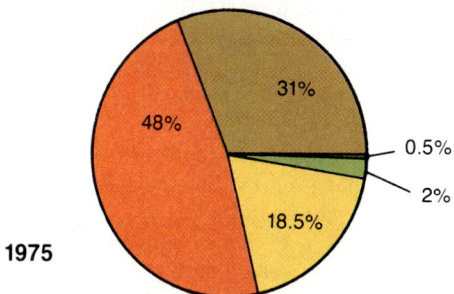

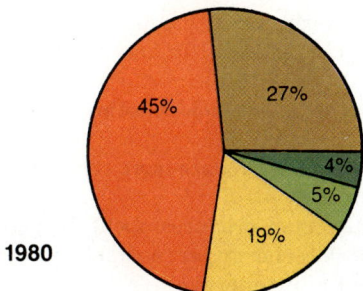

Coal

Oil

Natural gas

Hydroelectric and geothermal power

Nuclear power

What changes have taken place in the use of energy sources since 1925?

1925

- 83%
- 13%
- 3%
- 1%

1950

- 60%
- 28%
- 10%
- 2%

1975

- 31%
- 48%
- 18.5%
- 2%
- 0.5%

1980

- 27%
- 45%
- 19%
- 5%
- 4%

Conservation

Most nations agree that conservation should play a huge role during the transition. Conservation will allow us the time to examine all new sources, to weigh the advantages and disadvantages of each, and to find those sources that do not scar the environment.

Some futurists believe the world is capable of getting through the transition period by developing renewable resources. With such development, they say, the world will never have to worry about energy again.

Keeping Facts in Focus

1. Why has the United States lagged behind in breeder construction?
2. Describe how the United States plans to make it safely through the energy transition period.
3. What answers would you give to the key questions on page 256?

Working with Ideas

1. Why is it important for the United States to develop alternative energy sources during the transition period?
2. Why must trade-offs and compromises be an important part of any decision about the use of nuclear energy or other energy sources in the present and in the years to come?

Unit 7 REVIEW WORKSHOP

Test Your Geographic Knowledge

A. Tell whether each statement is true or false.
1. The sun is about 93,000 miles from the earth.
2. North Africa and the Middle East have huge oil reserves.
3. The Bay of Fundy is in the Soviet Union.
4. Anthracite is the most valuable of all coals.
5. Nuclear energy is used to boil water, which produces steam.
6. Brazil uses the wind to generate more than 80 percent of its electricity.
7. Breeder reactors can get 70 times more energy out of uranium than can standard reactors.
8. The energy delivered by the sun in one hour is enough to supply energy needs for a full year.
9. Geothermal energy sources are found in only a few places around the world.
10. Most nations of Europe have decided to hold off on the construction of breeder reactors.

B. Identify each of the following energy forms.
1. I rely on uranium.
2. Sometimes I take the form of a geyser.
3. As I move in and out twice a day, my energy can be used to produce electricity.
4. My energy is released when fossil fuels are burned.
5. My energy changes light, air, and water into food.
6. My energy was used in early times to pump water and grind grain in mills.
7. My potential energy is turned into motion energy with the help of huge dams.

Apply Your Reading Skills

A. See how well you can use the energy terms listed below. First, look them up in the Glossary at the back of this text to make sure you understand the meanings. Then, supply the correct missing word in each sentence.

thermal	renewable	bioconversion
fusion	radiation	evaporation
fission	hydrogen	gasification

1. The energy in the sun comes from a process called _____.
2. Solar energy is an example of a _____ resource.
3. The burning of garbage as a fuel is an example of _____.
4. The process that changes water into water vapor is called _____.
5. The process of converting coal into a clean gas or liquid petroleum is called _____.
6. _____ atoms are very common on earth.
7. Most of today's nuclear power is generated from the splitting, or _____, of uranium atoms.
8. Another term used to describe heat energy is _____ energy.
9. Two forms of energy found in the sun's _____ are heat and light.

B. Use encyclopedias, books, magazines, or other reference tools to do research on ONE of the following topics.
1. What simple fuel- and electricity-saving measures can families take to conserve energy in their homes?
2. What is gasohol, and can its use help the world conserve crude oil?
3. What progress has science made with electric automobiles?
4. What are the advantages and disadvantages of solar heating?

Apply Your Geographic Skills

Use the world energy production and consumption graph on page 261 to answer the following questions.
1. Which world regions consume more energy than they produce?
2. Which regions produce more energy than they consume?
3. Which region produces about 13 percent of the world's energy yet consumes only about 2 percent of its energy?

THE WORLD ENERGY PICTURE

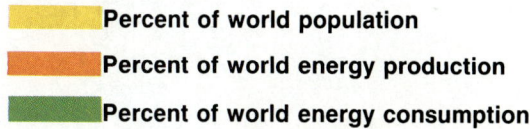

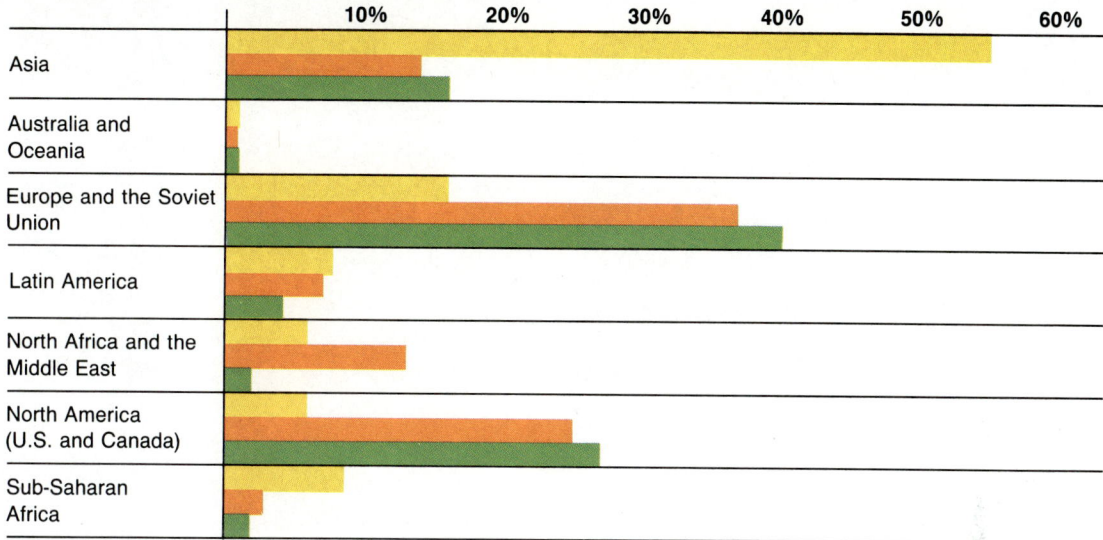

■ Percent of world population
■ Percent of world energy production
■ Percent of world energy consumption

	10%	20%	30%	40%	50%	60%

Asia

Australia and Oceania

Europe and the Soviet Union

Latin America

North Africa and the Middle East

North America (U.S. and Canada)

Sub-Saharan Africa

4. Which region has over 50 percent of the world's population yet consumes less than 20 percent of its energy?
5. Which region has less than 10 percent of the world's population and consumes more than 25 percent of its energy?
6. Which region leads the world in energy consumption?
7. Which region leads the world in energy production?

Apply Your Thinking Skills

A. Often facts may be used to prove a statement true or false. Sometimes, however, we do not have the facts needed to make a final decision. Read each statement below. Then, using the facts presented in this unit, tell whether you think each statement is true, false, or doubtful (a statement is doubtful if facts are lacking).
  1. Most of the earth's energy can be traced back to the sun.
  2. Most of today's major fuels are renewable resources.
  3. At present consumption rates, our fossil fuels will eventually be used up.
  4. Electrical energy can be generated from steam.
  5. In the future, each person will have a more equal share of the energy used on earth.

B. Use your imagination and the facts you know about energy to write a story about one of the following possibilities.
  1. The earth runs out of fossil fuels in 1982.
  2. Plutonium is stolen from a nuclear power plant. The thief plans to build a nuclear bomb.

Discuss These Points

Beginning in the 1970s, there was much controversy concerning nuclear power. Some people feel it is the world's most logical energy source. Others do not want any type of nuclear power. They believe the risks are far too great. What are some of the risks? Do you think these risks are worth taking? On the whole, do you believe nuclear power is safe? Why or why not?

Expand Your Geographic Sights

Asimov, Isaac. *How Did We Find Out About Energy?* (Walker and Company, 1975).

Boyd, Waldo T. *The World of Energy Storage* (Putnam, 1977).

Pringle, Lawrence. *Energy: Power for People* (Macmillan, 1975).

Smith, Norman F. *Energy Isn't Easy* (Putnam Publishing Group, 1984).

Yates, Madeleine. *Earth Power: The Story of Geothermal Energy* (Abingdon, 1980).

Unit 8 The World's Population

If the world's more than 4 billion people were to stand side by side, they would form a line long enough to circle the equator about 100 times.

Lesson 1 World Population Growth

Reading Focus

1. Sometimes in your reading, you will come across a word that is unfamiliar. You know that you can use a glossary or a dictionary to find a word's meaning. But often you can find the meaning of a word from its *context*— from the sentence or paragraph in which it is used. Some words are defined within a sentence. ("Population *density* is a figure that tells us the average number of people living on a square unit of land.") Sometimes other words in a sentence, or in another sentence, help you to understand a word. Can you tell what *urbanization* means from these sentences? *The movement of people from rural to urban areas has been taking place for thousands of years. In recent years, however, urbanization has taken place at a faster rate than ever before.* Look for such context clues when an unfamiliar word stumps you.

2. As you read this lesson, look for answers to these key questions:
   a. What is doubling time?
   b. What major change in the early period of world population growth made it possible for people to begin to live longer than 20 to 30 years?
   c. What two major sets of changes brought about a rapid increase in world population in about the last 300 years?

Vocabulary Focus

doubling time	*birth rate*	*death rate*
rate of natural increase		

Once, there was a collector of statues. Let us call the collector Thera. Thera resolved, after getting her first statue, to double her collection each year. From 2, in the second year, her collection grew year by year to 4, 8, 16, 32, and so on. By the 30th year, Thera had a total of 536,870,192 statues. After three more years and three more doublings, the collection totaled more than 4 billion statues. One more doubling and Thera's collection would top 8 billion.

Thera's story may seem unlikely, but it has a purpose. It is a capsule story of the growth of population. Earth now has a total of over 4 billion people, a figure that took tens of thousands of years to reach. Often, the doubling time—the period from one doubling of the earth's population to the next—took several thousands of years. But in the last 300 years or so, the doubling time has decreased sharply. The last doubling of the population took place between 1930 and

WORLD POPULATION GROWTH

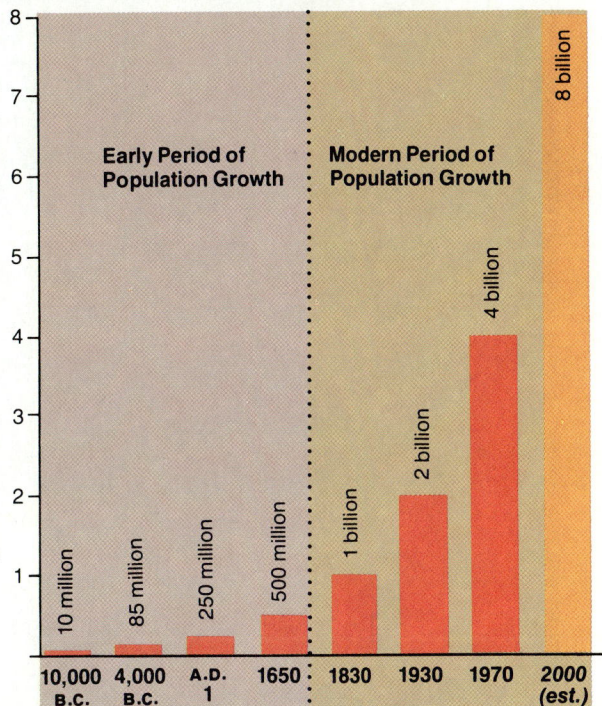

By the year 2000, the population of São Paulo, Brazil, may be about 26 million. It is now about 10 million.

1974, a period of 44 years. The next doubling time might be even less.

World Population—Then and Now

We can only guess at the actual number of people living on earth at any given time for much of the distant past. By using estimates for the past, along with the more accurate counts of modern times, however, we get a picture of world population growth over time. The graph on this page tells the number story. Refer to this graph often as you read about population changes through the ages.

Early population growth. The early period of population growth covers all the years up to A.D. 1650. Before the first date shown on the graph, people lived as hunters and gatherers. Their food supply was often limited and always uncertain. Few people, if they survived birth or childhood, lived longer than 20 to 30 years. In the long period before 10,000 B.C., there were years when births and deaths balanced each other and no population growth took place. There were other years when deaths outnumbered births and population decreased. But there were enough years within this period when births outnumbered deaths to register a slow but continual increase in the population. What was the estimated world population about 10,000 B.C.?

Between 10,000 and 4,000 B.C., an important change took place that made

265

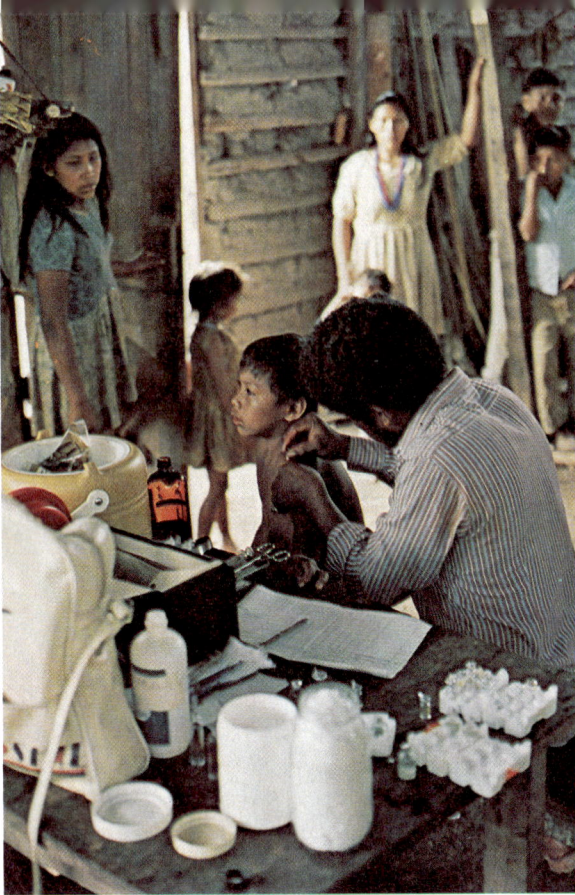

While others wait their turns in a Venezuelan health clinic, a child receives a vaccination.

it possible for more people to live longer. During this period, many people began to cultivate plants and tame wild animals. With the development of agriculture and its eventual spread across a large part of the world, the food supply became less limited and less uncertain. With more food available, there were fewer infant deaths. Adults lived longer than 20 to 30 years. As yearly increases mounted, the population doubled three times within a period covering 6,000 years—an average doubling time of 2,000 years. What was the estimated population of the world about 4,000 B.C.?

Four thousand years later, by the year A.D. 1, the population doubled twice to reach 250 million. Doubling time again averaged 2,000 years. But by A.D. 1650,

when the population doubled once again, the doubling time was shorter. It was only 1,650 years.

Modern population growth. Between 1650 and 1974, the earth's population doubled three times. Each time the doubling time was shorter. What was the doubling time between 1650 and 1830? Between 1830 and 1930? Between 1930 and 1974? When is the next doubling of the earth's population likely to take place? The decreasing amount of time that it has taken the world's population to double means that the population in the modern period has been increasing at a rapid rate. Compared to past totals, the world's population is becoming extremely large.

We can link the rapid and large increases in the world's population in a little more than 300 years to two major sets of changes: improving health practices and industrialization. With better health practices, people in many parts of the world began to increase their life spans to a greater degree than in the early period. At the same time, industrialization made it possible for many people to produce more goods, making it easier for them to meet their needs and wants. Both sets of changes worked together to bring about a decreasing number of deaths and a rapid increase in the total number of people.

Improving health practices began in the 1700s with a series of discoveries about the spreading of diseases. Throughout the early period of population growth, such diseases as smallpox, diphtheria, scarlet fever, and cholera had killed millions of people each year.

Improvements in waste disposal, water treatment, personal hygiene, and food preservation led to the discovery that the spread of most diseases could be controlled. Other discoveries brought about the development and use of vaccines and new techniques in surgery, health care, and sanitation.

Industrialization began with the discovery that steam could be harnessed and used to run machinery. As societies industrialized, other changes followed. Cities expanded. Improved tools and new methods of agriculture greatly increased the food supply. Modern processing techniques reduced food waste and spoilage. Faster and better methods of transportation increased trade and made resources more available to more groups of people. As levels of education rose and methods of communication improved, these changes spread from their centers of origin to other parts of the world.

A Closer Look at Population Today

So far we have been looking at population growth for the world as a whole. To get a clearer picture of population growth today, we need to look more closely at the differences that exist in population growth among nations. One of the first differences we need to look at is the rate of natural increase.

Rate of natural increase. Two things are important in figuring rates of natural increase. These are birth rates and death rates. The birth rate is the number of births there are each year for

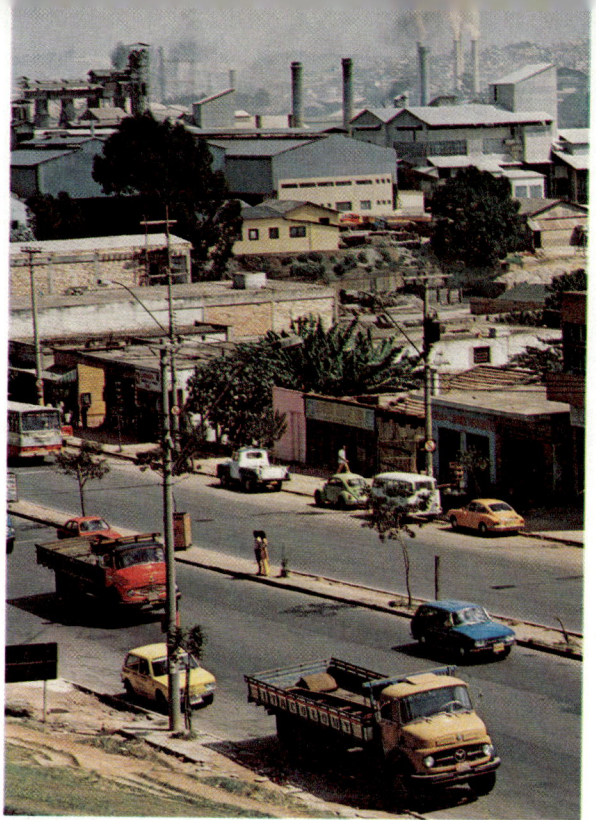

What are some of the benefits the industries of Belo Horizonte have given to Brazil's people?

every 1,000 people. The death rate is the number of deaths for every 1,000 people. The difference between the birth rate and the death rate, expressed as a percent, is the rate of natural increase. For example, say the birth rate of a nation is 20, and its death rate is 10. The difference between the figures is 10. The rate of natural increase is 1 percent (10 is 1 percent of 1,000).

The rate of natural increase is used to determine future doubling time. A rate of natural increase of 2 percent or more is considered high. At 2 percent, the population will double in 35 years. If the rate is higher, the population will double in less than 35 years. Most developing nations have a rate that is 2 percent or higher. Most industrialized nations have a rate of natural increase that is lower than 2 percent.

POPULATION STATISTICS: A SELECTED LISTING

Selected Nations	Population (in millions)	Birth Rate	Death Rate	Rate of Natural Increase	Doubling Time (in years)
Argentina	30.1	26	9	1.6%	45
Australia	14.5	16	7	0.8%	90
Austria	7.6	13	12	0.1%	—
Bangladesh	99.6	46	20	2.8%	26
Japan	119.9	13	6	1.0%	72
Libya	3.7	48	13	3.4%	21
Mexico	77.7	34	5	3.3%	22
Nigeria	88.1	50	19	3.2%	23
Turkey	50.2	34	11	2.3%	31
United Kingdom	56.0	13	12	0.0%	—
United States	236.4	16	9	0.7%	103
Western Samoa	0.2	37	7	3.0%	24
The World	**4,737.0**	**28**	**11**	**1.8%**	**40**

Based on 1981 figures

Rates of natural increase and doubling times for different nations are listed on the chart on this page. First examine the birth and death rates for the world. What is the world's present rate of natural increase? At this rate, what is the doubling time for the world's population?

Now look at the birth and death rates for the United Kingdom. Because its birth and death rates are the same, it has a zero rate of natural increase. Its population is neither increasing nor decreasing. Should its birth rate fall below its death rate, its population would decrease. It would have a negative rate of increase. Such is the case for Austria, another nation listed.

Finally, study the rates and doubling times of other nations on the chart. Which nations have rates of natural increase that are higher than 2 percent? What is the doubling time for each of these nations? What is the rate of natural increase for the United States? At this rate, how long will it take for our population to double? Based on their rates of natural increase, which nations on this chart would you consider industrialized nations? Developing nations?

The World Health Organization

The World Health Organization (WHO), an agency of the United Nations, aims to bring all people to the "highest levels of health." This is a tall order, one not yet achieved, but one that has helped to lessen the suffering of many people, especially those in developing nations.

In fulfilling its mission, WHO has worked with government planners, hospital staffs, teams of scientists seeking the causes of diseases, people in rural communities, and urban slum dwellers. By training health workers and by starting educational programs, WHO has improved the nutrition and the sanitation practices of many groups of people. By conducting mass inoculations in high-risk areas, the agency has helped to control the spread of such diseases as smallpox, influenza, and diphtheria. By spraying the breeding grounds of insects and other pests with chemicals, it has helped to limit the spread of other diseases, such as malaria, snail fever, and river blindness.

One of WHO's most effective programs has wiped out smallpox, a viral disease that usually killed one of every four of its victims. Those it did not kill, it left with disfiguring facial scars and, sometimes, blindness.

In 1967, WHO declared all-out war on the disease. At that time, over 130,000 cases were reported. Each year thereafter, the number of reported cases of smallpox declined. In 1978, only two cases were reported, and these cases occurred because of an accident in a laboratory where samples of the smallpox virus were stored for research. The victory over smallpox has been proclaimed. But WHO will remain ever watchful, in case the smallpox virus attacks in some future age. In the meantime, WHO will direct its energies toward other killer diseases.

Stages of population growth. As a nation develops, it appears to move through four stages of population growth.

In the *first* stage of growth, a population has high birth rates and high death rates. Before the 1700s, people everywhere were in this stage of growth, one marked by small increases in population over long periods of time. Today, there probably is no nation that has all of its people in this first stage. However, there may be isolated groups of people within some nations who still remain in the first stage. Where these people exist, they live mainly as hunters and gatherers, or they farm using the same kinds of tools and methods their ancestors of long ago used.

The developing nations of Asia, Africa, and Latin America have moved into the *second* stage of population growth. Birth rates remain about as high as those in the first stage. Death rates, however, have begun to fall. As the rates of natural increase rise to 2 percent or higher, population totals increase rapidly. About half of the population is under 15 years of age. Less than 3 percent of the people may be over 65. Although some industrialization has taken place, the majority of people still practice subsistence farming.

Most of the fully industrialized nations of the world have moved through the second into the *third* stage of population growth. Birth rates fall, and death rates remain low. The rates of natural increase range from less than 2 percent to about 1 percent. As the rates of natural increase fall, the number of young people in a population decreases. At the same time, the number of people over 65 grows.

At different times in history, a few nations have entered the *fourth* stage, which is marked by even lower birth and death rates than those in the third stage. Fourth stage nations have rates of natural increase lower than 1 percent. Birth rates may be even with death rates, or the birth rates may be lower than the death rates. A nation may enter this stage because war, disease, or famine has taken many lives. Still another cause may be the decision on the part of many couples to have no children. Whatever the cause or combination of causes, a nation in this stage may, after a few years, return to either the second or third stage of population growth.

Keeping Facts in Focus

1. What is the birth rate? The death rate? The rate of natural increase?
2. How do birth rates and death rates compare when a nation has a zero rate of natural increase? A negative rate?
3. What answers would you give to the key questions on page 264?

Working with Ideas

If you were a leader of a developing nation that had high birth and death rates and 70 percent of its people engaged in subsistence agriculture, what steps would you take to bring your nation into the next stage of population growth? What factors might limit your nation's entry into the next stage?

Lesson 2 Population Patterns and Groups

Reading Focus

1. Interpreting generalizations and the facts they contain is an important reading skill. Few generalizations are true of all people, in all places, at all times. So generalizations about people usually need to be qualified, or limited, by such words or phrases as *may, many, usually, on the average,* etc. For example, a generalization in this lesson reads as follows: *In some developing nations, fewer than 50 percent of the people can read or write.* Notice the word *some* and the phrase *fewer than 50 percent. Some* means more than one. *Fewer than 50 percent* is a phrase that covers any percentage between 0 and 49. Watch for other generalizations as you read this lesson. Take time to interpret them carefully.

2. As you read this lesson, look for answers to these key questions:
   a. What fact do graphs or maps about population distribution make most evident?
   b. What two kinds of population movements influence the distribution of population over periods of time?
   c. What are some of the groups into which a population can be divided to determine its characteristics?

Vocabulary Focus

demographer	*migration*	*statistic*
ethnic	*urbanization*	*immigration*
literacy	*population growth rate*	*emigration*
population density		

In the world's journey into the future, more and more people are concerned with the direction in which population growth is taking them. Will there be enough land to support growing populations? Will resources be sufficient? What will be the most pressing future need? Will it be food supply, education of the young, or care of older people?

To find answers to these questions, more and more governments are keeping records that population experts called demographers can use to discover population characteristics. Governments collect data about where people live, what ages and sex they are, what races or ethnic groups they belong to, what languages they speak, and what

Population for Regions of the World

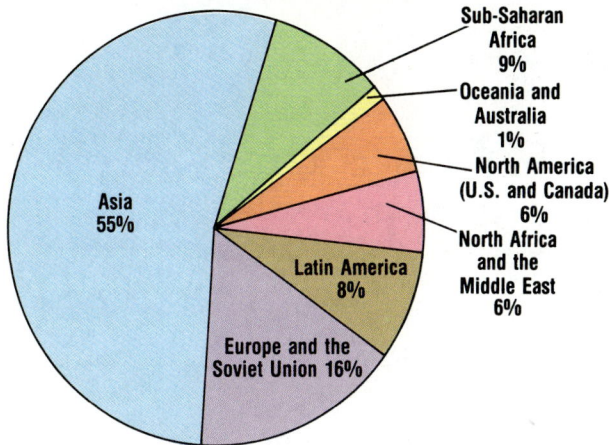

Sub-Saharan Africa 9%
Oceania and Australia 1%
North America (U.S. and Canada) 6%
North Africa and the Middle East 6%
Asia 55%
Latin America 8%
Europe and the Soviet Union 16%

Land Area for Regions of the World

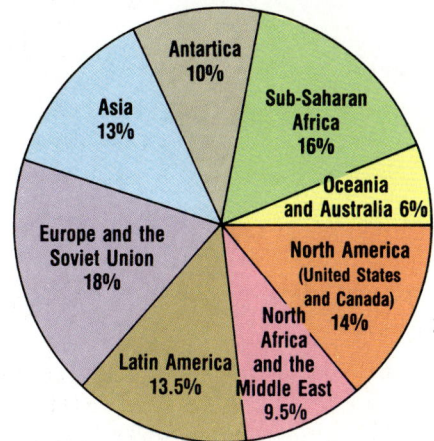

Antartica 10%
Asia 13%
Sub-Saharan Africa 16%
Oceania and Australia 6%
North America (United States and Canada) 14%
Europe and the Soviet Union 18%
North Africa and the Middle East 9.5%
Latin America 13.5%

religions they practice. They also collect data about incomes and occupations, about literacy (the ability to read and write), and education levels. Experts use the data to divide people into groups with the same general characteristics.

By mapping data on similar groups, the experts discover patterns of distribution. By charting and graphing data, they discover relationships and make comparisons. Their findings give national governments and world organizations the "road maps" they need in planning for the future.

Population Distribution

Graphs or maps can show the distribution of population. Studying these forms makes one fact most evident: population is distributed unevenly over the earth's land surfaces.

Distribution by regions. People make their homes in every geographic region except Antarctica. The graph above shows the percentage of the world's population living in different regions. Another graph to the right shows the per-

centage of the world's land area each region contains. By comparing the information on these two graphs, you begin to understand even more about the unevenness of population distribution. For example, Asia has 55 percent of the world's people and 13 percent of its land. What percentage of the world's land does North America have? What is its percentage of the world's people? What comparisons can you make for each of the other inhabited regions?

Distribution and density. The map on pages 272–273 shows both distribution and density. Population density is a figure that tells us the average number of people living on a square unit of land. We find this figure by dividing the number of people living on a given amount of land by the area of that land.

For example, the tiny European nation of Andorra has a population of 45,000. Its area is 466 square kilometers. Its population density is 97 persons per square kilometer. Belgium, another European nation, has an area of 30,562 square kilometers, But it has a population of 9,872,000. Belgium's population

NORTH

AMERICA

Vancouver
Seattle
Minneapolis Detroit
Chicago Toronto
St. Louis
New York
San Francisco Washington, D.C.
Los Angeles
Dallas
ATLANTIC
Houston
Monterrey
Havana
Guadalajara
OCEAN
Mexico City
Caracas

PACIFIC
Bogotá

OCEAN
SOUTH
Lima
AMERICA

World Population
Distribution

Santiago
Rio de Janei
São Paulo

Montevideo
Buenos Aires

Persons	
per sq mi	per sq km
0–5	0–2
5–50	2–20
50–100	20–40
over 100	over 40

POPULATION DENSITY BY REGION

	Total Population	Persons Per Square Km
Asia	2,594,808,000	128
Europe and the Soviet Union	765,400,000	28
Latin America	393,526,000	20
North Africa and the Middle East	271,649,000	20
Sub-Saharan Africa	426,669,000	19
North America (U.S. and Canada)	261,555,000	14
Australia and Oceania	23,477,000	3

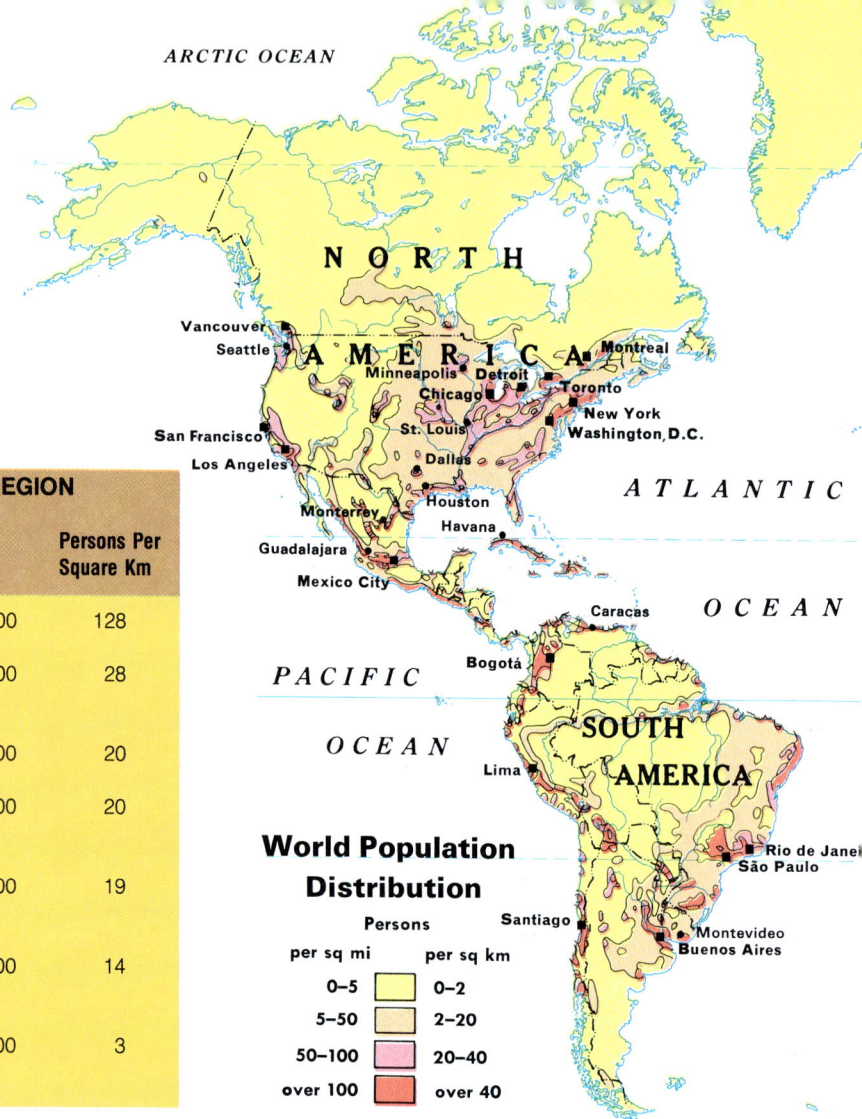

density of 323 persons per square kilometer is much greater than Andorra's.

On the map on pages 272–273, population densities range from very low to very high. Use the map to find the areas where population densities are very low. Find the areas where population densities are very high. What is the population density of your home area?

Where people live often depends on such things as the steepness of the land, soil fertility, precipitation, and climate. For the most part, few people live in those parts of the world that have steep, rugged highlands, vast stretches of des-

ert, or temperatures below freezing for much of the year.

Population density is generally higher in those areas that have mostly level land or gently rolling hills, climates with enough seasonal or yearly precipitation, and growing seasons long enough to produce a food supply.

Three large areas with very high population densities are Europe, India, and East Asia. Smaller high-density areas are generally found along seacoasts, especially where rivers empty into well protected harbors. Others are found along river valleys, at places where two

Arctic Circle

Stockholm
Glasgow
Leningrad
Moscow
Novosibirsk
London
Berlin
Paris
Warsaw
E U R O P E
Madrid
Rome
Istanbul
A S I A
Shenyang
Casablanca
Teheran
Tashkent
Peking
Seoul
Tientsin
Tokyo
Osaka
Alexandria
Delhi
Chungking
Shanghai
Cairo
Karachi
P A C I F I C
Tropic of Cancer
A F R I C A
Calcutta
Hong Kong
Bombay
Madras
Manila
Bangkok
O C E A N
Ho Chi Minh City
(Saigon)
Lagos
Singapore
Equator
Jakarta
Kinshasa
Surabaya
I N D I A N

O C E A N

AUSTRALIA
Tropic of Capricorn
Johannesburg
Sydney

Melbourne
© Follett Publishing Company

■ Selected urban areas with more than three million people

• Selected urban areas with less than three million people

or more rivers meet, or where natural resources are abundant. In most of these places, people have established cities.

Population Movements

Two population movements influence population patterns over a period of time. One is migration, the movement of people from continent to continent or from nation to nation within a continent. The other is urbanization, the movement of people from rural areas to urban areas within nations.

Migration. The many migrations that have taken place throughout history have helped to give many populations a mixture of races, ethnic groups, languages, and religions. Migrations decrease the population of one place and increase the population of another. Migration, then, affects the rate at which the population of a nation grows. The difference between birth and death rates in a population is only the natural rate of increase. To figure the total population growth rate of a nation, the effects of migration must also be taken into account.

Demographers—Sorters of Vital Statistics

Birth, death, marriage, the contracting of a communicable disease, employment—these and other important human events become vital statistics when they are recorded by governments. Governments record many vital statistics as they happen. They collect other vital statistics in periodic surveys called censuses.

Demographers study vital statistics to discover the changes that are taking place in populations. In studying statistics, they sort them into groups based on such factors as age, sex, place of residence, income, race, national origin, or level of education. Demographers tabulate, graph, or map statistics. They discover patterns and relationships among the patterns. Their work enables government officials to make predictions about future growth, to draft legislation, or to take care of health, housing, and education needs of the people. Many business people also use the services of demographers to determine the best locations for stores, factories, office buildings, or such sport and entertainment centers as tennis clubs, campgrounds, and movie theaters.

The opening lesson of each unit in the second part of this book contains a National Profiles chart, which lists statistics for each independent nation in the region. One of the statistics listed is the population growth rate. This figure includes the effects of migration on the rate of natural increase. For example, the chart on page 268 lists the rate of natural increase for the United States as 0.6 percent. The National Profiles chart for Unit 15 shows that the United States has a population growth rate of 0.9 percent. Migration accounts for the difference between the two rates.

Check the appropriate National Profiles chart for the population growth rate of each of the other nations listed on the chart on page 268. Compare the population growth rate with the rate of natural increase. Which nations are gaining population through immigration (people entering the country)? Which nations are losing population through emigration (people leaving the country)? Is there any nation for which the population growth rate and the rate of natural increase are the same?

Urbanization. The movement of people from rural to urban areas has been taking place for thousands of years. In recent years, however, urbanization has taken place at a faster rate than ever before. In 1920, only 10 percent of the world's people lived in cities with populations greater than 100,000. Today, such cities hold 40 percent of the world's people.

The number of large cities is also growing. In 1950, there were 71 cities with populations of one million or more. By 1975, there were 181. In general, 70 percent or more of the people in industrialized nations live in urban areas. In developing nations, the urban population is generally about 30 percent. However, urbanization in many developing nations is increasing rapidly.

Education of the young, employment, and care of the elderly are vital needs in all nations. In a village in Zaire (top), children learn skills they will need as adults. Unemployed people in India (middle) often compete for only a few job openings. In Japan (bottom), three generations or more often live in one household.

Urbanization at a rapid pace causes problems for many of the world's cities. Most people who leave rural areas for the cities come looking for jobs and a better standard of living. They come, however, with little education. Some may be illiterate (unable to read or write). When they come to the city, they find they are not qualified for the jobs that will pay them well. They either remain unemployed, or they take the jobs that no one else wants.

As city populations increase, so do demands for housing, water, electricity, sewage disposal, schooling, and medical care. Cities in industrialized nations have difficulties meeting these demands. Cities in developing nations have even greater difficulty providing for the people who think that in the cities they will find the fulfillment of their dreams.

Population Groups

Studies that separate a population into different kinds of groups reveal social, economic, and political relationships. Such studies help to classify nations as industrialized or developing and populations within nations as young or old, urban or rural, literate or illiterate, and rich or poor.

Population profiles. Dividing the population of a nation into age and sex groups provides a profile of the population. Study the population profiles on this page. The horizontal bars show the percentages of the male/female population in each age group.

Mexico's population profile is quite different from that of the United States or Sweden. Mexico's profile has the shape of a pyramid and shows that the population is growing at a rapid rate. About 18 percent of the people are under 5 years old. About 66 percent of the people are under 24. Mexico has a very youthful population.

The shape that the profile of the United States takes shows that its population is growing very slowly. Compared with Mexico, the United States has a greater percentage of people over 50. About 41 percent of its people are 24 or younger. And of this group, only about 7 percent are under 5.

POPULATION PROFILES

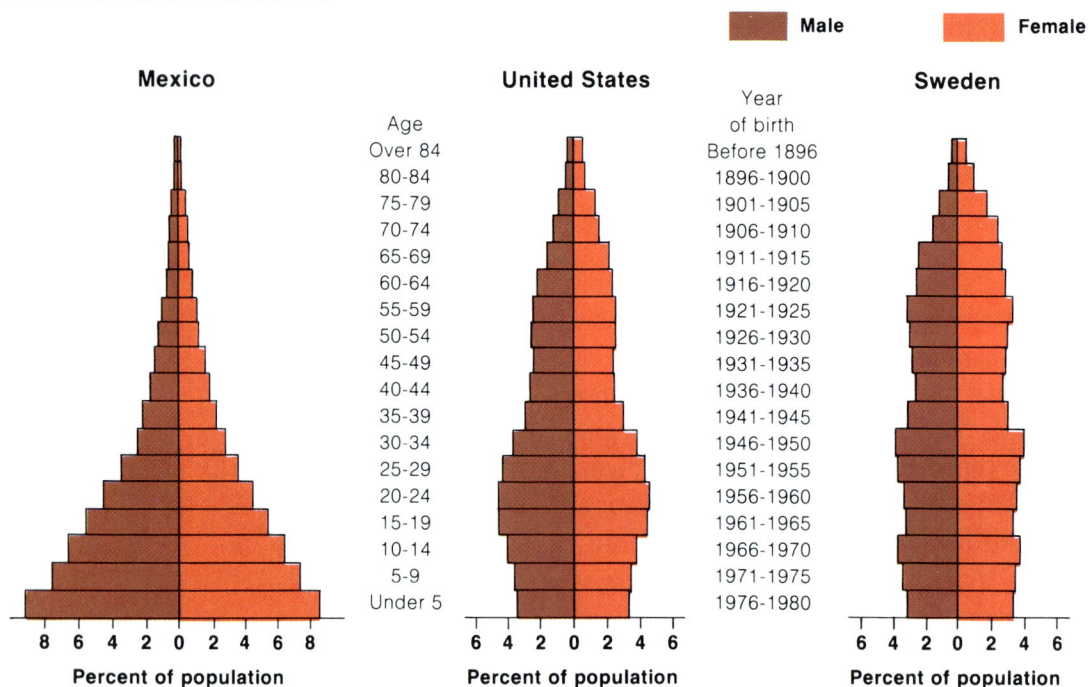

Age	Year of birth
Over 84	Before 1896
80-84	1896-1900
75-79	1901-1905
70-74	1906-1910
65-69	1911-1915
60-64	1916-1920
55-59	1921-1925
50-54	1926-1930
45-49	1931-1935
40-44	1936-1940
35-39	1941-1945
30-34	1946-1950
25-29	1951-1955
20-24	1956-1960
15-19	1961-1965
10-14	1966-1970
5-9	1971-1975
Under 5	1976-1980

Mexico — Percent of population

United States — Percent of population

Sweden — Percent of population

Male   Female

Sweden has a profile that shows the population has remained about the same since 1921. It has roughly equal numbers of people in almost all age groups under 60 years of age.

Urban/rural patterns. Keeping track of changes in urban/rural groupings shows the degree of urbanization that is taking place throughout the world or within a nation. The percentage of people living in the urban areas of a nation is another statistic listed on the National Profiles charts in Part Two.

Literacy. Most industrialized nations have a high degree of literacy (90 percent or higher). In some developing nations, fewer than 50 percent of the people can read or write. Literacy rates for most nations are also listed in the National Profiles charts.

Income. Another valuable statistic is per capita income—the average yearly income of each person. People in developing nations usually have per capita incomes of less than $2,000 in U.S. currency. Which nations listed on this page might be developing nations?

Nation	Urban	Rural
Argentina	86%	14%
Australia	85%	15%
Austria	55%	45%
Bangladesh	11%	89%
Japan	76%	24%
Libya	53%	47%
Mexico	65%	35%
Nigeria	20%	80%
Turkey	47%	53%
United Kingdom	76%	24%
United States	79%	21%
Western Samoa	21%	79%

Keeping Facts in Focus

1. Which land region has over half of the world's population?
2. What is population density?
3. On what kind of land is population density usually very low? On what kind of land is it usually higher?
4. What is migration? Urbanization?
5. What is included in determining the population growth rate that is *not* included in determining the rate of natural increase?
6. About what percentage of the people of industrialized nations live in urban areas? Of developing nations?
7. What answers would you give to the key questions on page 270?

Working with Ideas

If you could see into the future of a previously uninhabited continent, one similar in many ways to North America, where might you expect large cities and high population densities to develop? Where might you expect population densities to be very low? In a nation's first century of development, which might you expect to be greater—immigration or emigration? Why?

Nation	Literacy Rate	Per Capita Income (in U.S. dollars)
Argentina	94%	$4,610
Australia	99%	$10,087
Austria	98%	$10,995
Bangladesh	25%	$117
Japan	99%	$8,947
Libya	50%	$7,600
Mexico	74%	$2,273
Nigeria	25%	$827
Turkey	70%	$1,096
United Kingdom	99%	$8,620
United States	99%	$11,725
Western Samoa	90%	$770

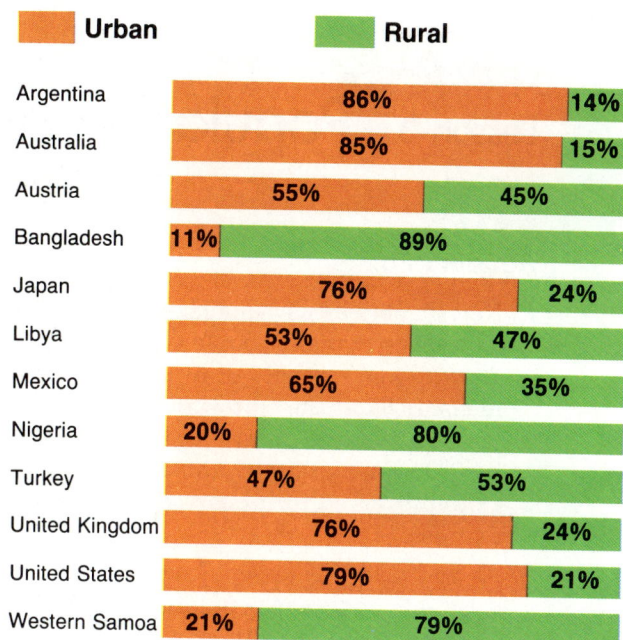

Lesson 3   Population and the Future

Reading Focus

1. Newspapers or newsmagazines often use editorial cartoons to express an opinion or to make people aware of a problem. Sometimes, cartoons tell a story to make people think about important political, economic, or social ideas. Look at the editorial cartoon on page 279. Study it carefully. What opinion about population growth do you think the cartoonist is expressing? Do you agree with this opinion? Why or why not? Bring to class other examples of editorial cartoons. See what ideas or opinions they seem to express.

2. As you read this lesson, look for answers to these key questions:
   a. What changes in population growth and distribution might take place by the end of the 20th century? What changes might take place in birth rates? In death rates? In urbanization?
   b. What are some environmental problems related to population growth?
   c. What are some things that are being done on the political level to offset population problems? On the economic level? On the social level?

Vocabulary Focus

trend   *forecast*   *life expectancy*

The cartoon on page 279 expresses an opinion held by some demographers. The opinion is this: population pressures on the earth may well put our world out of whack. The cartoonist is saying that soon—20, 35, 50 years from now, perhaps even in your lifetime—there will be too many people and too few resources to meet population needs.

Population Trends

Let us look at some of the trends that are emerging from present population patterns.

Growth and distribution. If its present growth rate stays nearly the same, the world's population is expected to reach 8 billion as the 20th century ends. Even if birth rates decline, the lowest population forecast for the year 2000 is 5.9 billion people.

By the year 2000, 12 nations will join the ranks of the 16 nations that, at the present time, have populations greater than 50 million. The Philippines, Iran, Thailand, South Korea, Zaire, and Colombia are likely to be among the 12.

Of the 9 nations that presently have populations between 50 and 100 million,

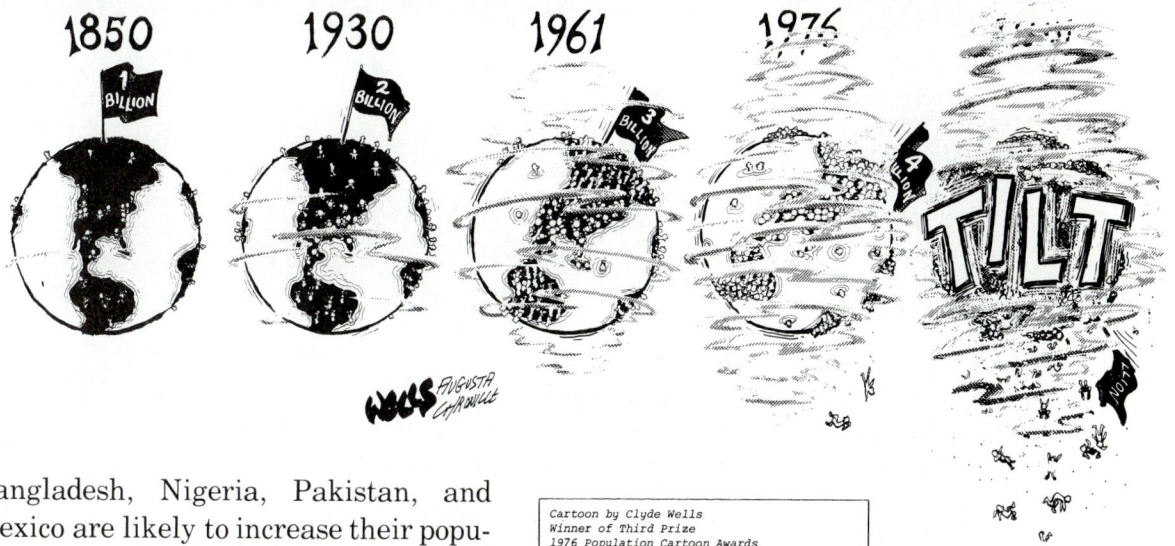

1850   1930   1961   1976

Cartoon by Clyde Wells
Winner of Third Prize
1976 Population Cartoon Awards
Reprinted by permission of Augusta Chronicle

Bangladesh, Nigeria, Pakistan, and Mexico are likely to increase their populations to more than 100 million. Nations already in this category and their current rankings are: (1) China, (2) India, (3) the Soviet Union, (4) the United States, (5) Indonesia, (6) Brazil, and (7) Japan.

Based on the 1982 census, China has a population of over one billion. India, before the year 2000, is likely to increase to one billion its present population of 686 million. Because of their higher growth rates, the populations of Indonesia and Brazil are expected to increase faster than those of the Soviet Union, the United States, and Japan.

New towns or cities may spring up where new deposits of minerals or energy resources are found. Wasteland may be turned into productive land as technology improves. There will be some movement to areas that are now sparsely populated. But on the whole, there will be little change in the unequal distribution of the world's people.

Birth rates. One hopeful trend is a lowering of birth rates in many of the world's developing nations. In about two thirds of the developing nations, the overall birth rate has fallen by about 15 percent since 1960. Some experts expect to see an even greater lowering of the birth rate in these nations as the 20th century comes to a close.

People and motor vehicles jostle for space in a narrow traffic lane in Jaipur, India. What other problems can rapid urbanization or industrialization cause in a developing nation?

279

What has happened in Japan is likely to happen in many developing nations. Women will enter the work force in greater numbers as their education levels rise and age-old customs change. The result—a lowering of world birth rates.

Death rates and life expectancy. Life expectancy is the average number of years a person can be expected to live. Between the 1960s and 1970s, average life expectancy rose from about 42 to about 54 years in developing nations and from about 65 to about 71 years in industrialized nations. Many people are living longer because they have better living conditions, more nutritious diets, and improved medical services. Fewer children are dying at birth. More children survive into their teens and beyond. By the year 2000, the world's population of people over 60 years of age is expected to be about 580 million. This is almost double the 1970 figure of about 300 million.

Urbanization. By the end of the 20th century, cities with populations greater than one million may number over 400—up from 181 in the mid-1970s. More than 250 of these cities will probably be located in the developing nations.

One United Nations study, based on 1975 population figures, made the following forecasts for the year 2000. São Paulo's population would increase from 10 to 26 million; Mexico City's, from 11 to 32 million; Cairo's, from 7 to over 16 million; Lagos', from 2 to over 9 million; and Bombay's, from 7 to 19 million. The world's urban population will rise to about 50 percent.

Problems and Prospects

Rapidly expanding populations put pressures on the environment, on individuals, and on governments.

Environmental problems. As more and more people use the resources of the environment, unwanted changes take place. For example, loss of plant cover or poor farming practices speed erosion. The landscape changes, floods increase, and, in some cases, deserts are enlarged.

People and the Environment

Sound vs. Noise—Setting Limits

Car horns blare. Factory whistles shriek. Sirens wail. Tires screech. Jet planes thunder overhead. Metal rasps against metal. Groups of people talk and laugh, shout and yell. Outdoors, in the home, in the office, in school, in the factory—almost everywhere—the air is filled with sounds of many kinds.

Sounds are high-pitched or low-pitched, loud or soft, long-lasting or short. They may be periodic or continuous, soothing or irritating. High-pitched, long-lasting, and loud sounds are harder for people to bear than low-pitched, short, or soft sounds. Periodic sounds that start, stop, and start again in a seemingly endless cycle can be more irritating than a continuous sound. Unfamiliar sounds are more disturbing than familiar sounds. Finally, nearness to the source makes some sounds more disagreeable than others. Several of these factors, even one of them, make some sounds more intense than others. It is a high level of intensity that turns sound into noise.

The intensity of sound is measured in decibels. A 0 decibel sound can barely be heard. A sound of 10 decibels is 10 times greater. But a 20 decibel sound is 100 times greater than a sound of 10 decibels and 1,000 times greater than a 0 decibel sound. The human ear can bear the intensity of most sounds or noises measuring up to 100 decibels without damage to hearing, if heard for only a short time. But sounds at this level and even at lower levels, heard for a long length of time or at close range to the source, can damage hearing. Sounds above the 100 decibel level are almost always damaging at close range. Sometimes the damage is permanent. At other times, it is only temporary.

Recognizing the damaging effects of noise, some governments have passed laws to limit and control noise. In the United States, the Noise Control Act passed in 1972 sets standards for acceptable noise levels for different activities. It also provides that certain safety measures be taken to protect the well-being of workers. People who guide jet planes on the ground wear protective earmuffs, as do many other workers who have jobs near high-intensity noises. Where

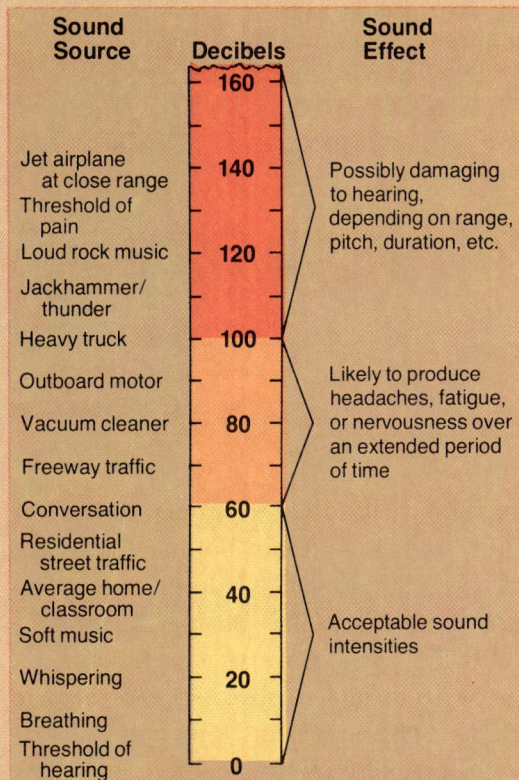

Sound Source	Decibels	Sound Effect
	160	
Jet airplane at close range	140	Possibly damaging to hearing, depending on range, pitch, duration, etc.
Threshold of pain		
Loud rock music	120	
Jackhammer/ thunder		
Heavy truck	100	
Outboard motor		Likely to produce headaches, fatigue, or nervousness over an extended period of time
Vacuum cleaner	80	
Freeway traffic		
Conversation	60	
Residential street traffic		
Average home/ classroom	40	
Soft music		Acceptable sound intensities
Whispering	20	
Breathing		
Threshold of hearing	0	

it is possible, the law requires the addition of noise-reducing parts to machines. It also requires the installation of special materials to rooms and buildings to absorb sound.

Besides damaging hearing, noise has other effects that are not easily measured. Noise is irritating. It makes people tired and nervous. Noise is distracting. It makes concentration difficult. It also lowers the ability of many people to work well.

Noise pollution does not harm the air, the water, or the land. What it does is harm the ability of people to live in harmony with nature and with each other. In its own way, noise is as serious a problem as other forms of pollution. Because it is a problem, it needs a solution.

An official of Malawi's Ministry of Agriculture shows farmers the proper way to apply fertilizers to their fields. Programs like this are helping many developing nations to meet food needs.

In heavily populated areas, forms of land, water, and air pollution are increasing to dangerously high levels. The alarming frequency of high levels of pollution robs nature of the time it needs to renew itself, disturbs the life cycles of living things, and even affects weather and climate.

Other problems. Food, clothing, shelter, good health, and jobs are among the most important needs of individuals. Yet, in overpopulated areas of the world, there are many people who cannot adequately meet these needs. Unmet needs may lead to social unrest within nations and conflicts among and between nations.

Future prospects. Solutions to the problems of population growth depend on the efforts of many people the world over. People working on the political level encourage government leaders to plan according to population needs. Scientists, bankers, and others working on the economic level spread industrialization and its benefits. Others working on the social level educate young and old, fight injustice, and treat physical and mental ills.

Keeping Facts in Focus

1. What two countries may have populations of one billion or more by the year 2000?
2. What is the definition of *life expectancy?*
3. What answers would you give to the key questions on page 278?

Working with Ideas

1. How might the life of a thirteen-year-old boy or a thirteen-year-old girl change as a nation passes from the first to the second stage of population growth?
2. Senior citizens—people 65 years of age or older—are the fastest-growing group in the U.S. population. How do you think our government should plan to meet the needs of this age group?

Unit 8 REVIEW WORKSHOP

Test Your Geographic Knowledge

A. Select the TWO items in each list that are a part of the definition of the first item.
   1. **rate of natural increase:** (a) birth rate, (b) migration rate, (c) death rate, (d) literacy rate.
   2. **population density:** (a) total number of square units of land in an area, (b) total number of people in an area, (c) number of cities in an area, (d) kinds of occupations.
   3. **population growth rate:** (a) income groups, (b) rate of natural increase, (c) age groups, (d) immigration and emigration.
   4. **developing nation:** (a) low rate of natural increase, (b) high growth rate, (c) high literacy rate, (d) low per capita income.
   5. **industrialized nation:** (a) low growth rate, (b) high rate of natural increase, (c) high literacy rate, (d) low per capita income.

B. Tell whether the following statements are true (T) or false (F).
   1. Population increased from about 10 million in 10,000 B.C. to about 500 million in A.D. 1650.
   2. The lower the natural growth rate is, the longer it takes a population to double.
   3. Doubling time is the number of years it will take for a population to decrease to half its size.
   4. Population is increasing faster in most industrialized nations than in most developing nations.
   5. For a nation to have a zero rate of natural increase, its birth rate must be the same as its death rate.

Apply Your Reading Skills

A. Study the editorial cartoon on this page. Then follow the directions in each part below.
   1. Answer these questions:
      a. As a symbol, what does a skull usually represent? What does it represent on this cartoon?
      b. As a symbol, what do the wheat-like plants represent?

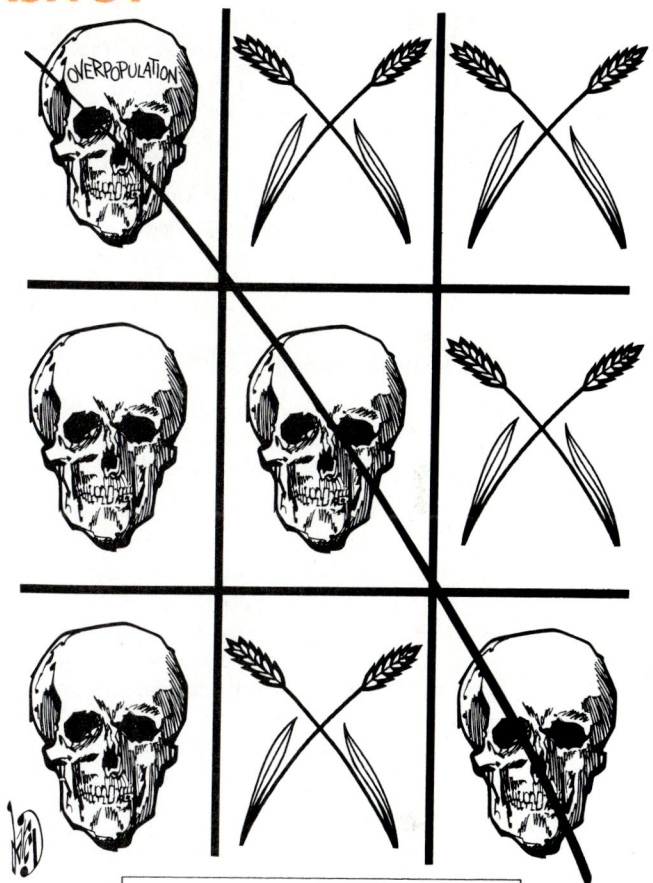

Cartoon by Tom Darcy
Winner of First Prize ○ 1976 Population Cartoon Awards
Reprinted by permission of Newsday

   c. In the cartoon, who is the winner of the game of tic-tac-toe?
   2. Which of the following statements describes the opinion of the cartoonist most fully and adequately?
      a. Four wheat fields are not enough to support five people.
      b. Every year, three out of five people die because of famine.
      c. Unless population growth declines, there will be too many people and not enough food.

B. Use encyclopedias, books, and other reference tools to do research. Then prepare a report on ONE of the following topics.
   1. How have cities changed through time?
   2. What might cities be like in the future?

283

3. What part did migration play in the development and expansion of the United States?

4. What are some of the problems of migrant farm workers?

5. What are some of the effects of forced migration on people uprooted from their homelands?

6. What were some of the notable events in the development of languages?

7. What might the life of women be like in the year 2000?

Apply Your Geographic Skills

A. Using statistics from the chart on page 268, construct a bar graph showing birth rates, death rates, or the rate of natural increase for at least five of the nations.

B. Study the population distribution and land use maps for North America on page 30 of the Atlas. Compare population distribution with land use. Then use the graphic-relief map on page 27 of the Atlas and the climate map in Unit 5 to discover relationships among population distribution, land use, physical features, and climates. Use these relationships to explain why some areas of North America are heavily populated and others are not.

C. Use the population profiles on page 276 to answer the following questions:
   1. What percentage of the population of Sweden are males between the ages of 25 and 29 years old?
   2. What percentage of the population of Mexico are females between the ages of 15 and 40?
   3. What age group is the largest in Mexico? In Sweden? In the United States?
   4. Which has a higher percentage of its population over 60 years of age—the United States or Sweden?

Apply Your Thinking Skills

Using statistics you gather from the charts or graphs of this unit, write a statement that supports each generalization stated below.
1. North America has a lower population density than Latin America.

2. Population increases in the modern period of population growth were greater and took place within a shorter time than in the early period.

3. A nation that has a low literacy rate generally has more than half of its population living in rural areas.

4. A low per capita income often indicates that a nation has not reached a high level of industrialization.

Discuss These Points

1. Make some predictions about what you think life in the United States will be like in the year 2000. Begin with predictions about population growth and movement that are based on present data or trends. Then consider the impact of these predictions on such things as technology, travel, land use, food supply, energy resources, employment, and education to make other predictions.

2. In the United States, some people think zero population growth is a worthy goal. What advantages or disadvantages can you see to zero population growth?

Expand Your Geographic Sights

Drummond, A. H., Jr. *The Population Puzzle* (Addison-Wesley, 1973).

Heaps, Willard A. *Wandering Workers: The Story of American Migrant Farm Workers and Their Problems* (Crown, 1968).

Lens, Sidney. *Poverty—Yesterday and Today* (Crowell, 1973).

May, Charles P. *The Uprooted* (Westminster Press, 1976).

Muller, Jorg. *The Changing City* (Atheneum, 1977).

Pringle, Laurence. *One Earth, Many People: The Challenge of Human Population Growth* (Macmillan, 1971).

Stwertka, Eve and Stwertka, Albert. *Population: Growth, Change, and Impact.* (Watts, 1981).

Switzer, Ellen. *Our Urban Planet* (Atheneum, 1980).

Thomson, David. *Language* (Silver Burdett, 1975).

Tripp, M., Ed. *Woman in the Year 2000* (Arbor House, 1974).

Part Two

People and Their Regional Environments

Part One of your text examined physical patterns of the earth—on a worldwide scale. Part Two will explore the earth on a smaller scale, by regions. A region is a part of the earth that has one or more characteristics that are shared throughout its area. Physical features are part of any regional environment. But different patterns of human life—or culture—also help to create distinct and different regions of the world. History, religion, language, and use of the physical environment all combine to create unique cultural regions.

Part Two of this book will introduce you to the main cultural regions of the world and the people who have brought them to life. In your study of these units, you will examine both cultural and physical influences that have shaped the environments of various regions of the world. You will explore each region's potential for agriculture and for industry. You will also discover how people of the various world regions depend on one another for trade, political support, and cultural identification. Just as physical geography can teach us how to live in harmony with nature, regional geography can point the way to understanding and peace among the peoples of the world.

Unit 9 Europe and the Soviet Union

Ruins in Athens stand as symbols of the great civilization of ancient
Greece that flourished about 2,500 years ago.

Europe and the Soviet Union
Political-Relief Map

⊛ National Capitals   • Other Cities

SCALE

Miles 250 500 1000 1500
Kilometers 500 1000 1500 2000
One Inch—About 1000 Miles

Lesson 1 Exploring Europe and the Soviet Union

Reading Focus

1. With this unit, you begin your study of different regions of the world. In the first lesson of each regional unit, you will find a National Profiles chart. These charts give special facts about every independent country in each world region. Look at the chart of National Profiles for Europe and the Soviet Union, pages 290–292. What eleven facts does the chart give for each country? Be sure you know what each term at the top of each column means. You have already studied the meanings of *population density* and *population growth rate* in Unit 8. *Literacy rate* means the percentage of each country's population that can read and write. Notice the dagger (†) before *GNP* at the top of the last column. Now look for the dagger in the footnote at the bottom of the chart. What does *GNP* mean? Remember that each *GNP* is shown in *millions* of dollars. *Per capita income* means the average yearly income for each person in a country. These are actual figures on the chart, again in U.S. dollars.

The map labels:

ARCTIC OCEAN

East Siberian Sea

Ambarchik

70°

Arctic Circle

Yakutsk

Magadan

60°

Bering Sea

Kamchatka Pen.

Sea of Okhotsk

Aleutian Islands (U.S.)

50°

Vladivostok

Sea of Japan

KOREA N. S.

JAPAN

PACIFIC OCEAN

40°

30°

135° 150° 165° ©FPC

Europe is a small continent—only Australia is smaller. But Europe is the birthplace of ideas and ways of life that have spread all over the world. Beginning in the 1400s, Europeans explored the seas and developed a lasting trade with other continents. In their trade and travels, Europeans spread their ideas about government, science, art, and religion far and wide.

The Physical Setting

Europe lies right in the middle of the land hemisphere. Its central location favored the early development of transportation and trade. This location helps explain Europe's worldwide importance today.

Location. Find Europe on the political map of the world on pages 10–11. What three hemispheres is it in? What ocean separates North America and Europe? Europe lies mostly in the middle latitudes. What is the latitude of mainland Europe's most northerly point, the North Cape of Norway? The southern tip of Spain on the Strait of Gibraltar is mainland Europe's most southerly point. What is its latitude? Find where these latitudes would be in North America. Which extends further north, Europe or the United States (excluding Alaska and Hawaii)?

Use the map on this page to help you locate the boundaries of Europe. The Ural mountain range and river separate Europe from Asia on the east. The Caucasus (kô′kə·səs) Mountains form another boundary of Europe. Between what two seas do the Caucasus stretch? What sea separates Europe from Africa?

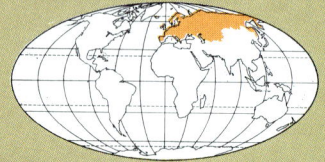

National Profiles: *Europe and the Soviet Union*

Independent Nation	Area/ Population	Population Density/ §Growth Rate	*Capital/ Largest City and Population	Urban Population/ Literacy Rate	Chief Exports (in order of importance)	†GNP (in millions of U.S. dollars)/ Per Capita Income
Albania	28,749 sq km 2,906,000	101 per sq km 2.1%	*Tirana 198,000	34% 75%	fuels, ores, metals, fruits & vegetables, raw materials	$2,150 $820
Andorra	466 sq km 45,000	97 per sq km 5.4%	*Andorra 10,900	N/A 100%	tobacco	N/A N/A
Austria	83,916 sq km 7,579,000	90 per sq km 0.1%	*Vienna 1,531,346	55% 98%	iron & steel, machinery, timber, textiles, chemicals	$68,890 $10,995
Belgium	30,562 sq km 9,872,000	323 per sq km 0.1%	*Brussels 144,000	95% 98%	iron & steel, gem-stones, textiles,	$85,420 $8,628
Bulgaria	111,852 sq km 8,969,000	80 per sq km 0.3%	*Sofia 1,070,000	60% 95%	machinery, motor vehicles, agricul-tural products, fuels	$35,300 $3,963
Czechoslovakia	127,946 sq km 15,466,000	121 per sq km 0.3%	*Prague 1,182,000	66% 99%	machinery, manufac-tured goods, fuels, raw materials	$147,100 $9,550
Denmark	42,994 sq km 5,112,000	119 per sq km 0.1%	*Copenhagen 645,198	84% 99%	meat, dairy products, machinery, textiles & clothing, chemicals	$56,400 $11,016
Finland	336,700 sq km 4,873,000	14 per sq km 0.5%	*Helsinki 483,051	60% 99%	wood pulp & paper, ships, machinery, iron & steel	$49,100 $10,124
France	551,670 sq km 54,872,000	99 per sq km 0.4%	*Paris 2,296,945	73% 98%	machinery, motor vehicles, food, agricul-tural products, iron & steel, textiles	$542,000 $9,996
Germany, East	108,262 sq km 16,718,000	159 per sq km 0.0%	*East Berlin 1,145,743	77% 99%	machinery, chemi-cals, metals	$165,600 $9,903
Germany, West	248,640 sq km 62,387,000	247 per sq km −0.2%	*Bonn 292,200 West Berlin 1,879,100	92% 99%	machinery, motor vehicles, chemicals, iron, & steel, textiles & clothing	$658,400 $10,682
Greece	132,608 sq km 9,984,000	75 per sq km 0.9%	*Athens 885,136	58% 95%	fruits & vegetables, petroleum products, clothing, iron & steel, tobacco	$38,600 $3,959

§Growth Rate: Includes natural increase and migration.

†GNP: Gross National Product, or the total value of all the goods and services produced by the people of a country within one year.

N/A: Data not available

National Profiles: *Europe and the Soviet Union*

Independent Nation	Area/ Population	Population Density/ §Growth Rate	*Capital/ Largest City and Population	Urban Population/ Literacy Rate	Chief Exports (in order of importance)	†GNP (in millions of U.S. dollars)/ Per Capita Income
Hungary	92,981 sq km 10,681,000	115 per sq km −0.1%	*Budapest 2,067,000	54% 98%	machinery, food, motor vehicles, chemicals, iron & steel	$65,200 $6,901
Iceland	102,952 sq km 239,000	2 per sq km 1.0%	*Reykjavik 85,782	88% 99%	fish & fish products, animal products, aluminum	$2,200 $9,322
Ireland	68,894 sq km 3,575,000	52 per sq km 1.2%	*Dublin 525,360	52% 99%	dairy products, live animals, textiles, chemicals, clothing	$17,000 $5,667
Italy	301,217 sq km 56,998,000	189 per sq km 0.3%	*Rome 2,840,259	69% 93%	machinery, motor vehicles, food stuffs, metals, textiles	$347,000 $5,314
Liechtenstein	168 sq km 27,000	161 per sq km 1.7%	*Vaduz 4,980	N/A 100%	postage stamps, metal products, chemicals, furniture	N/A $16,900
Luxembourg	2,590 sq km 366,000	141 per sq km 0.1%	*Luxembourg 80,000	68% 100%	iron & steel	$3,400 $9,289
Malta	313 sq km 356,000	1,137 per sq km −0.5%	*Valletta 14,527 Sliema 20,100	94% 83%	clothing, textiles, ships, printed matter	$1,140 $3,499
Monaco	1.5 sq km 28,000	18,667 per sq km 1.2%	*Monaco 28,000	100% 99%	foreign trade included with France	N/A N/A
Netherlands	33,929 sq km 14,437,000	425 per sq km 0.4%	*Amsterdam 700,759 *The Hague 456,886	88% 99%	food, machinery, chemicals, petroleum products, natural gas	$137,300 $9,807
Norway	323,750 sq km 4,145,000	13 per sq km 0.4%	*Oslo 450,386	44% 100%	crude oil, natural gas, metals, pulp & paper, fish	$56,200 $13,600
Poland	312,354 sq km 36,887,000	118 per sq km 0.9%	*Warsaw 1,500,000	57% 98%	machinery, fuel transport equipment, chemicals, food	$186,800 $5,160

National Profiles: *Europe and the Soviet Union*

Independent Nation	Area/ Population	Population Density/ §Growth Rate	*Capital/ Largest City and Population	Urban Population/ Literacy Rate	Chief Exports (in order of importance)	†GNP (in millions of U.S. dollars)/ Per Capita Income
Portugal	94,276 sq km 10,045,000	107 per sq km 0.4%	*Lisbon 812,400	30% 80%	textiles & clothing, cork & cork products, canned fish, wine, timber	$23,400 $2,237
Rumania	237,503 sq km 22,683,000	96 per sq km 0.5%	*Bucharest 1,861,007	50% 98%	machinery & transport equipment, petroleum products, chemicals, meat, clothing, timber	$104,800 $4,238
San Marino	62 sq km 23,000	370 per sq km 1.6%	*San Marino 3,000	92% 97%	postage stamps, wine, furniture, wool	N/A N/A
Spain	505,050 sq km 38,435,000	76 per sq km 0.5%	*Madrid 3,188,297	74% 97%	iron & steel, machinery, motor vehicles, fruits & vegetables, textiles, footwear	$179,700 $4,880
Sweden	448,070 sq km 8,335,000	17 per sq km 0.0%	*Stockholm 1,544,454	83% 99%	machinery, motor vehicles, paper, iron & steel, metal ores, chemicals	$81,000 $10,285
Switzerland	41,440 sq km 6,477,000	156 per sq km 0.2%	*Bern 145,000 Zurich 375,000	51% 99%	machinery, chemicals, watches & clocks, metal products, textiles, food stuffs	$95,600 $14,270
United Kingdom	243,978 sq km 56,023,000	230 per sq km 0.0%	*London 6,696,008	76% 99%	machinery, motor vehicles, petroleum, chemicals	$482,700 $8,620
U.S.S.R. (Soviet Union)	22,274,900 sq km 274,860,000	12 per sq km 0.9%	*Moscow 8,200,000	62% 99%	petroleum, natural gas, iron & steel, wood, agricultural products, machinery	$1,715,000 $6,352
Vatican City	0.4 sq km 1,000	228 per sq km 0.1%	*Vatican City 729	100% 100%	postage stamps	N/A N/A
Yugoslavia	255,892 sq km 22,997,000	88 per sq km 0.7%	*Belgrade 1,300,000	39% 85%	machinery, electrical equipment, transport equipment, fuel, manufactured goods, chemicals	$53,900 $2,370

§Growth Rate: Includes natural increase and migration.

†GNP: Gross National Product, or the total value of all the goods and services produced by the people of a country within one year.

N/A: Data not available.

Countries. Unlike the continents of North America, South America, and Africa, Europe and Asia are not almost entirely surrounded by water. The Ural mountains separate the continents near the middle of the huge Eurasian landmass. One country, the Union of Soviet Socialist Republics, lies on both sides of these mountains. The U.S.S.R. is partly in Europe and partly in Asia. Since its ways of life are linked more to Europe than to Asia, you will study the Soviet Union in this unit as a part of Europe.

Europe is a continent of diversity. The total land area of Europe is only slightly larger than that of the United States. Yet within Europe, there are more than thirty-three countries and more than twenty-five official languages. There are also thirty-two different kinds of money in use in Europe.

There are also great differences in the sizes of Europe's countries. Part of the world's largest country, the Soviet Union, lies in Europe. Vatican City and Monaco, the world's smallest countries, are also in Europe. Vatican City lies within the city of Rome. Find Monaco on the map on page 288.

The countries of Europe are diverse in still another way. Some are islands, and some are completely landlocked. Still others have at least one border on a sea. Find examples of all of these types of countries on the map on page 288.

Landforms. The landforms of Europe also show a great variety. The map of natural regions on this page shows the four main divisions of European land.

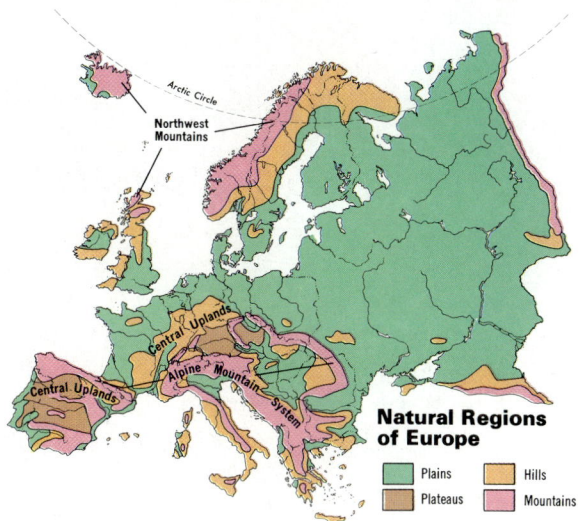

Natural Regions of Europe

Plains · Plateaus · Hills · Mountains

The northwest mountains. Compare the map of natural regions with the graphic-relief map of Europe on pages 16–17 of the Atlas. Notice that the northwest mountains separate Sweden and Norway and send a sprinkle of hills through the United Kingdom and Ireland. What region in the United Kingdom seems to be the most mountainous?

The northwest mountains are the remains of the oldest mountains in Europe. Over time, they became eroded. But through some later volcanic activity, they were raised again. Iceland is a volcanic island formed during this uplifting. (Find Iceland on the map on pages 16–17.) In general, the mountains of the northwest are not very high, but they are very rugged.

The plains. The plains of Europe stretch from the Atlantic coast all the way to the Ural Mountains and beyond. What countries lie within the plains area?

The European plains have many areas of rolling land, low hills, and

293

Rolling plains with a patchwork of crops typify much of the Polish landscape. Poland is a Communist nation, yet most farms are privately owned.

gentle rises. While the plains of the north are frozen for much of the year, the plains further south have growing seasons long enough for farming.

The central uplands. The mountains and plateaus spreading across the European plain form another system of very old mountains. Like the mountains in the northwest, these are built of hard, ancient rocks, rich in minerals.

During their long life, the mountains of the central uplands have become eroded. They are relatively easy to cross. Deep rivers and large stretches of plains separate these mountains into several distinct masses. The passes between these mountains encouraged road building and aided the early development of central Europe.

Alpine mountain system. The youngest and steepest of Europe's mountains belong to the Alpine system. The Alps themselves are only a part of this rugged system. Other Alpine ranges in southern Europe are the Pyrenees (pir′ə·nēz), the Apennines (ap′ə·nīnz), the Dinaric (də·nar′ik) Alps, the Pindus, and the Caucasus. The Carpathians, curving through eastern Europe, are also part of the Alpine system. Look at the graphic-relief map on pages 16–17. In what countries do you find each of these ranges?

Unlike the mountains of the central uplands, the Alpine ranges are difficult to cross. They are steep and rugged, much like the Rocky Mountains in North America.

Seas, rivers, and lakes. On a map, Europe looks like a puzzle of lands and seas, small and large peninsulas, and a scattering of islands. Land and sea fit into each other deeply in Europe. Look at the graphic-relief map on pages 16–17. Name three of the seas that form Europe's north and northwest coast. The deep bays and large peninsulas make Europe's coastline extremely long. Its

Little farming is done in the rugged terrain of the Austrian Alps. Most Alpine communities have grown up in the valleys, many as tourist resorts.

jagged line has made many natural harbors that have encouraged European shipping and fishing.

The oceans and seas are very important to all of Europe. Much of inland Europe lies no further than 300 miles (480 kilometers) from a seacoast. These parts of Europe have easy access to seas by the many river systems.

Find Europe's longest river, the Volga, on the map on pages 16–17. The Ural River and the Volga both empty into the Caspian Sea. Also find the Dnestr (nēs′ tər) and the Dnepr (nē′pər) rivers. Both empty into the Black Sea. The Daugava (dou′ gə · və) River in the northwestern part of the Soviet Union flows into the Baltic Sea. The Northern Dvina (də vē′ nä) River empties into the White Sea. Eastern Europe's important rivers are the Elbe, Oder, and Vistula. The Elbe empties into the North Sea, while the Oder and Vistula rivers flow into the Baltic Sea.

In the central part of Europe, the longest river is the Danube. The Danube starts in the Black Forest of Germany, then flows through Austria, Hungary, and Yugoslavia. It continues eastward, forming the border between Bulgaria and Rumania.

The Rhine River is the busiest waterway in western Europe. Its source is in the Swiss Alps. Follow its course to the North Sea on the map on pages 16–17. Through which countries does it flow? Between what two countries is the Rhine a boundary?

Europe has relatively few square miles of freshwater lakes. Check the map on pages 16–17. Most of Europe's lakes are in Finland and northwestern Soviet Union.

Climate. You have already seen that Europe and parts of North America lie within the same latitudes. Would you expect their climates to be similar? Turn to the map of world climates on pages

200–201. Compare Europe's climates with those in North America at the same latitudes. Which are generally warmer, European or North American climates? The waters surrounding Europe help moderate the climate. They help keep Europe warmer in winter and cooler in summer than parts of North America at the same latitude.

The North Atlantic Current, in the Atlantic Ocean, warms Europe's western shore. Westward winds carry this warmth even to inland spots. Except in northern Europe, most of Europe's mountains run in an east-west direction. They form little barrier to the westerly winds.

The winds carry moisture as well as warmth to inland Europe. Turn to the precipitation map on page 20. Note that the inland countries, such as Hungary and Czechoslovakia, receive about the same amount of precipitation as London, which is much closer to the sea.

The seas of Europe make the day-to-day weather very changeable. The difference in land and sea temperatures makes air masses above Europe unstable. Sudden brief storms are common. But they are seldom violent in Europe's moderate climate.

Population distribution. Roughly one sixth of the world's people, about 765 million persons, share the region of Europe and the Soviet Union. Europe's average population density is 28 persons per square kilometer (109 per sq m). Some areas in Europe have far fewer people than this average. Others have far more. Netherlands, for example, has an average of 425 per square kilometer (1,003 per sq m). Its population density helps make Europe's population density the second highest in the world.

Land use. The map on page 21 of the Atlas shows how people in various parts of Europe use the land. Europeans have taken good advantage of their land and sea resources. They have used the many harbors to develop shipbuilding and fishing industries. Using careful farming methods, they have cultivated almost one third of the land. In other parts of the world, less than one fifth of the land is cultivated. Finally, Europeans have used their resources to develop many other productive industries. Canals connect Europe's many rivers for easy transportation of goods. Many railways and highways also carry goods and people across Europe's wide plains and passable mountains.

Keeping Facts in Focus

1. In what hemispheres is Europe located?
2. What is Europe's longest river? Into what sea does it empty?
3. What river forms part of Europe's eastern boundary?
4. What is Western Europe's busiest waterway?
5. To what system do the youngest and steepest of Europe's mountain ranges belong?
6. What ocean current warms Europe's western shores?
7. In what country or countries is each of the following located: the Dinaric Alps, the Pyrenees, the Apennines, the Pindus, and the Caucasus?
8. How are the Alpine ranges different from the mountains of the central uplands? What mountain range in North America is similar to the Alpine system?

Many Europeans, especially in small towns, feel strong ties to the customs of their ancestors. On a Scottish island, the drone of a bagpipe heralds a wedding; in Tirol, Austria, a crisp brass band plays a stirring march.

The People's Heritage

The plains of Europe supported early farming peoples over 7,000 years ago. By 5,000 years ago, farming had spread to almost all parts of the continent.

At about this time, urban civilization in Europe began to develop around the Aegean Sea. Later, Romans built an empire that spread through Europe and nearby parts of the world. The Greek and Roman cultures developed laws and ways of government that have helped shape our present societies. Diverse peoples were united under Roman rule and under the widespread Christian religion. Although the Roman Empire fell to northern tribes in A.D. 476, many of the people remained linked by their religion and culture.

But the link was not unbreakable. Quarrels over religion, territories, and government soon dominated much of

the area's history. Christianity split into diverse groups. Empires split into separate kingdoms. Many distinct nations emerged, each wanting to preserve its own culture and unique ways of life. Frequent wars shifted the balance of economic and political power back and forth among nations.

In the first half of the twentieth century, the nations of Europe were in frequent conflict. Two world wars started in Europe, the first in 1914 and the second in 1939. Many cities were bombed and destroyed; many Europeans lost their lives. Even after the Second World War, European nations were divided over political and economic policies—especially Communism. Today, Europe is still split between Communists and non-Communists.

Amid the divisions, however, the force of industrialization was creating many

297

Western Europe
Northern Europe
Mediterranean Europe
Eastern Europe and the Soviet Union

ATLANTIC
OCEAN

North Sea
Baltic Sea

E U R O P E

Black Sea
Caspian Sea
Adriatic Sea
Mediterranean Sea
Aegean Sea

ASIA
AFRICA

compete with other nations of the world. Although tensions still exist, European nations are finding economic cooperation necessary to their survival.

Some people worry that the movement toward economic cooperation threatens Europe's rich diversity. They fear that local customs and regional differences will disappear. But others believe that Europe's diversity is rich enough to survive modern pressures.

Europe's cultural divisions. Each country in Europe is unique. But certain shared patterns of culture unite the countries of Europe into larger regions. The map on this page shows the four main cultural divisions of Europe. In the lessons that follow, you will be exploring each of these regions. As you continue to read this unit, look for the differences and similarities in Europe's four cultural regions.

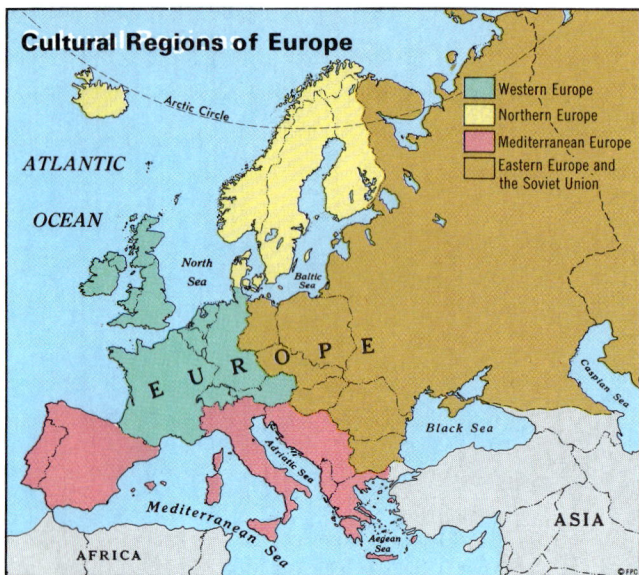

similarities among European countries. During both the 1800s and the 1900s, industrial centers attracted many people looking for work. As people moved into the cities, they often left behind local ways and customs. Regions at one time self-sufficient began to depend on imported raw materials for industry. Advanced methods of communication and transportation brought more regions of Europe in touch with each other and the world. The search for new markets spread the effects of industrialization as the European standard of living rose steadily.

Industrialization also created problems. It created the need for raw materials that led some countries to build up colonial empires. For a while, industry's need for cheap workers led to many unfair labor practices, including child labor. Industry also polluted the air and water, and helped use up a dwindling supply of energy and other resources. More and more, European countries must rely on each other's resources to maintain their rate of production and

Keeping Facts in Focus

1. Around what sea did urban civilization in Europe first begin to develop?
2. How are we indebted to the early Greek and Roman cultures?
3. What are Europe's main cultural divisions?
4. What problems were created by European industry's need for cheap labor? What were some other problems caused by industry?
5. What political philosophy divides Europe?
6. What answers would you give to the key questions on page 289?

Working with Ideas

1. How have Europe's physical features contributed to its cultural diversity?
2. Most of Europe was once united politically under the Roman Empire. What are the drawbacks to a united Europe today? What are the advantages?

Lesson 2 Western Europe

Reading Focus

1. Before reading this lesson, thumb through the pages and look carefully at the photographs. Getting an overview might help you focus your later reading more sharply. Then, on a separate sheet of paper, write down the headings in this lesson. What do they tell you about Western Europe? Do you think it is a highly developed region? Finally, take a close look at the maps. Take time to get familiar with the names of countries, cities, and bodies of water in this region. Then as you read about them, you'll be able to picture where they are on a map.

2. As you read this lesson, look for the answers to these key questions:
   a. In what ways is Western Europe a region of soft contrasts?
   b. Why is Western Europe good for farming?
   c. What are some reasons for Western Europe's industrial strength?
   d. What is the purpose of the Common Market?

Vocabulary Focus

productive	*cooperative*	*terraced*
migrant	*reclaim*	*tariff*

Western Europe is made up of eleven countries. Using the map on this page, name all of Western Europe's countries. Now turn to the chart of National Profiles on pages 290–292. Which Western European country is largest in area? Which has the largest population? What is the Western European country with the largest per capita income? What is its capital city?

The region of Western Europe is filled with soft contrasts. Within its small area, it contains all four types of European landforms. But the most common landscape in Western Europe is a wide stretch of rolling plains.

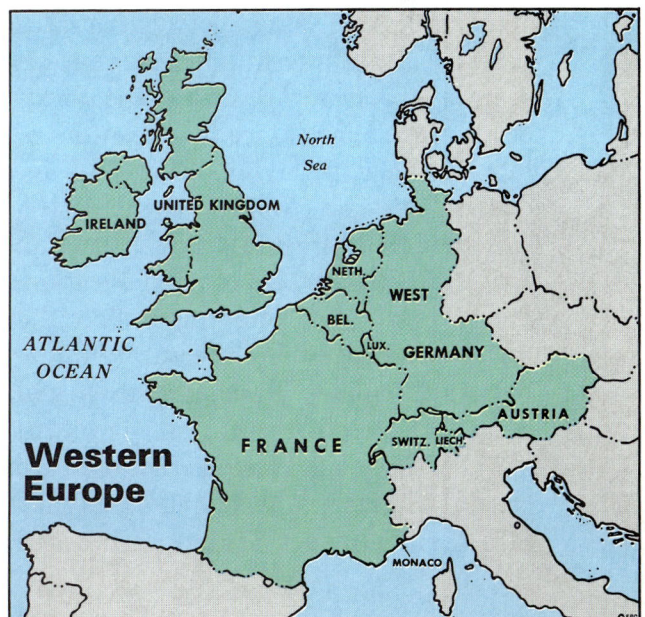

Western Europe

In northern France, as elsewhere in Western Europe, machines increase efficiency and help farmers achieve high crop yields.

The seasons of the year are also softly contrasting. Seasons change, but neither the winters nor the summers bring extreme changes. The precipitation in Western Europe is moderate in all seasons. But it is abundant enough to support a rich agricultural region. The plains of Western Europe have the most productive farmland on the continent.

Perhaps the sharpest contrast in Western Europe is that between country life and city life. While local traditions remain strong in the rural areas, Western Europe's cities are huge centers of bustling modern living. Enormous industries attract people to the cities, which offer a large variety of jobs. The cities are also centers of art, education, and science. New buildings reflect the styles of modern architecture. In the outlying regions, though, castles and cathedrals from centuries ago still stand. Cities of Western Europe are careful to preserve their history as they face the challenges of the present and the future.

A Rich Agricultural Region

The yields from Western European farmlands are among the highest in the world. Many farmers keep their lands fertile with chemicals and regular rotation of crops. Many also use modern, efficient machinery. On the large farms, groups of migrant workers, many of them from Mediterranean countries, help plant and harvest the crops. When one job is finished, they move on to other areas looking for more work.

On the smaller farms, family members usually work the fields. Sometimes one or two hired hands help out. Many of these small farmers also use advanced machinery. Most farmers in Western Europe own their own land, although some do rent from larger landowners. The average size of Western European farms is fairly small—about thirty acres (twelve ha).

Since World War II, the governments of Western European countries have encouraged small farmers to work in larger groups. Some farmers have formed cooperatives, or business agreements, with other farmers to share their costs, machines, and profits. In Western Europe, farmers in cooperatives share the cost of harvesting and selling their products. This sharing cuts down on costs for individual farmers.

There are still, however, some very small farms in Western Europe. On these, some of the family members work the farm. Their farming provides fruit, vegetables, and dairy products for the family's own use. It also provides some small crops that can be sold for extra cash. But the other family members work in nearby factories, earning the rest of the money for the family's support. Life for these part-time farmers is a blend of farming and industrial ways.

Grains and dairy farming. The map on page 302 shows the major crops for each area of Western Europe. The plains of southern England, central France, Belgium, and West Germany produce large crops of grain, especially wheat, barley, and rye. Sugar beets and potatoes are also important crops in Western European plains.

The cool, damp climate of Ireland and other countries in Western Europe helps dairy farming thrive. Dairy farms in these countries have produced some of the world's best breeds of livestock. People in the Netherlands also have thriving dairy farms. But they have had to overcome special problems to make their land manageable.

Almost 40 percent of the Netherlands was once under the shallow water of the

AVERAGE WHEAT YIELD
(Metric tons per hectare)

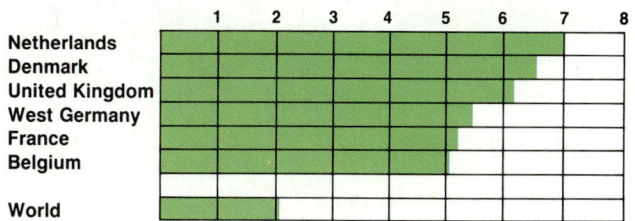

Country	Yield (metric tons per hectare)
Netherlands	~6.5
Denmark	~6
United Kingdom	~6
West Germany	~5.5
France	~5
Belgium	~5
World	~2

North Sea. Over 800 years ago, people began to reclaim this land from the sea. They built dikes and dams to stop the flooding. They also built canals into which the seawater drained. In winter, the water often rose above flood level. Windmills were built to pump it off the land. In the summer, water often fell to below the level needed for crops. Windmills would then pump it back onto the land.

Motors driven by electricity and other means of power have now replaced windmills in the Netherlands. Much of the reclaimed land has a rich soil that is ideally suited for farming. Only 6 percent of the Dutch labor force is involved in farming. But with efficient methods, Dutch farmers produce food, flower bulbs, seeds, and garden plants for home use and for export.

Farming in the hills and mountains. The upland and mountain regions of Western Europe have cooler climates and poorer soils than the plains. But farmers can earn good incomes from some special crops. In the foothills, grapes grow well on carefully terraced and fertilized land. Look at the map of

301

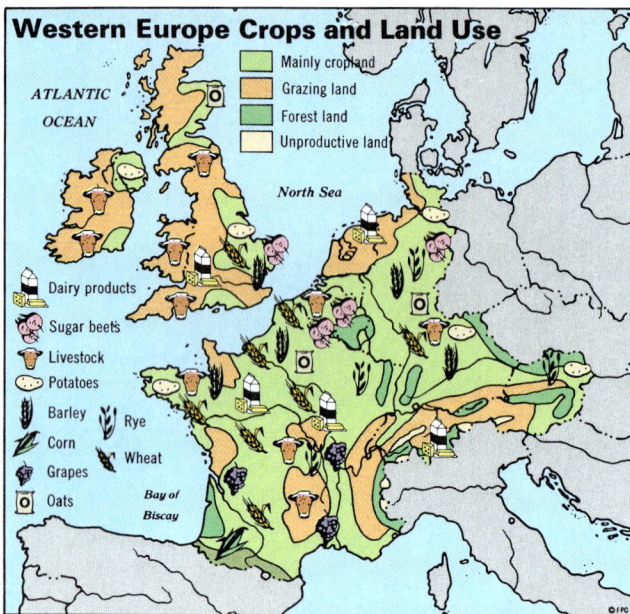

Western Europe Crops and Land Use

Mainly cropland
Grazing land
Forest land
Unproductive land

ATLANTIC
OCEAN

North Sea

Dairy products
Sugar beets
Livestock
Potatoes
Barley   Rye
Corn   Wheat
Grapes
Oats

Bay of
Biscay

© FPC

agricultural products. What country grows large amounts of grapes?

Sunny valleys in the mountain regions support truck farming and fruit growing. Higher areas around these valleys are good for raising cattle and sheep. Both meat and dairy foods are important products of these regions. The countries of Western Europe, especially France, produce a great variety of famous cheeses. Farmers in the Swiss and Austrian Alps also produce dairy foods. The outstanding scenic beauty of these areas attracts tourists year round. Alpine farmers, then, often add to their income with the tourist trade.

Abundant Mineral and Fuel Resources

Agriculture in Western Europe provides many jobs and products. But it was industry that made Europe rise to world leadership. Western Europe's rich mineral resources and well-developed systems of transportation favored industrial growth. These, combined with such modern production methods as the steam engine, helped turn Western Europe into an industrial giant.

Today, Western Europe can boast a wide variety of mining and manufacturing industries. France, England, and Germany contain some of the busiest industrial centers in the world. One reason for the dense industry in these areas is a concentration of mineral and

These Basque workers, near the Pyrenees in southern France, depend on sheep raising for much of their income. Isolated from the rest of the country, many Basques speak only their own native language, which is unrelated to French.

302

fuel resources. Look at the mining and manufacturing map on this page. Which Western European countries have good supplies of iron ore? What industrial fuel is abundant in England?

Mineral and fuel deposits alone are often not enough for building industrial regions. Other factors, including transportation, greatly affect the growth of industry.

Industrialization in the Ruhr Valley. Find the Ruhr region on the mining and manufacturing map on this page. Before 1860, the Ruhr Valley, named for the river that runs through it, was used mostly for grazing. Few people lived there. Today it is the most densely populated area of Germany. In some ways, the changes in the Ruhr Valley are typical of the changes the industrial revolution brought as it spread to other parts of Europe as well.

It was long known that the Ruhr Valley had some of the largest coal deposits in Europe. The coal could be used as a fuel in the production of steel from iron ore. But there was no iron ore

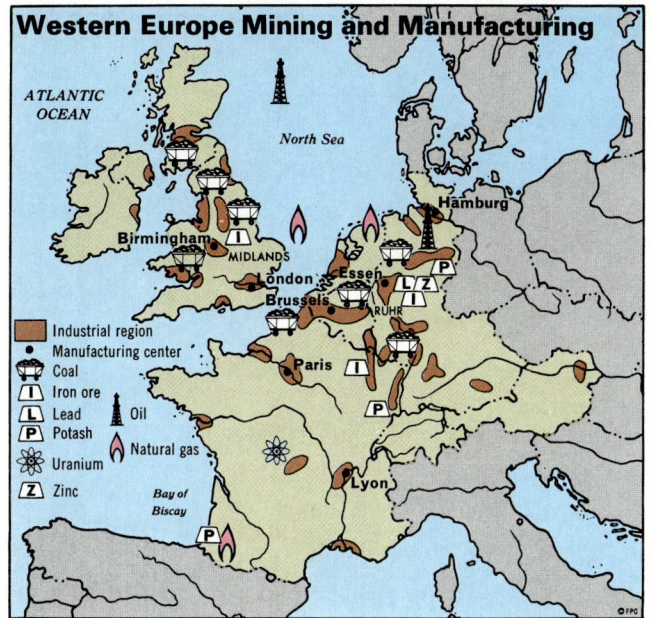

Western Europe Mining and Manufacturing

ATLANTIC OCEAN

North Sea

Hamburg

Birmingham

MIDLANDS

London   Essen

Brussels   L/Z

RUHR   P

Paris

Lyon

Bay of Biscay

P

- ▢ Industrial region
- • Manufacturing center
- Coal
- I Iron ore
- L Lead   Oil
- P Potash
- ❋ Uranium   Natural gas
- Z Zinc

in the area. Before 1860, transportation was so poor that iron ore could not be shipped into the Ruhr Valley.

Then governments began to build railroads and canals. Iron ore and other materials started coming into the Ruhr Valley from the iron-rich region near the Moselle River. The Ruhr Valley began to attract steel mills and factories. Plentiful jobs attracted millions of people. Huge urban centers sprang up, and they continue to thrive today.

European textile industries grew rapidly after the development of the steam engine in the 1700s. An automated spinning machine in a Liechtenstein plant is typical of the equipment today's workers operate.

WATERPOWER

(billions of kilowatt-hours per year)

amount produced
estimated reserves

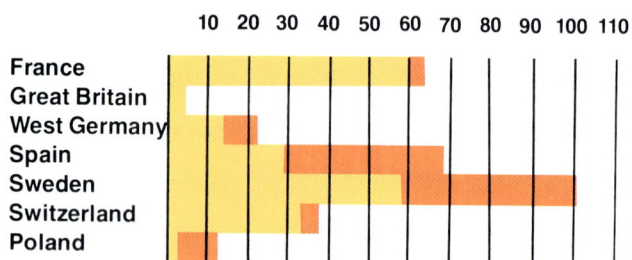

	10 20 30 40 50 60 70 80 90 100 110
France	
Great Britain	
West Germany	
Spain	
Sweden	
Switzerland	
Poland	

The story of the Ruhr Valley shows an important concern of European industry: cooperation. Without the exchange of resources among countries, production would lag. Cooperation is becoming more and more important for Western European nations. The metal industries of France, Belgium, Luxembourg, and Germany, for example, depend on one another in various ways. The Ruhr area provides coke to its neighbors. France and Luxembourg send iron ore to Belgium and Germany.

The Common Market. In addition to exchanging materials, Western European countries cooperate in another way. While remaining separate political units, these countries have worked out a way to share economic opportunities. In 1958, six Western European nations founded an organization called the European Economic Community, or the Common Market. Its purpose was to open up free trade among Western European countries. Today, Belgium, Denmark, France, West Germany, United Kingdom, Ireland, Italy, Luxembourg, Greece, and the Netherlands are all members of the Common Market.

Member countries charge no tariffs on imports from other members of the Common Market. Thus, they have created one big market for all member nations. They also share laborers.

In 1967 the European Coal and Steel Community (ECSC) and the European Atomic Energy Commission (EURATOM) joined with the Common Market to form European Communities (EC). With the same membership as the Common Market, the EC has political union as its final goal.

Protecting the environment. Europe faces a dwindling coal supply. Researchers are trying to develop ways to tap other energy sources. Pollution has also become a serious problem. Since 1952, when a heavy smog killed about 4,000 Londoners, strict laws have controlled pollution in England. But elsewhere in Western Europe, pollution remains a serious concern.

Keeping Facts in Focus

1. How have the Dutch reclaimed land for farming and other uses from the sea?
2. Why was good transportation essential to steelmaking in the Ruhr Valley?
3. What countries are Common Market members today?
4. What are two of the environmental problems facing industry today?
5. What answers would you give to the key questions on page 299?

Working with Ideas

1. How has the Common Market improved trade among its member nations?
2. What geographic conditions account for Western Europe's agricultural richness?
3. Why might industrial development in Europe be of concern to environmentalists?

Northern Europe

Reading Focus

1. The chart of National Profiles (pages 290–292) gives you some information about European countries. To learn more, you might want to make your own National Profiles chart for the Northern European countries. Follow the pattern of the chart on page 290. But change the headings to Climate(s), Land Use, Precipitation, and Growing Seasons. The Atlas maps on pages 20–21 and the climate map on pages 200–201 will supply the information you need. Here is an example to help you get started.

Countries	Climate(s)	Land Use	Growing Seasons	Precipitation
Denmark	Marine	Farming	5–7 mos. in the east 7–9 mos. in the west	20–40 inches

Now list Finland, Iceland, Norway, and Sweden. You can refer to your chart as you read this lesson.

2. As you read this lesson, look for the answers to these key questions:
   a. What features of Northern Europe show its glacial heritage?
   b. What are some problems in Northern European farming?
   c. How did the good supply of trees in Northern Europe help industry grow?
   d. What benefits do Northern Europeans receive from their governments?

Vocabulary Focus

fiord	*navigable*	*profit*	*retirement*
rapids	*arable*	*smelt*	

Over 11,000 years ago, glaciers covered most of the area that is now Northern Europe. Gradually, the Ice Age drew to a close. As the glaciers melted, they left behind a rugged environment, a harsh challenge to early settlers. In some areas, the glaciers left deposits of infertile soils, unsuitable for farming.

The land that was suited to farming was often scattered with boulders and gravel. The glaciers also left many areas of swamp and marsh, usable only if they were drained.

The people of Northern Europe have met these challenges with energy and hard work. They have made their region

Northern Europe

a center of agriculture, industry, and trade. They have also developed efficient governments that provide many benefits for their people.

Look at the map of Northern Europe on this page. What five countries make up Northern Europe? Today these countries are all independent nations. But they share a history that links them together.

The Heritage of the Northerners

Thousands of years ago, Northern Europe, sometimes called Scandinavia, was settled mainly by people moving in from lands further south. As the glaciers melted and the climate grew warmer, more people moved into the region. These new arrivals settled in the more northerly regions. As the population grew, the limited land resources of the area became a problem. Fishing had always been a source of income for the early Scandinavians. But by the late 700s, Northern Europeans had to depend on the sea even more. They developed a culture built around seafaring.

Some Northerners were seagoing traders. Others were raiders and con-

querors. They traveled both east and west seeking new fortunes and settling new areas, including Iceland and the Soviet Union. They also traveled to France and England, sometimes destroying coastal cities and stealing from them.

The raids continued for about 200 years, between A.D. 700 and 900. During this period, the Scandinavian homelands, in close touch with their raiding warriors, received many new ideas and much wealth from the conquered lands. Northerners who remained in Scandinavia built thriving farms, and the standard of living rose steadily.

Western Europe's influence began to reach Scandinavians by the end of the 900s. Many Northerners were converted to Christianity. Raiders found that they could become wealthy just by trading instead of conquering new areas. Many stayed on in villages they had conquered and became merchants. Some blended in with the Western Europeans. Others kept their ties to the homeland. Trade through these merchants helped strengthen the economy of their homelands. On this firm foundation, later generations of Scandinavians built lively, independent nations.

The glacial heritage. The natural features of many parts of Northern Europe help explain why seafaring was so successful. Look at the graphic-relief map on page 16 of the Atlas. Compare the coastlines of Norway and Sweden with those of other European countries. The highly irregular coastlines of Scandinavia show the glacial heritage of the area. The glaciers carved deep fiords, or

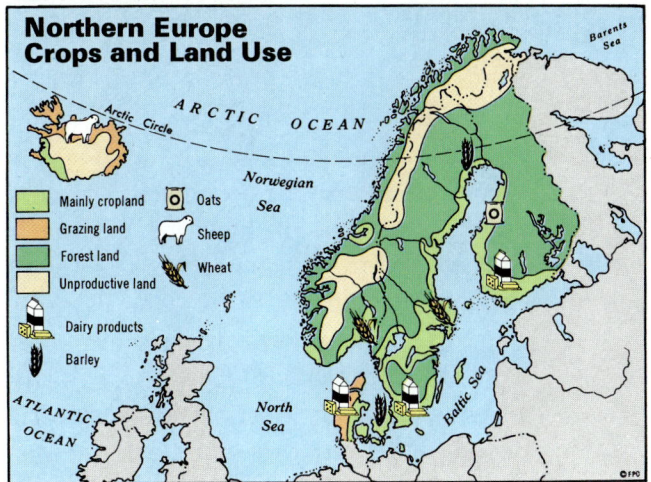
Northern Europe Crops and Land Use

How do you think the people living near this Norwegian fiord make their livings?

Rivers are also abundant, especially in Sweden and Norway. The sources of many of these rivers are in the Kjölen (chə(r)l'ən) Mountains. Check the map on pages 16–17 to find this mountain range. Rapids and waterfalls keep many of the rivers in Sweden and Norway from being entirely navigable. But the downward rush of these clear mountain rivers provides an abundant source of waterpower that can be harnessed to generate electricity.

Limited Agricultural Resources

The map on this page shows the main uses of land in Northern Europe. In which Northern European country is the land used mainly for crops and grazing? What is most of the land in Northern Europe used for? Where is the arable land in Norway? As you can see from the map, the importance of farming in Northern Europe varies from country to country. Farmland covers only about 5 percent of Norway. But in Denmark, over 50 percent of the land is used for farming.

Farmers and herders in Lapland, the northern part of Scandinavia, face a

U-shaped valleys, that the sea later filled. These fiords form natural harbors that provide easy docking for ships. They also offer shelter from storms at sea.

The glaciers also left behind many lakes. Many of the lakes are connected. Look at the map on pages 16–17. One of the Northern European countries has earned the nickname Land of a Thousand Lakes. Which country do you think has earned this nickname?

special problem. The oceans help moderate the climate of most of Northern Europe. But the areas north of the Arctic Circle still have a very long winter. Turn to the map of growing seasons on page 21 of the Atlas. How short is the growing season in most of northern Scandinavia? In this northern region, daylight disappears almost entirely for the whole month of December. Spring comes slowly here. The few farmers of this region live mainly in the river valleys. The short growing season limits what these farmers can produce. Often they must look for other sources of income to make a good living. Seasonal jobs in forestry and fishing often provide this extra income.

Meeting the challenges. Farmers elsewhere in Northern Europe have had to overcome other problems. In stony areas, they removed the boulders and used them to build fences and houses. In swampy areas, they drained the soil. In Denmark Danes, like the Dutch, have claimed lands from the sea. With hard work, Northern Europeans have found ways to use their land to the fullest.

Most of the farms in Northern Europe are small. Their average size ranges between 10 and 35 acres (4 to 14 ha) each. In some parts of Denmark, though, the farms are as large as 50 to 100 acres (20 to 40 ha) each.

To make up for the small size of the farms, farmers in Northern Europe have formed cooperatives. These help keep costs down. Cooperative farming is even more important in Northern Europe than it is in Western Europe. Unlike most Western European farmers, farmers in Northern Europe often share their equipment and profits. In Denmark, cooperatives are especially important. They handle the processing for about 90 percent of Denmark's pork products and dairy foods.

Dairy foods and livestock. The most important agricultural products of the Northern European countries are dairy foods and livestock. All five countries produce enough to export some cheese, butter, and other dairy products. Each also grows enough wheat, rye, barley, and oats to be self-sufficient in grains. Sweden, with somewhat more favorable conditions, even exports wheat.

Denmark, though, leads the Northern European countries in agricultural production. In fact, next to Dutch agriculture, Danish agriculture is the most efficient in all of Europe. Scientific methods have helped Denmark produce high-grade dairy products and bacon. These and other agricultural products make up about 20 percent of Denmark's exports. No other Northern European country exports a higher percentage of agricultural goods.

Abundant Fishing Resources

The resources of the sea have helped make up for the limited farmland available to Northern Europeans. There are seas all around Northern Europe. Small fishing villages near the openings of the fiords line the North Sea. The Arctic Ocean attracts thousands of cod fishers each spring. Scandinavians harvest the fish of the North and Baltic seas and the Arctic Ocean. Haddock and herring are abundant in these waters.

Iceland is one of the main fishing nations in the world. The warm North Atlantic Current waters meet the cold waters of the Arctic near the coast of Iceland. Their meeting stirs up the ocean's depths. In the mixing waters, there is plentiful food for large schools of fish. Fish and fish products make up about 70 percent of Iceland's exports. These help the country pay for goods that must be imported.

Norway also exports fish. Its total catch is about 2.5 million metric tons a year. Norwegians have fishing boats that are among the most modern in the world. Their fish-processing methods ensure a high-quality product.

Some of the cod caught in the waters off Norway's Lofoten Islands are salted and set out to dry.

Forests—Another Abundant Resource

As you saw from the map on page 307, forests are typical of the landscape in much of Northern Europe. Forests cover over 70 percent of Finland and make up much of Sweden as well. Scandinavian forests form one of Europe's largest reserves of softwood.

The trees are usually cut down during winter. When the rivers thaw, the trunks are floated downstream to the sawmills near the harbors. These mills supply wood products for two of Scandinavia's most famous industries: furniture and paper making.

Scandinavian forests are carefully managed. Long-range planning helps balance the rates of cutting and planting in Northern Europe's forests. Over half the wooded areas are managed by the government or by large industrial concerns. Small owners of the remaining forests are also careful to preserve their supply of trees. Scandinavians have earned a reputation for preserving all their resources carefully. They are also leaders in fighting pollution with strict laws.

Creative Use of Industrial Resources

Look at the map of mining and manufacturing on the next page. Scandinavia has a generally good supply of mineral resources. The abundance of iron ore in Sweden favored the early development of the steel industry there. But there was one major drawback. Neither Sweden nor its neighbors had a supply of coal. Coal and its by-product, coke, were commonly used to smelt iron ore and produce steel. Without coal, the

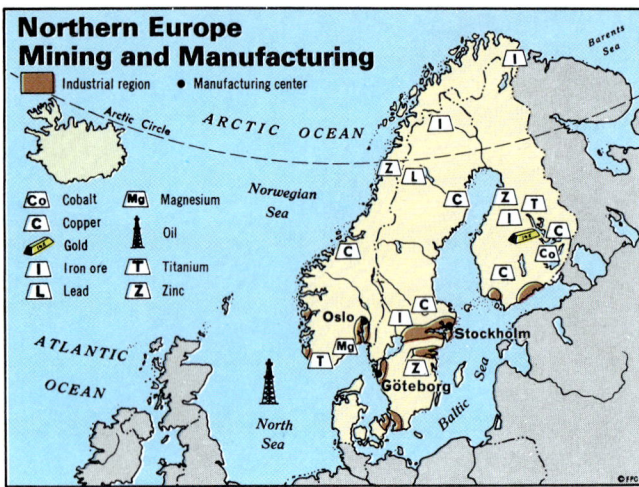

Northern Europe
Mining and Manufacturing

■ Industrial region   ● Manufacturing center

Co Cobalt   Mg Magnesium
C   Copper   Oil
Gold   T   Titanium
I   Iron ore   Z   Zinc
L   Lead

Swedes had to find another way to smelt iron. They turned first to their rich supply of timber. Timber in the form of charcoal was used as fuel to smelt iron ore into a very pure steel.

Later, water became the main source of energy. Scandinavia's many waterfalls and rapids became cheap sources of electricity. This new form of energy helped Northern European manufacturing grow by leaps and bounds. Electricity soon replaced charcoal in the Swedish processing of high-grade steel.

Major industrial products. Using waterpower for energy, many other industries have developed in Scandinavia. A number of industries depend on the region's supply of wood. Forests cover over 60 percent of Finland, so wood

pulp and paper lead Finland's list of exports. Paper products and timber are among the chief exports of Sweden.

Machinery and transportation equipment are other important Scandinavian exports. Norwegian shipbuilding is famous the world over. Swedish automobiles have also earned a worldwide fame for their dependability and fuel economy. Danish craft workers have found ways to turn their country's clay deposits into fine china. Other Scandinavian craft workers produce expertly made furniture.

Although Iceland is primarily a fishing nation, it does have mineral deposits from which it produces aluminum. It also has reserves of steam with which it generates electricity cheaply. Denmark lacks many minerals. Denmark, however, has begun to tap oil deposits beneath the North Sea's floor. Norway's offshore deposits have produced one fourth of its government income, making it an important oil-rich nation.

Sharing resources. As in other regions of Europe, the Northern European countries are finding cooperation more and more necessary to support their industries. Denmark needs pulp and

People in Iceland have tapped underground steam for energy. Power from underground reserves heats and lights the homes and offices of Iceland's only large city, Reykjavík.

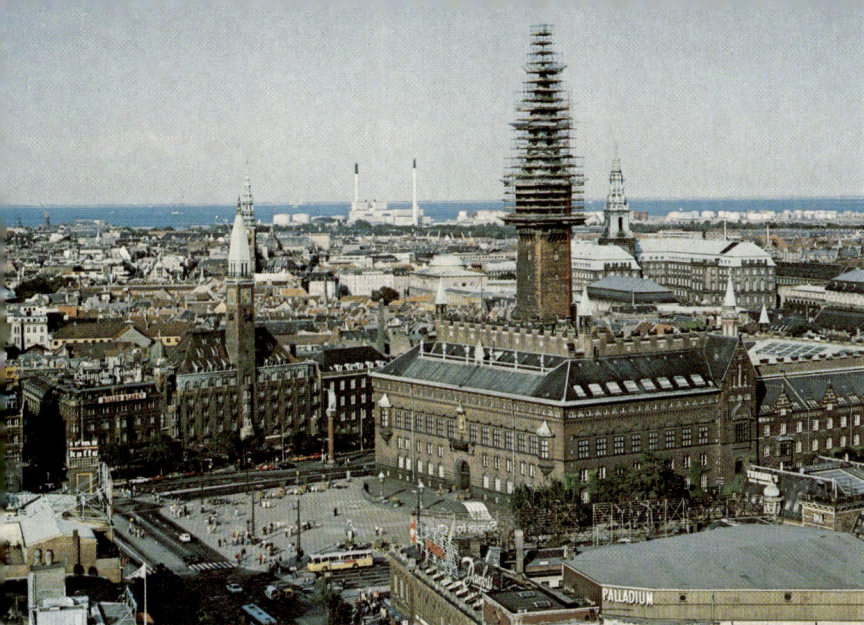

Copenhagen, the capital city of Denmark, grew to its present size partly as a result of its good location on world trade routes.

timber from Sweden and Norway. These countries in turn need Denmark's meat and dairy products. Sweden and Denmark need the electricity that Norway can produce. And Norway needs the machinery made in other countries. The Scandinavian countries share a common labor market like that in Western Europe. They have also formed the Nordic Council. Representatives from Scandinavian countries meet in this council to work out ways the nations can cooperate more fully.

A high quality of life. Scandinavians in both city and country enjoy a clean and sparsely populated environment. The Scandinavian countries are among the world's most prosperous nations. One fifth of Sweden's families, for example, own country homes where they can spend weekends and vacations. The majority of Northern Europeans live in cities, many of which are centers of learning, culture, and industry.

The governments of Scandinavian countries help provide a high quality of life. Their social security systems offer such services as complete medical care,

long vacations, free schools, and generous retirement funds. For these benefits, the Scandinavians pay high taxes.

In general, Scandinavians have planned carefully for the future. They have built solid economies and have found ways to share their wealth fairly among their people.

Keeping Facts in Focus

1. How do mountainous terrains help create good sources of waterpower?
2. What is a farm cooperative?
3. What Northern European country depends most heavily on its fishing industry?
4. What are some Scandinavian industries based on wood resources?
5. How do Northern European countries cooperate with one another?
6. What answers would you give to the key questions on page 305?

Working with Ideas

1. The map on page 307 shows that some parts of Northern Europe are considered agriculturally unproductive. In what ways might this land be considered productive?
2. If you were an iron and steel manufacturer, in which Northern European country would you be willing to build a new plant? Why?

311

Lesson 4 The Mediterranean Lands

Reading Focus

1. Before reading the entire lesson on the Mediterranean lands, read the first three paragraphs carefully. Through whose eyes do we see the Mediterranean lands in the second paragraph? Whose view do we see in the third paragraph? You may notice that the two views seem to conflict. But this conflict does not mean that either view is wrong. Instead, it suggests that both views are limited. Often, careful readers will have to balance differing points of views such as these. To achieve this balance, readers must try to figure out what explains the differing viewpoints. How might you explain the different views we see here of the Mediterranean lands? Can you put them together for a balanced view of the area?

2. As you read this lesson, look for the answers to these key questions:
   a. In what ways is the Mediterranean an unstable environment?
   b. What ways have Mediterranean farmers found to grow grains?
   c. Why has Mediterranean industry grown more slowly than industry in Western Europe?
   d. What is the general standard of living in Mediterranean lands?

Vocabulary Focus

malaria	*citrus*	*solar*
deforested	*fallow*	*technology*

The Mediterranean countries attract travelers from all parts of the world. On the map of Mediterranean Europe on page 313, find and name all the countries of the region.

Travelers to these countries delight in the beauty and history of the area. Rocky mountains and blue waters frame old, quiet villages. Ruins from ancient Greece and Rome remind visitors of the Mediterranean's rich history. Churches and museums, especially in Italy, house some of the world's most beautiful works of art. In the open markets, tour-ists find colorful stands of sun-ripened fruit. In the small shops, they find beautiful pottery and leather goods.

But many people who live in the Mediterranean countries see a different view of their region. They see tiny plots of abandoned farmland in poor villages. They see crowded trains carrying tired construction workers, miners, and factory workers back to their run-down homes. They see their friends taking jobs as waiters or house servants in Northern or Western Europe. And they see the contrast between the glories of

Mediterranean Europe

the past and the problems of today. Once prosperous and powerful, the Mediterranean countries today suffer from a number of economic difficulties.

A Difficult Environment for Farmers

Look at the graphic-relief map on page 16. Very little of the land in southern Europe is plains. The Mediterranean lands are mountainous and closely interlocked with the sea. In some places, high cliffs tower above the warm, salty Mediterranean waters. Elsewhere, the land slopes more gradually to the shore.

The slope of the land helps explain some of the area's problems in farming. Swift-running waters carry silt down the slopes to the narrow plains along the coasts. The silt deposits help make soil in these low-lying plains fertile. But they also cause problems. In some places, the silt builds up into small hills. These hills prevent the waters from draining to the sea. In rainy seasons, the rivers overflow and turn the nearby plains into marshlands. The marshes become breeding grounds for mosquitoes. Malaria is a constant threat.

Under the careful control of the Roman Empire, the plains were drained safely. But after the empire fell in A.D. 476, the neglected plains turned to swamps again. Draining the marshes again would have meant risking malaria. So many people moved into the hills to look for farmland. Here they faced other problems. The land they had come to farm was barren and badly eroded. Early farmers had first settled the hills about 3,000 years ago. They destroyed the oak and olive forests by cutting trees for timber and fuel. They allowed their herds of goats to destroy the grass covering. Then mountain rivers took over and eroded the exposed soil.

But farmers moving from the marshes found a way to make this land produce crops. They built terraces edged with stone to stop the erosion. And they harnessed the rivers to provide irrigation for the dry, deforested land.

But for about a thousand years after the Roman Empire fell, these hilly regions were prey to many invaders. Farmers were not always able to keep up their hard work of managing the land. During these times, swamps spread, monuments fell to ruins, and trade came to a standstill. It took hard work to overcome the effects of this neglect. And it takes hard work from today's farmers as well as careful government programs to protect the region's fragile environment.

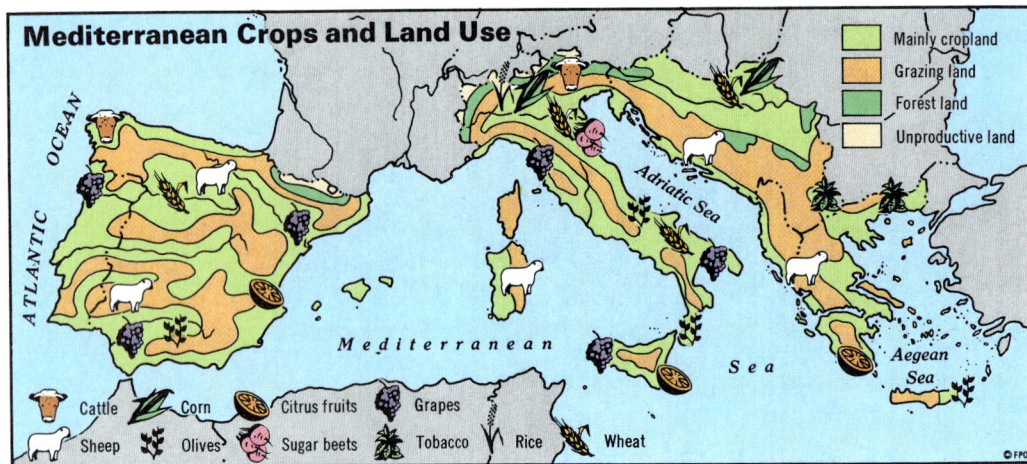

Mediterranean Crops and Land Use

Mainly cropland
Grazing land
Forest land
Unproductive land

ATLANTIC OCEAN

Adriatic Sea

Mediterranean

Sea

Aegean Sea

Cattle · Corn · Citrus fruits · Grapes
Sheep · Olives · Sugar beets · Tobacco · Rice · Wheat

© FPO

A region of small farms. Most of the farms in the Mediterranean are small. Farming methods are often old-fashioned, but they are not costly. Most farmers grow enough to feed their own families.

The map of agricultural production on this page shows the major crops of the Mediterranean countries. Trees bearing olives and citrus fruits, such as oranges and lemons, grow well in Mediterranean lands. Grapevines are also well suited to the climate, with its frequent periods of drought. Turn to the climate map on pages 200–201. During what season is drought likely? The olive trees and grapevines have long roots that can reach water deep underground. Vineyards in Spain, Portugal, and Italy have produced famous wines for centuries. Spain and Sicily are well known for their citrus fruits. And olive trees grow throughout the Mediterranean lands. Oil from olives is widely used for cooking in this region.

Grains are less able to withstand the periods of drought. Still, Mediterranean farmers have found a way to produce some grain crops. One year they plant grain in a field. The next year they let the field lie fallow. During the fallow year, the soil can store some of the rainwater for the next year's crops. Also during this year, flocks of sheep graze the land. Their manure enriches the soil. The sheep also supply meat, wool, and leather to the farmers.

Success and failure in Mediterranean farming. In some areas, farming has grown to a large scale. New techniques for draining the plains and irrigating the valleys have helped increase production in some areas. The Po Valley in northern Italy is one such region. Farmers there grow rice, cotton, sugar beets, and vegetables of all kinds. Nearby canning and preserving industries process whatever is not sold fresh. Dairy farming in the Po Valley helps make cheese another important product. The coastal plains of Spain are also highly productive. Citrus fruits, peaches, pears, and apricots grown there are sold throughout Europe. Check the map of growing seasons on page 21 of the Atlas. What is the length of the growing season on the southern coast of Spain?

In other areas, however, farms are failing. Many discouraged farmers leave their small plots and poor villages. They head for the cities where they hope to find jobs. Many people from Sicily,

Mediterranean Europe Mining and Manufacturing

Industrial Region • Manufacturing center

Milan
Zagreb
Belgrade
Barcelona
Naples

ATLANTIC OCEAN
Adriatic Sea
Mediterranean Sea
Aegean Sea

B Bauxite   C Copper   Natural gas   Oil   Uranium
Coal   I Iron ore   Su Sulfur   Tu Tungsten   Z Zinc   ©FPO

northern Greece, Portugal, and Yugoslavia become migrant workers in Germany, France, and Scandinavia. Often they make regular trips back to their villages. Sometimes, they return home with many years' savings. Back at home, they try their hand at farming once again.

Limited Fuel Resources for Industry

Look at the map of mining and manufacturing on this page. The Mediterranean lands have an adequate supply of minerals. But they lack good resources for fuel. What countries produce coal?

Most countries in southern Europe depend heavily on waterpower for energy. Portugal and Greece have some lignite, a type of coal. Italy has some natural gas. The mining of uranium may help Spain develop nuclear power. There is a small amount of oil under the Mediterranean Sea. But the general lack of fuel resources has slowed down the growth of industrial development in Mediterranean countries.

Only northern Italy has clearly joined the industrialized world. Abundant waterpower provides the energy for the area's thriving industries. Look at the map on page 16 of the Atlas. What

Many farmers in Cádiz, a southern province of Spain, grow a mixture of grapes, fruits, nuts, and grains. The patchwork of crops along the river Guadalete shows diversified farming.

Venice, Italy, located on the Adriatic Sea and near large rivers, was one of the great trading centers of the Middle Ages. Today, during an annual festival, gondoliers dress in the fashions of medieval merchants.

mountain range is the source of northern Italy's many rivers?

Over half of Italy's people live and work in the Po Valley. It is the most densely populated region in the country. Industries in the Po Valley produce cars, typewriters, farm machinery, and textiles of many kinds. The cities of Genoa, Turin, and Milan make up the industrial center of Italy. Turin leads the region in the production of automobiles. Genoa is the country's busiest port. Milan is the banking center of Italy. Workers in other areas of northern Italy produce chemicals, ceramics, glassware, and fine leather goods. There are also plants for processing the foods grown in the irrigated valleys nearby.

The more southerly areas of the Mediterranean region cannot compete with northern Italy in production. Portugal and Greece have shipbuilding industries. Yugoslavia exports some furniture. But these countries are mainly agricultural. They do have many workers and a varied supply of minerals. But they lack the energy resources needed to develop industry.

The Mediterranean climate may help provide new sources of energy. These would aid the area's future industrial growth. Solar batteries, for example, might tap the abundant sunshine of the region. Energy from the wind might be harnessed in modern windmills. But the level of technology is not very high in Mediterranean lands. Without advanced technology and money for development, industrial growth is likely to be slow.

Many tourists and religious pilgrims travel to Vatican City every year (above). The unusual architecture of Antonio Gaudí (left) also attracts many visitors in Barcelona, Spain.

Social, Political, and Environmental Problems

The Mediterranean lands face many problems. Where industry has developed, pollution threatens the air and water. Problems in land use make farming difficult. Early overfishing of Mediterranean waters has greatly reduced the supply of fish. Great numbers of poor people increase political problems. Look carefully at the chart on pages 290–292. Compare the gross national products of Mediterranean countries with those of other European nations. Now compare their literacy rates. Do you think there tends to be a relationship between a Mediterranean country's wealth and its percentage of people who can read and write?

Communism is gaining support, especially in Italy. Many people are attracted by its promise of shared wealth and social benefits. Albania and Yugoslavia now have Communist governments. Cooperation between Communists and non-Communists in the Mediterranean area is very limited. Although Mediterranean lands share certain problems, there is no one organization for cooperation among these nations.

It will be many years before the problems of the Mediterranean countries are resolved. Meanwhile, tourists continue to flock to the beautiful lands. The money spent by tourists is a large source of income for Mediterranean countries. It may one day help provide funds for much-needed development of problem areas in the Mediterranean lands.

317

Patterns of Government in Europe and the Soviet Union

All European countries have some form of representative democracy. Most elect two national assemblies, similar to our Senate and House of Representatives. The job of these parliaments is to propose and vote on new laws. Most European governments also have a chief official whose job is similar to that of our president. But the practice of democracy differs from country to country in Europe. The economic system of each country determines how its political system actually works in practice. The chart shows the basic differences between the practices of Communist and non-Communist governments in Europe.

COUNTRIES	ECONOMY	ROLE OF GOVERNMENT
Non-Communist countries Many countries in Western, Northern, and Mediterranean Europe, including Austria, Belgium, Finland, France, Italy, Norway, Sweden, Switzerland, United Kingdom, West Germany.	Mixture of free enterprise and socialism. Individuals have the right to compete for profits in most industries. Government does regulate some public services, such as transportation and communications.	Government exists to protect the rights and freedoms of all individuals. Voters choose candidates from a number of political parties to represent a variety of views in government. Government does own and operate some businesses. It also imposes taxes to provide social benefits, such as education and welfare for the poor.
Communist countries All countries in Eastern Europe and some countries in Mediterranean Europe, including Albania, Bulgaria, Czechoslovakia, East Germany, Hungary, Poland, Soviet Union, and Yugoslavia.	Individuals cannot compete for profits. All wealth is shared among the people according to their needs. Government owns and operates most or all businesses.	Government exists to protect the interests of the whole society, not individuals. There is only one political party—the Communist Party—that has any real power, so voters have few or no choices. Government receives and distributes all wealth. Party leaders make most economic decisions. Government provides all social services.

Keeping Facts in Focus

1. Which countries are a part of the culture region of Mediterranean Europe?
2. What are some important Mediterranean crops?
3. What answers would you give to the key questions on page 312?

Working with Ideas

1. How does farming in Mediterranean Europe differ from farming in Western Europe?
2. Which nations on the National Profiles charts on pages 290–292 have the lowest literacy rates? How might a low literacy rate affect the economic development of a nation?

Lesson 5 Eastern Europe and the Soviet Union

Reading Focus

1. Details in a reading selection serve a very important purpose. They help support, or verify, general statements and main ideas. Turn to page 320. Read the three paragraphs under the heading **Increasing farm production** carefully. The first paragraph states two main ideas. What are these ideas? The next two paragraphs offer details to support these ideas. Which main idea do the details in the second paragraph support? What details support the other main idea? Recognizing how details relate back to main ideas and help prove a point will sharpen your reading skills.
2. As you read this lesson, look for the answers to these key questions:
   a. In what ways are Eastern and Western Europe alike? In what ways are they different?
   b. How did Communism change farming and industry in Eastern Europe?
   c. Why has light industry grown slowly in Eastern Europe?
   d. What is daily life like under Communist rule?

Vocabulary Focus

ethnic	*loess*	*light industry*
intensive farming	*heavy industry*	*appliances*

From plowing to welding, the countries of Eastern Europe and the Soviet Union offer many different ways to earn a living. Their people come from a great variety of ethnic groups. Within this variety, however, there is some unity. Each of these countries has a Communist government and economy. The degree of government ownership of farms and industries varies from country to country. Still, Communism joins these nations into an economic unit distinct from other parts of Europe. Turn to the map on page 320. Name all the countries that form Eastern Europe.

Comparing Eastern and Western Europe

The regions of Eastern Europe and Western Europe are similar in several ways. On the whole, people in both Eastern and Western Europe use the land in much the same way. A look at the graphic-relief map on pages 16–17 will help explain why. Like Western Europe,

Eastern Europe and the Soviet Union

ARCTIC OCEAN

Arctic Circle

UNION OF SOVIET SOCIALIST REPUBLICS (U.S.S.R.)

Baltic Sea

EAST GERMANY

POLAND

CZECHOSLOVAKIA

HUNGARY

RUMANIA

BULGARIA

Black Sea

Caspian Sea

ASIA

©FPC

Eastern Europe is mainly a region of plains. There are more mountains in Eastern Europe, but plains are still the main landform.

Most of Eastern Europe has a continental climate. More northerly parts of the region have a subarctic climate. The climate of the countries in Western Europe is marine. Rains in Eastern Europe are usually heaviest in the summer, while rains in Western Europe are usually heaviest in the winter.

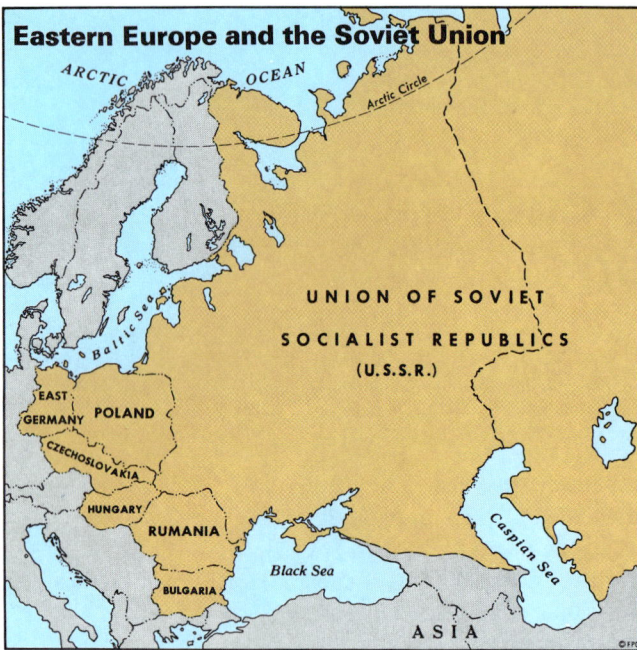

Communism and Agriculture

Most farms in Western Europe are privately owned. In many Eastern European countries, however, most of the land is owned by either the government or a number of families together. After the Communists took over, one of their first goals was to join small farms into larger, more efficient units. On state-owned farmland, farmers receive wages for their work. On collective farms, a number of families share the farm's earnings. Each family still owns a small patch of land to grow food for its own use or to sell for extra cash.

Many Eastern Europeans, especially the Polish, were against the move toward collective farms. In fact, the Polish government finally let the collectives break up. Now, 85 percent of all Polish farmland is privately owned. Other Eastern European governments, however, have not allowed most lands to return to private ownership.

Increasing farm production. Since World War II, agricultural production in

Near Tashkent, Soviet cotton farmers work on a collective farm. Cotton grown in the Asiatic region of the Soviet Union has coarse fibers and is used mainly for fillings and blankets.

После cotton is harvested, cleaned, and dried, it is ready for storage. On this state farm near Tashkent, workers use modern machines to move the cotton.

Eastern Europe has greatly increased. The production of wheat, for example, has doubled. But Eastern Europe is still not as productive as Western Europe. Several reasons explain Eastern Europe's lower production.

First, Eastern European farmers had never used the land as thoroughly as their western neighbors. They lacked a tradition of intensive farming. Second, industry in Eastern Europe was slow to develop. As a result, farming machines and chemicals for fertilizers are in shorter supply. Finally, early policies of the new Communist governments did not always work. Communist party leaders far from the farms were often unable to make good decisions about production and marketing. Appointed mainly for political reasons, many of these leaders had little skill in farming and management.

All of these factors are now changing. Farming is becoming more varied and intensive. Fruit and vegetable crops are getting larger. Dairy foods and new industrial crops, such as tobacco, are adding to Eastern Europe's yields. Industry has also grown. It is now producing more farm machines and chemical fertilizers. Government schools have educated farmers about new methods of irrigation. And the government has allowed local farmers to make more decisions about the collectives. It has also made prices and taxes more favorable to farmers. With all of these changes, Eastern Europe's agricultural production is growing quickly.

Chief Farm Products

The Soviet Union is one of the major wheat-growing countries of the world. In recent years, however, drought in the Soviet Union greatly reduced crops. The Soviet Union now imports some of its wheat from the United States and Canada. The map of agricultural products on page 322 shows that wheat is also grown in many other parts of Eastern Europe. Czechoslovakia and southern Poland, with their rich loess soils, produce a great quantity of wheat. Corn and wheat are the main crops of the plains of the Danube. Where else is corn grown? Eastern Europe grows well over half of Europe's supply of corn. Barley is another important grain crop, especially in Czechoslovakia.

■	Mainly cropland
■	Grazing land
■	Forest land
■	Unproductive land
🌻	Sunflower seeds
🌸	Sugar beets
	Barley
	Corn
	Grapes
	Livestock
	Oats
	Potatoes
	Rye
	Wheat

land and the Soviet Union. These countries lead the world in producing rye and potatoes. In fact, Poland grows more potatoes than all the other Eastern European countries combined.

Grapes and sunflowers are important cash crops in the southern portions of the Soviet Union and the Danubian plains. Grapes are used in making the region's many varieties of wine. And oil from the sunflowers is used throughout Eastern Europe as cooking fat.

Communism and Industry

Industrial development often depends on a strong, stable government to support it. During the years when Western Europe was industrializing, the govern-

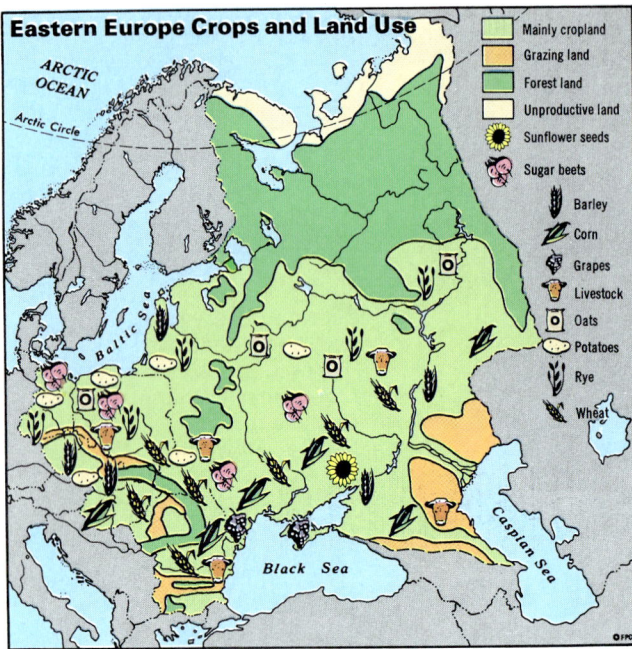

A variety of other crops. Rye and potatoes grow well in the sandy plains of the central and northern regions of Po-

The new Siberian railroad, which runs over miles of mountains, rivers, and blocks of ice, connects Moscow with Siberia's rich mineral resources.

ments of Eastern European countries were very unstable. Some industry developed in Eastern Europe, especially after World War I. But most areas remained largely agricultural.

When the strong Communist governments came to power, a new industrial age was born in Eastern Europe. The governments took over the existing industries. They also launched carefully planned programs to build up industry in underdeveloped regions. This effort has helped people in all regions of a country earn a fairly equal living. Czechoslovakia especially has achieved great industrial strength.

In the Communist countries of Eastern Europe, the government controls industrial planning and decisions. At first, the Communist plans stressed heavy industry. This included mining, metals, chemicals, construction materials, and electrical power. To support these industries, the governments devoted lots of money to improving transportation routes. Railways are especially important for shipping freight. Improved roads are making trucking more reliable.

Diverse mineral and fuel resources. The map on this page shows mineral and fuel resources in Eastern Europe and the Soviet Union. Most countries have fairly small deposits of a number of useful minerals. Two regions, however, have a great abundance of coal. These are the Silesian coalfield (in southwestern Poland and northeastern Czechoslovakia) and the Don Basin (Donbass) in the Soviet Union. Find these areas on the map of mining and

Eastern Europe Mining and Manufacturing

manufacturing. The Silesian coalfields have a reserve as great as that of the Ruhr fields in Germany. The Donbass reserves are also huge. In Siberia (that part of the Soviet Union that lies in northern Asia), there is another large deposit. There, one thousand miles east of the Urals, the Kuznetsk Basin is rich in coal.

All Eastern European countries have some iron ore. But they depend on the Soviet Union for high-quality ore. Look at the mining and manufacturing map. Find the iron ore deposits in the Ural region near Perm. Iron ore from this region is often exchanged for coal from the Kuznetsk Basin.

Rumania is the third largest producer of oil in Europe. It also produces large amounts of natural gas. Rumania uses some of these resources in its growing chemical industry. But it has sufficient supplies to export some oil and gas. Oil fields near the Ural Mountains, along with smaller fields in the Ukraine, supply the rest of the region's needs.

Recent Events in Eastern Europe

1917—Communists come to power in Russia

1924–38—Power struggle occurs in Soviet Union; Stalin becomes dictator; thousands of political prisoners are executed

1939—World War II begins; Germany signs a nonagression pact with the U.S.S.R.

1941—Germany attacks the U.S.S.R.

1945—World War II ends with most of Eastern Europe in the control of Soviet troops

1946–48—Communists come to power in Bulgaria, Rumania, Poland, Hungary, and Czechoslovakia; Yugoslavia develops close ties with the West

1956—Revolt against Soviet control fails in Hungary

1965—Rumania's new constitution stresses independence from the U.S.S.R.

1968—Revolt against Soviet control fails in Czechoslovakia

1980–82—Communist government in Poland checks efforts of union workers to gain better working conditions

1983—Nobel Peace Prize awarded to Lech Walesa, leader of the outlawed Polish labor union, Solidarity. Martial law lifted in Poland.

Trade among the Communist nations. Eastern European nations have formed an organization of economic cooperation similar to the Common Market in Western Europe. Their organization is called the Council for Mutual Economic Assistance (COMECON). Under COMECON, each nation is supposed to specialize in certain kinds of production. COMECON has not been entirely successful, partly because the Soviet Union has so much more economic power than the other member nations.

Most Eastern European countries depend on the Soviet Union for about a third of their foreign trade. Hungary's dependence on the Soviet Union is slightly greater than this average. Bulgaria's is far greater. About half of Bulgaria's trade is with the Soviet Union. Rumania, on the other hand, depends on the Soviet Union for only about 15 percent of its trade.

Most of the rest of Eastern European trade is among other members of COMECON. The Soviet Union, however, trades widely with Western countries. Twenty-five percent of its trade is with the West. Western nations supply many advanced technological products to the Soviet Union. Poland and Rumania also have significant trade with some Western European countries and even with the United States. But the bulk of Eastern European trade is among COMECON countries, with the Soviet Union at the center.

A slow change to light industry. The Communist leaders of Eastern Europe believe that heavy industry is the basis of industrial growth. They have centered their efforts, then, in building up heavy industry. Products from heavy industry have helped rebuild the many areas of Eastern Europe that were destroyed in World War II.

Now, much of that work is done. With their basic needs met, Eastern Europeans are beginning to look for a greater variety of consumer goods. Light industries, such as clothing, are beginning to grow. But the fashion of clothing and other personal goods changes very quickly. For this reason, central government planning in these industries is not always possible. Lately, the strict planning of the central leaders has relaxed. Local managers have more freedom to decide the products their plants will make. As a result, daily life in Eastern Europe and the Soviet Union is growing

This exhibit in the Kosmos Pavilion displays advanced technology in spacecraft. The Soviets have a world-wide reputation for their achievements in technology.

more like life in Western Europe. Foods, fashions, and household appliances in Eastern Europe are becoming more like those of Western Europe.

Life Under Communist Rule

Communist governments exercise strict control over the lives of their citizens. But some people think there is too much governmental control. These people feel deprived of many basic human rights and freedoms. Artists, writers, members of religious groups, and others cannot express their ideas or their beliefs without government censorship. People who oppose the government are often imprisoned or silenced in other ways.

Workers cannot easily change jobs or move to a different city. The Communist governments provide health care and old-age benefits for workers and their families. But personal income, except for high party and government officials, is low. After basic needs are met, many workers have little left to spend on items that make their lives happier or easier.

Keeping Facts in Focus

1. What are the main climates in Eastern Europe and the Soviet Union?
2. What ways have the Communist countries used to increase farm production?
3. Where are the major coal deposits in Eastern Europe and the Soviet Union?
4. Which country is the third largest European producer of oil?
5. What is COMECON?
6. How much foreign trade in most Eastern European countries is with the Soviet Union?
7. What answers would you give to the key questions on page 319?

Working with Ideas

1. What are similarities between COMECON and the Common Market? What are the differences between the two?
2. From time to time, the Soviet Union has purchased tons of grain from the United States. How does such a purchase benefit U.S. farmers? How does it benefit the Soviet Union?
3. How does heavy industry differ from light industry?

Unit 9   REVIEW WORKSHOP

Test Your Geographic Knowledge

A. Choose the letter of the item that correctly completes each statement.
1. The Northern European country most dependent on its fishing industry is (a) Sweden, (b) Iceland, (c) Denmark.
2. A heavily industrialized area of Italy is the (a) Po Valley, (b) Ruhr Valley, (c) Don Basin.
3. COMECON is an organization to promote economic cooperation among (a) Western European countries, (b) Northern European countries, (c) Eastern European countries and the Soviet Union.
4. Western Europe's busiest waterway is the (a) Volga, (b) Rhine, (c) Danube.
5. Western Europe's climate is (a) marine, (b) continental, (c) subarctic.
6. The Eastern European country that allows the most private ownership of farms is (a) Czechoslovakia, (b) Bulgaria, (c) Poland.
7. The culture region in which Portugal is located is (a) Western Europe, (b) Mediterranean Europe, (c) Northern Europe.
8. The longest river in Europe is the (a) Danube, (b) Volga, (c) Don.

B. Tell whether each statement is true or false.
1. Europe is in the middle of the land hemisphere.
2. Netherlands is one of the most densely populated countries of Europe.
3. Fiords in Scandinavia are the results of volcanic activity.
4. The Ural Mountains and the Ural River serve as the boundary between Europe and Asia.
5. Eastern Europe's agriculture is more productive than Western Europe's.
6. Most communist governments have stressed light over heavy industry.
7. Industry grew rapidly in Scandinavia once coal was discovered there.
8. Rumania is the third largest oil producer in Europe.

Apply Your Reading Skills

A. Skim over the lessons in this unit and find the dates for each of the following events in European history. Then arrange these events in the order in which they occurred.
1. The Roman Empire fell.
2. The First World War began.
3. The Common Market was established.
4. The Northerners raided Europe and other lands.
5. Urban civilization began to develop around the Aegean Sea.
6. The industrial revolution began.

B. Use library resources to help you write a report on ONE of the following topics.
1. What is the political relationship between Northern Ireland and Ireland? Between Northern Ireland and the United Kingdom? In what ways is religion a political issue in Northern Ireland today?
2. What is the difference between Eastern Orthodox and Roman Catholicism? In what European countries are many people followers of Eastern Orthodox traditions?
3. What in the Communist way of thinking conflicts with religion? How widespread is government action against religious practice in the Communist countries?

Apply Your Geographic Skills

A. Use the chart of National Profiles on pages 290–292 to help you answer the following questions.
1. Name the three European countries whose capital cities are not their largest cities.
2. Name the two European countries with no rural population. What does their urban population tell you about these countries' population densities?
3. What are the two European countries with the lowest literacy rates? How does their percentage of urban population compare to that of other European countries?

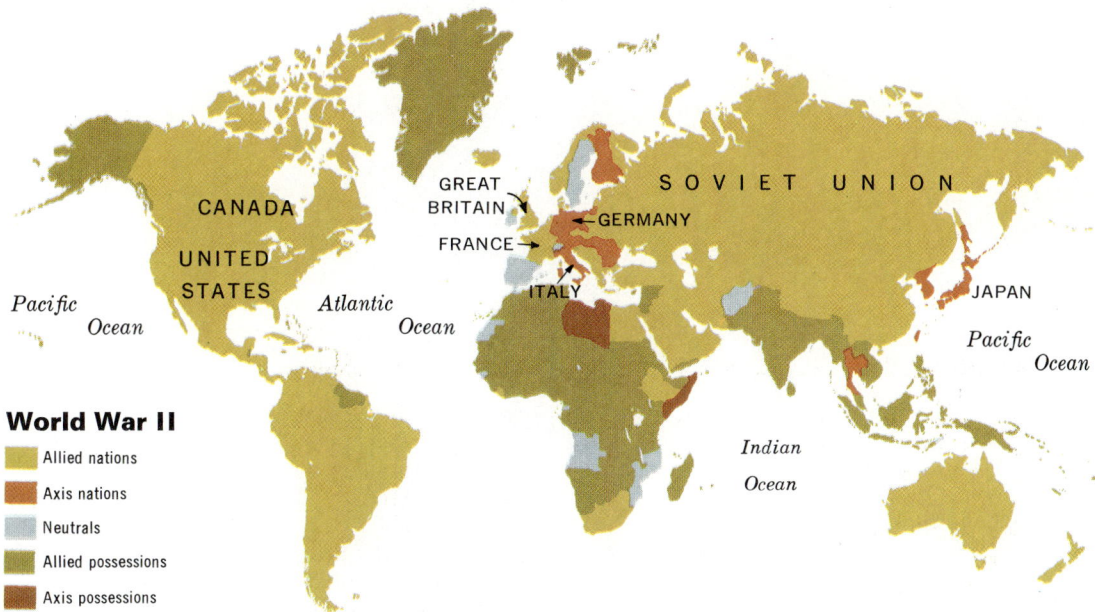

World War II

- Allied nations
- Axis nations
- Neutrals
- Allied possessions
- Axis possessions

Map labels: CANADA, UNITED STATES, GREAT BRITAIN, FRANCE, GERMANY, ITALY, SOVIET UNION, JAPAN, Pacific Ocean, Atlantic Ocean, Indian Ocean, Pacific Ocean

4. What information on this chart might help you determine which countries are least likely to suffer from industrial pollution?

B. Answer the following questions by using the map on this page and comparing it to the map on page 288.
   1. What Mediterranean countries were neutral (that is, neither Allied nor Axis nations) in World War II?
   2. Which Northern European country was an Axis nation? Which was neutral?
   3. Was Rumania an Allied or an Axis nation?
   4. What Western European countries remained neutral?
   5. Was Czechoslovakia an Axis nation?

Apply Your Thinking Skills

Each of the following descriptions applies to one of Europe's four main cultural regions. Decide which region each describes.
1. This region is sparsely populated. Its natural features are well suited to fishing and shipping. People in this region pay high taxes, but they enjoy many government benefits and services.
2. All countries in this region have a Communist government. This region grows more corn and potatoes than any other area in Europe. Much of the farming is done on collective farms. The industrial age in this region began after World War II.

3. This region has many ruins from Greek and Roman culture. Few areas of this region are highly industrialized. On the whole, the region lacks good fuel sources, except waterpower. Grapes and olives grow well in this region's warm climate.

Discuss These Points

Cities in Europe grew rapidly after the industrial revolution. Many people decided to give up their lives in the countryside and move to the city. Many people are still making that decision today. In general, what advantages does city life have over life in the countryside? What drawbacks does city life have? Compare and contrast daily life in the city and country, using specific examples you are familiar with. Which style of life is most appealing to you?

Expand Your Geographic Sights

Armitage, Paul. *The Common Market* (Silver, 1978).

Dornberg, John. *Eastern Europe: A Communist Kaleidoscope* (Dial, 1980).

Kingsbury, Robert. *Atlas of European Affairs* (Praeger, 1964).

Rothkopf, Carol Z. *East Europe* (Watts, 1972).

Strauss, Richard. *Coal, Steel, Atoms & Trade: The Challenge of Uniting Europe* (Coward, 1962).

The Soviet Union (Time: Life, 1984).

Unit 10 Asia

The mountainous terrain of Luzon, an island in the Philippines, has been made suitable for rice farming by the technique of terracing. With more than two billion people to support, Asian lands must be used to their greatest potential.

Exploring Asia

1. To understand the meaning of a new word, it is sometimes necessary to turn to a dictionary. Other times, however, the surrounding words and sentences provide helpful clues. In this lesson, the word *affluent* appears in paragraph two under the heading **People and problems** (page 336). Read the first three sentences of this paragraph. There are two important clues in these lines that help define *affluent*. The first is in the second sentence. The word *even* suggests that *affluent* is like *comfortable* but stronger. The second clue comes when the phrase "on the other hand" sets up a contrast. We learn that *starving* and *homeless* people are unlike *affluent* people. Putting the two clues together, how would you define the word *affluent?* Watch for other context clues as you read. Make sure to check your dictionary or glossary for words you do not understand.

2. As you read this lesson, look for the answers to these key questions:
   a. What physical features contribute to the isolation of Asia's people?
   b. What are the monsoons?
   c. What problems does Asia's huge population create?
   d. How did Europeans change the patterns of Asian life?

Vocabulary Focus

interior	*affluent*	*surplus*
monsoon	*domesticate*	*luxury*

More people have lived in Asia than in any other region on earth. Over 2.5 billion people, more than half the world's population, live in Asia today. All three major races—Mongoloid, Negroid, and Caucasoid—make up Asia's population. Asia is also the world's largest continent. But its huge population is not spread evenly over its vast lands. Some parts of Asia's interior are too cold, too dry, or too mountainous to support people. So millions and millions of Asians must crowd together in the areas of usable land to make their livings there.

Asia is also the home of some of the oldest cultures on earth. For many centuries, Asians had little contact with other cultures. Even within Asia, high mountains often prevented the exchange of ideas and ways of life. Many Eastern customs, languages, religions,

Asia
Political-Relief Map

⊗ National Capitals ★ Other Capitals • Other Cities

SCALE

Miles 0 200 400 600 800 1000
Kilometers 0 400 800 1200 1600

Map labels (selected):

U. S. S. R. · MONGOLIA · Ulaanbaatar · Gobi Desert · Takla Makan · CHINA · Peking · Hwang Ho · Grand Canal · Yellow Sea · Shanghai · Yangtze River · Chongqing · Xi Jiang · Canton · MACAO (Port.) · HONG KONG (Br.) · Sea of Japan · N. KOREA · Pyongyang · Seoul · S. KOREA · JAPAN · Hokkaido · Honshu · Tokyo · Osaka · Yokohama · Kure · Shikoku · Kyushu · Ryukyu Is. (Japan) · Tropic of Cancer · PACIFIC OCEAN · ★ Taipei · TAIWAN · TRUST TERRITORY OF THE PACIFIC ISLANDS (U.S.) · Equator

AFGHANISTAN · Kabul · IRAN · PAKISTAN · Islamabad · Indus R. · Karachi · Delhi · New Delhi · NEPAL · Katmandu · Himalaya · Mt. Everest (29,028 ft.) · Lhasa · BHUTAN · Thimbu · Ganges R. · Ahmadabad · Calcutta · Dacca · BANGLA DESH · BURMA · Rangoon · LAOS · Hanoi · VIETNAM · OMAN · Arabian Sea · Bombay · INDIA · Deccan · Western Ghats · Plateau · Madras · Bay of Bengal · Vientiane · THAILAND · Bangkok · KAMPUCHEA (CAMBODIA) · Phnom Penh · Ho Chi Minh City (Saigon) · Mekong R. · South China Sea · Manila · PHILIPPINES

MALDIVES · Male · SRI LANKA · Colombo · INDIAN OCEAN · BRUNEI · Bandar Seri Begawan · Moluccas · MALAYSIA · Kuala Lumpur · SINGAPORE · Borneo · Celebes · PAPUA NEW GUINEA · Sumatra · INDONESIA · Jakarta · Java · AUSTRALIA · © FPC

and patterns of life are therefore very different from each other and from those in the West.

But Eastern as well as Western countries share the same basic goal. They all want to provide as good a life for their people as possible. Eastern countries, with less industry and more people, face a constant struggle to achieve this goal. While they are borrowing some Western ideas to help them, many Asians are eager to preserve their own ways of life.

The Physical Setting

The continent of Asia covers almost a third of the world's land. It spreads across many climate zones and many types of landforms. It is an area of sharp contrasts and varying resources. The maps in this book will help you become familiar with the physical features that have shaped life in Asia.

Location. Turn to the map of the world on pages 10–11 of your Atlas. Locate the continent of Asia. Between

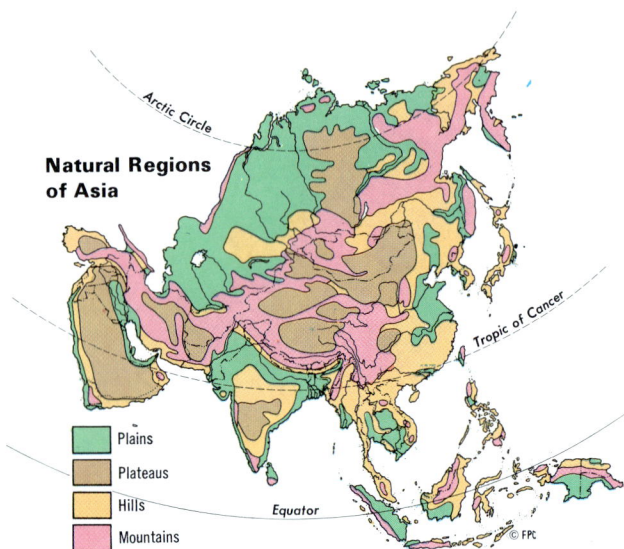

Natural Regions of Asia

- Plains
- Plateaus
- Hills
- Mountains

what latitudes does Asia lie? What ocean borders Asia on the north?

Countries. The map on page 15 of the Atlas shows all the countries on the continent of Asia. China and India are the two largest countries of Asia. The Maldives and Singapore, two island nations of Asia, are the smallest.

Compare the countries shown on page 15 with those shown in color on page 331. On the map in this lesson, part of the Soviet Union and the countries west of Pakistan are not shown at all or are not shown in color. You have already studied the Soviet Union as a part of Europe. You will study the countries west of Pakistan in Unit 12. These countries, although part of the Asian continent, form a cultural region called the Middle East.

Physical features. Compare Europe and Asia on the graphic-relief map (pages 18–19 of the Atlas). Which is more mountainous? Asia is, in fact, the most mountainous of all the continents. Asia's many mountains are high and rugged, making travel very difficult.

Most of Asia's mountains fan out from one rugged area called the Pamir (pə·mi(ə)r′) Knot. The Pamirs, the range at the heart of this area, are just north of Afghanistan and Pakistan. Find this range on the map on pages 18–19. What mountain ranges branch off to the northeast? What ranges extend to the east? The Himalayas, branching off to the southeast, are the highest mountains on earth. Find Mt. Everest in the Himalaya range. The highest peak in the world, Mt. Everest rises 29,028 feet (8,848 m) above sea level.

The map of Asia's natural regions (this page) clearly shows the pattern of mountain branches. Notice that the branches extend westward also. What ranges form the western ribs of this vast mountain system?

Deserts and plains. There are several large regions of deserts in Asia. In the west, much of Saudi Arabia is desert. There is also a desert area in the southern part of the Soviet Union. The large deserts in eastern Asia are the Takla Makan (täk′lə·mə·kän′) and the Gobi. The Takla Makan desert lies between the Tien Shan (tē·en′shän′) and Altun Shan (al·tōon′shän′), in northwestern China. Both of these mountain ranges block the moisture-bearing winds that could bring rain to the desert. Find the Gobi Desert, which stretches across Mongolia into China, on the map on pages 18–19. Also locate the Great Indian Desert.

Look at the map of natural regions above and locate the largest area of plains. The plains of the Soviet Union extend northward for many miles. But only those in the more southerly parts are suitable for farming. In the colder northern region, they form a zone of frozen tundra. Now locate the plains regions in northern India, in northern China, and in Southeast Asia. Notice the rivers running through these regions of plains.

Rivers. Although Asia's mountains create problems for travel, they also supply Asia with some of its most valued resources: the rivers. The rivers carry fertile soil, irrigate dry lands, and provide good transportation. Some of the most densely populated regions in Asia are those along its mighty rivers.

Three important rivers in India and neighboring lands have their sources in the mountains to the north. Find the Indus River on the map on pages 18–19. Into what sea does it flow? The other main rivers are the Ganges (gan'jēz) and Brahmaputra (bräm'ə p(y)ōo'trə). Using the map, name some of the big cities that lie along the Ganges.

The mountains also provide the sources for most of Southeast Asia's

Katanga Peak, near Mount Everest in Nepal, is part of the forbidding Himalaya range.

rivers. Trace the course of the Irrawaddy, Salween, and Mekong rivers through Southeast Asia on the map on pages 18–19.

The longest river in Asia is the Yangtze in China. The Yangtze flows eastward for 3,430 miles (5,520 km) and empties into the East China Sea.

In the dry lands of central India, farmers are using traditional ways to move water from a river to their farmlands. Can you describe the process pictured here?

National Profiles: *Asia*

Independent Nation	Area/ Population	Population Density/ §Growth Rate	*Capital/ Largest City and Population	Urban Population/ Literacy Rate	Chief Exports (in order of Importance)	†GNP (in) millions of U.S. dollars/ Per Capita Income
Bangladesh	142,500 sq km 99,585,000	699 per sq km 3.1%	*Dacca 3,458.602	10% 25%	jute & jute products, leather, tea	$1,0000 $117
Bhutan	46,600 sq km 1,417,000	30 per sq km 2.2%	*Thimbu 15,000	4% 5%	timber, fruits & vegetables, coal	$131 $109
Brunei	5,776 sq km 218,000	38 per sq km 3.9%	*Bandar Seri Begawan 51,000	76% 45%	crude oil, natural gas	$19,800 $27,000
Burma	678,600 sq km 36,196,000	53 per sq km 2.0%	*Rangoon 2,186,000	27% 78%	rice, teak, hard-woods, base metals, ores	$5,900 $180
China	9,600,000 sq km 1,034,907,000	108 per sq km 1.2%	*Peking 8,500,000 Shanghai 12,000,000	25% 75%	textiles & clothing, petroleum products, food, heavy machinery	$313,000 $308
India	3,136,500 sq km 746,388,000	248 per sq km 2.1%	*New Delhi 301,000 Calcutta 9,100,000	22% 36%	textiles & clothing, tea, gems, fish, iron & steel	$146,000 $209
Indonesia	1,906,240 sq km 169,442,000	89 per sq km 2.2%	*Jakarta 5,500,000	22% 64%	crude oil, natural gas, timber, rubber, coffee	$70,000 $440
Japan	370,370 q1 km 119,896,000	324 per sq km 0.6%	*Tokyo 8,300,000	76% 99%	machinery, motor vehicles, iron & steel, instruments, ships, textiles & clothing	$1,060,000 $8,947
Kampuchea (Cambodia)	181,300 sq km 6,118,000	34 per sq km 2.0%	*Phnom Penh 500,000	14% 48%	rubber, rice	$500 NA
Korea, North	121,730 sq km 19,630,000	161 per sq km 2.3%	*Pyongyang 1,283,000	60% 95%	lead & zinc, mag-nesite, metallurgical products	$16,200 $786
Korea, South	98,913 sq km 41,999,000	424 per sq km 1.5%	*Seoul 8,000,000	57% 90%	textiles & clothing machinery, footwear, iron & steel, ships	$70,800 $1,800

§Growth Rate: Includes natural increase and migration.

†GNP: Gross National Product, or the total value of all the goods and services produced by the people of a country within one year.

N/A: Data not available

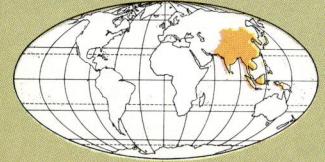

National Profiles: *Asia*

Independent Nation	Area/ Population	Population Density/ §Growth Rate	*Capital/ Largest City and Population	Urban Population/ Literacy Rate	Chief Exports (in order of Importance)	†GNP (in) millions of U.S. dollars)/ Per Capita Income
Laos	236,804 sq km 3,732,000	16 per sq km 2.3%	*Vientiane 200,000	15% 28%	timber, tin, ply-wood	$320 $90
Malaysia	131,313 sq km 15,330,000	117 per sq km 2.2%	*Kuala Lumpur 1,081,000	21% 48%	rubber, palm oil, tin, timber	$25,100 $1,750
Maldives	286 sq km 173,000	604 per sq km 3.0%	*Male 32,000	11% 36%	fish	$74 $462
Mongolia	1,564,619 sq km 1,860,000	1 per sq km 2.8%	*Ulaanbaatar 435,400	45% 80%	livestock, meat, wool	$1,200 $940
Nepal	141,400 sq km 16,578,000	117 per sq km 2.5%	*Katmandu 125,000	4% 20%	rice, jute & jute products, timber	$2,300 $148
Pakistan	803,000 sq km 96,628,000	120 per sq km 2.6%	*Islamabad 201,000 Karachi 5,103,000	28% 24%	rice, cotton, textiles & clothing, carpets	$27,000 $300
Philippines	300,440 sq km 55,528,000	185 per sq km 2.3%	*Manila 1,630,485	32% 88%	coconut oil, sugar, timber, copper, bananas	$39,000 $760
Singapore	583 sq km 2,531,000	4,341 per sq km 1.2%	*Singapore 2,334,400	100% 84%	petroleum products, machinery, rubber	$14,200 $5,745
Sri Lanka	65,600 sq km 15,925,000	243 per sq km 1.8%	*Colombo 1,262,000	22% 85%	tea, rubber, petroleum products	$4,400 $286
Thailand	512,820 sq km 51,724,000	101 per sq km 1.9%	*Bangkok 5,468,286	15% 84%	rice, sugar, corn, rubber, tin	$38,400 $800
Vietnam	329,707 sq km 59,030,000	179 per sq km 2.4%	*Hanoi 2,000,000 Ho Chi Minh City 3,500,000	19% 78%	agricultural products, coal, minerals, ores	$10,700 $189

PERCENT OF WORKING ASIANS IN VARIOUS OCCUPATIONS

Agriculture 61%

Mining & manufacturing 11%

Construction 3%

Transportation & communication 4%

Business & services 4%

Other 17%

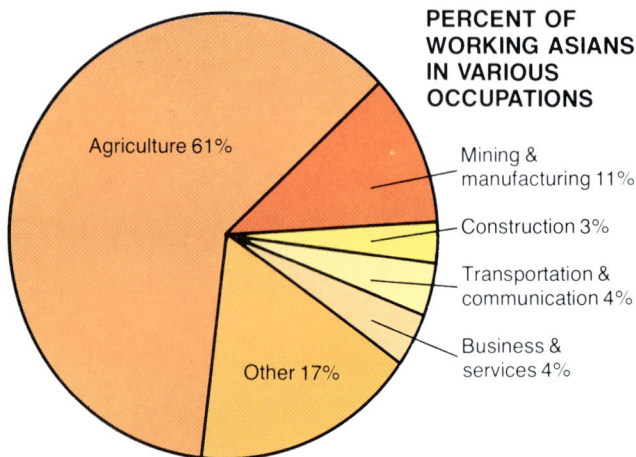

The Hwang Ho is another important river in eastern Asia. Also called the Yellow River, it flows across China for about 3,000 miles (4,830 km) and empties into the Yellow Sea.

Asia's seas. Many seas along the Asian coast create peninsulas and separate Asia's islands from the mainland. The Arabian Sea on the west and the Bay of Bengal on the east border India. The Sea of Japan separates Japan from the Soviet Union and the two Koreas. The South China Sea lies between the Philippines and Vietnam.

Climate. Turn to the map of climates on pages 200–201. Because Asia is so big, its range of climates is much wider than Europe's. Many factors influence Asia's climates. One of the most important is the monsoon. Monsoons are winds that blow in regular patterns according to the season. In the winter, the monsoons blow from the north, over land and to the sea. They bring with them dry, cool air. In the summers, however, the monsoons come from the south, starting this time over the warm seas. Summer monsoons bring much heat and rainfall to most of Southeast Asia. Sometimes the heavy rains cause flooding.

Turn to the map of precipitation on page 20 of your Atlas. Notice that not all areas of Asia benefit from the moisture-bringing summer monsoons. Compare the graphic-relief map (pages 18–19) with the climate map. What natural features might explain why much of central Asia remains dry, even in the summer?

Making a living in Asia. Over a third of the world's people live in China and India. Find these countries on the maps on pages 20–21. Where are their most densely populated areas? What is the yearly rainfall in these areas? How long is China's longest growing season? India's? What is the chief land use in China's most populous area? In India's? Which gets more rain on the average—Europe's farming areas or those of India and China?

The graph on this page shows that most Asians make a living as farmers. By using the chart of National Profiles on pages 334–335, you can tell other things about living in Asia. What are its chief agricultural exports? Which countries export industrial products?

People and problems. Asia's enormous population, already facing food shortages in some areas, is growing even larger. Look again at the chart of National Profiles on pages 334–335. What countries have the slowest population growth rates? How do the literacy rates and types of exported goods for these countries compare to those of other Asian countries? How do you think their standards of living compare?

The quality of everyday life in Asia varies from country to country. In Japan, for example, many people live comfortable, even affluent, lives. In some parts of India, on the other hand, homeless, starving people roam the streets of large and small cities. Asians in Burma, Indonesia, and Malaysia still live much as their ancestors did long ago. Throughout Asia's cultural region, however, people are trying to find new and better ways to develop their region's resources, to increase industrial development and increase employment opportunities, and to raise the standards of living for many.

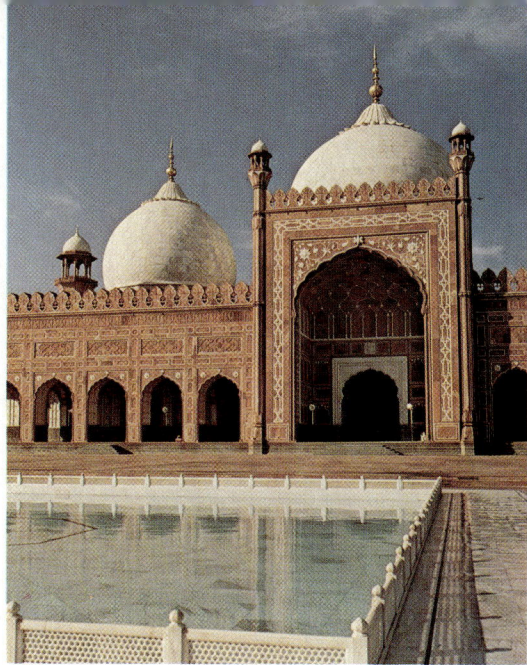

Islam, the Muslim religion, inspired this magnificent mosque in Lahore, Pakistan.

Keeping Facts in Focus

1. What is the name of the world's highest mountain peak? What is its elevation? In what mountain range is it found?
2. Into what body of water does the Indus River flow? The Ganges? The Brahmaputra?
3. What three major races make up the population of Asia? How much of the world's population lives in Asia?
4. How much of the world's land area does the continent of Asia cover?
5. Name three things Asia's rivers do to make them a valued resource.
6. In what occupation do most of Asia's people work?
7. What influence do the monsoons have on Southeast Asia?
8. What is the longest river in Asia? Through which country does it flow?

The People's Heritage

The ancient peoples of Asia developed some of the earliest and most advanced cultures in the world. Prehistoric farmers of Southeast Asia were the first to domesticate pigs, ducks, geese, and several kinds of fruits and vegetables.

By 4000 B.C., many farmers were living in the valleys of the Indus River and the Hwang Ho. The Asian farmers learned how to use the regular flooding of these rivers to irrigate their fields. Their earliest crops were mainly wheat and barley.

Soon, farmers in other regions of Asia learned how to irrigate their lands. In time, farmers adapted irrigation methods to rice farming. They were able to grow large crops from fairly small areas of land. Population in these regions became very dense.

Powerful states, or nations, arose in these densely populated areas. Because the farming methods were very efficient, people were able to grow more food than they needed. They were able to sell their surplus crops, and trade increased. Expanding trade helped build towns and cities where goods could be exchanged or transported.

Since not everyone had to work at farming, people began to seek other

Many Asian cultures have a long tradition of excellence in art. Dancers in Thailand (above) and Japan (lower right) perform in beautiful costumes. The rock-carved Hindu temple in India was both a technical and artistic feat.

occupations. Some became merchants. Others became artists and architects, creating splendid towns and monuments. Still others began to study and experiment with new ideas. Early Asian inventions include the decimal system, the chariot, the windmill, and possibly the first wheel. Early Asians also invented gunpowder, soap, the art of printing, metalworking processes, and advanced instruments for sailing. Through the years, Asians developed great civilizations with a high level of learning.

The coming of the Europeans. By the 1400s, Europeans were becoming interested in Asia's wealth and achievements. They built large ships to seek out a sea route to Asia. In time, European ships appeared in many Asian ports. At first, the Europeans wanted only to buy Asian goods—fine silks, dishes, spices, and other luxury items. But the Europeans soon saw that they could also sell their own goods to Asians.

Gradually, Europeans spread their influence over much of Asia. Sometimes they fought wars to win influence. At other times, they helped Asian rulers put down their local enemies. In return for their help, the Europeans then demanded special favors and trading rights. Sometimes Europeans used trickery to gain control of Asian governments and trading policies.

The upsetting of Asian life patterns. Once in control, Europeans flooded Asian markets with their own factory-made goods. The Asians willingly bought the cheaper European goods. As a result, many Asian craft workers were put out of business. Their handmade goods were too expensive to compete with the European goods.

Farmers also changed their ways. Instead of growing crops for local use, they began selling crops to Europeans. Asian cotton, silk, and jute began flowing to European factories. There, these raw materials were made into finished goods. Great amounts of Asian rice, tea, tin, and rubber also went to Europe. The Asian countries, once self-sufficient, became dependent on European trade.

Japan, however, followed a different course. Japanese rulers started early to build their own factories and industries. As a result, Japan did not have to depend on European factory goods. Its government was not controlled by European rulers. Japan even joined the Western nations in trying to control trade and governments of other Asian countries.

Independence for Asian countries. By the early 1900s, Asian countries were trying to free themselves from foreign influence. It has taken many years for all Asian governments to become independent. Many Asian countries are still not equal partners in the modern industrial world. They are trying to build up their industries. But they face several problems.

Asian nations have many mineral resources needed for industry. They also have enough people to supply workers for mines and factories. But they lack the machines, money, and other materials needed to develop these resources. Where industry has developed in Asia,

foreign industrial countries have often supplied the missing materials.

Industry in the primarily agricultural continent of Asia creates as well as solves some problems. For example, industry does not usually benefit all people in a country equally. The already skilled, educated, and wealthy city dwellers benefit the most from new industry. The people in the countryside often do not share in the industrial gains. Many of them remain poor.

As Asian leaders look to the future, they will have to find ways to balance the technology of the West with Asia's own ways of life. They will have to develop economic independence and avoid more foreign control. And they must find ways to share fairly Asia's slowly growing wealth with its people.

Keeping Facts in Focus

1. What features make the densely populated areas of Asia good to live in?
2. Why did many Asian craft workers go out of business when trade with Europe began?
3. How did Japan's course of history differ from that of other Asian nations?
4. What goals are Asian countries working to achieve?
5. What answers would you give to the key questions on page 330?

Working with Ideas

A rich deposit of petroleum has just been discovered in a developing Asian nation. The nation has an annual growth rate just under 3 percent, a literacy rate of 13.5 percent, and a rural population of 95 percent. Its government will soon receive billions of dollars for its oil. How should the nation use this money to improve the lives of its people?

Lesson 2 India and Its Neighbors

Reading Focus

1. Discovering the relationship between cause and effect can help clarify problems and sometimes even lead to solutions. The paragraph right before the heading **Efforts to modernize agriculture** on page 344 describes what happens if farmers cannot grow enough food to last the year. Read this paragraph carefully, looking for the cycle of causes and effects. As you can see, the effect of small crops is that farmers need to sell part of their lands and use the money to buy food. The next step in the cycle is that this effect becomes a cause for something else. Less land causes the farmers to grow even less food the next year. The cycle continues, cause becoming effect, until the farmers' lands are completely gone. As you read the whole lesson, look for other cause and effect relationships. Look for the causes of South Asia's problems. See if understanding these causes helps suggest any possible solutions or goals that the South Asian nations could work to achieve.

2. As you read this lesson, look for the answers to these key questions:
   a. How has South Asia's history shaped the area's problems?
   b. What are some reasons for the shortage of food in South Asia?
   c. What ways have South Asian farmers found to add to their incomes?
   d. How does South Asia's industrial development compare to that of Europe?

Vocabulary Focus

subcontinent	*millet*	*cottage industry*
plantation	*pesticide*	*subsidy*

The lands of the Indian subcontinent include Pakistan, India, Nepal, Bhutan, Bangladesh, and Sri Lanka. Look at the map on pages 18–19. What mountain ranges form a curved wall that separates these countries from the rest of Asia?

Despite these natural barriers, invaders long ago found ways to enter the Indian subcontinent. Many came through passes in the northwest curve of the mountain wall. Invaders brought their languages, religions, and other aspects of their cultures with them. Soon there were Muslims, Hindus, and Buddhists spread all over Indian lands. There were hundreds of languages, and few forces to unify the diverse peoples.

Colonization and Independence

In the 1500s and 1600s, Europeans began to arrive in India by sea. The Portuguese were the first to arrive. Later, the British came and found it easy to gain control of the divided people. India became a colony of the United Kingdom. Throughout the long years of British rule, Hindus and Muslims often fought among themselves for power. They also began to fight against the British. Indians, both Hindu and Muslim, began to demand their independence. During the 1930s, with the help of a leader named Mohandas K. Gandhi, Indians began to regain some control over their own affairs. But fighting between Hindus and Muslims continued.

Great Britain finally agreed to India's independence after World War II. Bloody riots between Hindus and Muslims then broke out. Leaders finally agreed that the only way to stop the fighting was to make separate Hindu and Muslim nations. India became a Hindu nation in 1947, and parts of India on the east and west sides of its northern boundary became Pakistan—a nation with a Muslim majority.

But the fighting continued. Hindus and Muslims clashed over control of Kashmir in the northern section of West Pakistan. People in East Pakistan later rebelled against West Pakistan. They wanted to govern their own affairs. In 1971, East Pakistan became the independent nation of Bangladesh.

The smaller countries. Both Nepal and Bhutan, two kingdoms in the northeast region of the subcontinent, also had

How do farmers on this tea plantation in Sri Lanka protect their sloped croplands from wind and water erosion?

to fight to keep their independence. Nepal, like India, was threatened by the British. But rugged Nepalese soldiers were able to defeat the British in 1816. Bhutan was also under some British control. From 1910 to 1949, the British controlled all of Bhutan's foreign relations. In 1949, when the British withdrew their control, Bhutan made strong ties to India. Bhutan had few foreign relations then until 1959, when China seized some of its land. Since then, Bhutan has depended even more on India for protection. But it has remained an independent kingdom.

Sri Lanka, an island off the southern tip of India, became an independent republic in 1972. The Portuguese and Dutch were the first to colonize this island, once called Ceylon. The British soon gained control and set up many rubber plantations in Ceylon. In 1948, Ceylon became an independent dominion of the British Commonwealth. A

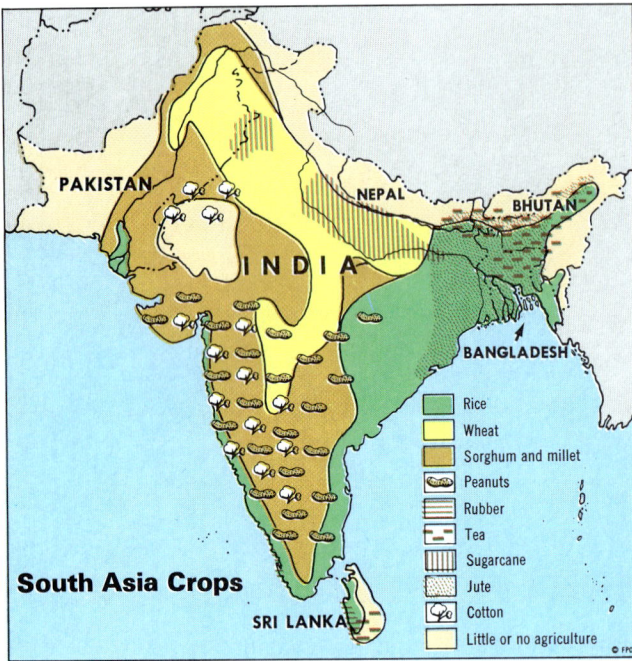

South Asia Crops

Legend:
- Rice
- Wheat
- Sorghum and millet
- Peanuts
- Rubber
- Tea
- Sugarcane
- Jute
- Cotton
- Little or no agriculture

new constitution in 1972 finally set up the independent republic of Sri Lanka.

The long history of invasions and foreign domination has left India and its neighbors with many problems. Differing languages, religions, and customs still divide the people. Years of forced dependence on European powers kept South Asian lands from building self-sufficient economies. Now, as independent nations, South Asian lands must work quickly to overcome the problems their history has left them.

Millions of people to feed. With each step forward, another problem forces South Asian progress to a halt. That problem is overpopulation. Almost one billion people live in India and its neighboring lands today. And the population is growing rapidly. By the year 2000, the area's population could reach 1.5 billion. India alone already has about one seventh of the world's people to feed. The Indian government has tried to find ways to slow down the

population growth. But it has not yet found a plan agreeable to all the people. As the population grows, so does the problem of starvation. Even today, daily life in many parts of South Asia is a constant struggle for survival.

A Land of Farmers

Most people on the subcontinent make their livings from farming. India grows more peanuts, sugarcane, and tea than any other world region. Bangladesh leads the world in jute production. (Jute is a plant from which twine and burlap are made.) The map on this page shows the area's other important crops and where they are grown. Compare this map to the precipitation map on page 20 of your Atlas. Which crops seem to need a very wet climate?

In drier areas with good soil, wheat and cotton are the major crops. Sorghum and millet can grow in dry areas with poorer soils. Even in the areas of good soil, farmers often plant sorghum and millet along with wheat. If there is a drought one year, at least the sorghum and millet will survive it. Peas, lentils, and beans are also important crops for these areas. Like sorghum and millet, they can survive droughts.

Controlling water supplies is one of the area's chief farming problems. Where there are great rivers, as in northern India, the farmers must control flooding during the wet monsoon summer. In these areas, dams and irrigation canals keep the waters under control. There is usually enough water to irrigate the lands even in the dry season. But elsewhere on the subcontinent, many rivers dry up in the winter. In some places, farmers drill wells to tap underground waters. Even these wells dry up sometimes. At present, India has developed only about 5 percent of its water resources. Until more resources are developed, farmers must work very hard to find enough water to survive the dry season.

Ancient farming practices. Lack of water is only one problem that limits the amount and kinds of food South Asian farmers can grow. Another is the lack of modern farming methods. Most South Asians farm their lands just as their ancestors did centuries ago. These farmers live in small, crowded villages near their tiny plots of land.

Most farmers own fewer than five acres (2 ha) of land. These five acres, however, are not generally contained in one stretch of land. Instead, farmers own small patches of land scattered around the area. For example, a family may own one acre of sandy soil for growing millet, beans, and sorghum. In a nearby swampy corner, the family may own some more land for growing rice. A patch of wasteland might be used for grazing and collecting firewood. Some farmers also own a small patch for growing vegetables, spices, coffee, and sugarcane.

Most of the farming is done by hand, and the work goes slowly. Preparing and planting a small field can take as long as two months. The hard ground must be broken by a single-bladed wooden plow. Only a few wealthy farmers can afford metal plows. Spreading manure by hand is sometimes a whole month's work. Farmers also harvest the crops by hand, often asking their neighbors to help. After harvest, some farmers use teams of bullocks to help with the threshing. But in many places, even the threshing is done entirely by hand.

After one crop is harvested, the work must begin again for a second crop. With careful rotation, some farmers can raise as many as three crops a year. Despite their hard work, however, their small lands and old fashioned farming

Members of many farming families in India add to their incomes by working in such cottage industries as spinning and weaving.

This fertilizer plant in Pakistan was built in a desert near the natural gas supplies needed for the manufacturing process. Customers come from nearby farms to buy the yield-increasing fertilizer.

methods limit the amount of food they can grow. Some farmers are not able to grow enough food to last until the next harvest.

If their food supplies run out, farmers may have to mortgage or sell part of their lands to buy food. Their remaining lands then produce even less food. Many farmers are forced to go deeper and deeper into debt. Some lose all their lands. Between 1960 and 1974, the percentage of landless farmers in Bangladesh rose from 17 to 40 percent. Studies predict that between 1970 and 1985, 4 million Pakistani farmers will have lost their lands. If these farmers cannot hire themselves out, they will join the unemployed. South Asian cities are already crowded with jobless farmers living in slums.

Efforts to modernize agriculture. To help solve some of these problems, South Asian governments are trying to develop better farming methods. With better irrigation systems, more land could be used for farming. More commercial fertilizers would increase yields. At present, cow dung is the chief fertilizer. But even that is in short supply, since it is also burned for fuel. Producing more fertilizer is an important goal of South Asian industries.

There have also been some efforts toward improving the varieties of seed the farmers use. But to grow well, these improved varieties often need special fertilizers and pesticides. Only a few wealthy farmers who can afford such chemicals have succeeded in increasing their yields.

Better farming methods will take a long time to develop. In the meantime, farmers must find other ways to earn money when their food runs out. South Asian governments are encouraging some farmers to develop cottage industries. These include weaving, spinning, toy and furniture making, and craftwork. Products from cottage industries are often exported, bringing extra cash to farmers. One third of India's total production of finished cotton cloth comes from its cottage industries. Farmers in the foothills of the Himalayas increase their incomes by finishing raw silk produced by silkworms. Although the income from cottage industries is not great, it helps farmers survive.

344

Developing Industry

After a late start, industry in the South Asian lands is growing. India is the world's largest producer of cotton cloth. Bombay and Ahmadabad, on India's west coast, are centers for the area's cotton industry. Pakistan produced few cotton products in the late 1940s. But now it produces enough cotton to make sizable exports. South Asian lands also export many products made from jute. Much of South Asia's foreign trade is with the United States, the Soviet Union, Japan, Great Britain, Hong Kong, and Saudi Arabia.

Heavy industries have been slower to develop than light industries. A look at the map of South Asia mining and manufacturing shows that India has the greatest share of natural resources for industry. The other countries lack industrial resources. Both Pakistan and Bangladesh, however, have built petrochemical plants to turn their supplies of natural gas into fertilizers and other chemicals.

India does have some heavy industry. Most of it developed after independence. The steel and iron mills near Calcutta are among the largest in the world. Foreign governments have helped India build other large steel mills. These are now under the control of India's government. About 70 percent of India's coal comes from the states of Bihar and West Bengal.

India also has some important engineering and chemical industries. Asia's largest chemical fertilizer plant is in India. Several factories for assembling cars and trucks are also helping to ex-

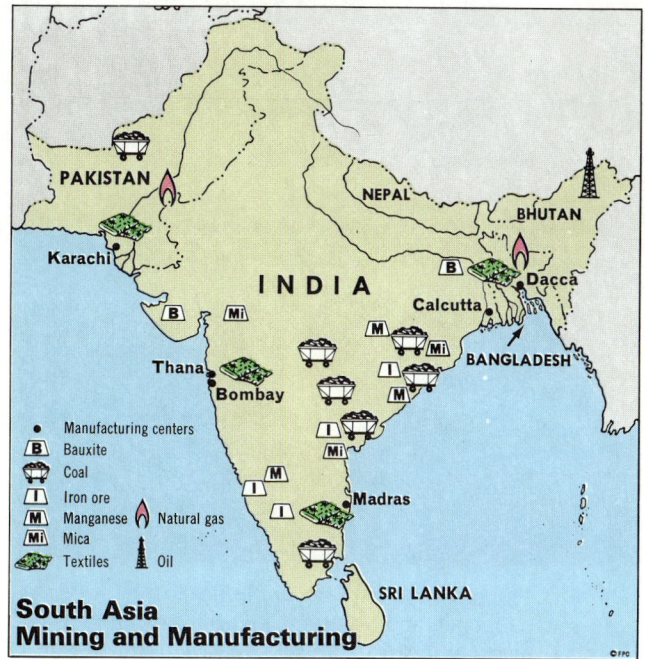

South Asia Mining and Manufacturing

Manufacturing centers •
B Bauxite
Coal
I Iron ore
M Manganese
Mi Mica
Textiles
Natural gas
Oil

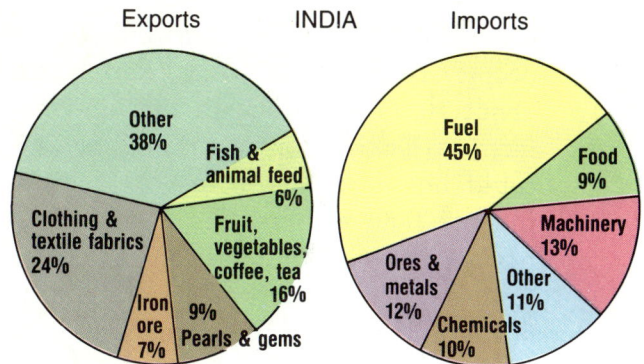

Exports INDIA Imports

Exports: Other 38%, Fish & animal feed 6%, Fruit, vegetables, coffee, tea 16%, Pearls & gems 9%, Iron ore 7%, Clothing & textile fabrics 24%

Imports: Fuel 45%, Food 9%, Machinery 13%, Other 11%, Chemicals 10%, Ores & metals 12%

pand industry in India. The government of India recognizes that industrial development is a key to solving many of the country's problems. So it offers subsidies to new industries and helps build industrial plants. It is also building dams along the rivers to convert waterpower into electricity. There are even a few nuclear power stations in India.

In 1971 industry in India provided jobs for only about 21 million people out of a total population of more than 600 million. By 1977 industry in India had

The steel industry is very important in India, especially in and near Calcutta. At this foundry in eastern India, workers cast steel for use in making trucks.

Poor people in Varanasi, India, wash clothes in the Ganges River. Some people in India lack not only modern conveniences but adequate food and shelter as well.

provided jobs for more than 130 million people. At the same time there were about 70 million workers and landless farmers looking for jobs that industry could not provide. As fast as India's industrial development has been, it has not been fast enough to keep pace with India's rate of population growth. Many of the landless and unemployed live in slums under conditions of extreme suffering and poverty.

Travelers to India's cities might not, at first, see the terrible problems. Twelve Indian cities have populations over one million. Most of these cities have two main sections. In one section, where the British used to live, houses are clean and rather large. Visitors to these sections would see strong Western influences in the modern buildings and big stores. Only the rich Indians can afford to live in these areas now. Some of these people have cars and telephones. The other section of the city is usually cluttered with small buildings,

has impassable roads, and is crowded with starving people. South Asian leaders face enormous challenges as they shape the futures of their nations.

Keeping Facts in Focus

1. What three South Asian crops can survive long periods of drought?
2. Are winters wet or dry in South Asia?
3. What are some cottage industries?
4. Which countries are the chief trading partners with South Asian nations?
5. How is India's history of British rule reflected in its cities today?
6. What answers would you give to the key questions on page 340?

Working with Ideas

1. How were the natural barriers that surround the subcontinent an advantage to its people? A disadvantage?
2. Why does a developing country often have a high percentage of its people working in agriculture?
3. What long-standing rivalries exist within and among nations on the Indian subcontinent that often influence political events?

Lesson 3 Southeast Asia

Reading Focus

1. Each lesson in this book presents a number of facts. From these facts, careful readers can form opinions. Sometimes, however, authors present their opinions along with the facts. For example, this statement is a fact: "In Malaysia, four fifths of the land used to grow rubber trees is owned by European or Chinese businesses." The following, however, is an opinion: "Malaysia should have more control over its own resources." Which of the following is a fact? Which an opinion? (a) The governments of Southeast Asia must find a way to reduce the population growth rate. (b) Southeast Asia's population is expected to reach 600 million by the year 2000.

   Very often, opinions are stated in sentences that contain key words, such as *should*, or *must*, or other words that express judgment. Balancing facts and opinions is one of the reader's most important tasks.

2. As you read this lesson, look for answers to these key questions:
   a. What different kinds of farming techniques are practiced in Southeast Asia?
   b. What changes did the plantations bring to Southeast Asia?
   c. How do Southeast Asia's mineral and fuel resources compare with those of South Asia?
   d. What are some of the benefits and drawbacks of foreign industries in Southeast Asia?

Vocabulary Focus

monarchy   *subsistence*   *commercial*   *capitalist*

The region of Southeast Asia stretches from Burma on the west to New Guinea on the east. Its total land area is about half that of the United States. Southeast Asia is a region of hilly peninsulas, islands, and shallow seas. Look at the map on page 331. Name the countries in this region.

The graphic-relief map, pages 18–19 of the Atlas, shows the diversity of Southeast Asia's landforms. There are stretches of wide plains, several of them watered by the region's mighty rivers. There are also long strips of rocky mountains, and rain forests thriving with trees and colorful flowers.

The people of Southeast Asia are also diverse. Throughout the area's history, people from China and India moved into Southeast Asia. Despite these two main influences from Indian and Chinese culture, the region shows little cultural or

Patterns of Government in Asia

Many Asian countries are trying to build modern societies free from Western economic influence. To do so, they need carefully worked-out plans for economic growth. Some countries have become communistic. In these, every economic activity is controlled by the government. Others have adopted capitalist systems similar in theory to those of Western European nations. In practice, however, some democracies have come under the rule of dictators, or authoritarians. The practices of Asia's main forms of government are outlined in the chart.

DEMOCRACY	AUTHORITARIAN DEMOCRACY	DEMOCRATIC CENTRALISM (Communism)
Several political parties compete for election to representative assembly (parliament). Assembly meets often to propose and vote on new laws. Chief executive is appointed by assembly. Government offers guidelines for capitalist economy, but otherwise does not control economy. Government protects such rights as free speech and press and right to compete for profits. Where practiced: Japan.	Several political parties exist, along with a democratic constitution. But the ruling party has the only real power, often maintained by the military. Economy is capitalist, but wealthy interests often get special government favors. Government is not always efficient at carrying out plans. Opposition is sometimes not tolerated. Government often censors individuals or groups opposed to its policies, sometimes imprisoning them without trial. Where practiced (selected countries): Indonesia, Malaysia, Philippines, South Korea.	Communist party has complete control over government. It chooses candidates for election to assembly. Assembly does not meet often. Government owns all businesses and makes all economic decisions. State also employs all workers and tells them where to work. Most criticism is censored. Individual rights are less important than national interests. Government desires people to shift their loyalty from family and religious ties to the state. Where practiced (selected countries): Kampuchea, China, Laos, Vietnam.

political unity. Recent years have brought much political unrest to the region, including the lengthy war in Vietnam. The United States played a great part in this war.

Except for Thailand, long an independent monarchy, the countries of Southeast Asia are newly independent. Each is now striving toward national unity. As they are developing their economic strength, the nations of Southeast Asia are finding abundant resources to help them.

A Farming Region

Most Southeast Asians are farmers. The maps on pages 20–21 of the Atlas provide some clues to the region's suitability for farming. The growing season is year round, and rainfall in Southeast Asia averages more than 80 inches (200 cm) a year. Most of the area's rainfall, however, comes in the summer months, when moisture-laden monsoon winds bring heavy rainfall.

Traditional farming. The map of land use on page 21 shows that much of the

Many Philippine people adopted Catholicism, the religion of the Spaniards who had colonial control over the Philippines until 1898.

This monk in Bangkok, Thailand, is one of the 245 million Buddhists in the world. Buddhism is the chief religion in Southeast Asia.

land in Southeast Asia is used for hunting, gathering, and traditional farming. Many Southeast Asian farmers have found effective ways to work the land without modern tools and machinery. These subsistence farmers live in wooded areas in the mountains. Lacking modern tools, farmers in these regions cannot clear the land by cutting down the trees. Instead, they burn the trees and use the ash as fertilizer for the land. Hoes and planting sticks are the only tools the farmers need to plant their crops. Chief among these crops is dry rice. It requires little water or fertilizer. Other crops are beans, corn, and manioc, a plant with a starchy root.

After a few years, however, the soil becomes exhausted. The farmer then moves to another area and once again clears the land by burning. It takes about fifteen years for the original field to become usable again. Farmers plan their movements so they return to a field every fifteen years.

Wet rice farming. For many years, most wet rice farming was done in terraces on the hills. Dams and irrigation canals gradually brought water down from the hilltops. Swamps and heavy floods had made the deltas at the bottom of the hills unusable. But when foreign nations began to occupy Southeast Asian lands, they brought advanced methods to reclaim the deltas. Careful drainage, irrigation, and flood-control systems turned the swampy deltas into good lands. Many farmers began to move into these delta regions. The population in these areas is still growing today.

Most farmers in the wet rice areas grow only enough for their own needs. In addition to rice, they grow coconuts, yams, beans, manioc, and vegetables. Some of the delta rice farming, however, is commercial. Study the chart of National Profiles (pages 334–335). What Southeast Asian countries export a great amount of rice?

349

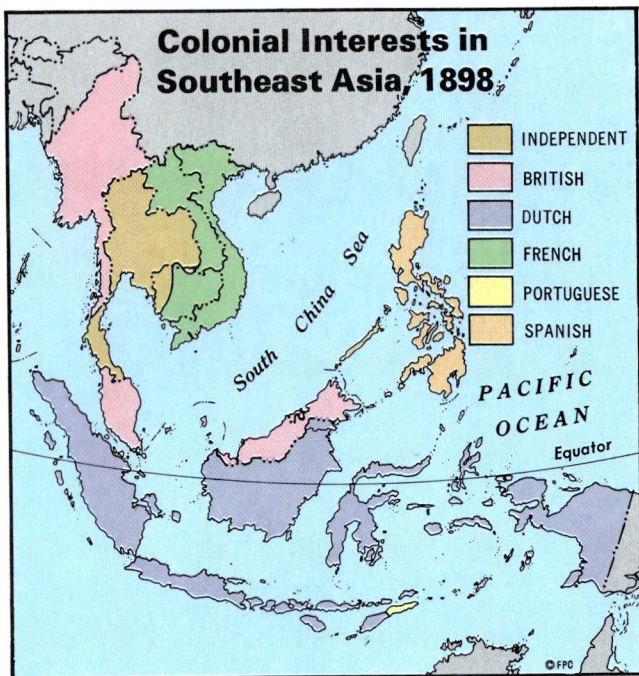

Colonial Interests in Southeast Asia, 1898

▨	INDEPENDENT
▨	BRITISH
▨	DUTCH
▨	FRENCH
▨	PORTUGUESE
▨	SPANISH

PACIFIC OCEAN

South China Sea

Equator

©FPC

Plantation crops. Southeast Asia is the world's major supplier of tropical plantation crops. Plantations are large estates worked by hired laborers who often live on the estates. Most plantations grew up during the years of European control of Southeast Asia. Its key location made the area an excellent site for exports going both east and west. Europeans set up large coconut and spice plantations. Later they introduced such new plantation crops as rubber and tea.

The large plantations helped Southeast Asia's economy grow. Europeans built much-needed roads, railways, and shipping lines to connect the plantations with seaports. The plantations even changed the way local farmers made their living. Many smaller farmers began to grow crops for export instead of for their daily needs.

Today, small farmers produce sizable amounts of Southeast Asia's exports.

Some governments in Southeast Asia are offering tax breaks and financial aid to encourage these small farmers to become even more productive. Foreign influence is still strong in these nations. In Malaysia, for example, four fifths of the area used to grow rubber trees is owned by European or Chinese businesses. Even in Indonesia, where the government owns all of the rubber plantations, much of the rice growing is still in foreign hands. Southeast Asian governments hope that helping small farmers increase their production will help the countries become less dependent on foreign businesses.

Forestry. The great demand for teak, mahogany, and other tropical hardwoods has helped forestry grow in Southeast Asia. Check the National Profiles chart (pages 334–335). In which Southeast Asian countries is timber a chief export product? The demand for some woods is so great that foresters are cutting huge numbers of trees. Many are not being replaced. Lands once forested are now barren and exposed to erosion. And the supply of valuable trees is dwindling. Many people feel that the short-term profits from forestry are not worth the problems careless cutting can cause.

The Mineral Boom and the Growth of Industry

Southeast Asian countries have only recently begun to develop their plentiful mineral resources. One half of the world's tin comes from Southeast Asia, most of it from Malaysia. Malaysia also supplies about 4 percent of the world's

Rice grown on terraces such as these on Bali is a chief crop of Indonesia.

bauxite. Seven percent of the world's chromite comes from the Philippines.

Southeast Asia also has reserves of tungsten, nickel, manganese, zinc, and lead. But more important than these is the area's rich supply of natural gas and oil. Some experts believe that Southeast Asia has the largest reserves of oil ever discovered. Soon after oil was discovered, European, North American, and Japanese oil companies moved in to control the oil regions. Their technologies helped develop the area's resources and provide jobs for many Southeast Asians. But some Southeast Asians believe they should have more control over their resources and industrial growth.

Southeast Asia also has about one eighth of the world's reserve of gas. Brunei (broo′ nī), on Borneo, already has the largest natural gas plant in the world. Similar plants are being built off the coasts of Sumatra and Borneo. As with oil production, foreign interests have a big share in Southeast Asia's natural gas industry.

Industry has grown rapidly in Southeast Asia. Most of the area's industry is geared to serving the needs of the mining companies and plantation owners. This type of industry is of lasting importance to the region's economy.

Another type of industry, however, is raising some problems for Southeast Asia. Assembly-line factories in the fields of electronics, toy making, wig making, and clothing are springing up all over Southeast Asia. These are owned mainly by foreigners, attracted to the area by the good supply of cheap labor. Southeast Asian governments have encouraged new industries by offering tax breaks during the first few years of business. They want to help

Southeast Asia Mining and Manufacturing

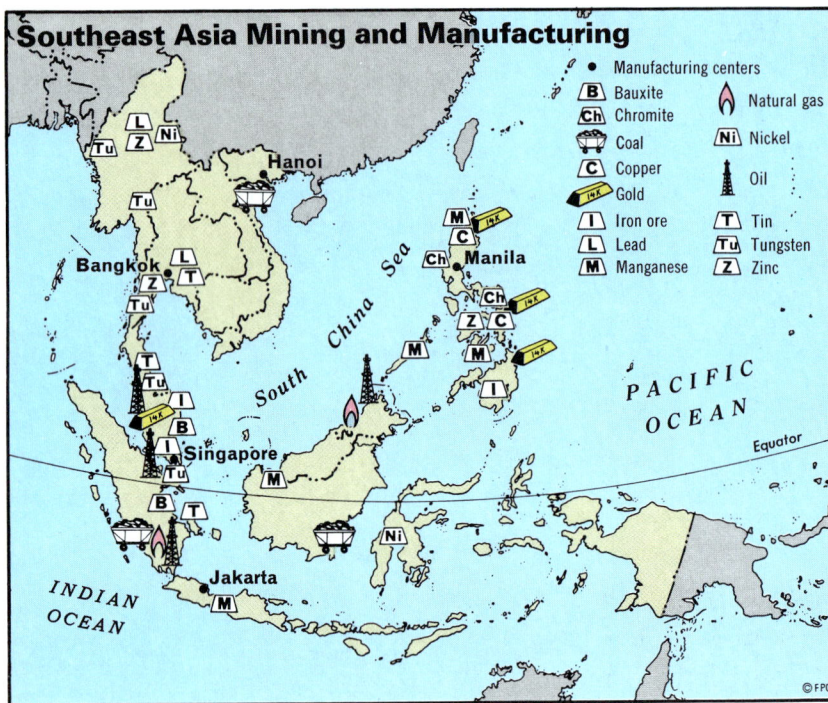

Legend:
- ● Manufacturing centers
- B Bauxite
- Ch Chromite
- Coal
- C Copper
- Gold
- I Iron ore
- L Lead
- M Manganese
- Natural gas
- Ni Nickel
- Oil
- T Tin
- Tu Tungsten
- Z Zinc

Map labels: Hanoi, Bangkok, Manila, Singapore, Jakarta, South China Sea, PACIFIC OCEAN, Equator, INDIAN OCEAN

©FPC

build strong industries that will lend lasting support to their nation's economy. But in many cases, foreign firms stay in an area only until their tax break runs out. Then they move to another area and start again. Their own profits soar, but Southeast Asian economies do not share the profits.

The population of Southeast Asia is growing rapidly. Its present population of 400 million is expected to reach 600 million by the year 2000. Many people believe that the foreign-owned, temporary businesses are not providing a good future for the growing population. They offer only temporary jobs and leave many people unemployed when they move to a new area. Many people believe Southeast Asian industry should look to the future instead of to short-term profits. They believe Southeast Asian industries must find ways to use Western technology and business ideas without giving up security and profits to foreign investors.

The growth of cities. Along with industries, cities have been growing quickly in Southeast Asia. Their populations have risen sharply in recent years, as educational and business opportunities have increased. Although most people still live in rural areas, some Southeast Asian cities are among the world's largest. Manila's population, for example, is nearly six million.

POPULATIONS OF SELECTED SOUTHEAST ASIAN CITIES

	1936	1960	1981
Phnom Penh, Kampuchea	103,000	450,000	500,000
	1947	**1960**	**1982**
Bangkok, Thailand	781,700	1,633,400	5,468,286
	1948	**1960**	**1980**
Manila, Philippines	1,366,800	2,135,700	1,630,485
	1931	**1961**	**1981**
Jakarta, Indonesia	533,000	2,933,000	5,500,000
	1947	**1957**	**1980**
Kuala Lumpur, Malaysia	176,000	316,200	1,081,000

Southeast Asia's cities are becoming crowded, clogged with many bicycles and some autos. Skyscrapers and shiny new buildings form modern skylines.

But industry in Southeast Asian cities has not grown fast enough to provide jobs for all the qualified people. Many young graduates cannot find jobs. And unskilled farmers coming into the cities from rural areas have an even harder time finding work.

Some Southeast Asians believe that Communism will help solve the problems of industry, unemployment, and growing population. Southeast Asia now has three Communist nations: Laos, Kampuchea (Cambodia), and Vietnam. They became Communist states only recently, after bitter civil wars in which some Western powers gave their support to the non-Communists. These countries are eager to develop economic independence, even from other Communist nations. However, after so many years of dependence on foreign powers and so many years of war, the move toward economic independence is slow.

Some Southeast Asian countries do not look to Communism as an answer for their problems. Since 1967, Indonesia, Malaysia, the Philippines, Singapore, and Thailand have been members of the Association of Southeast Asian Nations (ASEAN). This organization encourages independence from foreign powers and cooperation among member nations. Like the Common Market in Europe, ASEAN is working to reduce trade barriers among members in an effort to strengthen their capitalist economies.

SOUTHEAST ASIA'S TRADING PARTNERS

Percentage of Southeast Asia's Exports to Trading Partner	Trading Partner	Percentage of Southeast Asia's Imports from Trading Partner
25.9	Japan	22.4
14.6	Countries within Southeast Asia	13.3
3.8	Other Asian Countries	5.2
2.3	Middle East	14.2
21.2	United States	16.2
15.6	Western, Northern, and Mediterranean Europe	17.0
2.0	Eastern Europe and the Soviet Union	.6
3.2	Oceania and Australia	4.1
11.4	Remaining Nations	7.0

Keeping Facts in Focus

1. What Southeast Asian country has been an independent monarchy for many years?
2. When did delta rice farming become widespread in Southeast Asia?
3. How are Southeast Asian governments encouraging farmers of small plots to become more productive?
4. What foreign powers have large interests in Southeast Asia's oil industries?
5. Why were foreign investors attracted to Southeast Asia?
6. What answers would you give to the key questions on page 347?

Working with Ideas

Why might a foreign electronic manufacturer be interested in locating a factory in Southeast Asia? What drawbacks might such a manufacturer consider before making a final decision?

353

Reading Focus

1. China is undergoing dramatic change. China has begun to incorporate "incentives" and other capitalist ideas into its form of Communism. Since Mao (a former Communist leader) died in 1979, policy has shifted from strict internal discipline to practical development. The new leaders have made reforms in the agricultural system and increased incomes.

   All over China, giant posters with pictures and slogans guide the people's way of thinking. New posters replace those from Mao's time. Popular slogans from the past include:

   "Everyone is a soldier."

   "The precepts (teachings) of the Revolution should be more dear to a Communist than his own life."

   A slogan seen now is:

   "Time is money! Efficiency is life!"

   How do these slogans show the basic change that has gone on in China? What are some of the benefits and problems that might come with change in China?

2. As you read this lesson, look for the answers to these key questions:
   a. How do the four regions of China differ from one another in climate and land use?
   b. What initial changes in farming did the Communists make?
   c. What is the current farming system?
   d. What kinds of relations does China have with its close neighbors?

Vocabulary Focus

oasis	*commune*	*nomadic*
protein	*armaments*	

China's civilization is one of the oldest in the world. Its written history goes back to about 1500 B.C. For many centuries, Chinese ways of life changed very little. But the China of today has broken many of its ties with the past. When the Communists came to power in 1949, they began making sweeping changes. They even changed the country's name to the People's Republic of China. Opposed to these changes, many Chinese fled the mainland and now live on the island of Taiwan. On the mainland, however, Communism became a new way of life. China's strength grew rapidly. In the years since 1949, China has become one of the major world powers.

More people live in China than in any other country on earth. China's population of one billion includes about one fourth of the world's people. China's land area is also immense, the third largest in the world. But about two thirds of China's lands are only sparsely settled. Most of China's people live in the remaining third of the land. Look at the population map on page 20 of the Atlas. Compare it to the graphic-relief map on pages 18–19. What physical features help explain the distribution of China's population?

China's Great Regions

The map on this page shows the four main regions of China. Look at the graphic-relief map on pages 18–19. What mountain range in High China forms part of the country's southern border? Between these mountains and the Kunlun Mountains in the north of High China lies the wide Tibetan Plateau. In most parts of Tibet there is little vegetation. The mountain ranges keep out most moisture-bearing winds. In the southeast, however, some of the lower river valleys occasionally benefit from the summer monsoons. In these areas, some agriculture is possible, although the growing season is short. Most people in High China live in this milder southeast region.

Dry China also has high, rugged mountains that keep out rain-bearing winds. The region of Dry China includes

Four Regions of China

the vast Takla Makan and Gobi deserts. In the region of Sinkiang, most people live near oases where there is enough water for some farming and grazing. Since 1949, the Communist government has sent Chinese workers to Sinkiang to introduce new methods of irrigation. It has also sent engineers and other skilled workers to develop mining and manufacturing in that region.

North and east of Sinkiang, in Inner Mongolia, most of the land is used for grazing. There is some irrigated land along the Hwang Ho. But the high salt content of the soil limits farming to only a few small areas. The region does, however, have good reserves of coal and iron that help support the steel industry in China.

Wheat China and Rice China are the most densely settled areas of China. Turn to the climate map on pages 200–201. What are the climates of Wheat China and Rice China?

Wheat China has large stretches of flat land suitable for grain crops. Farms in Wheat China are the largest in the country. The Hwang Ho (Yellow River) is the heart of the farming region in

In 1953, the Chinese government combined small fields into collective farms to help increase yields. In 1958, the collectives were combined to form huge communal farms. Communal farmers are paid in wages.

Wheat China. In northwestern China, the Hwang Ho flows through a dry, treeless land covered with loess. The loess, a fine yellow soil formed from dust blown from the dry lands of Mongolia, washes into the Hwang Ho and gives the river its name.

In the northeast, Wheat China stretches into a plains area where many farmers grow soybeans. The area also has good supplies of coal and iron ore and is one of China's main centers of heavy industry.

Rice China is very different from Wheat China. Rice China is hilly. It has a long growing season, but the climate is humid. The Yangtze waters the northern half of Rice China. Unlike the Hwang Ho, the Yangtze is very manageable. It is an important transportation route, navigable year round. More people live in the Yangtze Valley than in the whole of the United States. Shanghai, at the river's mouth, has over ten million people crowded into its urban area.

In the southern part of Rice China, there are about 2,000 people per square mile (800 per sq km) of arable land. In the Xi River delta, most farmers grow two crops of rice a year.

Another part of Rice China is a region between the Yangtze and Xi rivers. It is a hilly region with short, swift rivers whose valleys support some farming. Along the coast, most people live in small fishing towns.

The Need for Efficient Farming

Only about a tenth of China's vast lands are suitable for farming. With such limited lands, farming must be highly efficient to support a population as large as China's. For many years, farming methods were not efficient enough. Hunger and starvation were widespread. Now, however, most Chinese people have enough to eat. Beef is scarce in China, since there is not enough land to spare for grazing. But many Chinese farmers raise pigs. Pigs can feed on scraps of garbage, so they do not take up valuable land for grazing. Rice and other grains make up about

This farming commune is near Peking. In what region of China is Peking?

These workers are producing textiles, one of China's chief exports.

two thirds of the Chinese diet. Some farmers raise fish in natural or artificial ponds for protein.

The Chinese government has put much effort into finding ways to increase agricultural production. For example, the government employs farmers in their off-seasons to build dams, ponds, canals, and dikes. These help conserve water and expand irrigation systems. To increase yields, the government is developing industries to produce chemical fertilizers. To combat erosion, the government has encouraged the planting of trees. Many trees in China were cut long ago and burned for fuel. Reforesting will help supply valuable timber and make more land usable for farming.

One of the biggest changes the Communists made was to combine small farms into larger communes. They created about 24,000 communes, each one self-sufficient. When the farmers were not working the land, they were put to work in newly built local industries. The government did away with all private property. It set up child-care centers so women would be free to work. The gov-

ernment hoped the communes could develop ways to mass-produce the area's important goods.

The communes continued to operate but they were not completely successful. Farm production increased, but not enough to feed the expanding population. The commune system did help develop small industries that can benefit farmers.

Beginning in 1979, the communes were replaced by an "agricultural responsibility system." Now smaller economic units, some the size of a household, make contracts with the government to deliver a set amount of produce for a fixed price each year. After that goal is met, peasants can sell any surplus at a profit. This system has greatly increased all agricultural production and brought more money to the peasants.

Along with other major changes in the 1980s, China has realized that in order to insure its economic future, it must control population growth. Already the world's most populous country, China has taken severe steps to

China's Foreign Trade Partners

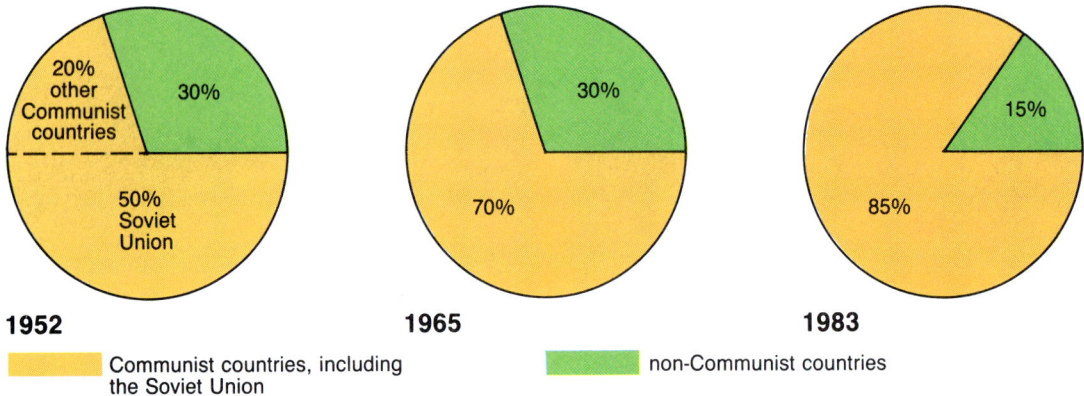

1952	**1965**	**1983**

Communist countries, including the Soviet Union — non-Communist countries

curb continued population growth. The government has taken measures to enforce what it considers the ideal family size. Couples are encouraged to have only one child and are penalized for having more. This policy has already begun to affect the population growth rate.

The Development of Industry

China long ago developed a strong handicraft industry. It remains strong to this day, supplying many of the country's consumer goods. Foreigners introduced larger-scale industries in China around 1900. Most large-scale industries in pre-Communist China centered on textiles. By the mid 1930s, China was producing one fifth of the world's cotton. The Japanese introduced steel, iron, and engineering industries during the period before World War II, when they controlled a part of northern China.

After World War II and the civil war that brought the Communists to power,

the Chinese had to rebuild the industries that foreigners left behind. Like the Soviets, Chinese Communists put most of their efforts into heavy industry, such as machine building and armaments. For a while, the Soviets even helped build up China's industry. But after a dispute in the mid-1960s, the Soviet Union withdrew its aid. Most of China's trade is now with non-Communist countries.

Coal is China's most important source of energy. It has more than 500 billion metric tons of coal reserves. In 1980 China exported more than 10 billion tons of coal. It also exported 300,000 barrels of oil daily. China's other minerals include tin, titanium, talc, tungsten, iron ore, and vanadium.

Behind China's industrial development lies a huge program of education. Before the Communists came to power, only about 10 percent of China's people could read or write. Look at the chart of National Profiles on page 334. What is China's current literacy rate?

China needs trained scientists, engineers, and technicians to help the country to greater economic strength. But the government is eager to preserve a balance between countryside workers and industrial leaders. For this reason, the government requires even the highly trained specialists to spend regular periods working at manual labor. The leaders hope this measure will prevent splits among the Chinese. They also hope it will ensure that policy makers are well informed on local problems and needs.

China's Neighbors

The Mongolian People's Republic. The Mongolian People's Republic covers the Gobi desert and its grassy margin. Mongolia shares much of its history with China. China has sometimes tried to control Mongolia. But the present government has strong ties to the Soviet Union.

Mongolia has long been a land of nomadic herders. It has more livestock per capita than any other country. The government, aided by the U.S.S.R., has

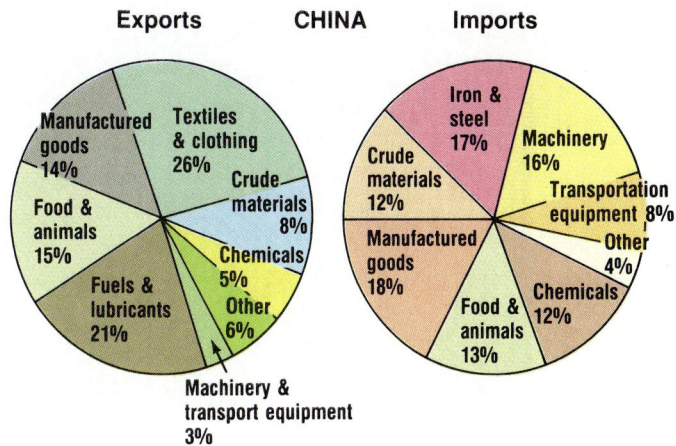

Exports   CHINA   Imports

Exports pie chart:
- Manufactured goods 14%
- Textiles & clothing 26%
- Crude materials 8%
- Chemicals 5%
- Other 6%
- Machinery & transport equipment 3%
- Fuels & lubricants 21%
- Food & animals 15%

Imports pie chart:
- Iron & steel 17%
- Machinery 16%
- Transportation equipment 8%
- Other 4%
- Chemicals 12%
- Food & animals 13%
- Manufactured goods 18%
- Crude materials 12%

tried to limit the wanderings of these herders and set up stable communities. It has also set up some diversified agriculture. In addition, it has helped develop some industries, most of them related to processing animal products.

Korea. Korea is a hilly peninsula with some plains on the west. Like Mongolia, Korea has often been pulled between two stronger countries. It has been dependent at times on either China or Japan. In 1953, as a result of the Korean War, Korea was divided in two.

Non-Communist South Korea is flatter, warmer, and more moist than North

These workers, part of a commune production team in Sian, are producing bricks for use in housing construction. Men and women share work equally in China.

359

Hong Kong, an island off the southeast coast of China, is a British colony. Under a 1984 agreement, between China and Great Britain, Hong Kong will be returned to China in 1997. The 5.4 million people will be guaranteed certain rights for 50 years. Plywood is one of the products of the city of Pusan, South Korea (right).

Korea. It therefore produces more agricultural goods. Two crops of rice, barley, sweet potatoes, vegetables, and fruits can be grown each year. South Korea has few mineral resources. But recent Japanese and American investments in South Korea have helped light industry grow.

Communist North Korea is suitable for growing some rice, corn, barley, and millet. It has much more industry than South Korea. Its more abundant resources, along with the strong Communist stress on industry, make it a highly industrialized nation.

Keeping Facts in Focus

1. Which two regions of the People's Republic of China are the most densely populated?
2. What are some differences between the Hwang Ho and the Yangtze River?
3. How do most people in the Mongolian People's Republic make their living?
4. What answers would you give to the key questions on page 354?

Working with Ideas

Why do Chinese leaders require that all city workers spend some time working in the countryside? Why do you think Chinese leaders do not require countrysiders to spend some time working in the city?

360

Lesson 5 Japan

Reading Focus

1. Most lessons in Part Two of this book contain brief histories of the areas under study. To keep the historical events straight in your mind, you might find it helpful to keep a list of dates as you read. For example, the first two paragraphs in this lesson contain two dates. Start your list with these. Make sure you leave enough room between the dates to add any others that might come along in the lesson.

   1542—Japan made first contact with the West

   1853—Commodore Matthew Perry arrived in Japan and trade opened with the West

   Remember to watch carefully for dates as you read the lessons. By remembering important dates, you can compare events in one country with events taking place at the same time in another country.

2. As you read this lesson, look for the answers to these key questions:
   a. How is Japan's population spread among its various islands?
   b. What makes Japan's agriculture so efficient?
   c. How has industry in Japan grown despite the country's general lack of resources?
   d. What are Japan's important industrial cities?

Vocabulary Focus

tremor	*tilling*	*commuter*
persimmon	*duster*	

In 1542, a Portuguese ship on its way to China was blown off its course. It landed in Japan instead of China. For the first time in its history, Japan made contact with the West. Europeans found a highly developed civilization in Japan, one that shared many achievements and cultural features with its neighbor, China. At first, Europeans were welcome in Japan. But soon Japanese leaders began to fear that the Europeans would try to make Japan a colony. To prevent domination, Japan cut off almost all contact with foreign nations for about 200 years.

During these years, however, some Japanese heard of the great advances the European countries had made in science and industry. Some began to feel that Japan could learn much from these nations. So when Commodore Matthew Perry, an American official, arrived in

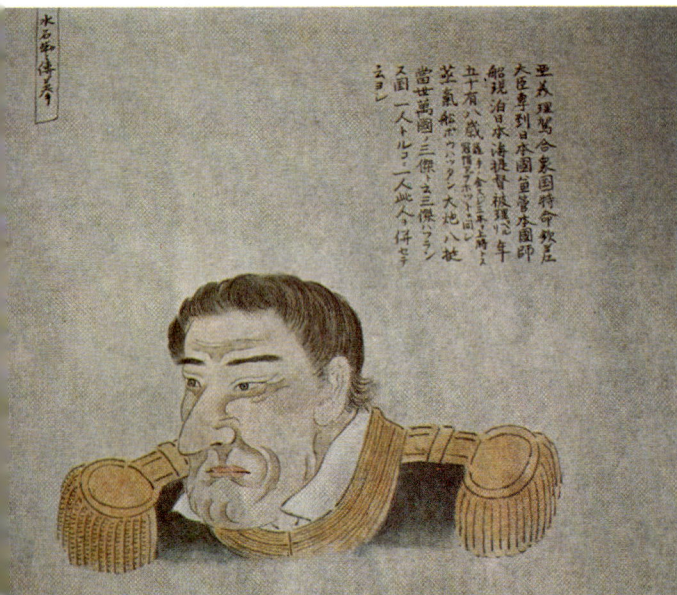

A Japanese painter exaggerated the unfamiliar European features of Matthew Perry (above). The Peace Dome in Hiroshima (right) is a reminder of the World War II bombing of Japan.

1853 asking Japan to open its doors to the United States, he was not turned away. Soon Japan had trade agreements with thirteen other nations as well.

Once the doors were opened, the Japanese and Europeans exchanged many ideas. The West learned much from the Japanese culture and drew great economic benefit from Japanese trade. Japan also learned much. Experts from England supervised the building of railroad and telegraph lines. French advisers helped the Japanese set up a new legal system. Americans showed them how to set up a modern school system, organize a national postal service, and improve their agriculture. Germans in Japan started a medical service and trained army officers.

Borrowing ideas and money from some Western nations, the Japanese built steel mills, shipyards, electric power plants, and textile factories. By 1900, Japan had become the most advanced industrial nation of Asia.

But Japan was not self-sufficient. It lacked most raw materials for industry and had to depend on imports from other countries. In an effort to achieve more economic independence, Japan began to build an empire. It wanted to control mineral-rich colonies that could support its industry. In the 1930s, Japan seized some lands from China. During World War II, Japan grew more powerful. In 1941, Japan attacked the U.S. naval base at Pearl Harbor, and the United States entered the war. The United States and its allies waged a fierce struggle in the Pacific, finally winning some important victories in Japanese-held territories. In 1945, the United

States dropped atomic bombs on two Japanese cities. Japan finally surrendered, and its empire fell.

But Japan made a quick recovery from the war. It established good relations with its former enemies. It turned from war to rebuilding its industries. Its strength now provides prosperous lives for its people. Japan is once again an important world power.

A Necklace of Islands

Japan is made up of four main islands and over 3,000 smaller islands. Look at the map of the world on pages 10–11 of the Atlas. What North American country lies within the same latitudes as Japan? Now turn to the climate map on pages 200–201. Do the climates of the North American region in the same latitude correspond to those of Japan?

Despite the similarity in climates, Japan receives more rain than the eastern coast of the United States. Like many other Asian countries, Japan receives heavy monsoon rains, especially in July. In the winter, the dry monsoons from the north bring sunny and mild weather.

Look at the map on page 331. Name the four main islands of Japan. What sea separates them from Korea and the Soviet Union? Japan also includes the Ryukyu Islands. What sea separates this chain of islands from China?

Japan's environment made the region difficult to tame. Nearly four fifths of Japan's land is covered with hills and mountains. Much of this land is unsuitable for farming. Volcanoes, some still

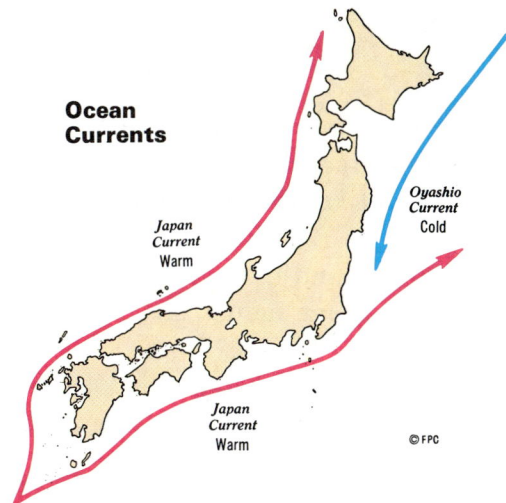

Ocean Currents

Japan Current Warm

Oyashio Current Cold

Japan Current Warm

© FPC

Which would you expect to be warmer, southern Japan or northern Japan? Check your answer on the map on pages 200–201.

active, are scattered throughout the mountains. Earthquakes are frequent in Japan. Each year, there are about 1,500 measurable quakes. But most of them are only mild tremors. Japan's rivers are generally swift and shallow, not well suited to navigation.

There are, however, stretches of plains that are good for farming. Most of the good land is near the sea. The biggest area of plain is on Honshu, Japan's largest island. Tokyo and Yokohama lie in this large plain. Good prospects for both agriculture and industry make this region one of the most densely populated areas in the world. Almost 80 percent of Japan's people live on the island of Honshu.

Hokkaido, the second largest island, also has some good lands for farming. The most extensive farmlands lie in the southwest. There are also some plains on the eastern coast. Irrigated rice, sugar beets, and white potatoes are

Hokkaido's important crops. Only about 5 percent of Japan's people live on Hokkaido. About 11 percent of the Japanese people live on Kyushu. The coal deposits on this island have helped make it one of the great industrial regions of Japan.

Shikoku is Japan's smallest island. Most of its farmland lies along the coastline of the Inland Sea.

Land Use in Japan

Japan is the world's seventh most populous country. Its land area is slightly smaller than that of California. But Japan's population is a little more than half that of the United States.

Highly efficient agriculture. In terms of their productivity, Japanese farmers are more efficient than the farmers of any other country in Asia. Only about one fifth of Japan's land is suitable for agriculture. So each square kilometer of land must support about 1,500 people. Japanese farmers supply about 70 percent of their country's food needs, and Japan's yield per acre is the second highest in the world. Ninety percent of Japan's 14 million cultivated acres are devoted to food crops.

Half of Japan's farmland is used for growing rice. Most areas support two rice crops a year. Other crops include wheat, barley, white and sweet potatoes, and soybeans. Japanese farmers also grow apples, pears, oranges, tangerines, persimmons, and most common vegetables.

Some of Japan's crops are grown for industrial use. For example, mulberry trees are grown for their leaves, which are used to feed the silkworms that supply raw silk for the textile industry. Flax, hemp, and reeds are other industrial crops. Their materials are used to make mats that serve as floor and wall coverings in many Japanese homes.

Many part-time farmers work in urban offices and factories during the week. On the weekends, they return to their family farms to help with farming tasks.

Japanese farmers use many small machines such as tilling machines, electric pumps, sprayers, and dusters. Japan's swift-running rivers are a good source of cheap electrical power. Even farming areas that are most distant from large cities have electricity to supply power for electric lights and the use of household appliances.

Forestry and wood industries. About two thirds of Japan is forested. The forested hills are patchworks of different kinds and sizes of trees, including cedars, oaks, maples, and pines. The Japanese use their wide variety of woods in many ways, but they control the cutting and the planting of trees carefully.

Because Japan cannot keep up with the world demand for wood, it has invested in forests overseas. Japanese firms are logging timber from forests in Indonesia, South America, and Alaska. With these foreign investments, Japan continues to be an important world supplier of wood for commercial and industrial uses.

Japan's Fishing Industry

Japan is one of the world's greatest fishing countries. The Japanese lead the world in total tons of fish caught and in tuna fishing. They are second only to the United States in salmon fishing and rank equal with the U.S.S.R. in whaling. Japanese fishing boats travel to all parts of the world. Many of the most modern vessels are equipped to process the day's catch and to keep it refrigerated. These ships often are away from their home ports for months at a time. The Japanese export frozen, canned, and dried fish.

They also sell lobsters, cultured pearls, and crab meat to other countries.

A Modern Industrial Giant

Despite its size, Japan has become a modern industrial giant. Modern industry began in the 1890s with the use of machines to manufacture silk, a fabric that had a heavy worldwide demand then. Money from exported silk helped finance the development of other industries. Soon Japan was importing cotton and manufacturing cotton goods. Japanese businesses sold these goods all over the world. Japanese silk still

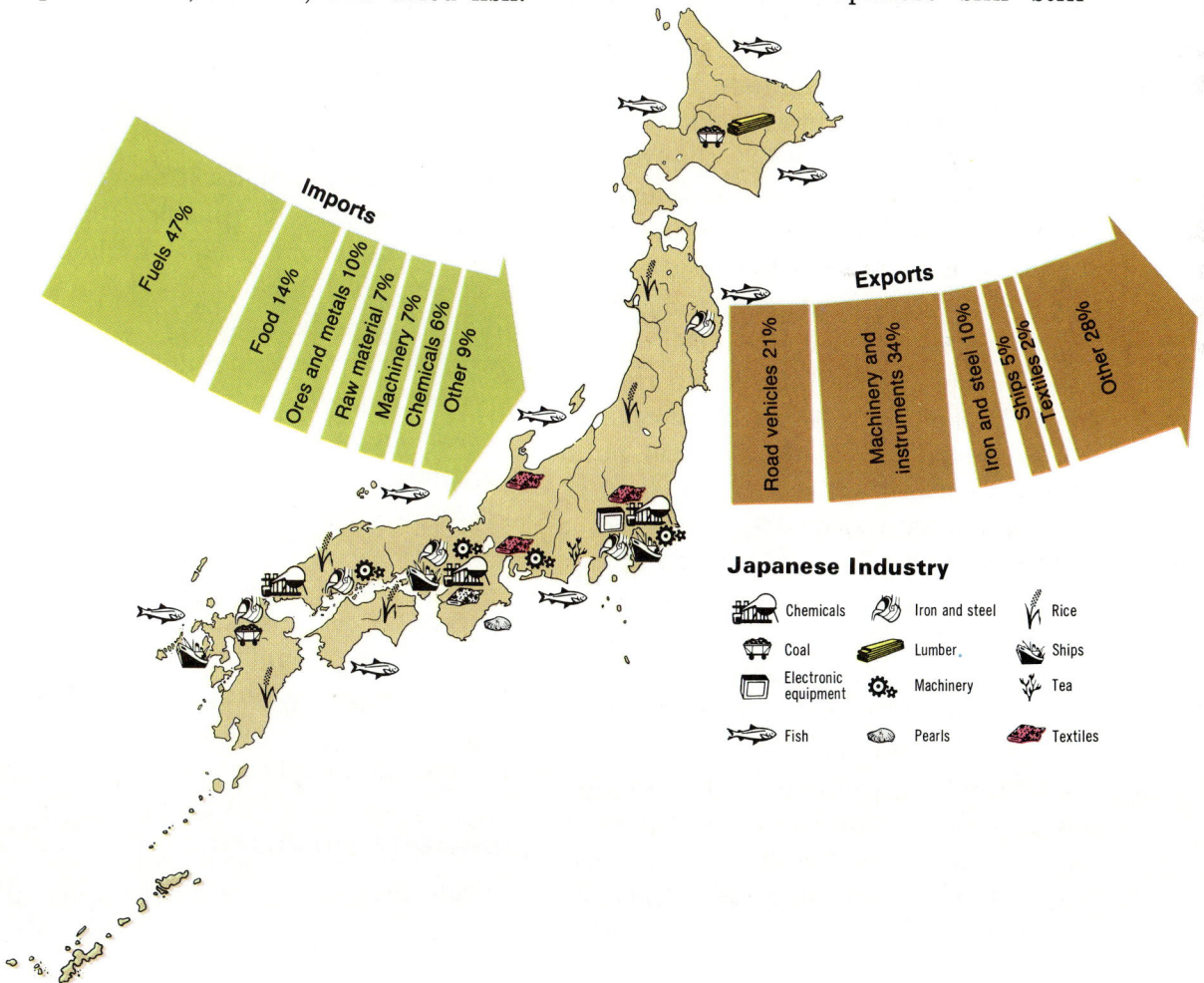

Imports
Fuels 47%
Food 14%
Ores and metals 10%
Raw material 7%
Machinery 7%
Chemicals 6%
Other 9%

Exports
Road vehicles 21%
Machinery and instruments 34%
Iron and steel 10%
Ships 5%
Textiles 2%
Other 28%

Japanese Industry

Chemicals	Iron and steel	Rice
Coal	Lumber	Ships
Electronic equipment	Machinery	Tea
Fish	Pearls	Textiles

makes up about 60 percent of the world's supply. But the manufacture of silk has declined considerably since the development of synthetic fabrics.

Heavy industry and the manufacture of electrical goods have grown rapidly in Japan. Japan, however, has very few mineral resources with which to support its manufacturing industries. Lead, zinc, and sulfur are the only minerals produced in large enough amounts to meet Japan's manufacturing needs. Japan imports most of its industrial raw materials and great amounts of fuel. Trade with the United States, Canada, Australia, Saudi Arabia, and other countries is therefore very important to Japanese industry.

With the help of imported raw materials, Japan has become the world's leading producer of cars, television sets, video-cassette recorders and other major electronic equipment. Japan ranks second in steel production, turning out about 16 percent of the world's steel supply. Japan now has a reputation for making high-quality goods.

Industrial growth has helped build large, modern cities. Japan's urban population has grown from 50 percent in 1955 to 76 percent in 1981. Tokyo is the world's second largest urban area, with a population approaching 17 million. Other metropolitan areas with populations of 2 million or more are Osaka, Nagoya, and Kitakyushu. The streets of many of Japan's large cities, like those in European or American cities, are crowded with cars and people. High-speed commuter trains connect Tokyo with its suburbs. Other

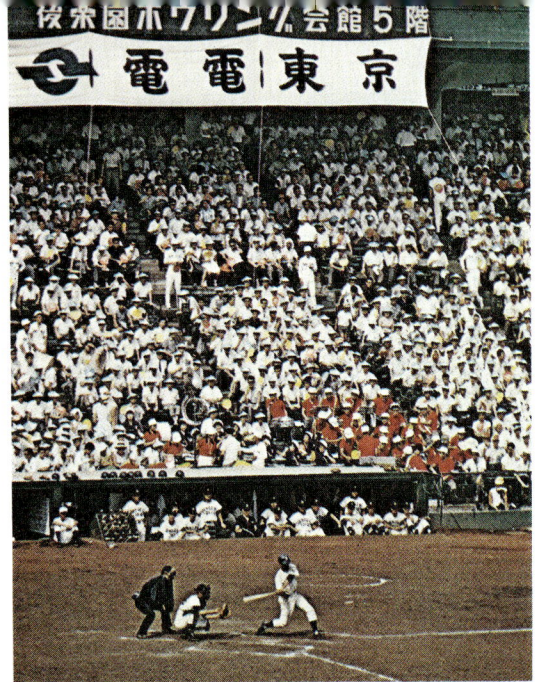

Japan has adopted many Western ways. Baseball is the nation's most popular sport.

high-speed trains shuttle Japanese travelers between Tokyo, Yokohama, and other cities.

Few buildings in Japanese cities are higher than five or six stories because Japan has about 1,500 earthquakes each year. Buildings of this size are less likely to experience severe damage if a strong earthquake should strike.

Japan's many industries provide a variety of jobs for its people. Many Japanese workers stay with one employer their whole lives. Because of their loyalty, Japanese workers seldom have to worry about being fired. Employers, in turn, reward their workers with lifelong benefits. Many factories employ women, but few women in Japan hold positions of importance in the business world.

Education in Japan

The government requires that all Japanese children attend at least nine

Almost 3 million people live in the great port city of Yokohama, just south of Tokyo on the Pacific Ocean. Japan's location allows easy trade with both North America and the Asian mainland.

years of school. Many Japanese students also go on to higher schooling. But students must compete with one another for entrance into the more than 340 colleges and universities in Japan. The examinations are difficult, so most Japanese students prepare very hard for them. Japan has more of its young people in college than any other country except Israel and the United States.

Problems and Progress

On the whole, Japan's recent history makes its story a successful one. But Japan's success has had its price. One problem is pollution. Another problem is Japan's heavy dependence on other countries for raw materials to feed its industries. Despite these problems, Japan's development since World War II has given its people a high standard of living. In both agriculture and industry, the Japanese people have found

ways to make much out of little. This ability will surely be of value as the Japanese continue to search for solutions to the problems their progress has brought.

Keeping Facts in Focus

1. What are the four main islands of Japan? Which island is the largest? The most densely populated?
2. How much of Japan's land is covered with hills and mountains?
3. How does Japan's agricultural yield per acre compare with that of other countries?
4. Which of Japan's crops are used mainly for industry?
5. Why is trade so important to Japan?
6. What answers would you give to the key questions on page 361?

Working with Ideas

1. Why can Japan be considered a model of efficiency?
2. Why can Japan never be entirely self-sufficient if it wishes to have a highly developed economy?

Unit 10   REVIEW WORKSHOP

Test Your Geographic Knowledge

A. On a separate sheet of paper, fill in the word or words that best complete each statement.

1. Most of Asia's mountains fan out from the _____.
2. The two large deserts in eastern Asia are the _____ and the _____.
3. The longest river in Asia is the _____.
4. In the winter, the monsoons blow from the _____, over land and to the sea.
5. _____ and _____ are the two most populous countries in Asia.
6. The one Asian country that kept out foreign control was _____.
7. _____ and _____ clashed in India, until finally India was divided and Pakistan became a separate nation.
8. _____ is the world's largest producer of jute.
9. Indian cities still reflect the times when India was a colony of _____.
10. Europeans in Southeast Asia set up large _____ to grow such crops as rubber and tea.
11. One half of the world's tin comes from Southeast Asia, most of it from _____.
12. The Communist countries of Southeast Asia are _____, _____, and _____.
13. There are now over _____ people in China.
14. Japan lacks _____ for industry and has to depend on imports from other countries.
15. Japan's biggest trading partner is the _____.

B. Tell whether each statement is true or false.

1. Asia's population is spread evenly over its lands.
2. The Irrawaddy, Salween, and Mekong rivers are in Southeast Asia.
3. Monsoon rains reach all parts of Asia.
4. Most Asians live in cities and work at industrial jobs.
5. Bangladesh was once a part of Pakistan.
6. ASEAN is an organization for economic cooperation among Communist Asian nations.
7. Most of China's foreign trade today is with other Communist nations.

Apply Your Reading Skills

Below are some of the new words introduced in this unit. Some of the words refer most accurately to agriculture. Some refer to natural features, and others refer to economic matters. Make a chart with these three headings: *Agricultural Terms, Economic Terms, Natural Feature Terms.* Place each of these words under the appropriate heading.

tilling	capitalist	interior
oasis	subcontinent	monsoon
mortgage	plantation	affluent
domesticate	pesticide	luxury
persimmon	subsidy	subsistence

Apply Your Geographic Skills

The graphs on page 369 show population and food supplies of India and Japan. Use these graphs to answer the following questions.

1. Which country has the denser population, Japan or India?
2. In which country is half the land suitable for farming?
3. Is the daily average calorie consumption in Japan greater than the necessary minimum?
4. In which country does animal protein make up about 41 grams of daily intake?
5. On the average, how many grams of all proteins are available to the Japanese daily? How many are available to the Indians?
6. Between what years did India's rice production drop from about 53 million metric tons to 45 million metric tons? How does Japan's rice production during these years compare to India's?
7. In what year did India's population reach 500 million? In what year did Japan's population first exceed 100 million?
8. Which do you think is harder to predict for the future—rice production or population growth? Why?

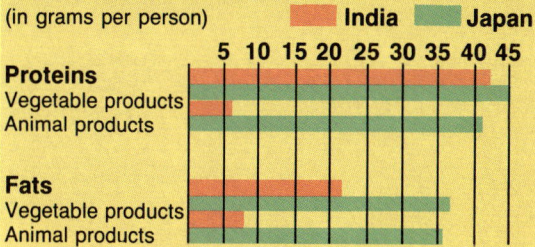
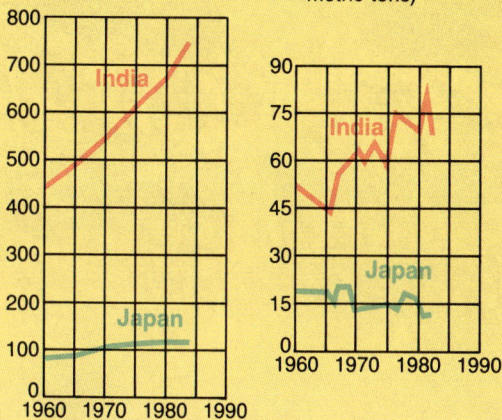
Apply Your Thinking Skills

A. Below is a list of some of the smaller nations in Asia. Each of these smaller nations has strong historical and cultural ties to one of the three powerful nations in Asia—China, India, or Japan. Match each of the smaller nations below with the more powerful nation to which it has the strongest cultural or historical ties.

1. Bangladesh	A. China
2. North Korea	B. India
3. Sri Lanka	C. Japan
4. South Korea	
5. Vietnam	
6. Nepal	
7. Mongolian People's Republic	
8. Laos	
9. Bhutan	
10. Pakistan	

B. Arrange these historical events in the order in which they occurred.
1. India becomes an independent republic.
2. ASEAN is formed.
3. Matthew Perry arrives in Japan.
4. East Pakistan becomes the independent nation of Bangladesh.
5. The United States drops bombs on two Japanese cities.
6. Mohandas K. Gandhi leads the Indians in resistance to British rule.
7. Japan attacks Pearl Harbor.
8. Communists come to power in China.

Discuss These Points

In the early 1970s, a group of people in the Philippines that had lived in complete isolation for thousands of years was seen by foreigners for the first time. These people are called the Tasaday. Until they were seen, they had had absolutely no contact with any other culture on earth. Experts predict that once the Tasaday make contact with the modern world, their own culture will fade away within thirty years. Many people believe that the Tasaday should be left in complete isolation so they will not be able to adopt new ways of life that would destroy their ancient culture. Others believe the Tasaday should be given a chance to learn about other ways of life and decide which they prefer. What do you think? Do you think the Tasaday would be able to judge the new ways of life presented to them? Should the Tasaday be left alone completely? Give reasons for your views.

Expand Your Geographic Sights

Davidson, Judith. *Japan: Where East Meets West* (Dillon, 1983).

Loescher, Gil and Loescher, Ann D. *China: Pushing Toward the Year Two Thousand* (Harcourt, Brace, Jovanovich, 1981).

Poole, Frederick K. *Southeast Asia* (Watts, 1973).

Roland, Albert. *Profiles from the New Asia* (Macmillan, 1970).

Sarin, Amita V. *India: An Ancient Land, A New Nation* (Dillon, 1984).

Unit 11 Australia and Oceania

Under these blue waters off the northeast coast of Australia is the Great Barrier Reef, a favorite spot for deep-sea divers from all around the world.

Lesson 1 Exploring Australia and Oceania

Reading Focus

1. There are many ways to preview a lesson. You can thumb through the pages and look at illustrations and section headings. You can skim quickly to get a general idea of what you will be studying. You can also use the Atlas and other maps in the text to preview a lesson. This lesson, for example, is about Australia and Oceania. Before reading the lesson, look carefully at the map on page 373. Note that Oceania is made up of Melanesia, Micronesia, and Polynesia. Next, find Australia and the three regions of Oceania on the world political map (pages 10–11). Then find Australia and the islands of Oceania on the world climate map (pages 200–201), world vegetation map (pages 206–207), and the special-purpose maps on pages 20–21 of the Atlas. These maps will give you good background information that will help you as you read.

2. As you read this lesson, look for the answers to these key questions:
   a. What are Australia's three major land regions? Which is the most densely populated?
   b. What are the three main island types in Oceania? How was each formed?
   c. When did Europeans first arrive in the Pacific world?
   d. Why were Australia and Oceania sought after as colonial possessions?

Vocabulary Focus

artesian well   *coral*   *typhoon*

In the southwestern waters of the Pacific Ocean lie a large landmass—the continent and country of Australia—and thousands of scattered islands. The island region, called Oceania, is made up of three main island groupings—Micronesia, Melanesia, and Polynesia. Find Australia and the three island groups of Oceania on the world graphic-relief map (pages 12–13). Which island group lies mainly to the east of the International Date Line?

The Physical Setting

Look closely at the map on page 373. Between what latitudes does the region of Australia and Oceania lie? Is New Zealand north or south of the Tropic of Capricorn? At what latitude do the Gilbert Islands lie?

Independent countries. Australia, covering an area almost as large as the mainland United States excluding Alaska, is by far the largest country in the

Australia and Oceania
Political-Relief Map
⊛ National Capitals   ● Other Capitals   • Other Cities

CHINA · East China Sea · JAPAN · TAIWAN · South China Sea · PHILIPPINES · BRUNEI · MALAYSIA · INDONESIA · Timor Sea · Darwin · Arafura Sea · New Guinea · PAPUA NEW GUINEA · Port Moresby · Bismarck Archipelago · SOLOMON ISLANDS · Honiara · Coral Sea · Great Barrier Reef · Great Dividing Range

MICRONESIA · Mariana Islands (U.S.) · Guam (U.S.) · TRUST TERRITORY OF THE PACIFIC ISLANDS (U.S.) · Caroline Islands (U.S.) · Wake Island (U.S.) · Marshall Islands (U.S.) · MELANESIA · NAURU · Yaren · Tarawa · Gilbert Islands · KIRIBATI · TUVALU · Funafuti · VANUATU · Port-Vila · New Caledonia (Fr.) · FIJI · Suva · TONGA · Nukualofa

PACIFIC OCEAN · International Date Line · Honolulu · Hawaii (U.S.) · Christmas Island · Phoenix Islands · Line Islands · Tokelau Is. (N.Z.) · WESTERN SAMOA · Apia · American Samoa · Wallis & Futuna (Fr.) · Cook Islands (N.Z.) · Niue (N.Z.) · POLYNESIA · Society Islands (Fr.) · Tuamotu Archipelago · Marquesas Islands (Fr.) · French Polynesia · Pitcairn Is. (Br.) · Easter Island (Chile)

UNITED STATES · MEXICO · Tropic of Cancer · Equator · Tropic of Capricorn

AUSTRALIA · WESTERN AUSTRALIA · Great Sandy Desert · NORTHERN TERRITORY · SOUTH AUSTRALIA · QUEENSLAND · Brisbane · NEW SOUTH WALES · Perth · Adelaide · VICTORIA · Canberra · Sydney · Melbourne · Darling R. · Murray R. · Tasman Sea · Auckland · North I. · NEW ZEALAND · Wellington · Christchurch · South I. · So. Alps · TASMANIA · Hobart · INDIAN OCEAN

SCALE · Miles 0 250 500 1000 1500 · Kilometers 0 500 1000 1500 2000 · ©FPG

southwestern Pacific. The island of Tasmania, south of the Australian continent, is one of Australia's six states. Other large countries in this region are New Zealand in Polynesia and Papua New Guinea in Melanesia. Papua New Guinea occupies the eastern half of the island of New Guinea, the second largest island in the world. The western half of New Guinea is part of Indonesia.

The National Profiles chart on page 374 lists all the independent countries in the southwestern Pacific region. Except for Australia, all are part of one of the three island groupings of Oceania. Find each independent nation in its island grouping on the map on this page. In which island group do you find Tonga?

Australian landforms. Australia is the world's flattest continent, although it does have several mountainous regions. The map on page 375 shows Australia's three main land regions: the Eastern Highlands, the Central Lowlands, and the Western Plateau.

The Western Plateau makes up about three fifths of Australia. Look at the map on page 24 of the Atlas. The Macdonnell and the Hamersley ranges rise above the level of the plateau, which also includes several deserts. Rainfall is light, rivers are few, and lack of water has limited the development of much of this region.

Rainfall is also light over much of the Central Lowlands. But this region has other important sources of water. The Murray River drains a large area of the Central Lowlands. The Murray and its branches form Australia's largest river system and provide irrigation water

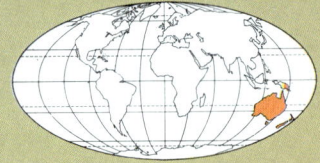

National Profiles: *Australia and Oceania*

Independent Nation	Area/ Population	Population Density/ §Growth Rate	*Capital/ Largest City and Population	Urban Population/ Literacy Rate	Chief Exports (in order of Importance)	†GNP (in) millions of U.S. dollars)/ Per Capita Income
Australia	7,692,300 sq km 15,462,000	2 per sq km 1.3%	*Canberra 251,000 Sydney 3,280,000	85% 99%	coal, wool, wheat, iron ore, beef	$153,000 $10,087
Fiji	18,272 sq km 686,000	38 per sq km 2.1%	*Suva 65,000	37% 80%	sugar, copra	$1,850 $1,852
Kiribati	684 sq km 61,000	90 per sq km 1.6%	*Tarawa 22,148	N/A 90%	phosphates, copra	$36 $630
Nauru	21 sq km 8,000	381 per sq km 1.3%	*Yaren N/A	N/A 99%	phosphates	$155 $21,400
New Zealand	268,276 sq km 3,238,000	12 per sq km 1.1%	*Wellington 342,000 Christchurch 321,000	83% 98%	meat, dairy, fish, wool	$24,000 $7,090
Papua New Guinea	475,369 sq km 3,353,000	7 per sq km 2.8%	*Port Moresby 123,624	13% 32%	copper, coffee, cocoa, coconut products, fish	$2,000 $650
Solomon Islands	29,785 sq km 263,000	9 per sq km 3.7%	*Honiara 19,200	9% 60%	copra, timber, fish	$110 $460
Tonga	977 sq km 106,000	108 per sq km 2.0%	*Nukualofa 19,000	N/A 95%	copra, bananas, coconut & coconut products	$50 $520
Tuvalu	26 sq km 8,000	308 per sq km 1.6%	*Funafuti 2,200	19% N/A	copra	$4 $570
Vanuatu	14,763 sq km 130,000	9 per sq km 2.7%	*Port-Vila 15,100	N/A 20%	copra, fish, meat	N/A N/A
Western Samoa	2,849 sq km 162,000	57 per sq km 0.9%	*Apia 35,000	21% 90%	copra, cocoa, timber	$130 $770

§Growth Rate: Includes natural increase and migration.

†GNP: Gross National Product, or the total value of all the goods and services produced by the people of a country within one year.

N/A: Data not available

for farmland. Artesian wells are the other source of water in the Central Lowlands. These underground reserves of water reach the surface under their own pressure. Most of the water from these wells is too salty for human use or for irrigating farmland, but it can be used for watering livestock. Find the Great Artesian Basin on the map on page 24.

The Eastern Highlands region contains over half of Australia's population. Locate the Great Dividing Range, which runs along Australia's eastern coast, on the map on page 24. The highest of these ancient mountains are in the south. Their heights, however, do not match those of Europe's Alps or of other younger ranges of the world. A warm, moist climate favors the narrow coastal plain that lies between the mountains and the Pacific Ocean.

The Eastern Highlands region contains some of Australia's best lands for farming and grazing, as well as its largest cities. Australia's longest river, the Darling, also has its source in the Eastern Highlands region.

Landforms of Oceania. The Pacific islands nearest to Australia and the Asian mainland are called continental islands. These include New Zealand and most of the Melanesian islands. Continental islands are the tops of sunken mountain ranges. Thousands of years ago, before the level of the sea rose, land bridges connected these islands to the continental mass.

Further into the Pacific are other island types. One is the high island. High islands are formed by volcanoes that

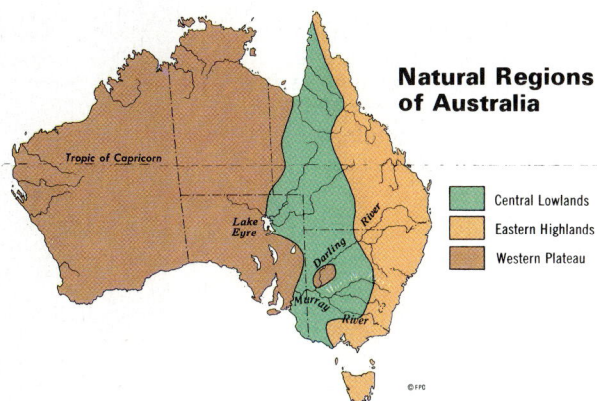

Natural Regions of Australia

Central Lowlands
Eastern Highlands
Western Plateau

build up layers of lava. Over the years, these layers rise above the water's surface. The Hawaiian Islands are high islands, as are Tahiti in the Society Islands and several of the southern Cook Islands.

Another island type, the low island, also has a volcanic core. But the volcanic layers of low islands remain under the water's surface. Tiny sea animals, called coral, attach themselves to the volcanic core. When these animals die, their limestone skeletons remain, forming a coral reef. Layers of limestone build up until the reef reaches the water's surface. Sand and dirt collect on the protruding reef and a new island is formed. Low islands, also called coral islands, rarely rise more than ten to twenty feet (3 to 6 m) above the sea. The Gilbert and Marshall Islands are examples of low, or coral, islands.

Climate and land use. The world climate map on pages 200–201 shows the various climate zones in the southwest Pacific region. Most of the Pacific islands lie within the Tropics, so they

Bora Bora Reef (right) in Polynesia is an island formation called an atoll. Atolls are coral reefs that encircle a lagoon. What island type is shown below?

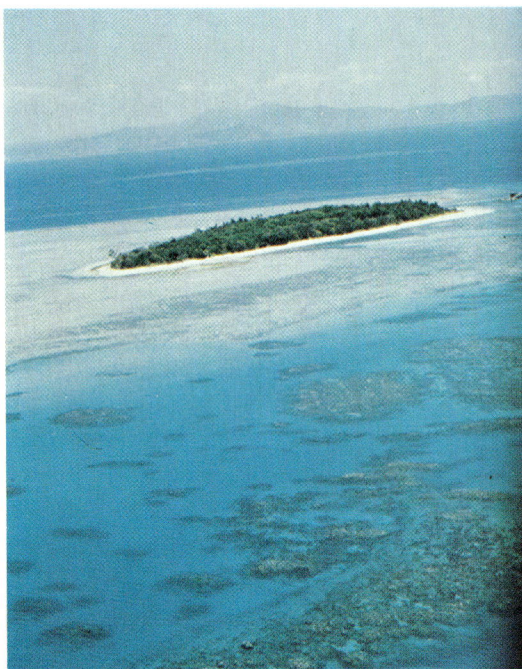

enjoy warm climates year round. Most have a wet and a dry season, although some of the low islands receive light and irregular rains. The islands are prey to tropical storms, called typhoons, that can cause very serious damage.

Most people in the Pacific islands make their livings from agriculture. With only a few exceptions, the islands of Oceania lack most resources needed for industry. Australia and New Zealand, on the other hand, have sufficient resources for many important industries, including both mining and manufacturing. But even these countries, by far the most industrialized in the region, depend heavily on agriculture for their income.

The People's Heritage

The first settlers in the Pacific world probably came from Asia more than 30,000 years ago. Some settled in what is now New Guinea; others spread south to Australia. Still others migrated eastward, settling Melanesia, Polynesia, and Micronesia. Isolated from the rest of the world, early peoples in the southwest Pacific depended on hunting and traditional food gathering for survival.

Europeans began to explore the Pacific world in the early 1500s. By the end of the 1700s, European sea captains had visited most of the island groupings and begun claiming colonies. Sailing under the Dutch flag, Abel Tasman sighted

High islands, such as the one pictured here in the French Society Islands, usually have fertile soils that can support lush vegetation.

New Zealand in 1642. Britain's Captain James Cook made a number of voyages between 1768 and 1779. He explored New Zealand and made the first British colonial claims in Australia. In 1840, the native Maori (maoõ(ə)r′ē) people of New Zealand signed a treaty with Britain. On the basis of this treaty, the British established colonial claims on New Zealand as well.

In the late 1800s, Western interest in the Pacific was at a peak. Developers from many Western nations were attracted to Australia because of its mineral wealth, especially its gold. The strategic location of the Pacific islands also gave them importance as military bases and as ports where steamships could refuel. France, Germany, Spain, Great Britain, and the United States all competed for control of the islands.

In the early 1900s, Australia and New Zealand gained their independence from Great Britain. After World War II, when their value as military bases lessened, a number of the smaller islands in the Pacific also gained their independence. The cultures of nations in the Pacific world, however, have been permanently influenced by the Western powers that controlled them in the past.

Keeping Facts in Focus

1. What three main island groupings make up Oceania?
2. Which country is also the world's flattest continent?
3. How high do low islands usually rise above the sea?
4. Tell whether each of the following is an example of a continental, high, or low island: Gilbert Islands, New Zealand, Tahiti.
5. What foreign power colonized Australia and New Zealand?
6. What two sources of water besides rainfall are usually available in the central lowlands of Australia?
7. What answers would you give to the key questions on page 372?

Working with Ideas

1. What effects has the Great Dividing Range had on the settlement and development of Australia?
2. Use the maps on pages 78, 80, and 86 to discuss the influence of the earth's inner forces on the formation of Oceania.

377

Lesson 2 Australia

Reading Focus

1. For over 100 years, Australia was a colony of Great Britain. You have already read about countries in Asia that were colonies at one time. Review what you learned about these former colonies, especially the impact of colonization on the native peoples and their struggles since independence. As you read this lesson, look for similarities and differences between Australia and former colonies in Asia. In future units, you will learn about former colonies in other regions of the world. What you learn about Australia in this lesson will help you understand how a colonial past can affect present-day economies in countries all over the world.

2. As you read this lesson, look for the answers to these key questions:
   a. Who are the Aboriginals?
   b. What products make up most of Australia's agricultural exports?
   c. What percentage of Australia's export earnings comes from minerals?
   d. How have Australia's patterns of trade changed since World War II?

Vocabulary Focus

Aboriginals	*merino*	*mutton*	*station*
revenue	*land grant*	*refrigerated*	

For thousands of years, the dry continent of Australia supported a native population of well under half a million. These earliest Australians, called Aboriginals, lived nomadic lives dependent on hunting and gathering. In 1788, when the first European settlers arrived in Australia, there were about 500 different Aboriginal tribes.

As the European population of Australia grew, the Aboriginal population declined. Today there are only about 140,000 Australians of Aboriginal descent, although their numbers are starting to grow again. Some live on reservations in the interior, following the traditions of their ancestors. Others work on Australia's large sheep and cattle ranches or in the cities. The Australian government recognizes Aboriginals as citizens. But few Aboriginals enjoy the same standard of living as most other Australian citizens.

Indeed, the standard of living for most Australians is notably high. The Australian government spends about 25 percent of its revenue on welfare, providing many generous benefits to its people. These include medical care, retirement pensions, and unemployment

Many Aboriginals, following age-old traditions, strip bark for use in making shelter.

insurance. Most people can afford to live comfortable lives in clean, modern cities and towns. Despite the forbidding nature of much of their land, Australians have found ways to develop a stable and prosperous economy. And the government is now planning programs that will give the first Australians, the Aboriginals, more opportunities to share in this prosperity.

Limited Agricultural Possibilities

The earliest Europeans in Australia settled mainly near Sydney on the continent's east coast. There they found farming conditions barely adequate. Since the Great Dividing Range prevented westward expansion, the settlers had to find ways to use their land to the fullest. They needed a better source of income than crop raising to support their growing populations.

The importance of livestock. Near the end of the 1700s, in a town near Sydney, a British officer named John Macarthur began to experiment with improving sheep. Using Spanish merino sheep that were used to hot, dry climates, Macarthur and others developed a breed that could produce a high quality of wool. The results of these experiments not only solved the early settlers' problems but also laid the foundation for much of Australia's future economic growth.

The wool industry began to flourish. After 1813, when explorers found a pass through the Great Dividing Range, settlers spread westward. Eager to develop Australia's interior, colonial governors gave land grants to people willing to move west. Many of these settlers took herds of merinos and other sheep with them. By the early 1830s, Australia was exporting large amounts of high-grade wool to Britain. Today Australia is the leading wool producer, supplying almost a third of the world's wool. It also exports large amounts of mutton and lamb.

With the introduction of refrigerated transport in the 1880s, cattle became increasingly important in Australia's economy. Australia's 22.5 million beef cattle graze on huge ranches, called stations. Find the many cattle grazing areas on the map on page 380. Most dairy farms are along the eastern and southwestern coasts, where abundant rains make good pasturelands. About half the products from Australia's highly mechanized dairies are for export.

Grains and other crops. While wool is Australia's chief agricultural product, wheat is the second most important. In

This valley vineyard is in New South Wales. Where else in Australia can farmers grow grapes?

Australia grows enough grain to export large quantities, including about two thirds of its total wheat yields. Much of Australia's exported wheat goes to the People's Republic of China.

Other Australian crops vary from one climate zone to another. In the tropical zones, farmers grow pineapples, bananas, and sugarcane. Apples, Australia's most important fruit crop, as well as berries and peaches, grow well in the temperate zones. In the area of irrigated farmlands along the Murray river system, farmers grow rice, citrus fruits, vegetables, and grapes.

fact, many farmers who raise sheep also grow wheat. After a wheat harvest, the sheep are brought in from their grazing lands on the hillsides and fallow fields. They feed on the stubble of the newly harvested wheat and fertilize the land with manure. Look at the map on this page and find areas where both wheat and sheep are farmed. Sometimes other grains, such as barley and oats, are also grown on sheep and wheat farms. Some of the grains are used for livestock feed.

Abundant Industrial Possibilities

Although Australia is one of the world's largest exporters of agricultural products, more people are employed in manufacturing, mining and service industries than in farming. Look at the National Profiles chart on page 374. What percentage of Australia's population lives in urban areas?

Mining industries. Australia's earliest mining operation grew out of the discovery of copper in 1842. By the late 1800s, gold, silver, lead, zinc, and tin

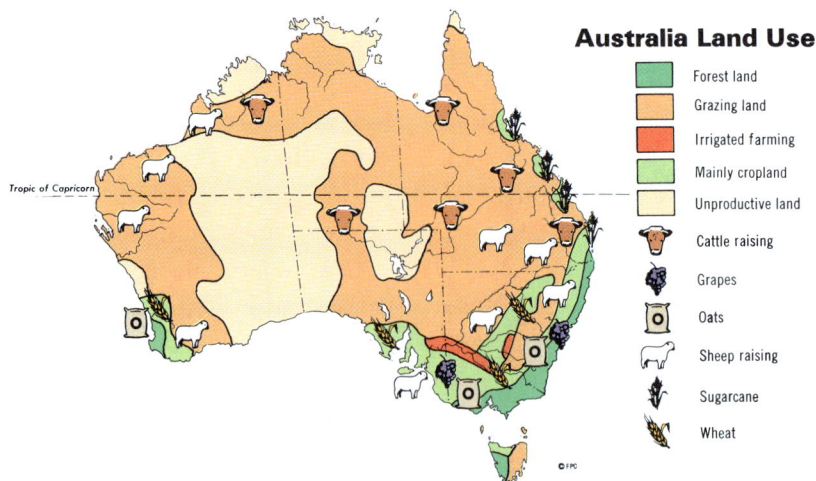

Australia Land Use

- Forest land
- Grazing land
- Irrigated farming
- Mainly cropland
- Unproductive land
- Cattle raising
- Grapes
- Oats
- Sheep raising
- Sugarcane
- Wheat

Tropic of Capricorn

The Snowy Mountains Hydroelectric Scheme in New South Wales brings coastal water inland for both power and irrigation.

COMPARING AUSTRALIA'S EXPORTS: 1960, 1982

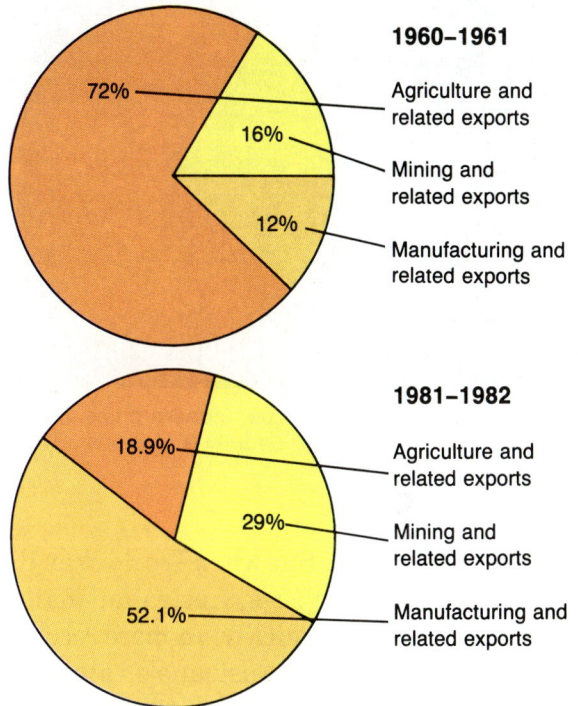

1960–1961

- 72% Agriculture and related exports
- 16% Mining and related exports
- 12% Manufacturing and related exports

1981–1982

- 18.9% Agriculture and related exports
- 29% Mining and related exports
- 52.1% Manufacturing and related exports

had also been discovered in abundant reserves. But it was the discoveries, after 1950, of oil, natural gas, coal, nickel, iron ore, uranium, and bauxite that have had the greatest impact on Australia's economy. Australia is now the world's largest producer of bauxite. It produces enough oil to meet about 70 percent of its own needs. Minerals now make up over one fourth of Australia's export earnings. Many of Australia's minerals go to Japan as raw materials for industry.

Manufacturing industries. At one time a colony, Australia had long depended on imports for many of its man-

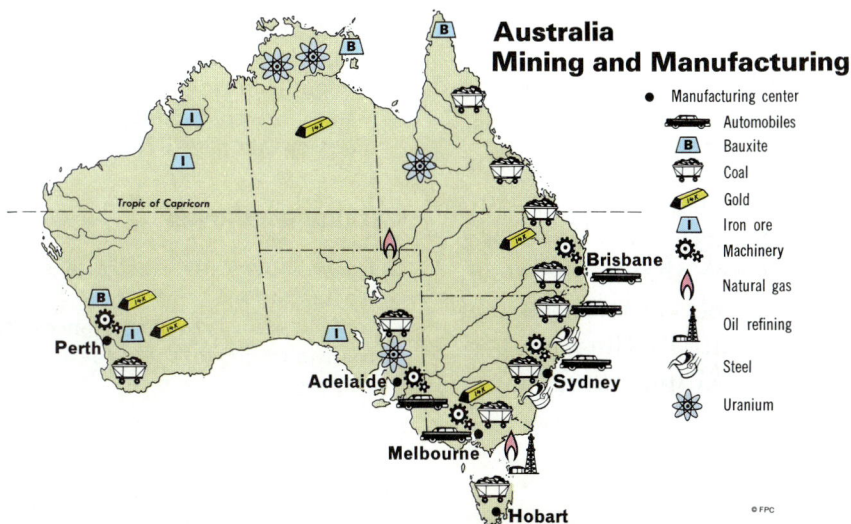

Australia Mining and Manufacturing

- ● Manufacturing center
- Automobiles
- B Bauxite
- Coal
- Gold
- I Iron ore
- Machinery
- Natural gas
- Oil refining
- Steel
- Uranium

Perth, Adelaide, Sydney, Brisbane, Melbourne, Hobart

Tropic of Capricorn

© FPC

Australia has several large, modern cities. Melbourne's skyline (left) rises above the Yarra River. Sydney (right) has a world famous opera house.

ufactured goods. But when World War II cut off British imports, Australian manufacturing grew quickly. In more recent years, income from Australia's valuable mineral exports has helped build even more manufacturing industries. As a result of its increased manufacturing, Australia now imports fewer manufactured goods and more materials for its own industries.

The direction of Australia's foreign trade has also changed. Since World War II, Australia has reduced both imports from and exports to Great Britain. Japan, the United States, China, and New Zealand are now Australia's chief trading partners.

Today, manufactured goods account for about 25 percent of Australia's gross national product and an increasing percentage of exports, mainly to New Zealand. The largest manufacturing industries are those that make steel, automobiles, machinery, chemicals, and electronic equipment. Textiles and processed foods are also important manufactured products. Most manufacturing is centered in Australia's six state capitals: Sydney, Brisbane, Adelaide, Hobart, Melbourne, and Perth.

Keeping Facts in Focus

1. What are some of the welfare services the Australian government provides?
2. Where were the earliest European settlements in Australia? When did the settlements begin spreading to other parts of the continent?
3. How did John Macarthur help shape Australia's economy?
4. What fruits do Australian farmers grow?
5. What are Australia's chief manufacturing industries? What conditions sped the growth of manufacturing since the 1940s?
6. What answers would you give to the key questions on page 378?

Working with Ideas

1. In what ways is Australia similar to the United States?
2. In what ways is the experience of the Aboriginals similar to the experience of Native Americans?

Lesson 3 Oceania

Reading Focus

1. Good ideas are based on solid facts. Sometimes, however, people form ideas before knowing any facts. For example, most people have an idea about what life is like on the Pacific islands. They picture swaying palms and hula skirts, gentle breezes and moonlit tides. Many people probably think that the islands are a tropical paradise, where nature provides abundantly for the natives and life is easy. This impression of the islands, formed with no solid basis in fact, is a fantasy. Try to be aware of any such impressions you might have about various regions of the world. As you read the material in this lesson and others, let the facts, and not your previously formed impressions, guide your geographic learning.

2. As you read this lesson, look for the answers to these key questions:
   a. In what ways is New Zealand similar to Australia? In what ways is it different?
   b. What are the chief sources of cash income on the Pacific islands?
   c. What evidence of Western influence is visible in the Pacific world?
   d. How do Melanesians, Micronesians, and Polynesians differ?

Vocabulary Focus

iron sands	*copra*	*taro*
silica sands	*sago flour*	*breadfruit*

Of all the Pacific islands, New Zealand has most in common with Australia. Like Australia, New Zealand has a population made up mainly of Europeans, mostly of British descent. The official language of both New Zealand and Australia is English. Both countries depend heavily on livestock for export products. Both countries have growing manufacturing industries and a high percentage of urban dwellers. And both countries have advanced social welfare programs that help maintain their high standards of living.

But there are also important differences between New Zealand and Australia. For example, farming is easier in New Zealand. Abundant rainfalls, especially on the west coasts, combine in some places with good soils to create rich farmlands and pastures. Look at the map on page 24. East of the Southern Alps on South Island is a large stretch of alluvial plains, ideal for grazing sheep and growing grains.

About one third of New Zealand's land is too mountainous for use. But the temperate climate makes the other two

383

New Zealand Products and Land Use

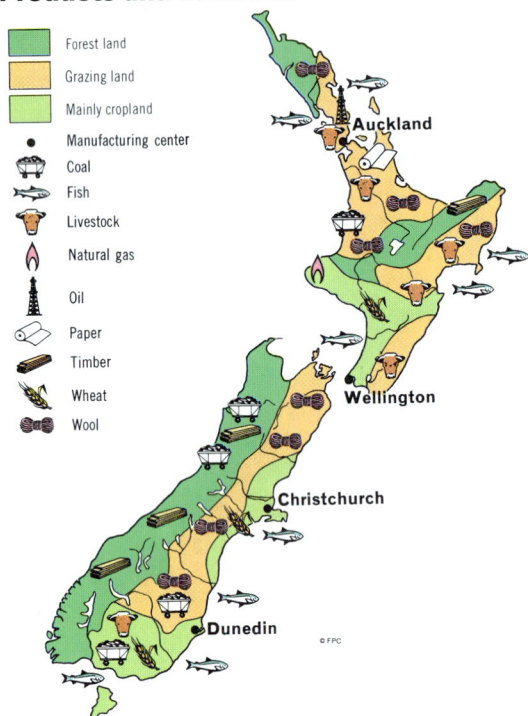

Forest land

Grazing land

Mainly cropland

• Manufacturing center

Coal

Fish

Livestock

Natural gas

Oil

Paper

Timber

Wheat

Wool

Auckland

Wellington

Christchurch

Dunedin

© FPC

thirds usable year round. About half of the usable land supports crops or sown pastures. The rest has thick commercial forests and grasses for grazing. New Zealand is second only to Australia as an exporter of wool. Its dairy and meat products make up 45 percent of its foreign earnings. It is self-sufficient in wheat production, and also grows barley, maize, peas, and potatoes for local use. New Zealand also exports some apples, pears, and seeds.

While New Zealanders benefit from better farming conditions, their mineral resources are less abundant than Australia's. There are only small deposits of copper, silver, gold, and tungsten. New Zealand does, however, have good sup-

plies of coal, iron ore, and natural gas. Local iron sands are used in New Zealand's government-run steel industry off the west coast of North Island. Silica sands are found on South Island.

Manufacturing industries are growing, although not as quickly as in Australia. Agricultural products still make up over three fourths of New Zealand's exports. The main manufacturing industries are food processing, automobile assembly, wood products, textiles, and machinery. New Zealand's good supply of waterpower, coal, and underground steam helps power these industries.

Life in the Pacific Islands

While most New Zealanders live in cities, people on the other Pacific islands live mainly in small fishing or farming villages. The villages are usually located near good fishing areas along the ocean shore or near inland rivers and lagoons. The average village has about two or three hundred people. Most villagers raise their own food, make their own clothing, and build their own homes.

Following the traditions of their ancestors, many island villagers feel strong community ties. They share in the work of the village, and they often share their goods as well. Until recently, few people earned cash incomes, so they had to depend on the exchange of goods to fill their needs.

Even today, cash is scarce in many island villages. What cash there is comes from the sale of such products as coconuts, bananas, and sugarcane to

Once called the Navigator Islands, Polynesia is still well known for the excellent sailing skills of many of its people.

Western export companies. Coconuts are the most important island product. Countries all over the world import copra—dried coconut meat—and coconut oil from the Pacific islands. In the growing cities and towns, however, cash is the chief means of exchange. Mining industries, military bases, and tourism are the main sources of cash in these areas.

Western influence. Although many islanders still follow traditional ways, Western influence is obvious in some places. Industry has developed on a few of the islands. Foreign-owned mines in New Caledonia make it the world's third largest producer of nickel. Foreign investments in Papua New Guinea are developing the world's largest copper mine. Phosphates from bird deposits on Nauru make mining very important on this island as well. Money from the sale of phosphates has earned the people of Nauru one of the highest per capita incomes in the world.

But industrial problems come hand in hand with industrial progress. The waters around Nauru are now polluted. Many fish are dying. The phosphate reserves will be depleted by the late 1980s. There are no clear plans for developing economic resources to replace the income from phosphates.

Westerners brought industrial goods as well as industry itself to some of the Pacific islands. Many islanders are moving to cities and other areas where industrial jobs and goods are available. Some cities are becoming very crowded. Papeete, the capital of Tahiti, now has slums and traffic jams.

The flow of tourists from Western nations also influences island life. Tourism can be a good source of income. Some islands, such as Fiji, Tahiti, and the Cook Islands, encourage the tourist industry. Other islanders, however, feel that tourists and their accommodations will spoil the islands' natural beauty and destroy the local customs.

385

Patterns of Government in Australia and Oceania

Most independent nations of the Southwest Pacific have adopted democratic systems of government patterned after those of the Western nations that at one time controlled them. Most are also members of the South Pacific Commission, an organization founded in 1947 by Australia, France, Great Britain, The Netherlands, New Zealand, and the United States, to promote the economic and social progress of Oceania. Through this commission, the Pacific islanders have received much Western guidance in agriculture, technology, health care, and education.

But there is a growing desire among the islanders for less dependence on Western nations. With the goal of reducing economic dependence on foreign nations, several countries of Oceania formed the South Pacific Forum in 1971. This organization is composed only of countries in the South Pacific region. It established the South Pacific Board for Economic Cooperation. This board creates programs that help members rely more on one another and less on foreign powers for economic support. Although most islands still have strong economic ties to Western powers, the South Pacific Forum is an important step in the direction of an economically as well as a politically independent Oceania.

INDEPENDENT DOMINIONS OF THE BRITISH COMMONWEALTH	NONINDEPENDENT REGIONS
British crown is ruler in name only. Chief executive, usually Prime Minister, and Cabinet have responsibility to carry out laws. Tonga, a constitutional monarchy, has a King as chief executive. Parliament, usually popularly elected, has power to make laws. In Western Samoa, tribal chiefs are the present legislators; when their seats become vacant, successors will be chosen by remaining legislature. Most countries have several active political parties. Government involvement in economy varies from country to country. New Zealand's government has much control over economy; Australia's somewhat less. Where practiced (selected countries): Australia, Fiji, New Zealand, Tonga, Western Samoa.	Status of nonindependent regions varies. Some, such as American Samoa and Midway Island, are territories or possessions administered by one foreign power. Others, such as New Hebrides and Phoenix Islands, are administered by several foreign powers. While the islanders themselves have little voice in these administrative governments, they are active in the village governments, which are often headed by tribal chiefs. Other nonindependent regions are self-governing territories that have chosen to keep formal ties with their foreign powers for military and economic support. Complete independence is a goal for some of these regions. Where practiced: Throughout Oceania (see map, page 373).

Differences Among the Island Groups

Although the general pattern of life is similar throughout most of Oceania, each island grouping is distinctive in some ways. Melanesians, Polynesians, and Micronesians have differing physical features as well as unique customs and traditions.

Melanesia. Melanesia means "black islands." The island grouping got its name from the skin color of most of its inhabitants. Many of the islands of Melanesia have thick forests and tall mountains. Rain forests, tropical diseases, insects, and rugged landforms have kept out many foreigners. As a result, Melanesia has remained more isolated than

The people of Oceania are distinctive physically as well as culturally as these pictures of a Melanesian (left), a Polynesian (center), and a Micronesian (right) clearly show.

other parts of Oceania. Especially away from the coast, Melanesian tribes still follow ancient ways of life. Within the Melanesian islands, however, there is an extensive trading system. Melanesians trade such goods as shells, sago flour, clay pots, and feathers with villagers on other islands.

The continental islands of Melanesia are favored with adequate soil and a good climate for farming. Melanesians therefore depend less on fishing and sailing than other Pacific islanders. Most Melanesians are subsistence farmers who grow sweet potatoes, yams, taro, bananas, squash, and pumpkins.

Polynesia. Polynesia means "many islands." Polynesians have the lightest skin of the three Pacific races.

Unlike the Melanesians, Polynesians are excellent sailors, fishers, and shipbuilders. They also practice farming. Taro and coconuts are the chief cultivated crops. While some of the farming is at a subsistence level, there are also large commercial plantations in Polynesia. Plantation crops and tourism supply Polynesians with the money to buy manufactured goods.

Micronesia. Micronesia means "small islands." Of Micronesia's 2,000 small islands and atolls, only about 100 are inhabited. The islands have meager resources. Most of the farming is for subsistence. The subsistence crops include breadfruit, bananas, taro, and yams. Coconuts are a main export crop.

Keeping Facts in Focus

1. How do farming conditions in New Zealand compare with those in Australia?
2. What minerals and fuels does New Zealand have?
3. What are New Zealand's main industries?
4. What are the main cash crops of the Pacific islands?
5. Where in Oceania is nickel mined? Where is the world's largest copper mine?
6. What answers would you give to the key questions on page 383?

Working with Ideas

1. What does New Zealand have in common with Japan? What differences are there?
2. What problems are the most pressing ones for the Pacific islands' peoples to overcome?

Unit 11 REVIEW WORKSHOP

A. On a separate sheet of paper, write the word or words that best complete each statement.
1. Oceania is made up of Micronesia, Melanesia, and _____.
2. The three largest countries in the southwest Pacific region are Australia, Papua New Guinea, and _____.
3. New Hebrides and New Caledonia are part of _____, one of the three large island groups of Oceania.
4. Australia's three land regions are the western plateau, central lowlands, and _____.
5. Of Australia's three land regions, the _____ has the fewest sources of available water.
6. Although the water from them is too salty for many uses, _____ are still valuable sources of water in the central lowlands of Australia.
7. Some of Australia's best farming and grazing lands as well as its largest cities are located in the _____ region.
8. Limestone from the skeletons of small sea animals, called _____, forms the upper portion of low islands.
9. Tropical storms, called _____, are a serious threat to some Pacific islands.
10. The strategic location of the Pacific islands made them attractive possessions for _____ reasons.
11. Early European settlements in Australia were mainly along the continent's _____ coast.
12. John Macarthur's experiments with _____ sheep laid the foundation for much of Australia's future economic growth.
13. _____ is Australia's second most important agricultural product.
14. Australia is the world's largest producer of _____, which is used in making aluminum.
15. Japan, the United States, China, and _____ are Australia's chief trading partners.

B. Tell whether each statement is true or false.
1. Farming is easier in Australia than in New Zealand.
2. New Zealand's dairy and meat products make up nearly half of its foreign earnings.
3. Underground steam helps power New Zealand's growing manufacturing industries.
4. Most New Zealanders live in small farming villages.
5. Coconuts are the most important product of the Pacific islands.
6. The world's largest copper mine is being developed in New Caledonia.
7. Mining phosphates is the chief economic activity on Nauru.
8. All Pacific islanders welcome the tourist trade for the money it brings to the islands.
9. Melanesians depend less on fishing and sailing than other Pacific islanders.
10. Micronesians are a mixture of Polynesians, Melanesians, and Malayan groups.
11. The rich soils of Micronesia support many export crops.

Apply Your Reading Skills

Use library resources to write a report on ONE of the following topics:
1. World War II battles in the southwest Pacific region.
2. The history, activities, and goals of the Australian government's Department of Aboriginal Affairs.
3. The history of Hawaii before statehood.
4. Traditional sailing methods of the Polynesian islanders.

PEOPLE ENTERING AND LEAVING AUSTRALIA

Years	People Entering			People Leaving			Net Increase
	Men	Women	Total	Men	Women	Total	
1979	86,430	75,780	162,200	56,060	47,840	103,900	58,300
1980	89,360	80,940	170,300	49,880	43,480	93,360	76,940
1981	109,060	95,720	204,780	45,760	40,360	86,120	118,660
1982	111,290	99,360	210,650	46,870	40,830	87,690	122,960
1983	91,650	81,090	172,740	52,600	44,690	97,290	75,450

Apply Your Geographic Skills

Partly because of a rising unemployment rate, the Australian government has reduced the number of immigrants it will admit to the country each year. Study the first table on this page to learn more about immigration patterns in Australia's recent past. Use the table to help you decide whether each of the following statements is true or false.

1. There was a steady decline in the number of people entering Australia between 1979 and 1983.
2. There was a steady decline in the number of people leaving Australia between 1979 and 1983.
3. In 1983, significantly fewer people entered Australia than in 1982.
4. In general, more women than men enter and leave Australia.
5. In 1983, Australia's population showed the smallest net increase resulting from arrivals and departures.
6. In each year between 1979 and 1983, more people entered Australia than left it.

Apply Your Thinking Skills

Using the employment figures in the second table on this page, decide whether workers in each of the following occupations would have a fairly easy time finding a job in Australia or a fairly difficult time. On a separate sheet of paper, write an E next to those who could easily find jobs, and a D by those whose job opportunities have recently decreased.

1. Carpenter
2. Social worker
3. Radio announcer
4. Factory worker
5. Truck driver
6. Miner

EMPLOYMENT PATTERNS IN AUSTRALIA*
(In thousands of persons)

	1981	1982	1983
Mining	91	96	97
Manufacturing	1,262	1,245	1,151
Construction	489	470	419
Transport and storage	345	367	356
Communications	123	141	133
Community social services	1,005	1,021	1,030
Other	3,062	2,410	3,075
TOTAL	5,956	5,742	5,571

*Excludes people employed in agriculture. defense, and private domestic service.

Discuss These Points

After World War II, many islands in Micronesia became part of the U.S. administered Trust Territory of the Pacific Islands. Under the terms of the trusteeship, the United States had the responsibility of helping the islanders prepare for future independence. What kinds of assistance do you think the islanders needed to become ready for self-government? What benefits do you think developing regions such as Micronesia receive under the trusteeship of a world power? What drawbacks might there be under such a trust agreement?

Expand Your Geographic Sights

Australia (Time-Life, 1985).

Ellis, Rennie. *We live in Australia* (Watts, 1984).

Heyerdahl, Thor. *Fatu-Hiru: Back to Nature on a Pacific Island* (Doubleday, 1975).

Kaula, Edna M. *The Land and People of New Zealand* (Lippincott, 1972).

May, Charles P. *Oceania: Polynesia, Melanesia, Micronesia* (Nelson, 1969).

Unit 12 North Africa and the Middle East

Clusters of green vegetation near oases stand out against the desert lands of north central Algeria.

Exploring the Middle Lands

Reading Focus

1. Photos are an important and enjoyable addition to a book. Very often, they communicate more effectively than words. Before reading this lesson, take a close look at all the photographs included in it. Try to learn as much about the Middle Lands as you can from the photos. To organize what you learn, make a list of everything you can conclude about life in the Middle Lands from the pictures. Look for such features as landforms, climate, occupations, and ways of life. As you read the lesson later, compare your list to the written information in the text. Were your conclusions accurate?

2. As you read this lesson, look for the answers to these key questions:
   a. Some cities in the Middle Lands have been called "gifts of the rivers." Why do you think this is a good description?
   b. How has the discovery of oil changed life in the Middle Lands?
   c. Which conquerors had the most lasting impact on people in the Middle Lands?
   d. What is the basis of the conflict over Palestine?

Vocabulary Focus

wadi   *strategic*   *nationalism*

The region known as the Middle Lands extends eastward from the Atlantic coast of Mauritania and Morocco across North Africa to Afghanistan. The people in the Middle Lands share many historical and cultural experiences. Some of the world's earliest civilizations and three of the world's major religions developed in the area. These cultures and religions had a worldwide influence. Today the rich supply of oil has given many countries within the region the power to influence the economies of oil-starved countries throughout the rest of the world.

The Physical Setting

The Middle Lands are divided into two smaller regions. The first of these is North Africa. The vast desert lands of the Sahara separate North Africa from the rest of the African continent, a region known as Sub-Saharan Africa. You will study Sub-Saharan Africa in Unit 13.

The second smaller region within the Middle Lands is often referred to as the Middle East. All the nations shown in color on the map on page 393 that are not in North Africa are a part of the

North Africa and the Middle East
Political-Relief Map

Middle East. Locate Egypt on the map on page 393. Although Egypt is a part of North Africa, it has close political ties with the countries of the Middle East and is often considered a part of it.

The people of the Middle Lands share an environment where water distribution is a problem. The land is lush with greenery where water is abundant, yet it is dry and forbidding where water is in short supply.

Mountains and deserts. The nations of the Middle East and North Africa occupy about 9 percent of the world's land surface. But their population is only about 6 percent of the world's total. The many mountains and deserts make the Middle Lands unable to support a large population. Look at the political–relief map on this page. Which region is more mountainous, the Middle East or North Africa? What mountains are in the northern part of Morocco?

The Middle Lands also have large areas of desert. The largest is the Sahara, which stretches across much of northern Africa. Deserts also form much of the Arabian Peninsula. The Syrian Desert extends from Jordan and Syria into Iraq. In the south, the Rub'al Khali (rōōb′al·käl′ē) Desert covers most of Saudi Arabia. Refer to the map above. Through which countries does the Libyan Desert extend?

Desert winds have carved strange shapes in the Algerian Sahara south of the Ahaggar Mountains.

National Profiles: *North Africa and the Middle East*

Independent Nation	Area/ Population	Population Density/ §Growth Rate	*Capital/ Largest City and Population	Urban Population/ Literacy Rate	Chief Exports (in order of Importance)	†GNP (in) millions of U.S. dollars)/ Per Capita Income
Afghanistan	647,500 sq km 14,448,000	23 per sq km 1.9%	*Kabul 891,750	15% 12%	fruits & nuts, natural gas, carpets	$2,800 $200
Algeria	2,460 sq km 21,351,000	77 per sq km 3.1%	*Algiers 2,200,000	45% 46%	crude oil, petro-leum products	$42,900 $2,142
Bahrain	596 sq km 409,000	686 per sq km 3.9%	*Manama 121,986	80% 40%	petroleum products, aluminum	$4,000 $10,000
Cyprus	9,251 sq km 662,000	72 per sq km 1.3%	*Nicosia 121,500	42% 89%	food, wine, cement	$2,172 $3,342
Egypt	1,000,258 sq km 47,049,000	47 per sq km 2.7%	*Cairo 5,084,463	49% 40%	crude oil, cotton & cotton goods	$30,800 $690
Iran	1,647,240 sq km 43,820,000	27 per sq km 3.1%	*Tehran 4,496,159	49% 48%	crude oil, petroleum products, carpets	$66,500 $1,621
Iraq	445,480 sq km 15,000,000	34 per sq km 3.3%	*Baghdad 3,205,645	66% 70%	crude oil	$30,000 $2,150
Israel	20,720 sq km 3,855,000	185 per sq km 1.5%	*Jerusalem 398,000	87% 88%	diamonds, citrus fruit, clothing, chemicals	$22,200 $5,612
Jordan	96,089 sq km 2,689,000	28 per sq km 3.8%	*Amman 1,232,600	60% 70%	fruits & vegetables, phosphates	$4,900 $1,875
Kuwait	16,058 sq km 1,758,000	109 per sq km 6.2%	*Kuwait 60,400 Hawalli 152,300	88% 71%	crude oil, petro-leum products	$27,600 $25,850
Lebanon	10,360 sq km 2,601,000	251 per sq km 0.1%	*Beirut 1,100,000	64% 75%	paper, food & beverages, chemicals, machinery	$4,100 $820
Libya	1,758,610 sq km 3,684,000	2 per sq km 5.2%	*Tripoli 858,000	53% 50%	crude oil	$26,500 $7,600
Mauritania	1,085,210 sq km 1,623,000	1 per sq km 2.0%	*Nouakchott 250,000	23% 17%	iron ore, fish	$720 $400
Morocco	490,200 sq km 23,565,000	48 per sq km 2.9%	*Rabat 435,510 Casablanca 1,371,330	42% 28%	phosphates, fruits & vegetables, textiles & clothing	$15,200 $640

§Growth Rate: Includes natural increase and migration.

†GNP: Gross National Product, or the total value of all the goods and services produced by the people of a country within one year.

N/A: Data not available

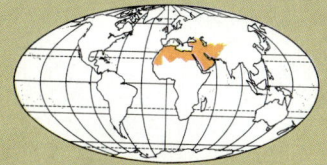

National Profiles: *North Africa and the Middle East*

Independent Nation	Area/ Population	Population Density/ §Growth Rate	*Capital/ Largest City and Population	Urban Popula- tion/ Literacy Rate	Chief Exports (in order of Importance)	†GNP (in) millions of U.S. dollars)/ Per Capita Income
Oman	212,380 sq km 1,009,000	5 per sq km 3.1%	*Muscat 50,000	7% 20%	crude oil	$6,300 $6,828
Qatar	10,360 sq km 276,000	27 per sq km 3.3%	*Doha 190,000	86% 40%	crude oil	$7,900 $27,790
Saudi Arabia	2,331,000 sq km 10,794,000	5 per sq km 3.3%	*Riyadh 1,793,000	67% 52%	crude oil, petro- leum products	$120,000 $14,117
Syria	186,480 sq km 10,075,000	54 per sq km 3.4%	*Damascus 1,142,000	48% 50%	crude oil, cotton, textiles, tobacco	$18,400 $1,957
Tunisia	164,206 sq km 7,202,000	44 per sq km 2.6%	*Tunis 1,000,000	52% 62%	crude oil, phosphates, clothing	$8,700 $1,183
Turkey	766,640 sq km 50,207,000	65 per sq km 2.1%	*Ankara 1,877,775 Istanbul 2,772,708	45% 70%	cotton, textiles, tobacco, wheat, fruits, nuts, metals	$53,806 $1,096
United Arab Emirates	82,880 sq km 1,523,000	18 per sq km 10.3%	*Abu Dhabi 449,000	72% 56%	crude oil	$30,000 $30,000
Yemen, North	194,250 sq km 5,902,000	30 per sq km 2.7%	*Sanaa 277,800	37% 15%	cotton, coffee	$3,800 $740
Yemen, South	287,490 sq km 2,147,000	7 per sq km 2.9%	*Aden 365,000	33% 25%	petroleum products	$792 $430

Climate. Most of the countries in the Middle Lands have a desert climate. Within this climate zone, the average annual rainfall is less than 10 inches (25 cm). The areas with heavier rainfalls are in the more mountainous lands of the Middle East. Turn to the precipitation map on page 20. How do the mountains help explain the pattern of precipitation in the Middle Lands?

Summers are hot in most of North Africa and the Middle East. In some places, temperatures may reach as high as 135° F (58° C). Across much of North Africa winters are cooler than summers. In the interior highlands of Turkey, Iran, and Afghanistan, temperatures often drop below freezing.

Rivers, floods, and desert waters. Turn to the map on page 393. How

ATLANTIC OCEAN

EUROPE

Black Sea

Caspian Sea

SOVIET UNION

TURKEY

Mediterranean Sea

TUNISIA

CYPRUS

SYRIA

LEB.

AFGHANISTAN

MOROCCO

Suez Canal

ISRAEL

IRAQ

IRAN

ASIA

JORDAN

KUWAIT

ALGERIA

LIBYA

EGYPT

Persian Gulf

SAUDI

BAHRAIN

QATAR

MAURITANIA

Red Sea

ARABIA

UNITED ARAB EMIRATES

OMAN

Middle Lands
Petroleum Production and Shipping

N. YEMEN

S. YEMEN

INDIAN OCEAN

©FPC

- Oil fields
— Pipelines
Major Sea Routes

many rivers can you find? The Tigris and Euphrates rivers have their sources in the mountains of Turkey. Into what sea does the Nile River flow?

Both the Nile and the Euphrates were sites for early civilizations in the Middle Lands. Their regular floods helped fertilize the nearby land so that farming could thrive. The Nile flows in a wide valley with steep sides. Each year at the beginning of the tropical rainy season, the waters of the Nile rise. Because of the steep-sided valley, the floods do not spread far from the river.

The Euphrates, in contrast, flows through a wide, open plain. Over the centuries, deposits of silt and mud have raised the level of the riverbed. Like the Hwang Ho in China, the Euphrates flows several feet above the level of the surrounding plain. In the spring, snow melting from the mountains swells the waters of the Euphrates. Without a steep-sided valley to contain them, the flood waters spread over many miles. Sometimes, thousands of people lose their homes to the floods.

Although there are few main rivers in the Middle Lands, there are many short, irregular rivers in the deserts. These are called wadis. For most of the year, wadis are nothing but eroded riverbeds, completely dry. But when there is a sudden cloudburst, a wadi might fill up all at once with rushing waters. Within a few hours, the riverbed will once again be dry. The water from the wadi sometimes filters down hundreds of feet underground. It might be years before another cloudburst will fill the wadi again.

The wadis are not dependable as sources of water. In fact, their sudden floods have sometimes damaged nearby towns. A better source of desert water is the oasis. Oases are fed by springs. In some cases, the water in these springs is from rains that fell many years ago. If the water in an underground spring gets

396

The Nile River flows through Egypt in a wide valley. The waters give life to a narrow strip of land along the banks.

used up, it may be many years more before the supply is restored.

Land use. The generally dry climate and desert lands limit the amount of agriculture the Middle East and North Africa can support. Look at the map of land use on page 21. What explains the snake-like strip of farmland in Egypt? Other areas of good farmland lie along the Tigris and Euphrates rivers. Where else in the Middle East and North Africa is land used for farming? In what countries is land used for forestry?

The impact of oil. People in the Middle Lands have long known about their excellent supply of some minerals. Saudi Arabia, Algeria, and Turkey have some of the largest iron reserves in the world. The region also has plentiful supplies of copper, manganese, lead, salt, and potash. Oil, first discovered in Iran in 1908, has since been found in many other places in the Middle Lands. The map on page 396 shows some of the major oil fields of the Middle Lands. It also shows how oil is transported to other parts of the world.

Today, North Africa and the Middle East produce about 23 percent of the world's petroleum. Experts estimate that oil reserves in this region make up close to 70 percent of the world's total supply. The discovery of oil has brought wealth to many countries. Recently, it has also brought burdens of debt because of the drop in oil prices. Many countries are unable to pay for projects they began when high oil prices brought greater revenues.

Keeping Facts in Focus

1. What African countries are part of the Middle Lands?
2. What Asian countries are part of the Middle Lands?
3. What desert stretches across much of northern Africa?
4. What is a wadi? What is an oasis? Why is an oasis a better source of water than a wadi?
5. What is the average rainfall in a desert climate?
6. Where are the areas of heavier rainfall found in the Middle Lands?
7. Which rivers in the Middle Lands were sites for early civilizations?
8. What natural resource found in the Middle Lands has given the region great economic power throughout the world?

The People's Heritage

About 6,000 years ago, the Middle Lands supported some of the ancient world's greatest cultures. The Babylonian culture arose between the Tigris and Euphrates rivers. The Egyptian culture developed near the Nile River. Both cultures produced great works of art and science and had splendid cities. Other cultures, including the Sumerian, Phoenician, Aramaean, and Hebrew, developed in nearby lands. From these cultures came many ideas about religion, mathematics, and government that have shaped human thinking all over the world. The location of the Middle Lands at the crossroads of Europe, Africa, and Asia made it easy for ideas, as well as goods in trade, to spread far and wide.

The properous cultures of the ancient Middle Lands fell to invaders beginning in about 800 B.C. Invaders from many different regions sought control of the strategic trading positions occupied by the Middle Lands. One after another, Persians, Greeks, Romans, Byzantines, Arabs, and Turks claimed the Middle Lands as part of their empires. Each group of conquerors left behind monuments to their conquest. But none has been more lasting than the living "monuments" of language and religion left by the Arabs after their conquest in the A.D. 600s. The Arabic language and the religion of Islam have continued to play a unifying role in the daily lives of the people of the Middle Lands.

The Turkish Ottoman Empire, named for its founder Osman, had gained control of most of the Middle Lands by the late 1600s. Soon after that, the Middle Lands began to decline in their importance as a center for trade. Sea explorers had found an all-water route between Europe and Asia around the southern tip of Africa. Traders no longer needed the Middle Lands as crossroads. As the economy failed in the Middle Lands, political unrest grew. European countries, by then industrialized and seeking expansion, began to move into the politically weak Ottoman Empire.

Colonial interests in the Middle Lands. The Middle Lands once again became important to trade when the French in Egypt completed the Suez Canal in 1869. The canal connected the Mediterranean and Red seas, greatly shortening the water route between Europe and Asia. Both France and Great Britain wanted to control this important trade route. During World War I, the Europeans defeated the ruling Turks and divided up most of the Arab lands into French or British colonial territories. During the next 60 years, each of these territories demanded and finally achieved independence.

In 1945, seven newly independent countries formed the Arab League. Since then, fifteen other nations have joined. The purpose of the organization was to promote political and economic cooperation and settle disputes peacefully. But rivalries and a growing sense of nationalism in each separate country have limited the success of the Arab League. Controversy over Palestine, a former British territory, has further divided the Arabs. They all agree that

Arab rights in Palestine should be protected. But they disagree on the specific measures that would guarantee those rights to the people of Palestine.

Palestine had been the home of the Jews until about 2,000 years ago. Romans drove many Jews out of Palestine in the first century A.D. Some remained in Palestine, keeping up Jewish traditions under the rule of various empires, including the Arab and Turkish. Those who fled made new homes in countries all around the globe. But many longed to return to their former homeland.

After World War I, Great Britain agreed to the formation of a new Jewish homeland in Palestine. But during the many years following the Roman invasion, Arabs had settled in Palestine and claimed it as their homeland. When the Jews began to return to Palestine, they found the Arabs unwilling to share their land. For many years, Jews and Arabs fought over this land. In 1947, Great Britain asked the United Nations to work out a solution. The UN proposed that Palestine be divided in three parts. One part would be an Arab state; a second would be a Jewish state. The third area, the city of Jerusalem, would be under international control. Under this plan, Israel became an independent state, a new homeland for Jews, in 1948.

Many Arabs did not approve of the UN plan. Numerous fights between Arabs and Jews broke out. An all-out war broke out in 1967, during which Israel seized control of more lands. Another war flared in 1973. Israeli and Arab leaders are still trying to work out solutions to their conflicting claims.

Thousands of Arabs fled from Palestine during the fighting in 1948. Many still live in crowded camps in Jordan, Syria and Lebanon.

Keeping Facts in Focus

1. What was the site of the ancient culture of Babylonia?
2. What was the site of the ancient culture of Egypt?
3. What are some other cultures that arose in nearby lands?
4. What language and what religion unite most of the Middle Lands today?
5. Which western European countries had colonial interests in the Middle Lands?
6. When was the Suez Canal completed? How did its completion change life in the Middle Lands?
7. What is the Arab League?
8. What answers would you give to the key questions on page 392?

Working with Ideas

1. How has oil affected the countries of the Middle East?
2. What makes peacekeeping so difficult in the Middle Lands?

Lesson 2 Agriculture in the Middle Lands

Reading Focus

1. The first section of this lesson describes the nomadic way of life. This way of life is probably very different from your way of life. To get a better understanding of nomadic ways, imagine that you are a member of a nomadic tribe. Try to imagine what a whole day would be like as a desert nomad. What would you do when you woke up? Where would you wash? What foods would you eat and what clothes would you wear? What jobs would keep you busy during the day? What would you do for fun? What is it like to sleep in the desert? A good imagination adds color and life to reading. Trying to put yourself in other people's shoes can help develop a good imagination and a better understanding of different ways of life.

2. As you read this lesson, look for the answers to these key questions:
   a. How do the terms "the desert and the sown" describe making a living in the Middle Lands?
   b. What are some modern irrigation techniques in the Middle Lands?
   c. What is life like on a Middle Eastern farm?
   d. What are the chief crops of the Middle Lands?

Vocabulary Focus

land reform	*moshav*	*regulate*
kibbutz	*communal*	*bilharzia*

Geographers often use two words to describe the Middle Lands: "the desert and the sown." Where water is unavailable, desert tribes lead their herds of goats and sheep from pasture to pasture in a regular seasonal cycle. Wherever there is enough water to irrigate fields, farmers have sown the land and set up stable agriculture. Both ways of life are thousands of years old in the Middle Lands.

The Desert

Nomadic herding is one of the few possible ways to make a living in the unwatered regions of the Middle Lands. Some nomads live mainly in the mountainous areas. In the spring, these nomads leave their winter grazing grounds in the foothills and head up into the mountains. As the mountain snow melts, it leaves fresh grass for the sheep and goats to feed on. On the return trip in the fall, the herds feed on stubble from harvested fields.

Other nomads live mainly in the flat desert areas. During the dry season, they set up tents near oases. When the rainy season comes, they move their goats and camels to better pastures. Sometimes they move into the winter pastures left behind by the mountain nomads. Although these pastures are too harsh for sheep, camels can survive easily on them.

Nomads have few belongings besides their tents, cookware, clothes, and rugs. Sheep, goats, and camels provide the nomads with such needs as milk, cheese, wool, leather, and dung for fuel. Nomads trade wool or a sheep for other needs, such as knives, flour, and coffee.

Nomads like their life with its patterns of movement. When the day's work is done, they sit in front of their tents to discuss their next move. During the hottest part of the year, they travel at night to avoid the sun. Although they move regularly from place to place, nomads also do some farming on the side. Once a year, they sow some wheat or millet near a water hole. Six months later, they return to harvest their crop. They know that other nomads will not touch their crop while they are gone. Nomads are loyal to their own tribes and respectful of others.

Nomads know no national boundaries. They travel wherever they can find good pasture. Some governments in the Middle Lands have tried to settle nomads in one permanent place as full-time farmers. Most nomads resist such government efforts. Some, however,

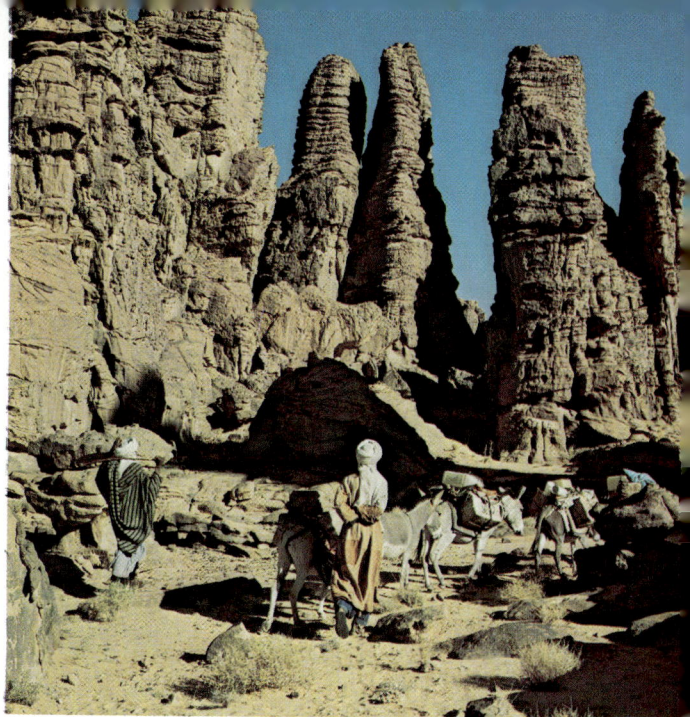

Rock columns are the eroded remains of a sandstone plateau in southern Algeria.

have taken temporary jobs in the newly developed oil fields. With their earnings, they can afford year-round irrigation on their farmlands. But most still prefer the nomadic way of life. Until the governments can build up pastures in farming areas, the nomads will most likely continue to move their herds to whatever pastures can support them.

The Sown

For centuries, farmers in the Middle Lands have been seeking new and better ways to irrigate their lands. In earliest times, farmers probably carried water from rivers or wells to their fields in buckets. Through the years, farmers developed other irrigation techniques, some of which are still used today.

Modern irrigation plans. In recent years, governments in the Middle Lands have developed huge projects to use all

401

available water supplies to the fullest. For example, Israel's water-distribution plan has tripled the country's supply of usable water. The Israeli government built a 150-mile (250-km) pipeline to carry water from the Sea of Galilee in the north to the Negev Desert in the south. As a result, hundreds of farm settlements in the Negev can now survive. The government also drained about 15,000 acres (6,000 ha) of marshland near the Jordan River and turned them into fine farmlands.

The Aswan High Dam in Egypt is one of the largest irrigation projects in the world. The dam stores enough water to irrigate one million acres (400,000 ha) of now unused land. Another 100,000 acres (40,000 ha) can be irrigated year-round so that they can produce two or three crops a year. Water from the Aswan High Dam will make it possible for Egypt to greatly increase its agricultural production.

In Syria, the government has encouraged private investors to develop irrigation systems. As a result, irrigated land has increased threefold in the past twenty years. The government itself has also developed irrigation projects. Between 1960 and 1965, the Syrian government spent nearly 40 percent of its money on irrigation. One of its biggest projects opened up 200,000 acres (80,000 ha) of new land, serving 50,000 farmers.

Political problems have slowed down Jordan's irrigation plans. The government wants to use water from the Jordan River to irrigate land that can now be used only for herding. But the river crosses several political boundaries.

Until all countries on the river can work together, Jordan will have to use a costly system of its own that works against neighboring networks.

Farming Life in the Middle Lands

Most farmers live in small villages in houses built of sun-baked mud brick. Farmers usually have very little furniture. Often they sit and sleep on woven mats. They usually eat flat bread, vegetables, cheese, and yoghurt made from goats' milk. Meat is usually saved for special occasions. Many villagers weave rugs from sheep or camel wool dyed in bright colors. Each village has its own pattern of colors and designs. Rugs from the Middle Lands, especially Iran, are sold all over the world.

All around the village, the farmlands spread out. Each owner's plot usually extends only as far as the farmer can comfortably walk after a hard day's work. The best land is divided into small plots for grain and other crops. The farmers share the poorer lands for use as pastures.

In general, the farmers use few machines. Most families have only an ox or a donkey to pull the plow, turn the waterwheel, and carry loads. Other work is often done by hand.

Land reform programs. Until recently, few farmers in the Middle Lands owned their own farms. Irrigation systems often require the work of many people. Few farmers were able to irrigate their lands on their own. Instead, they depended on wealthy landlords who could maintain the networks and

Following traditional ways, women in Iran (left) make flat bread together. On an Israeli kibbutz (right), farmers use modern methods and machinery.

make sure each farmer took only a fair share of water. In exchange for the water, some farmers had to give as much as four fifths of their crops to the landlord.

Recently many governments in the Middle Lands have changed these conditions. Through a series of land reform programs, they have broken down the large estates and divided them among landless farmers. In Egypt, for example, a single landowner may own no more than seventy acres (28 ha) and no fewer than seven (2.8 ha). Most of the other countries have similar laws regarding land distribution.

Farmers' cooperatives. To increase production, Middle Eastern governments have encouraged cooperative farming. In a cooperative, each family owns a small section of several larger fields. One field might be for cotton, one for rice or corn, and one for clover. In each of these, the farmers are responsible for taking care of their portions. But because each large field has only one crop, some big jobs can be done by machines. The farmers share the cost of the machines and use them for such jobs as spraying insecticides, spreading fertilizer, and harvesting.

Cooperative farming in Israel is especially efficient. There are two kinds of collective farms in Israel: the kibbutz and the moshav. On the kibbutz, the villagers share all property. Farmers eat their meals together. Children are raised in a communal nursery. All farmers are equal partners and they all perform certain assigned jobs. On the moshav farms, each family owns its own land, usually about twelve acres (4.8 ha). The families are free to make their own farming decisions. But they join with other farming families to purchase supplies and sell their crops through a government agency. Both the kibbutz and the moshav make Israeli farming highly efficient.

Percent of Population Earning a Living from Agriculture in Selected Countries of the Middle Lands			
Afghanistan	69.0%	Jordan	28.9%
Algeria	30.0%	Lebanon	17.0%
Egypt	50.0%	Libya	21.3%
Iran	33.0%	Morocco	50.0%
Iraq	30.0%	Saudia Arabia	28.0%
Israel	6.5%	Turkey	61.0%

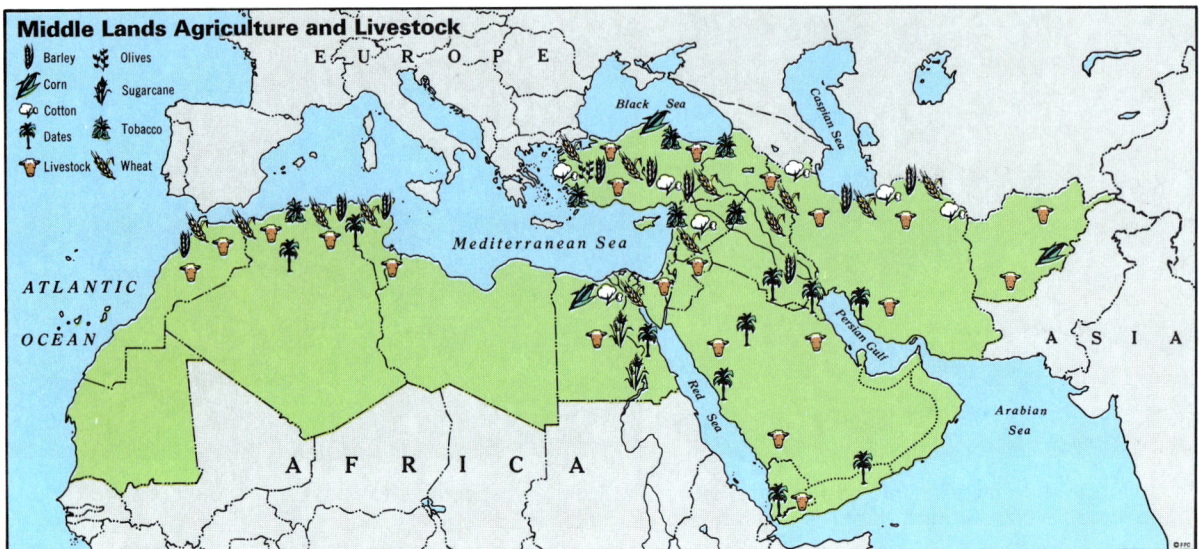

Barley   Olives
Corn   Sugarcane
Cotton   Tobacco
Dates   Wheat
Livestock

EUROPE

Black Sea

Caspian Sea

Mediterranean Sea

ATLANTIC
OCEAN

Red Sea

Persian Gulf

ASIA

Arabian
Sea

AFRICA

A variety of crops. The map on this page shows the chief crops of the Middle Lands. Wheat, barley, and rye grow well in the northerly regions, while corn grows well in the south.

Wheat is a staple food in the Middle Lands. It can be made into bread or pancakes or ground coarsely and served like rice. Barley, hardier than wheat, is the chief grain in Libya, Iraq, Morocco, and parts of Iran. Farmers grow rice in the Nile delta and other well-watered valleys in Morocco, Syria, Turkey, and Israel. Some governments limit rice growing because it requires so much water. Corn is Egypt's chief cereal crop.

Egypt also grows some of the world's finest cotton. Cotton fields cover one fifth of Egypt's land. Turn to the chart of National Profiles on page 394. Which other Middle East countries export cotton? All countries in the Middle Lands grow at least enough cotton for their own use. Most of the cotton grows on large plantations. Village farmers can sometimes earn extra money by working on cotton plantations at harvest time.

The Middle Lands export a great amount of fruits. One area in southern Iraq alone supplies 80 percent of the world's dates. Algeria also produces dates, known for their high quality. Israel and the North African countries produce and export large amounts of citrus fruits and grapes. North Africa also exports large supplies of olives.

Moroccan wool is dyed colorful shades and hung out to dry. Where in Morocco do people use the land for grazing sheep and other livestock?

404

Patterns of Government in the Middle Lands

Until recently, many of the countries in the Middle Lands were colonies of European powers. Their own independent governments are young, and many are in a state of gradual change. Most governments in the Middle Lands, however, share one common feature. Except in Israel, a Jewish state, the government laws in the Middle Lands often reflect the religious laws of Islam. Political leaders are, therefore, often regarded as religious leaders as well. Since the 1979 revolution in Iran, Islamic law is the official law of the land. Iran's leader, the Ayatollah Khomeini, is a religious leader whose influence spreads beyond Iran. Israel's government leaders have strong ties to Judaism. But Israel's civil laws are generally kept separate from religious laws. The various types of governments in the Middle Lands are described in the chart on this page. Although they differ in name, the various types of government share many similar practices.

REPUBLICS	MILITARY REPUBLICS	KINGDOMS	EMIRATES
President is chief executive and is often elected by popular vote. President usually has more power than in Western republics. President can often make laws by decree, sometimes without approval of assembly. Usually only one political party is officially recognized. Israel's government is an exception, closely resembling Western republics. Economies are mixture of free enterprise and socialism. Where practiced (selected countries): Afghanistan, Algeria, Egypt, Israel, Lebanon, Mauritania, Tunisia.	President has great power. Sometimes executive power is shared by three or four men. Military leaders play important role in advising president. Usually several political parties exist, but president's party has the greatest power. Economies are mixture of free enterprise and socialism. Government owns and operates most industries. Where practiced: Iraq, Libya, Syria, Yemen.	King has great power. King appoints cabinet members and sometimes even lawmakers. Some kingdoms elect lawmakers to assembly by popular vote. But king has power to disband assembly. Number of political parties varies from kingdom to kingdom. Most, however, have only one or no political parties. Most kingdoms have mixture of free enterprise and socialist economies; governments usually control some big industries. Where practiced (selected countries): Jordan, Kuwait, Morocco, Saudi Arabia.	Very similar to kingdoms. Emir is chief executive, often chosen from ruling family. Emir appoints cabinet and has power to disband assemblies. Sometimes emir rules by decree. In United Arab Emirates seven emirs share governing power, each with control over his own region. Where practiced (selected countries): Bahrain, United Arab Emirates.

Many farmers in the Middle Lands must still work very hard to bring water to their lands.

Progress and Its Price

With improved irrigation and larger, combined fields, farm yields have been growing. Egypt, for example, now has the world's largest yields per acre of corn and sugarcane. In the 1920s, Turkey had to import more than half its food. Today, it can grow almost all of the food it needs.

Year-round irrigation has also created some problems. The more water there is flowing through the soil, the more rapidly salt deposits build up. About three fifths of Iraq's irrigated soils already have dangerous levels of salt. There is a remedy, however. Underground drains can keep excess water from building up on the soil and depositing salt. But these drainage systems are costly. Few farmers can afford them.

The Aswan High Dam in Egypt offers a good example of another problem in modern irrigation methods. The dam holds back water from the Nile River and regulates the flow of water year round. But the dam also holds back silt and mud. When the Nile used to flood, it left rich silt deposits that fertilized the land. Now farmers have to use costly chemical fertilizers.

The river floods also served another important purpose. They killed the snails that carried a serious disease called bilharzia. Without the floods, the snails keep multiplying and spreading the disease to humans.

Settling nomads and decreasing hazards of new irrigation systems are only two of the important problems in the Middle Lands. Both require much hard work and costly, large-scale projects. Lasting solutions to these problems also require peaceful political conditions and close cooperation among neighboring nations. To many observers, these conditions still seem a long way off.

Keeping Facts in Focus

1. How do the movement patterns of mountain nomads differ from those of desert nomads?
2. How do farm settlements in Israel's Negev Desert get enough water to survive?
3. Why is Jordan's irrigation system so costly?
4. What are some fruits that the Middle Lands export?
5. What problems has the Aswan High Dam created? What problems has it solved?
6. What answers would you give to the key questions on page 400?

Working with Ideas

1. How is farming different in Israel than in other Middle Land countries?
2. Why have many farmers in the Middle Lands welcomed efforts at land reform?

406

Lesson 3 Industry in the Middle Lands

Reading Focus

1. There are many different categories of industry. One way to break down the main topic of industry is to separate different industries according to the specific goods they produce. For example, the furniture industry is different from the shoe industry because the product is different. Another way to divide industries is according to the *types* of goods they produce, rather than the specific goods themselves. By this method, both the shoe and the furniture industry belong to one type of industry—light industry. By the same reasoning, one factory that makes machinery and another that makes transportation equipment both belong to the category of heavy industry. Still another way to break down industry is according to how the goods are produced. For example, both factory workers in Egypt and villagers in Afghanistan weave cotton into cloth. The villagers are working in cottage industries, while the Egyptians are working in mechanized industry. As you read this lesson on industry, try to decide how many different categories each industry mentioned can belong to.

2. As you read this lesson, look for the answers to these key questions:
   a. How does industrial production in the Middle Lands today compare to production twenty years ago?
   b. What local industries do most countries in the Middle Lands have?
   c. Why are the oil-producing nations so important to world affairs?
   d. How is income from oil being used to improve life in the Middle Lands?

Vocabulary Focus

professional   *accessible*   *water distillation*

For thousands of years, farming and herding were the two most common ways of earning a living in the Middle Lands. Farming, however, often included small cottage industries as well. People in farming villages often spent much time making rugs and other textile products. By the 1400s, products from these and other cottage industries, such as metalwork, silverware, pottery, and leather, were in demand in Europe. Many Middle Eastern towns grew wealthy from selling these goods to Europe and Asia. Cottage industries still account for a high percentage of the income in the Middle Lands today.

A Late Start in Industrializing

Modern industry is still very new in the Middle Lands. Modern factories were not built in the area until the late 1940s. In Jordan and Syria today, only

In street markets such as the one in Acre, Israel (left), many handmade goods are sold. Many tools are also handmade, such as the bellows used in metalworking shown at the right.

14 and 23 percent, respectively, of the goods produced each year comes from industry. In Egypt, industry makes up 27 percent of the gross national product. In Israel, industry produces almost a third of the gross national product. These figures are still low compared to some Western industrialized nations. But industrial production in the Middle Lands is now three times higher than it was twenty years ago.

The map on page 409 shows the mineral and fuel resources of the Middle East and North Africa. The Middle Lands have a wide variety of resources. But they have only small supplies of coal and iron ore. The scarcity of these important resources has limited industrial production in the Middle Lands.

Most countries in the Middle Lands have, however, built up some modern industries to meet basic needs. They have built spinning and weaving mills to process their raw cotton. The textile industry in Egypt employs over 30 percent of the country's industrial workers. Most North African and Middle

Eastern countries also have cement plants, flour mills, and factories to make fertilizer, soap, and soft drinks. Furniture, shoes, stoves, and cigarettes are also produced locally. Israel is the only

Handwoven Iranian rugs bring a very high price, especially in foreign markets.

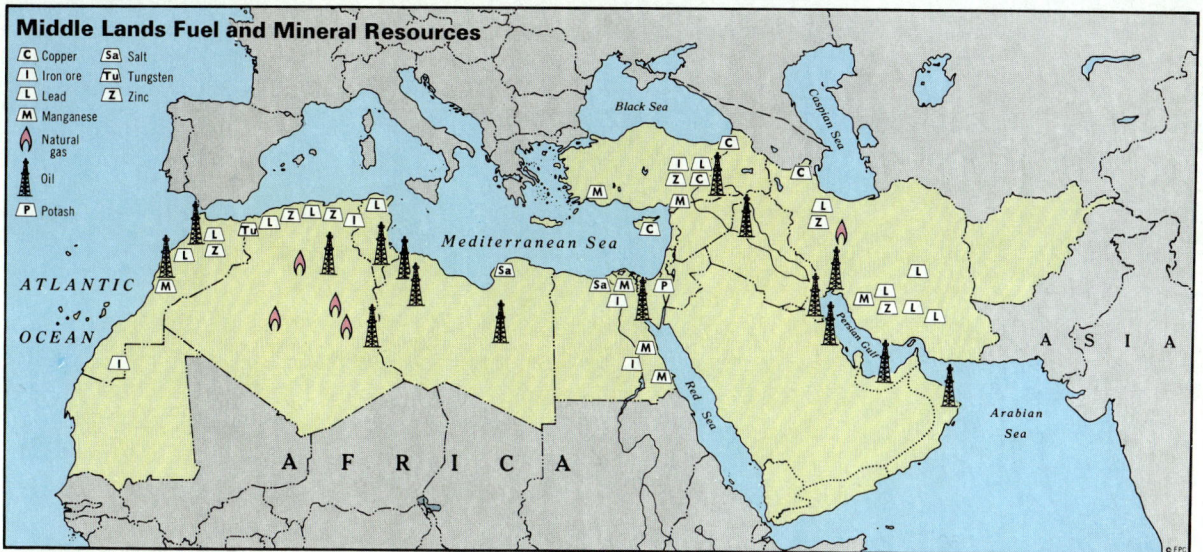

Middle Lands Fuel and Mineral Resources

C Copper
I Iron ore
L Lead
M Manganese
Natural gas
Oil
P Potash
Sa Salt
Tu Tungsten
Z Zinc

Middle Eastern country whose industrial products make up a high percentage of its exports.

Most countries in the Middle Lands need to import such heavy goods as machinery and transportation equipment. However, in Turkey, Iran, Israel, and Egypt, there are some foreign-owned assembly plants for making motors and electric machinery. There are plans to build similar factories in Lebanon, Syria, and Iraq.

Until recently, poor local transportation and a lack of energy resources slowed down the growth of industry. It was often cheaper to import needed goods from countries far away. But in the last fifteen years, governments in the Middle Lands have improved roads and built new power stations. Education has also improved. There are now many highly qualified professionals, especially in engineering. These professionals are eager to see their countries become advanced industrial nations. With their new-found wealth from oil, many countries in the Middle Lands are making rapid strides toward this goal.

The Oil Boom

The Middle East is a region that is extremely rich in oil. Iran, Saudi Arabia, and Kuwait are three of the world's top oil-producing countries. Certain natural conditions make the Middle Eastern oilfields especially productive. First of all, the oilfields in the Middle East are extremely large. Also, the oil in the ground is under very high natural pressure. For this reason, few wells and little pumping are required to bring the oil to the surface. In some cases, the crude oil rises by itself. Finally, Middle Eastern oil lies under easily accessible land. In Southeast Asia, on the other hand, the oil lies under dense jungles. In South America, oil must be pumped from forbidding swamps. Because Middle Eastern oil is in open country, production costs in this region are lower than in Southeast Asia or South America.

When Middle Eastern oil was first discovered, European companies quickly moved in. They succeeded in gaining the right to pump and market the oil.

The United States, East Germany, and France are the biggest buyers of Algerian oil. Algeria also exports great amounts of natural gas.

The Middle Eastern countries, as colonies of European powers, received only a small share of the oil wealth. After World War II, however, the independent Middle Eastern lands began to demand a bigger share. In 1951, the government of Iran took over all oil production. Since that move, Middle Eastern lands have won a greater and greater share of the wealth from oil. In 1960, Iran, Iraq, Kuwait, Saudi Arabia, and Venezuela formed the Organization of Petroleum Exporting Countries (OPEC). This organization succeeded in increasing Middle Eastern shares in oil wealth. In less than ten years, OPEC increased Middle Eastern shares from less than 50 percent to more than 80 percent.

Highly industrialized countries in other parts of the world became more and more dependent on Middle Eastern oil. Even the increased oil production in the North Sea could not meet the needs of Western Europe, Japan, and North America. The OPEC countries grew powerful as they recognized the world's need for their oil. In 1970, they were able to take complete control of both the production and pricing of Middle Eastern oil. Between October and December of 1973, the price of Saudi Arabian oil jumped by about 300 percent. But oil prices began to fall by the early 1980s. Many non-OPEC countries began producing more oil. OPEC does not have the same power that it had in the 1970s. This has caused problems between member countries as they try to determine how to regain some of their control over oil prices.

Profits and Progress

At first, the sudden wealth from oil benefitted only the few who were in a position of political power. Then governments began to expand the benefits that oil wealth brought to their countries. They took on major development and improvement projects. Nomadic grazing, poor farming, and fishing and pearl diving were the most common ways people in Kuwait could make a living. Now, the government operates a water-distillation plant built from oil money. For the first time in its history, Kuwait uses irrigation for agriculture. The city of Kuwait has become a large, modern city. Although its population is less than that of the new suburb of Hawalli, Kuwait remains the country's political and commercial center.

Saudi Arabia and Iran have also made vast improvements with oil income. In 1974, Saudi Arabia built more than 2,000 miles (3,200 km) of highways. It also built electric and water conservation plants and expanded its irrigation networks. Iran has built steel mills and machine-tool plants with its oil wealth. It has also spent a great deal of money on its program to teach its people to read and write.

The shifting balance of power. But just as the oil producing nations benefitted from the sudden wealth that oil revenues brought them, they began to incur major problems when oil prices started to fall in the 1980s. Some of the oil rich nations took on major industrialization and improvement programs and planned to finance them with oil revenues for years to come. Drops in oil prices have left these countries short of money to pay for these projects.

Not all countries in the Middle Lands have benefitted from sudden oil wealth. The countries with few or no oil reserves now have little power to influence affairs in the region. Egypt, industrialized fairly early, is not a major oil-producing country, but it has remained a politically influential power in the region because of its industrialization.

Even within oil-producing countries, some people remain outside the benefits of wealth. Kuwait and Saudi Arabia have started government programs that help distribute their wealth. But in Iran, little has been done since the Shah's exile to raise the subsistence level of most of Iran's people.

Oil wealth has undoubtedly brought many improvements to the Middle Lands. But there are still many unsolved problems that, for the present, make the Middle Lands a trouble spot for the rest of the world.

Keeping Facts in Focus

1. What are some products from cottage industries in the Middle Lands?
2. What important industrial resources do the Middle Lands lack?
3. What industry in Egypt employs 30 percent of the nation's industrial workers?
4. What three natural features make Middle Eastern oilfields especially productive?
5. What is OPEC?
6. What changes has wealth brought to Kuwait? To Saudi Arabia?
7. What answers would you give to the key questions on page 407?

Working with Ideas

1. Oil has been very important to the countries of the Middle East. What are some of the benefits? What are some of the problems? Which are greater, the problems or benefits?
2. Why is the political instability of the Middle Lands upsetting to peacekeeping efforts of world leaders?

Unit 12 REVIEW WORKSHOP

Test Your Geographic Knowledge

A. Choose the letter of the item that correctly completes each statement.

1. The Suez Canal separates (a) Turkey from Europe, (b) the Arabian Peninsula from Iran, (c) the Arabian Peninsula from Africa.
2. People living in the Middle Lands make up (a) 9 percent, (b) 6 percent, (c) 3 percent of the world's total population.
3. The desert covering most of Saudi Arabia is the (a) Syrian, (b) Sahara, (c) Rub' al Khali.
4. Oases are supplied with water by (a) wadis, (b) mountain rivers, (c) springs.
5. The ancient Babylonian culture grew up (a) along the Nile River, (b) along the Jordan River, (c) between the Tigris and Euphrates rivers.
6. The Arabs conquered the Middle Lands (a) about 6,000 years ago, (b) around 800 B.C., (c) in the A.D. 600s.
7. The Suez Canal connects the (a) Red Sea and Mediterranean Sea, (b) Persian Gulf and the Indian Ocean, (c) Red Sea and the Gulf of Aden.
8. Israel became an independent state in (a) 1948, (b) 1967, (c) 1973.
9. Thirty percent of Egypt's industrial workers are employed in the (a) engineering industry, (b) textile industry, (c) electronics industry.
10. The founders of the Organization of Petroleum Exporting Countries (OPEC) include all the following countries EXCEPT (a) Venezuela, (b) Iran, (c) Egypt.

B. On a separate sheet of paper, fill in the word or words that best complete each statement.

1. The three religions born in the Middle Lands are ———, ———, and ———.
2. Although Egypt is in North Africa, it is generally considered a part of ———.
3. Because of its steep-sided valley, floods from the ——— do not spread far from the river.
4. Short, irregular rivers in the desert are called ———.
5. The fertile land along the Nile River was the site of the ancient ——— civilization.
6. The Middle Lands produce about ——— percent of the world's petroleum.
7. The ——— Dam in Egypt is one of the largest irrigation projects in the world.
8. The kibbutz and moshav are examples of cooperative farming in ———.
9. Small supplies of ——— and ——— have limited industrial production in the Middle Lands.
10. Industry in Israel makes up about a ——— of that nation's gross national product.

Apply Your Reading Skills

Use library resources to write a report answering ONE of the following questions.

1. What is the Palestine Liberation Organization? How do its views on Israel differ from the views of other Arabs?
2. What forces turned the Sahara region into a desert?
3. What were the ancient civilizations in the Tigris–Euphrates valley like?
4. How are Islam and Judaism alike? How are they different?

Apply Your Geographic Skills

Use the map on page 413 to find answers to these questions.

1. In how many nations is Arabic the only major language?
2. How many nations have more than one major language?
3. In which two countries is Persian a major language?
4. In what countries is the Berber language spoken?
5. Are there more Jews or more Muslims in the Middle Lands?
6. What is the major language of the Jews?

Middle Lands
Major Religions and Languages

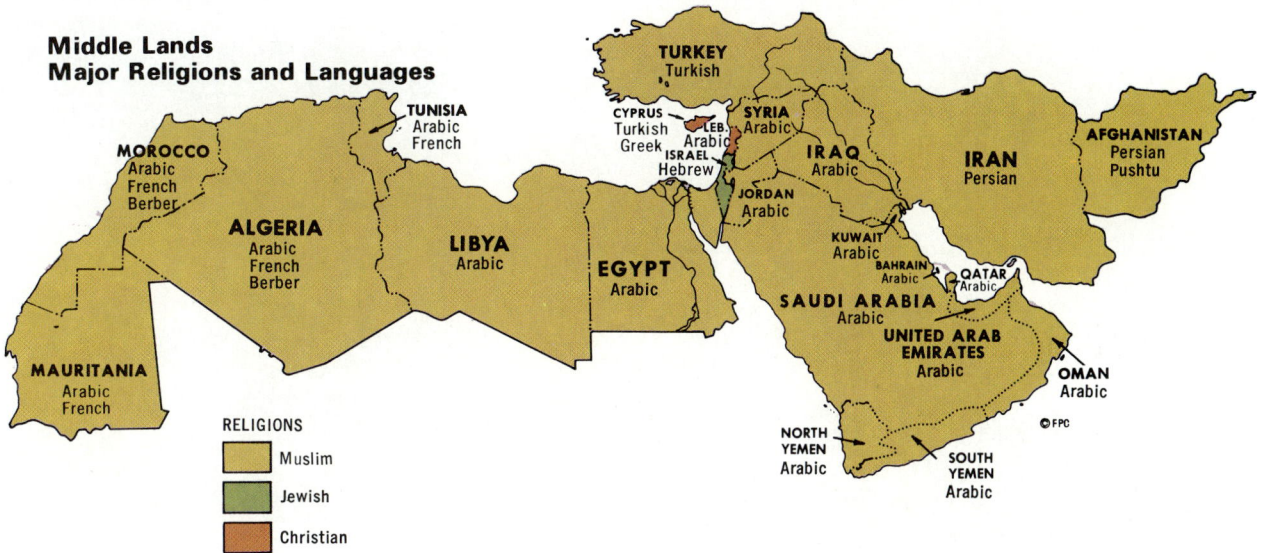

TURKEY Turkish

TUNISIA Arabic French

CYPRUS Turkish Greek

SYRIA Arabic

LEB.

IRAQ Arabic

IRAN Persian

AFGHANISTAN Persian Pushtu

MOROCCO Arabic French Berber

ISRAEL Hebrew

ALGERIA Arabic French Berber

LIBYA Arabic

EGYPT Arabic

JORDAN Arabic

KUWAIT Arabic

BAHRAIN Arabic

QATAR Arabic

SAUDI ARABIA Arabic

UNITED ARAB EMIRATES Arabic

OMAN Arabic

©FPC

MAURITANIA Arabic French

NORTH YEMEN Arabic

SOUTH YEMEN Arabic

RELIGIONS

- Muslim
- Jewish
- Christian

7. In what way is Lebanese culture similar to most other North African and Middle Eastern cultures? In what way is it different?
8. What European influence is evident on this map?
9. What is the only country in which Greek is a major language?
10. What makes the culture of Iran and Afghanistan different from that of their neighbors?

Apply Your Thinking Skills

A. Listed below are some of the problems facing countries in the Middle Lands. Arrange them in order of importance, listing what you consider the most important first.
 1. Lack of coal and iron ore for development of industry
 2. Lack of water
 3. Lack of regional cooperation
 4. Open hostilities between Jews and Arabs
 5. Diseases, such as bilharzia
 6. Unfair distribution of wealth
 7. Illiteracy

B. Write a short report explaining why you arranged the above items the way you did.

Discuss These Points

1. Some people in the oil-wealthy nations are using their money to buy controlling interests in foreign businesses. What are some of the advantages and disadvantages of these investments to the people living in the foreign countries? What, if anything, can or should the foreign goverments do to limit outside control of their businesses?
2. In the late 1970s, OPEC countries accounted for over 40 percent of the world's production of oil. By 1984 OPEC's share of world oil production had declined to about 23 percent. What effect has this had on OPEC's ability to control prices? What could be some of the reasons why OPEC's share of world production has decreased? Will this trend continue? Why or why not

Expand Your Geographic Sights

Cross, Wilbur. *Egypt* (Childrens, 1982).

Edmonds, I. G. *Allah's Oil: Mideast Petroleum* (Nelson, 1977).

Hicks, Jim. *The Empire Builders* (Silver, 1974).

Hicks, Sam. *Desert Plants and People* (Naylor, 1966).

Martin, Rupert. *Land and People of Morocco* (Macmillan, 1968).

Peretz, Don. *The Middle East* (Houghton Mifflin, 1973).

Spencer, William. *The Islamic States in Conflict* (Watts, 1983).

Worth, Richard. *Israel and the Arab States* (Watts, 1983).

Unit 13 Sub-Saharan Africa

Slight and irregular rainfall makes about two thirds of Kenya's land too dry to support many crops. In the dry regions, the Masai and other peoples graze cattle.

Lesson 1 Exploring Sub-Saharan Africa

Reading Focus

1. The region south of the Sahara is called Sub-Saharan Africa. The word *Sub-Saharan* breaks down into two parts. One part is *Sub-*, which is a prefix meaning "under" or "below." The other part is *Saharan*. Put together, the parts explain the word's meaning: "under or below the Sahara." Many other words begin with the prefix *sub-*. What do these words mean?

subarctic	*subnormal*	*substandard*
subfreezing	*subsoil*	*subway*

   If *subnormal* means below normal, what does *supernormal* mean? *Super-* is another prefix. It means "over, above, or in addition to." What does *supernatural* mean? Being able to break words down into their parts will help you understand new words. What are the meanings of some other prefixes you should watch for as you read?

2. As you read this lesson, look for the answers to these key questions:
   a. In what ways is the physical environment of Sub-Saharan Africa different from that of North Africa?
   b. What farming problems do Sub-Saharan Africans face?
   c. How does Sub-Saharan Africa's industrial development compare with that of other regions of the world?
   d. What outside influences played a strong part in the region's history?

Vocabulary Focus

tsetse fly	*industrialist*	*pan-Africanism*

The vast desert lands of the Sahara separate North Africa from the rest of the African continent. For centuries, the Sahara limited travel and trade between the lands to the north and south of it. As a result, the culture of central and southern Africa, which together are called Sub-Saharan Africa, is somewhat different from that of North Africa. In many ways, the physical environment is also different. Yet most of the countries in both parts of Africa share many of the same problems. Most countries in both regions are newly independent. And most are struggling toward the same goal: to enter the modern industrial world without losing their unique identities.

Sub-Saharan Africa

Political-Relief Map

⊛ National Capitals
• Other Cities

SCALE
Miles 0 200 400 600 800 1000
Kilometers 0 400 800 1200 1600

The Physical Environment

Africa is the second largest continent in the world. Only Asia is larger. Its great area includes a variety of landforms and climates that influence the way people can make their livings.

Location. You have already studied the region of North Africa as part of the Middle Lands. Sub-Saharan Africa includes all countries south of the North African region. Using the map above and its key, determine how far north and how far south of the equator the region extends.

Some of Africa's countries are islands in the Atlantic and Indian oceans. The largest island-nation is Madagascar. The smallest is the Seychelles. Find these island-nations on the map on pages 10–11. Both are located in the Indian Ocean.

National Profiles: *Sub-Saharan Africa*

Independent Nation	Area/ Population	Population Density/ §Growth Rate	*Capital/ Largest City and Population	Urban Popula- tion/ Literacy Rate	Chief Exports (in order of Importance)	†GNP (in) millions of U.S. dollars)/ Per Capita Income
Angola	1,245,790 sq km 7,770,000	6 per sq km 2.4%	*Luanda 475,300	21% 20%	crude oil, coffee, diamonds	$3,900 $501
Benin	115,773 sq km 3,910,000	34 per sq km 3.1%	*Porto-Novo 208,000 Cotonou 487,000	15% 20%	palm products, cotton, palm kernel oil, other agricultural products	$1,100 $294
Botswana	569,800 sq km 1,038,000	9 per sq km 3.6%	*Gaborone 72,000	16% 24%	diamonds, cattle, animal products, copper & nickel ores	$722 $910
Burundi	28,490 sq km 4,691,000	165 per sq km 2.8%	*Bujumbura 141,000	2% 25%	coffee, tea, cotton	$1,200 $272
Cameroon	475,400 sq km 9,506,000	20 per sq km 2.7%	*Yaoundé 435,000 Douala 637,000	29% 65%	oil, cocoa, coffee, timber	$7,000 $845
Cape Verde	4,040 sq km 300,000	74 per sq km 1.2%	*Praia 36,000 Mindelo 40,000	26% 37%	fish, bananas, salt	$142 $473
Central African Republic	626,780 sq km 2,585,000	4 per sq km 2.8%	*Bangui 375,000	41% 33%	cotton, coffee, diamonds, timber	$658 $273
Chad	1,284,640 sq km 5,116,000	4 per sq km 2.5%	*N'Djamena 303,000	18% 20%	cotton, meat, fish	$500 $110
Comoros	2,170 sq km 455,000	209 per sq km 2.8%	*Moroni 22,000	12% 15%	perfume oils, vanilla, vegetable oils, cloves	$90 $230
Congo	349,650 sq km 1,745,000	5 per sq km 3.0%	*Brazzaville 200,000	39% 50%	crude oil, timber, fertilizer	$1,800 $1,140
Djibouti	23,310 sq km 289,000	12 per sq km −8.8%	*Djibouti 200,000	74% 20%	cattle	$116 $400
Equatorial Guinea	27,972 sq km 275,000	10 per sq km 2.5%	*Malabo 25,000	54% 20%	cocoa, coffee, wood	$100 $417
Ethiopia	1,178,450 sq km 31,998,000	27 per sq km 2.3%	*Addis Ababa 1,408,000	13% 15%	coffee, hides & skins	$4,800 $141
Gabon	264,180 sq km 958,000	4 per sq km 3.9%	*Libreville 180,000	32% 65%	crude oil, timber, manganese ore, uranium & thorium ores	$3,500 $2,742
Gambia	10,360 sq km 725,000	70 per sq km 3.4%	*Banjul 40,000	16% 15%	peanuts & peanut by- products	$240 $370
Ghana	238,280 sq km 13,804,000	58 per sq km 3.2%	*Accra 1,045,000	31% 30%	cocoa, aluminum, wood, gold	$10,500 $420

§Growth Rate: Includes natural increase and migration.
†GNP: Gross National Product, or the total value of all the goods and services produced by the people of a country within one year.
N/A: Data not available

National Profiles: *Sub-Saharan Africa*

Independent Nation	Area/ Population	Population Density/ §Growth Rate	*Capital/ Largest City and Population	Urban Population/ Literacy Rate	Chief Exports (in order of Importance)	†GNP (in) millions of U.S. dollars)/ Per Capita Income
Guinea	246,050 sq km 5,579,000	23 per sq km 2.7%	*Conakry 575,000	19% 20%	bauxite, alumina, diamonds	$582 $102
Guinea-Bissau	36,260 sq km 842,000	23 per sq km 1.8%	*Bissau 109,500	24% 9%	peanuts, palm kernels, fish	$177 $98
Ivory Coast	323,750 sq km 9,178,000	25 per sq km 3.2%	*Abidjan 1,686,000	32% 24%	cocoa, coffee, timber	$7,700 $871
Kenya	582,750 sq km 19,362,000	28 per sq km 4.1%	*Nairobi 959,000	10% 47%	petroleum products, coffee, tea, sisal	$6,300 $316
Lesotho	30,303 sq km 1,474,000	49 per sq km 2.5%	*Maseru 75,000	5% 55%	wool, mohair, wheat	$569 $424
Liberia	111,370 sq km 2,160,000	19 per sq km 3.3%	*Monrovia 306,000	28% 24%	iron ore, rubber, diamonds, lumber	$800 $385
Madagascar	595,700 sq km 9,645,000	16 per sq km 2.7%	*Antananarivo 600,000	16% 53%	coffee, vanilla, cloves & clove oil, sugar	$3,200 $360
Malawi	95,053 sq km 6,829,000	71 per sq km 3.2%	*Lilongwe 130,000 Blantyre 250,000	8% 25%	tobacco, tea, sugar, peanuts	$1,340 $213
Mali	1,204,350 sq km 7,567,000	6 per sq km 2.3%	*Bamako 620,000	17% 10%	cotton, livestock, peanuts, grain	$1,000 $138
Mauritius	1,856 sq km 1,018,000	548 per sq km 1.6%	*Port Louis 146,884	44% 61%	sugar, textiles, clothing	$960 $890
Mozambique	786,762 sq km 13,413,000	17 per sq km 2.8%	*Maputo 785,000	9% 14%	cashew nuts, cotton, textiles, sugar, mineral products	$1,500 $150
Namibia (South West Africa)	823,620 sq km 1,111,000	1 per sq km 3.0%	*Windhoek 77,400	7% N/A	foreign trade included with South Africa	N/A N/A
Niger	1,266,510 sq km 6,284,000	5 per sq km 3.3%	*Niamey 300,000	13% 5%	uranium, livestock, fruits & vegetables	$2,000 $425
Nigeria	924,630 sq km 88,148,000	95 per sq km 3.4%	*Lagos 1,404,000	20% 25%	crude oil, cocoa	$74,000 $827
Rwanda	25,900 sq km 5,836,000	225 per sq km 3.3%	*Kigali 150,000	4% 37%	coffee, tea	$1,388 $270
São Tomé and Príncipe	964 sq km 89,000	92 per sq km 1.1%	*São Tomé 20,000	33% 50%	cocoa, copra, coffee	$30 $300
Senegal	196,840 sq km 6,541,000	33 per sq km 3.2%	*Dakar 978,553	30% 10%	peanuts, peanut oil, phosphates	$2,500 $410

National Profiles: *Sub-Saharan Africa*

Independent Nation	Area/ Population	Population Density/ §Growth Rate	*Capital/ Largest City and Population	Urban Population/ Literacy Rate	Chief Exports (in order of Importance)	†GNP (in) millions of U.S. dollars)/ Per Capita Income
Seychelles	404 sq km 66,000	163 per sq km 1.3%	*Victoria 23,000	37% 60%	cinnamon, vanilla, copra	$128 $1,330
Sierra Leone	72,261 sq km 3,805,000	53 per sq km 2.7%	*Freetown 375,000	25% 15%	diamonds, iron ore, coffee, cocoa	$1,200 $270
Somalia	637,140 sq km 6,393,000	9 per sq km 2.3%	*Mogadishu 400,000	30% 60%	livestock, hides bananas	$1,875 $375
South Africa	1,222,480 sq kn 31,698,000	26 per sq km 2.4%	*Pretoria 528,407 *Capetown 213,830 Johannesburg 1,536,457	48% 50%	wool, diamonds, corn, uranium, sugar	$73,600 $2,500
Sudan	2,504,530 sq km 21,103,000	8 per sq km 2.7%	*Khartoum 333,921	20% 20%	cotton, gum arabic	$7,100 $345
Swaziland	17,364 sq km 651,000	37 per sq km 3.0%	*Mbabane 33,000	15% 65%	sugar, asbestos, wood pulp, citrus fruits	$500 $880
Tanzania	939,652 sq km 21,202,600	23 per sq km 3.2%	*Dar es Salaam 1,400,000	20% 66%	coffee, cotton, fruits & vegetables, diamonds, tobacco, sisal, tea	$5,200 $281
Togo	56,980 sq km 2,926,000	51 per sq km 3.1%	*Lomé 283,000	15% 18%	phosphates, cocoa, coffee	$950 $340
Uganda	235,690 sq km 14,819,000	65 per sq km 3.2%	*Kampala 458,000	8% 52%	coffee, cotton, tea	$4,800 $280
Upper Volta	274,540 sq km 6,733,000	25 per sq km 2.5%	*Ouagadougou 200,000	8% 7%	cotton, livestock, oilseeds & nuts	$1,100 $169
Zaire	2,343,950 sq km 32,158,000	14 per sq km 2.9%	*Kinshasa 3,000,000	30% 15%	copper, cobalt, diamonds, petroleum	$3,400 $570
Zambia	745,920 sq km 6,554,000	8 per sq km 3.2%	*Lusaka 538,000	43% 54%	coper, zinc, cobalt, lead, tobacco	$2,900 $476
Zimbabwe (Rhodesia)	391,090 sq km 8,325,000	21 per sq km 3.0%	*Harare 686,000	20% 45%	chrome ore, gold tobacco, asbestos cotton	$7,100 $880

§Growth Rate: Includes natural increase and migration.

†GNP: Gross National Product, or the total value of all the goods and services produced by the people of a country within one year.

N/A: Data not available

Lake Natron in northern Tanzania lies within a Great Rift trench. The uplands in northeast Tanzania are rich in timber, although many of the forests were cleared in the early 1900s and replaced with coffee plantations.

A continent of plateaus. Much of Africa's interior consists of plateaus and hilly uplands. In southern and eastern Africa, high mountains rim the plateaus near the coast. Between the mountains and the coast, there is often a strip of coastal plain. Turn to the graphic-relief map on page 23. Trace the coastal plains from Senegal on Africa's west coast around the continent to Somalia on the east. Now turn to the map of land use on page 21. Are all of these plains used for farming? Where along Sub-Saharan Africa's coast are there areas of little-used land?

The Great Rift Valley. The plateaus in eastern Africa were long ago cracked by a shift in the land that created a huge fault. When the land shifted, a series of deep trenches formed between the cracked surfaces. Lava poured out and built up higher plateaus, sometimes topped by volcanic peaks. Find Mt.

Kenya and Mt. Kilimanjaro in eastern Africa on the map on page 417. These are two of the volcanic peaks formed when the land shifted. Which is the higher peak?

The series of trenches extend all the way from Syria in the Middle East to Mozambique in southern Africa. Together the trenches form the Great Rift Valley. In some places, the trenches are 40 miles (64 km) wide and 1,500 feet (454 m) deep. Two of Africa's largest lakes, Nyasa and Tanganyika, lie in Great Rift trenches. Find these lakes on the map on page 417. Lake Tanganyika, 420 miles (676 km) long, is the longest freshwater lake in the world.

To the west of the Great Rift Valley near the equator, the plateaus sink into a lowland area. The Zaire Basin is part of this lowland. On the whole, the plateaus of western Africa are lower than those in the east.

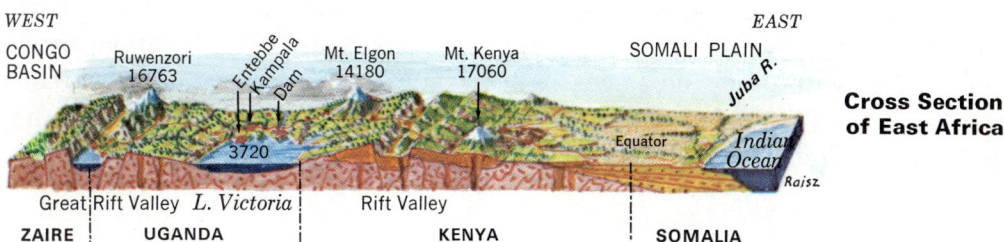

WEST

CONGO BASIN

Ruwenzori 16763

Entebbe
Kampala
Dam

Mt. Elgon 14180

Mt. Kenya 17060

SOMALI PLAIN

Juba R.

EAST

3720

Equator

Indian Ocean

Raisz

Cross Section of East Africa

Great Rift Valley   *L. Victoria*   Rift Valley

ZAIRE   UGANDA   KENYA   SOMALIA

The Kariba Dam on the Zambezi River supplies much-needed electrical power to the many copper industries in Zambia and Zimbabwe.

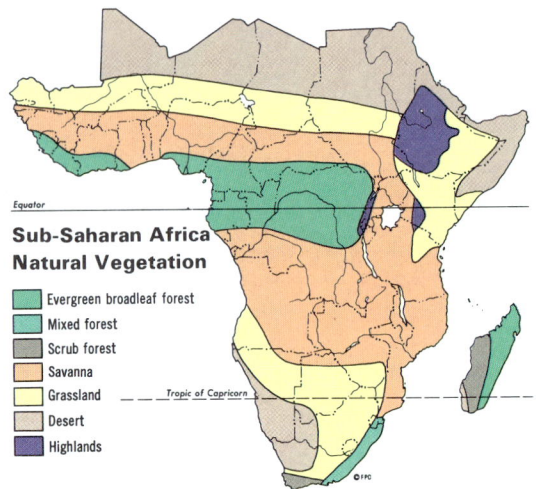

Sub-Saharan Africa Natural Vegetation

- Evergreen broadleaf forest
- Mixed forest
- Scrub forest
- Savanna
- Grassland
- Desert
- Highlands

Equator

Tropic of Capricorn

©FPC

Rivers. Throughout the continent, rivers beginning in the highlands drain outward toward the seas. Trace the routes of the Niger, Zaire, and Kasai rivers to the sea, using the map on page 417. Lake Victoria, Africa's largest lake, helps feed the Nile on its course to the Mediterranean Sea. Find the Limpopo and Zambezi rivers in southern Africa. Into what ocean do they empty?

Find Victoria Falls on the Zambezi River. Waterfalls are common in Africa, occurring when rivers rush over sudden drops in the level of land. Waterfalls make Africa's rivers difficult or impossible to navigate. They do, however, provide an excellent source of waterpower. Although much of it is untapped, Africa has over a third of the world's potential waterpower.

Climate. Refer to the map on pages 200–201. Are there any parts of Sub-Saharan Africa that are cold year round? Read the description of each of Africa's climate zones carefully. Are the differences among climates centered mainly around temperature or around rainfall?

Compare the precipitation map on page 20 with the map of natural vegetation (page 422). What vegetation grows in the areas with the heaviest rainfalls? What kinds of vegetation cover most of Madagascar?

Land use. The map on page 21 shows the chief ways Africans use their land. What is the most common land use in Sub-Saharan Africa? Where else in the Eastern Hemisphere is land used for traditional farming and grazing? Compare the map on page 21 with the map on page 417. In what countries is forestry important?

Now check the chart of National Profiles on pages 418–420. Which countries export a large amount of animal products (wool, meat, hides, and skins)? What do these exports tell you about the way people use their land?

In many parts of central and western Africa, the tsetse fly makes cattle herding almost impossible. The bite of the tsetse fly kills cattle and brings disease to humans. Scientists are trying to find ways to control the tsetse fly and keep it away from cattle. But at present, the tsetse fly limits herding in about a third of Africa's lands.

Over 70 percent of Sub-Saharan Africa's people work at farming. Farmers in Africa face several problems. In some parts of Africa, rainfall is unpredictable. In one year, rainfall might be very low. Crops might die under the blazing sun. In another year, rainfall might be so heavy that it washes away seeds and crops.

Another farming problem is Africa's general lack of rich soils. Most of Africa's soils lack humus, the decayed matter that enriches soil. Soil with little or no humus easily loses its fertility. Many Africans have found that they cannot use the same plot of land year after year, as European farmers can. They must therefore move from one plot to another in search of fertile soil.

Mineral resources and industry. Parts of Africa have some of the world's richest deposits of certain minerals. In some of these places, mining has become second only to farming in the number of people it employs. Other parts of Africa, however, have few mineral resources. Look at the map on this page. What parts of Sub-Saharan Africa have many mineral resources? Which have almost none? What important minerals and energy sources seem to be in short supply in Sub-Saharan Africa?

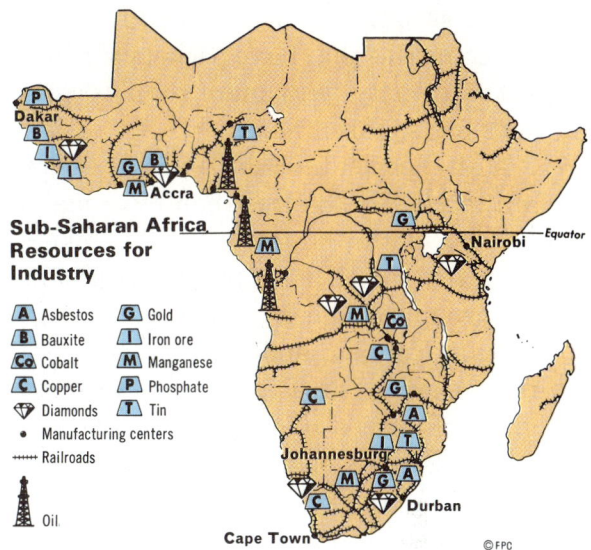

Sub-Saharan Africa Resources for Industry

A Asbestos
B Bauxite
Co Cobalt
C Copper
◇ Diamonds
G Gold
I Iron ore
M Manganese
P Phosphate
T Tin
• Manufacturing centers
••••• Railroads
Oil

Despite the continent's abundance of certain mineral resources, Africa has few manufacturing industries. The many unnavigable rivers make transportation very difficult, slowing down industrial growth. Coal and other energy resources are generally lacking or undeveloped. When European industrialists came to Africa in the 1800s, they built roads, railways, port cities, and power stations that helped change these conditions. Some African countries, such as South Africa and Nigeria, now depend on industry for about a third of their gross national products. On the whole, however, industry in Sub-Saharan Africa plays a smaller role in the economy of the region than does agriculture.

People of Sub-Saharan Africa. Less than 400 million people live in Sub-Saharan Africa. Turn to the map of population distribution on pages 272–273. What is the population density for most of Sub-Saharan Africa? Name the countries that have pockets of population density with more than 40 persons per square kilometer.

Most people in Sub-Saharan Africa are black. The Bantus (ban′tōōz) make up one of the largest groups of black Africans. The Nilotes (nī·lō′tēz), another group of black Africans, live in Kenya, Uganda, and parts of east central Africa. The Pygmies are a third group. Pygmies live mainly in the forests of Zaire, Congo, and Gabon. There are very few Pygmies in Africa, fewer than 200,000.

A fourth group of black Africans are the Khoisans (koi′sänz). These people, including the Bushmen and the Nama, live in the deserts of Botswana and Namibia. Believed to be the oldest group of African peoples, there are now only about 84,000 Khoisans in Africa.

Sub-Saharan Africa also has some non-black people. About five million people of European descent live in Africa, nearly two thirds of them in South Africa. Africa also has about one million people of Indian descent. These people settled in southeastern Africa when they left India and Pakistan in the 1800s.

Keeping Facts in Focus

1. What rank does Africa's size give it among the continents of the world?
2. What is Africa's largest island-nation? What is its smallest?
3. What is the Great Rift Valley?
4. How does the tsetse fly affect the way some Africans can make their living?
5. Name two problems facing African farmers.
6. Give two reasons for Africa's slow industrial growth.

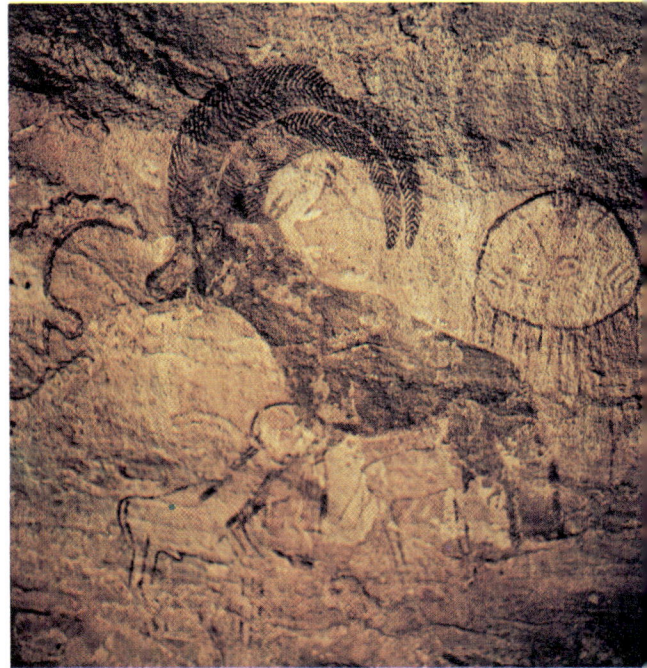

Rock paintings in southern Algeria record a time when the Sahara region still supported abundant plant, animal, and human life.

The People's Heritage

Some scientists believe that Africa may be the home of the first human beings on earth. They have found evidence that a Stone Age culture existed about two and a half million years ago in eastern Africa. Gradually, humans spread to other parts of Africa and to other continents.

By about 5000 B.C., people living in North Africa had begun to farm and domesticate animals. At that time, the Sahara was not a desert. It was an area of good farmland. But by 2500 B.C., the changes in climate had turned the Sahara into a desert. Farmers moved south in search of better lands. There they

Nearly half of all Nigerians are Muslims. Most of these live in the northern part of the country. This Muslim mosque is in the city of Kano. Most people in the southern part of the country are Christians.

grew rice and yams and kept herds of cattle.

The Sahara blocked most communication between northern and southern Africa. Some ideas, however, reached southern Africa through Egypt. One of these was the use of iron for tools. Early Africans south of the Sahara continued to migrate even further south, taking farming and the use of iron with them.

After the Muslim conquest in North Africa (A.D. 651-710), trade across the Sahara increased. This trade extended Muslim influence into the interior of Africa. Several empires, led by Muslim rulers, grew up in western Africa. Trade with North Africa made these empires wealthy.

Between A.D. 1000 and 1500, there were also large kingdoms in central and southern Africa. Most of these had little contact with the outside world. One of them, the Karanga kingdom, had some trade with foreign merchants on Africa's east coast. Although Africa's regular coastlines offer few natural harbors that encourage seafaring, strong sailing winds on the eastern coast helped early sea trade develop there.

Europeans in Africa. The winds and currents on Africa's west coast did not favor seafaring. But in the mid 1400s, the Portuguese built ships that were able to sail against these winds and currents. They were able to start trade routes to empires in south and west Africa. Africa's gold was the prize most Portuguese and European merchants sought.

Before long, however, slaves became another valuable prize. African rulers sold their enslaved prisoners of war to European traders. Many slaves died on the crowded, miserable voyage from Africa. Those who survived faced lives of forced labor in Europe and on plantations in the Americas. The slave trade finally began to decline in the 1800s.

425

The harbor of Mombasa in Kenya is one of the busiest ports on the east African coast.

Merchants then found that other African goods, such as palm oil and peanuts, could bring good profits in Europe.

Europeans began to tighten their control of African resources during the 1800s. Competing with one another for territories, they set up strong colonial rule in most of Africa. By 1914, only Liberia and Ethiopia were left as independent countries. In many places, especially South Africa, Europeans took the best land for themselves. The Africans could make only meager livings from the poor lands that remained. Europeans in other parts of Africa brought some benefits of their culture, such as education and medical care, to Africans. But resentment of colonial rule grew strong all over the continent.

It was not until after 1960 that African nations achieved independence. Some did so by fighting; some made peaceful agreements with the Europeans. Some African nations changed their names from those the Europeans gave them to names approved by Africans. National pride has grown very strong.

But problems continue. Many newly independent nations lack skilled workers and experienced leaders. Europeans still have economic control, and racial hatred and tribal conflicts have slowed progress in several African nations. Military governments have gained control in a number of African nations. And in recent years, drought has brought starvation and death to many people in parts of Africa.

Pan-Africanism, a belief that Africans can and should work together without foreign influence, is on the rise in Africa. The East African Community (EAC) and the Organization of African Unity (OAU) have been formed to encourage close cooperation. But it may be many years before Africa can solve the problems arising from its past and become a stable, secure region.

Keeping Facts in Focus

1. What turned the Sahara into a desert?
2. Which two African countries remained independent during Africa's colonial period?
3. What is pan-Africanism?
4. Explain how Africa's coastline affected trade.
5. What answers would you give to the key questions on page 416?

Working with Ideas

How did most nations in Sub-Saharan Africa benefit from colonial rule? What are some problems these nations face today because of colonial rule?

Grasslands and Forests in Sub-Saharan Africa

Reading Focus

1. Much of Sub-Saharan Africa is covered either with grasslands or with forests. The two areas are different enough from each other to be considered separate areas, but they are also related to each other. First, they usually border each other, yet one area merges into the other without a sharp change. Second, the people in each area look to the other area to supply some of their needs. While reading this lesson, look for other relationships between grassland areas and forest areas. Also watch for changes that have taken place in their relationships through the years. By putting together small pieces of information, you will be able to form interesting conclusions.

2. As you read this lesson, look for the answers to these key questions:
   a. How does the climate in the grasslands change from season to season?
   b. What products come from grassland farming?
   c. What is the life of grassland herders like?
   d. What replaced hunting and gathering as the main occupation in the forest areas?

Vocabulary Focus

savanna	gum arabic	plantain
baobab	phosphorus	cassava
acacia	kola nut	maize

Grasslands and forests cover the greater part of Sub-Saharan Africa. At one time, the grassland areas were more populated and wealthier than the forest areas. Farming and trade were difficult in the dense forests where the land was hard to clear. But in recent years, countries in the forest areas have built up their economies. They have developed seaports for trade with Europe. Today, both the grasslands and the forests play important roles in Africa's economic development.

Patterns of Life in the Grasslands

Look at the map of natural vegetation on page 422. Grasslands, both with and without trees, stretch all across Africa from the west coast to parts of the east coast. They form a curved arc around the evergreen broadleaf forest. How far south do grasslands extend?

Over the past few years much of the grasslands area has been turning into desert. There are many reasons for this; some are human causes and others are natural ones. This change has caused massive food shortages and widespread famine. Much work will have to be done to restore the grasslands.

There are two distinct seasons in the grasslands, the dry season and the wet season. The dry season usually lasts for four to seven months. January is dry north of the equator; July is dry to the south. During the dry season, many trees lose their leaves, and most rivers run dry. The grasslands become browner and browner. Farming comes to a standstill. People use the dry season to trade, build houses, work on arts and crafts, and visit friends.

The first sign of the wet season is usually the flight of migrating birds. When grasslands dwellers spot the birds, they know the rain clouds cannot be far behind. With much rejoicing, the people wait for the rains to start. With the rains, the land turns green, trees start to bloom, and children run out to splash in the showers. The farmers send their cattle to pasture away from the farm and get ready to start the new planting.

Grasslands close to the Sahara have thin grass that withers completely in the dry season. Trees are rare, although there may be a few thorny bushes. In wetter areas further from the desert, there are more trees. On the edge of the rain forest, grasslands are covered with waist-high grass and many trees. These areas on the edge of the rain forest are called savannas.

In recent years rainfall has been very scarce. This lack of water has caused a drought. While droughts are normal occurrences in Africa, the effects are worse because of the deepening poverty, growing population, and abuse of the land. One of the worst abuses of the land is "deforestation", the cutting down of most trees. Twenty years ago, 16 percent of Ethiopia's land area was covered with forests. Today only 3.1 percent is forest. Without trees the land loses its ability to hold moisture. The soil erodes.

Grassland trees. In these wetter areas, the baobab is a common tree. Its odd, barrel-shaped trunk is sometimes used to make houses. People also use the bark of the baobab to make rope and cloth. The fruit of this tree has a tasty pulp that people enjoy eating.

Mango trees also produce a sweet, tasty fruit. For many years, mangoes from Africa's grasslands bruised too easily to be shipped very far. But stronger varieties that resist bruising have recently been developed.

Another common tree is the acacia. The acacia produces gum arabic, a substance used in making glue, candy, and some drug products. The country of Sudan is the world's largest producer of gum arabic.

Farming in the grasslands. The main crops in the grasslands are those that can survive long dry spells: sorghum, wheat, and millet. Corn grows well close to the equator, where moisture is more abundant. Some farmers along the rivers also grow dry rice.

When hollowed out, the wide trunk of the baobab tree can be used as shelter. The fruit of the baobab, called monkey bread, grows to over a foot long.

Recent irrigation improvements in some areas have made wet-rice farming possible. Because of the increasing desert area most grassland farmers can not grow enough food foor their own use.

Peanuts and cotton are the main cash crops. Gambia, Senegal, Mali, Sudan, and Nigeria are the main producers of peanuts.

Cotton is an important crop in Sudan. The Sudan Cotton Board, a government agency, controls cotton production and irrigation in this country. The government shares some of the cotton profits with the farmers. The rest is used to fund the Cotton Board and other government programs that help develop farming. In Western Africa, foreign investors own most of the cotton plantations.

Attempts to improve farming. Farming in the grasslands is hard work. Because there is only one rainy season, farmers must work quickly to make the most of it. Big machines are seldom used because of the high cost of maintaining them. Primitive farming methods have hurt the soil. Because there are so many people to feed, there is constant pressure to keep the land in production. The soil has worn out.

Grassland herding. Some grassland tribes depend completely on cattle herding for their living. These people are nomads, living in temporary shelters on their moves between pastures. Other people, especially in mountainous areas, are less affected by drought. These people have permanent villages and do some farming and cattle raising. Animal manure fertilizes their land, so harvests are often large. Africa's herding tribes include the Fulani (fōō′län·ē) in West Africa and the Masai (mä·sī′) in East Africa.

Farming and herding in Ethiopia and Somalia. Most of Ethiopia lies within the grasslands region, although the desert is spreading (see map, page 422). But, for several reasons, Ethiopia's farming patterns are different from those of its neighbors. First of all, Ethiopia lies on a very high plateau, much of it over 10,000 feet (3,030 m) above sea level. Its altitude affects temperature and rainfall. Turn to the climate map on page 201. How does Ethiopia's

The Ethiopian highlands made invasion difficult and helped keep Ethiopia isolated for many years.

climate differ from that of its neighbors?

Ethiopia once differed from other African countries because the land was exceptionally rich in phosphorus and potash. Plentiful rainfall allowed a variety of crops to grow, including bananas, coffee, and dates. Today farmers have a difficult time growing basic crops. The desert has moved in on Ethiopia's grasslands. The result has been wide-spread famine.

Ethiopia and Somalia together form what is called the Horn of Africa. Find this jutting horn on the map on page 23. Most of Somalia lies at a much lower elevation than Ethiopia. Its climate is very hot and dry. Farming is possible only in or near oases. Nomadic herding is the main occupation. Many herders can make only a poor living from their camels, goats, and sheep. For many years, the herders bordering Ethiopia and the farmers living within it have had conflicting claims on the land. Conflicts continue today, often creating tense political conditions.

Grassland highways for trade. Before the forest areas developed seaports, goods from the forest had to be sent to other parts of the world over the grasslands. Even today, grassland traders handle many goods. Diamonds from

Sierra Leone, Ghana, and the Congo Basin still often go through grassland traders to reach the Middle East. Dried fish, kola nuts, and gold from the forest countries are also often shipped over grassland routes. And salt from mines in the Sahara reaches the forest through grassland markets.

During the rainy season, trade usually goes by waterways. In the dry season, trucks carry freight over dusty roads. Trade links have brought some unity to the diverse African tribes. Although most tribes have their own language, many people have learned a common language used to conduct trade.

The famine, which made the headlines in 1984, was the worst in Africa's history. Ethiopia, Mozambique, and Chad were the hardest hit. The rest of the world sent relief supplies. Also needed are solutions to the problems that caused the famine.

Keeping Facts in Focus

1. How do grasslands close to the forest differ from those near the desert?
2. What uses do Africans find for the baobab tree?
3. What grassland crops can survive long dry spells?
4. What are the main cash crops of the grasslands? Where are they grown?
5. What are some of the causes and effects of the spreading desert in Ethiopia?

Many people in the modern city of Antananarivo on Madagascar (left) do intensive farming within the city limits. In the forests of Uganda (right), people follow more traditional ways of living and farming.

Patterns of Life in the Forests

Humans moved into the forests when the Sahara was no longer able to support them. But for many years, the tangled forests were also unable to support a large population. Before iron tools were developed to clear the forests, hunting and gathering were the only ways forest people could make a living. Even after people had iron tools, farmers found that many crops could not survive in the forest's dark shade. After about A.D. 800, forest farmers learned of plantains (large, green, banana-like fruits) and cassava through contacts with the East. With these shade-tolerant crops, forest farming began to replace hunting and gathering. The tsetse fly made cattle herding impossible. But it also kept out enemy raiders, whose horses would have been killed by the fly.

Village farming. Farming is even harder in the forests than it is in the grasslands. Plants grow very quickly in the forests. Even after a field has been cleared, new weeds spring quickly to life and the clearing must begin again. Plantain, cassava, yams, and rice are the chief food crops of the forest areas. Maize, sweet potatoes, and pineapples also grow well here. Europeans probably introduced these crops in the 1500s.

The forests have long produced cash crops as well. One of the forest's oldest cash crops is kola nuts. Europeans introduced oil palm, cacao, coffee, and rubber to meet the needs of their markets. Farmers were able to grow these new crops without disturbing their traditional farming patterns.

Hunting and gathering in the forests. Most of the people in the rain forest are now farmers. Some, however, such as the Pygmies, are still hunters and gatherers. They do not live in permanent villages. Instead, they follow their animal prey in a hunt. Pygmies usually hunt with nets in groups of three or four

431

Zaire's Ituri Forest is the home of about 35,000 Mbouti Pygmies. This dense tropical forest has traditionally offered the Pygmies protection from hostile outsiders.

families. Their hunting techniques require close cooperation. The men spread the nets, which are sometimes 300 feet (90 m) long. Then the women and children chase the animals into the nets.

When they are not hunting, Pygmies gather fruits, nuts, berries, and wild honey. On these days, the forests ring with the Pygmies' call-and-answer songs.

Many Pygmies have regular trade with nearby villages. In exchange for honey and meat, the villagers provide the Pygmies with such goods as peanuts, yams, and metal tools.

Today, the Pygmies' forest home is threatened. Loggers and farmers are seeking new lands. The forests are being cleared in many places. Also, governments are trying to settle the Pygmies in permanent villages. But Pygmies are like nomads. They have no permanent homes. They view the forest as the giver of life and cross national boundaries at will. They honor the spirits that they believe guard their forest homes.

Forest countries today. A number of changes are taking place in forest countries today. On the average, per capita income has increased. The new wealth has come, in part, from a rise in industry, the modernization of port cities, and the development of an active sea trade with other world regions.

Keeping Facts in Focus

1. What are the chief food crops of forest areas?
2. Why is farming difficult in the forests?
3. What cash crops did Europeans introduce to Africa's forest areas?
4. What interests are threatening the Pygmies' forest home today?
5. Explain why forest countries have become wealthy in recent years.
6. What answers would you give to the key questions on page 427?

Working with Ideas

1. How does life on the grasslands differ from life in a rain forest?
2. Why might an anthropologist argue that every effort should be made by countries in Sub-Saharan Africa to preserve the way of life of Pygmies and other nomadic tribes?

Industry and Cities in Sub-Saharan Africa

Reading Focus

1. Facts are the building blocks of ideas. Ideas come from interpreting the information facts provide. Read the first paragraph under the heading **New Patterns of Life in Industrialized Areas** (page 436). One of the facts you learn from this paragraph is that usually it is the men who seek distant industrial jobs, leaving women and children behind. By interpreting this fact, you can start to form an idea about what position women generally hold in African life. Using this fact, discuss these questions. a) Do you think African men and women share equally in industrial life? b) Do you think African men regard women as weak and frail? c) Do you think there is a strong women's rights movement in Africa? d) How do you think industry has affected family life in Africa? Before you can be certain of your answers, you need more facts. But meanwhile you have formed the beginnings of an idea by interpreting one fact.

2. As you read this lesson, look for the answers to these key questions:
   a. Which is more important in Africa, mining or manufacturing?
   b. How do physical problems limit Africa's industrial growth?
   c. How do political problems limit Africa's industrial growth?
   d. In what ways is South Africa different from other African countries? In what ways is it similar?

Vocabulary Focus

hydroelectric	*refinery*	*apartheid*
exploit	*antimony*	

Most of the industry in Sub-Saharan Africa centers around mining. Much of it was developed by the Europeans, especially in the early 1900s. But even before the Europeans, Africans were mining gold in Ethiopia and Guinea, and tin in western Africa. In the 1860s, Europeans discovered diamonds in Kimberley, South Africa. Twenty years later, they discovered the huge gold deposits of the Rand in South Africa. These discoveries brought many interested Europeans to Africa. All over the continent, Europeans began exploring and mining for metals and other minerals. Soon they discovered copper, lead, zinc, and tin, often near the sites of older African mines.

Abidjan, the port capital of Ivory Coast, is an important center for trade and industry.

World Wars I and II weakened European economies. As a result, there was little new exploration in Africa. But after World War II, mineral industries in Africa soared. As they grew, cities also grew. Many former colonies became independent nations during these years. New ways of life appeared all over the continent. Expanding industries have brought both progress and problems to the people of Africa. As in other parts of the world, people in Africa are working out ways to balance the problems with progress.

Industrial Development

Look again at the map of industrial resources of Sub-Saharan Africa on page 423. What minerals are found near the city of Johannesburg? Near Dakar? Where is tin mined? Where is oil an important resource?

Copper. Uganda was once an important copper producer. But it has used up most of its copper reserves. Copper now accounts for about 4 percent of Uganda's exports. Zambia and Zaire are larger producers of copper. Zambia has about one fourth of the world's total copper supply. Money from British and American investors has helped develop Zambia's copper mines. A string of mines runs from Zambia into southeastern Zaire. These mines form what is called the Copperbelt. A city has grown up at each of the mining sites. Hydroelectric stations on the upper Zaire and Zambezi rivers power the machines for mining and smelting the copper.

Diamonds. South Africa is the world's largest producer of gem diamonds. It is also the third largest producer of less refined, industrial diamonds, after Zaire and the Soviet Union. Look at the map on page 423. Where else in Sub-Saharan Africa are diamonds found?

Most of Zaire's diamonds are mined near the city of Kananga. Find this city on the map on page 417. There are also diamond fields in Angola. Check the National Profiles chart on pages 418–420. What percentage of Angola's exports do diamonds make up?

Gold. For many years, gold was the leading export in countries near the Gulf of Guinea. Liberia and Ghana still produce some gold. But now most of Africa's gold comes from the Rand area in South Africa. The Rand, which includes the cities of Johannesburg and

The copper industry in Zaire was developed mainly by Belgians, who ruled the country for many years. Belgians still manage this plant, but Zairian workers now receive a share of the profits.

Pretoria, is the most important mining and industrial area in Africa. More than three million people live in the Rand.

Much of the Rand's gold has already been mined. But gold reserves there are still large enough to support mining for many more years. Some of South Africa's mines are the deepest in the world, reaching 11,000 feet (3,333 m) into the earth.

Coal and iron ore. In general, Africa lacks good supplies of coal. There is some coal, however, available in Zambia, South Africa, and Zimbabwe. Most of it lies near the surface, so it is easily mined. Coking coal from southwestern Zimbabwe and South Africa finds important uses in Zimbabwe's iron ore smelting industries. Other iron ore producers are Liberia and Sierra Leone.

Oil. Only within the last twenty years have some Sub-Saharan African countries discovered and exploited their oil resources. Nigeria is now a major oil exporter and has become a member of OPEC. Other oil producers are Gabon, Congo, and Angola. How do you think oil production has affected the economies of these countries?

Handicaps to full industrial development. Look again at the map on page 423. Railroads are fairly well developed in southern Africa. But elsewhere, especially in the forest countries, poor transportation remains a major handicap to industrial growth. Undeveloped sources of power, especially waterpower, are another problem. Conflicts within and between nations have also slowed down industrial growth. Political problems in Zaire, Angola, South Africa, Mozambique, and Nigeria have often centered around conflicting claims on these countries' mineral reserves and power sources. These conflicts make investments risky. Without invested money for development, production is sometimes limited.

Africa's handicaps limit manufacturing as well as mining. The major manufacturing plants are those that process the area's resources for export. These include copper smelters and refineries in Zambia, Zimbabwe, and Zaire, an aluminum plant in Ghana, and an oil refinery and rubber-processing plant in Nigeria. Most of Sub-Saharan Africa's exports go to European countries, the United

Workers assemble a drilling platform at Port Harcourt, Nigeria. To tap its petroleum reserves, Nigeria needs more skilled workers.

States, and Japan. In 1975, the European Economic Community (Common Market) and forty-six African, Caribbean, and Pacific countries made agreements to ease restrictions on trade among themselves. International trade within Africa is limited. Most African countries have extensive trade with the European powers that colonized them earlier.

Many larger cities also have manufacturing industries that produce goods for local use. Factories producing beer, furniture, and shoes are fairly common in African cities.

Only the country of South Africa can be called a fully industrialized nation. It lacks many of the handicaps that have limited African industry in general. South Africa has other problems, however. You will learn more about South Africa later in this lesson.

New Patterns of Life in Industrialized Areas

Mining and urban areas have provided many new jobs for Africans. People seeking these jobs often come from the grasslands, sometimes travelling hundreds of miles. The families they leave behind usually continue to farm the land in the traditional ways. Usually it is the men who seek the jobs; women and children stay behind.

The men usually work at industrial jobs until they have earned their desired amount of money. Many of these workers spend almost two thirds of their adult lives outside of their tribal areas. When they return, brothers or cousins often take their place in the industrial area.

Even while they are away, these migrant workers keep close ties with their tribe. Usually they form neighborhoods with other migrant tribesmen working in the same area. Sometimes they have a chief related to the tribal chief back home. They often speak their own language. But, at the same time, they are absorbing many aspects of their city's culture. They learn new job skills, new ways of living, and sometimes new languages. When they do return to their tribal homes, they bring these new ways with them.

Migrant workers often cross borders in search of work. Many miners in South Africa come from Lesotho, Botswana, Swaziland, and Malawi. Some people from Malawi and Zambia work in Zimbabwe. Sudan's cotton plantations attract workers from Niger and Chad. And Ghana and Ivory Coast draw workers from Nigeria and Upper Volta. Through these workers, money from the modern economy reaches the traditional farming areas. European-styled clothing and small home appliances also reach even the most remote villages.

Africa's cities. Some of Africa's large cities grew up as mining centers. Others were built by European powers as the seat of their colonial governments. The centers of these colonial cities consist of office buildings, stores where European goods are sold, and international banks and agencies. During Africa's colonial period, only white Europeans lived in this central area. Even now, many whites still live in these areas, along with the top government leaders and workers. Less important government workers live in outlying areas.

Many colonial governments developed systems of public education, medical care, modern sewage disposal, and subsidized housing in Africa. After independence, African nations took control of these costly services. The people hired to work them are usually those educated in European ways.

Slums are growing in Africa's cities. The cities cannot provide enough jobs to employ all who seek them. Many of Africa's unemployed are well-educated people. Their farming families worked hard to send them to school. If they return to the farm, they are sometimes regarded as failures. So they make a poor living at odd jobs, sometimes unloading ships, sometimes selling trinkets to tourists. Some grow vegetables in a vacant lot and sell them in the city market. Many of the slum dwellers spend most of their free time in the lobby of government buildings. There they wait for news about job openings within the government.

African governments are trying to clear the slums. Some have tried moving the slum dwellers to the cities' outskirts. Others have ordered them to return to their tribal villages. But many return to the cities. The recent drought in western Africa brought many more starving people into the cities. There they hoped to get some of the food donated from other countries. Many who survived the drought stayed in the city even after the drought was over.

Keeping Facts in Focus

1. What mineral discoveries brought many Europeans to Africa in the late 1800s?
2. Name three of Africa's chief mineral exports. Where are they mined?
3. What Sub-Saharan African nation is an OPEC member?
4. What handicaps limit Africa's industrialization?
5. Would you agree that Africa's migrant workers live in two cultures at once? Explain your answer.
6. Describe a typical African city that was once the seat of colonial government.

Patterns of Government in Sub-Saharan Africa

Before the Europeans arrived, patterns of government in Africa were very diverse. Some of the governments were kingdoms. In the forest areas, kings came to power by claiming to have spiritual and religious powers. In kingdoms in south and east Africa, cattle owners often ruled over the poorer farmers. Many grew wealthy from the farmers' labor. There were also areas that had no central government. In these areas, small, local groups had decision-making powers on local issues. At times, these small groups would meet to work out decisions affecting larger areas.

The arrival of the Europeans brought an end to most of these patterns of government. After the long years of colonial rule, newly independent African nations adopted governments based on European models. There is still some diversity among African governments. But they all share the feature of having one strong, central authority. The chart on this page shows the main types of African governments today.

INDEPENDENT REPUBLICS	MONARCHIES	MILITARY GOVERNMENTS
Most African republics have a very strong chief executive, usually a president. Most also have a popularly elected parliament and only one official political party. (Gambia and Senegal are exceptions; they have several active political parties). The amount of government involvement in the economy varies from country to country. Mozambique has a totally socialist economy; Ivory Coast encourages private as well as foreign investment and ownership. Most republics have a mixture of government-owned and private industries, although subsistence agriculture is the main occupation. Governments supply limited public services.

Where practiced (selected countries): Botswana, Cameroon, Gabon, Ivory Coast, Kenya, Liberia, Mozambique, South Africa, Tanzania, Zaire, Zambia. | Emperor or king has most of the political power. King sometimes rules by decree, with or without advice of council of appointed ministers. Kings can, and have, suspended constitutions, claiming even more power for themselves. Number of political parties varies. Government provides very few public services. The few industries are usually in private hands.

Where practiced: Lesotho, Swaziland. | Rulers came to power through military takeover of government. Most military governments have strong executives supported and advised by military leaders and armed forces. Most have only one political party; some have none. In those countries that have parliaments, ruling party leaders control parliament. In some countries, the parliament as well as the constitution has been suspended, leaving military leaders to rule by decree. Opposition is not tolerated. In some cases, such as Nigeria and Ghana, the military government has promised to return the country to civilian rule in a short time. Government usually owns most industries.

Where practiced (selected countries): Benin, Burundi, Congo, Ethiopia, Ghana, Nigeria, Somalia, Sudan. |

Dutch colonists set up vineyards in Cape Town, South Africa, in the 1600s. Wine is still an important product from the coastal region today.

South Africa, a European Island in Black Africa

No other African country shows its European influence as much as South Africa. Although the majority of its population is black, the whites living there hold most of the positions of power. South Africa is wealthier and more industrialized than any other African nation. Its towns and cities are very much like those in Great Britain and the United States. Its exports make up about a third of Africa's total export trade. Customers for South Africa's exports include Great Britain, the United States, Japan, Western European countries, and nearby African nations. South Africa produces over half of the continent's electric power. It has almost half of Africa's telephones and cars and has Africa's best transportation network.

Even South Africa's farming reflects European ways. Farmers raise cattle and sheep on very large ranches, sometimes as big as 10,000 acres (4,000 ha). Unlike most herders in the grassland areas, South African herders have permanent grazing lands.

The Mediterranean climate on South Africa's coast has encouraged such crops as citrus fruits and olives. The early foreign settlers brought their European farming methods with them, including techniques for processing. Jams, jellies, and wines are some of the food products from South Africa's coastal area. In the non-white areas, blacks use traditional farming methods that often result in low yields.

It was the area's great mineral wealth that first brought large numbers of Europeans to South Africa. Today, South Africa leads the world in the production of gold, platinum, gem diamonds, and antimony. It also has large supplies of other minerals, including uranium, manganese, and asbestos.

In an effort to reduce its need for imports, South Africa has also developed many manufacturing industries. Chief among them is smelting. South Africa also produces gasoline from coal. But it still depends heavily on the Middle East for most of its gasoline. Engineering and manufacture of machinery are also well developed in South Africa.

Crowded living and poor farming conditions in the areas reserved for blacks attract many South Africans to jobs in the gold mines (left). Black workers often live in housing provided by the mining companies (right).

Apartheid. While industry has made South Africa wealthy, not all South Africans have a share in the wealth. The nation is sharply divided between the whites and blacks. Apartheid (ə·pär′tāt), separation by race, is the law in South Africa. Blacks in South Africa make up most of the working class. But they are not allowed to vote. They are not allowed to live in "white only" areas. They may not own land except in places reserved for blacks. Most of the land in such reserves is unsuitable for farming or other purposes. Until 1979, blacks could not join labor unions or compete for better jobs. They still may not marry whites. The freedom of South Africa's other non-white citizens, including Asians and people of mixed races, is also strictly limited.

South Africa's white government is firm in its policy of apartheid. It encourages programs that keep development of black citizens separate from that of whites. For example, the South African government foresees a time when each of the ten "all black" reserves will become an independent country with self rule. The United Nations does not support this future for South Africa. It does not believe that South Africa will withdraw its control of these areas, even after their independence. Also, these areas are too poor to support large populations. For this program to work, the South African government would have to make huge investments to irrigate, fertilize, and industrialize these areas. Such investments seem unlikely.

In protest of South Africa's apartheid, many countries around the world refuse to have extensive trade with South Africa. Some athletes have refused to participate in international events in which a South African team will also compete. So far, outside pressures have not per-

Independent Namibia, once controlled by South Africa, now claims control over Walvis Bay, Namibia's only deep-water port. This peaceful inlet overlooks Walvis Bay's Namib Desert.

suaded the white South African government to abolish its policy of apartheid. But tensions within South Africa continue to mount. In June of 1976, riots broke out in several parts of the country. About 600 people were killed. Most of them were blacks.

The best of European influence in Africa has brought many improvements. These include disease control, public services, and greater industrial wealth. But the worst of European influence has brought racism and injustice. South Africa is not the only country with racial problems. Violence between blacks and whites has scarred the histories of Kenya, Zaire, and Zimbabwe. After the African majority (97 percent of the population) gained political control in Kenya, many of the Europeans fled the country. When political and racial tensions flared in Zaire in the 1960s, many Belgians also fled. Not until April 1980 did Zimbabwe become an independent nation under the control of its black majority.

Many people hope that Africans of all colors and backgrounds in whatever country they reside will one day be guaranteed their basic human rights.

Keeping Facts in Focus

1. What countries are South Africa's chief trading customers?
2. How does South African livestock raising differ from that in other grassland areas?
3. What is South Africa's chief manufacturing industry?
4. What is apartheid?
5. What answers would you give to the key questions on page 433?

Working with Ideas

Why do some South Africans support their country's policy of apartheid? Why do others oppose it?

Unit 13 REVIEW WORKSHOP

Test Your Geographic Knowledge

A. Choose the letter of the item that correctly completes each statement.

1. Much of Sub-Saharan Africa's interior is made up of (a) fertile plains, (b) deserts, (c) plateaus and hilly uplands.
2. The Niger River empties into (a) Lake Victoria, (b) the Gulf of Guinea, (c) the Mediterranean Sea.
3. The most common land use in Sub-Saharan Africa is (a) traditional farming and grazing, (b) forestry, (c) grazing.
4. Most Africans of European descent live in (a) Uganda, (b) Mozambique, (c) South Africa.
5. By 1914 all the following were under colonial control except (a) Zimbabwe, (b) Ethiopia, (c) South Africa.
6. Savannas are (a) cold, mountain areas, (b) grasslands near the desert, (c) grasslands near the rain forests.
7. The tree that produces gum arabic is the (a) mango, (b) acacia, (c) baobab.
8. The main cash crops of the grasslands are (a) peanuts and cotton, (b) sorghum and millet, (c) corn and wheat.
9. Farming in Ethiopia is being affected by (a) modern farming methods, (b) heavy rainfall and rich soil, (c) desertification, the advancing desert.
10. Dried fish, kola nuts, and gold are typical products from the (a) forest countries, (b) desert countries, (c) grassland countries.

B. On a separate sheet of paper, fill in the word or words that best complete each statement.

1. Cassava is an important forest crop because it can tolerate _____.
2. The Copperbelt runs from Zambia into southeastern _____.
3. In general, Africa lacks good supplies of _____, an important fuel source for industry.
4. _____ in western Africa is a major oil exporter and a member of OPEC.
5. Most of Sub-Saharan Africa's exports go to European countries, the United States, and _____.
6. In many large African cities, local manufacturers produce beer, furniture, and _____.
7. Some African cities grew up as mining centers; others were the seats of _____ governments.
8. South Africa produces over _____ of the continent's electric power.
9. The _____ climate on South Africa's coast supports such crops as citrus fruits and olives.
10. Separation by race, or _____, is the law in South Africa.

Apply Your Reading Skills

A. Use the National Profiles chart on pages 418–420 to name the country that each of the following phrases describes.

1. The most densely populated country
2. The country with the highest per capita income
3. The country most dependent on fish for its exports
4. The country whose capital city is Bangui
5. The country with the highest literacy rate
6. The country with the lowest population growth rate

B. Using library resources for information, write a report on one of Sub-Saharan Africa's capital cities. Khartoum, Dar es Salaam, and Nairobi have especially interesting histories. Try to cover both the problems and achievements of the capital you choose.

Apply Your Geographic Skills

Use the maps on page 443 to find answers to these questions.

1. What is the land surface like in Kenya's largest farming area?
2. What physical features help explain why there are also farming areas on Kenya's southeast coast?

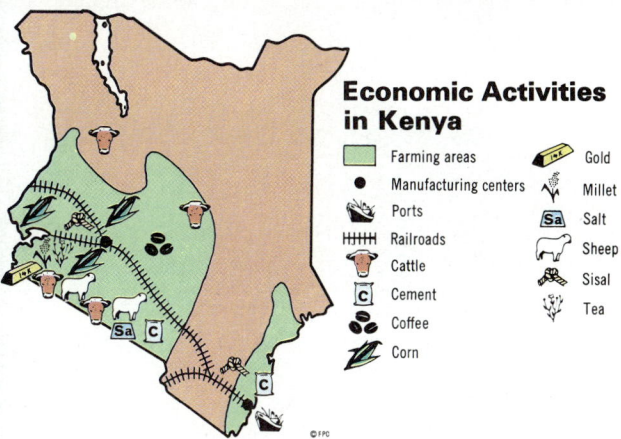

Economic Activities in Kenya

- Farming areas (green)
- ● Manufacturing centers
- Ports
- ┤┤┤┤ Railroads
- Cattle
- C Cement
- Coffee
- Corn
- Gold
- Millet
- Sa Salt
- Sheep
- Sisal
- Tea

©FPC

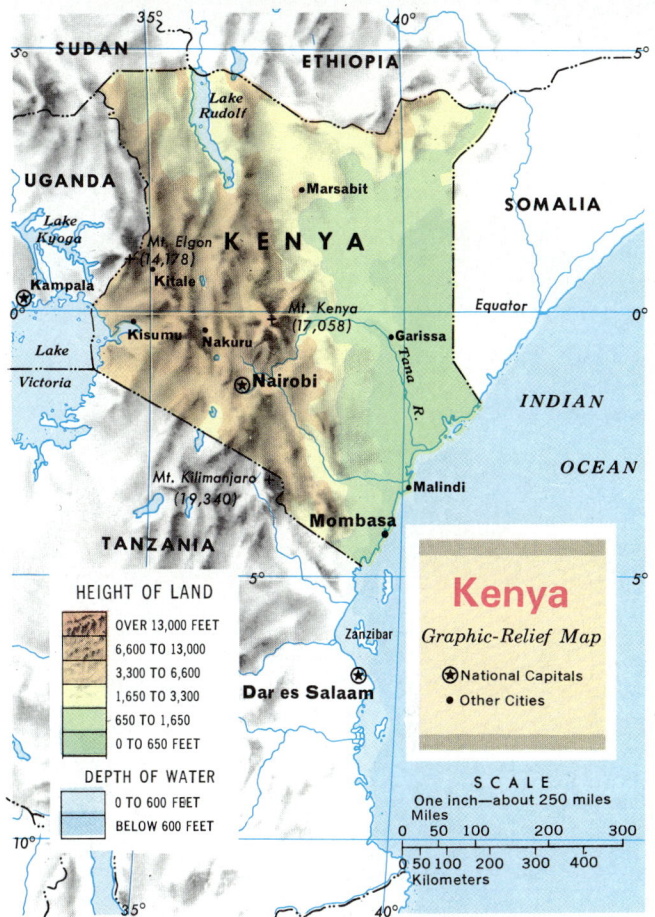

Kenya
Graphic-Relief Map

⊛ National Capitals
● Other Cities

HEIGHT OF LAND
- OVER 13,000 FEET
- 6,600 TO 13,000
- 3,300 TO 6,600
- 1,650 TO 3,300
- 650 TO 1,650
- 0 TO 650 FEET

DEPTH OF WATER
- 0 TO 600 FEET
- BELOW 600 FEET

SCALE
One inch—about 250 miles
Miles
0 50 100   200   300
0 50 100 200 300   400
Kilometers

3. What is the chief agricultural activity near Mt. Kenya?
4. In what other ways besides farming can Kenyans earn a living?
5. What two transportation routes connect Nairobi and the Indian Ocean?
6. Where are most of Kenya's mineral reserves —in the highlands or in the lowlands?
7. What possibilities for trade with other nations does Kenya's location offer?

Apply Your Thinking Skills

Listed below are some features of Sub-Saharan Africa. On a separate sheet of paper, indicate whether the phrase in column A or the phrase in column B better describes each feature.

Feature	Column A	Column B
1. soils	rich and fertile	poor in humus
2. rainfall	very light	generally unpredictable
3. waterpower	good but undeveloped supply	few available sources
4. minerals	abundant throughout the region	absent in some places
5. industry	centered on mining	centered on manufacturing
6. occupations	mainly farmers	mainly industrial workers

Discuss These Points

A British writer of the last century described Africa as the "white man's burden." He felt that it was the duty of developed civilizations to educate Africans in the ways of modern life and to convert them to Christianity. Many people, however, feel that it is wrong and disrespectful to impose foreign ways of life on people. But even many of these people recognize that the European colonists did achieve some important improvements in Africa, such as disease control and developed transportation routes. On the whole, do you think the improvements brought by the colonists outweigh the cultural problems that came with them? Do you think it is possible to bring such improvements to undeveloped regions without also imposing foreign values and ways of life? Discuss.

Expand Your Geographic Sights

Brooks, Lester. *Great Civilizations of Ancient Africa* (School Book Service, 1971).

Foster, Blanche. *East Central Africa* (Watts, 1981).

MacClintock, Dorcas. *African Images: A Look at Animals in Africa* (Scribner, 1981).

Murphy, E. Jefferson. *Understanding Africa* (Crowell, 1969).

Russell, Vivian. *Life in Africa* (Viking, 1976).

Trupin, James E. *West Africa: Ancient Kingdoms to Modern Times* (Parents Magazine Press, 1971).

443

444

Unit 14 Latin America

The empire of the Inca flourished in the Andes Mountains until the arrival of conquering Spaniards in the 1530s. The ancient city of Machu Picchu in Peru, an example of the great building and engineering skill of the Inca, attracts many visitors every year.

Lesson 1 Exploring Latin America

Reading Focus

1. The pronunciation key that introduces the Glossary at the back of your book will help you learn how to pronounce new words. The key lists the phonetic symbols that stand for certain common sounds. Use the key to discover which words these phonetic spellings represent:

   kôz′ wā   ə·lōō′ mə·nəm
   bôk′ sīt   hyōō·mid′ ət·ē

   The phonetic spellings may look difficult at first. But once you become familiar with the symbols, they will help you pronounce each new word you encounter. How would you write the phonetic spelling of the word *estuary*?

2. As you read this lesson, look for the answers to these key questions:
   a. What physical features of Latin America form barriers to development?
   b. What are the major river systems of South America?
   c. What features moderate Latin America's climates?
   d. What are Latin America's chief racial and cultural groups and in what order did they arrive in the region?

Vocabulary Focus

estuary	*llano*	*astronomy*
Pampas	*mestizo*	*dictator*

Parts of Latin America are in the midst of sweeping changes. Industry is helping to create modern cities and bring economic advancements. Some Latin Americans are also beginning to enjoy social and political advancements. Older ways of life are giving way to modern ideas.

But there are also many parts of Latin America that have remained isolated from modern changes. To this day, there are some people in Latin America who will live their whole lives without ever voting in an election, operating a ma-chine, or reading a newspaper. Many are poor farmers, working the sometimes difficult land much as their ancestors did. It may be many years before Latin America bridges the gap between modern city life and the traditional ways of the countryside.

The Physical Setting

Some of Latin America's physical features help explain the gap between traditional and modern ways. Many of these natural features work in combina-

Latin America
Political-Relief Map

⊛ National Capitals   ★ Other Capitals
• Other Cities

SCALE

Miles
0   200   400   600   800   1000

0   400   800   1200   1600
Kilometers

tion with one another to create areas of isolation and barriers to development.

Location. Look at the world graphic-relief map on pages 12–13. Latin America's most northerly country is Mexico. What river forms a border between the United States and Latin America? Latin America stretches south from this river to the southern tip of South America.

Countries. There are thirty independent countries in Latin America. Look at the map on this page. Which is the largest country? Some Latin American countries, such as Cuba, Jamaica, and Barbados, are islands. Find these countries on the map on this page. What bodies of water surround Cuba? The string of islands between Florida in the United States and Venezuela is called the West Indies.

447

National Profiles: *Latin America*

Independent Nation	Area/ Population	Population Density/ §Growth Rate	*Capital/ Largest City and Population	Urban Population/ Literacy Rate	Chief Exports (in order of Importance)	†GNP (in) millions of U.S. dollars)/ Per Capita Income
Antigua	440 sq km 80,000	181 per sq km 1.3%	*St. John's 25,000	34% 88%	clothing, rum, lobsters	$125 $1,650
Argentina	2,711,300 sq km 30,097,000	11 per sq km 1.6%	*Buenos Aires 2,908,000	86% 94%	meat, corn, wheat wool, hides	$130,000 $4,610
The Bahamas	11,396 sq km 228,000	20 per sq km 2.0%	*Nassau 138,500	58% 89%	pharmaceuticals, cement, rum	$1,400 $6,000
Barbados	430 sq km 252,000	586 per sq km 0.3%	*Bridgetown 7,600	37% 99%	sugar, sugar cane, electrical parts, clothing	$997 $3,977
Belize	22,973 sq km 158,000	7 per sq km 2.1%	*Belmopan 4,500 Belize 39,887	50% 80%	sugar, fish, molasses, citrus	$169 $1,120
Bolivia	1,098,160 sq km 6,037,000	5 per sq km 2.6%	*La Paz 881,400 *Sucre 66,300	33% 75%	tin, crude oil, lead, zinc, silver, tungsten	$5,600 $933
Brazil	8,521,100 sq km 134,380,000	16 per sq km 2.3%	*Brasília 1,776,908 São Paulo 7,032,547	56% 74%	coffee, iron ore, machinery, soybeans & soybean products	$295,000 $2,360
Chile	756,626 sq km 11,655,000	15 per sq km 1.5%	*Santiago 4,085,000	82% 90%	copper, metal ores, paper products	$23,600 $1,950
Colombia	1,139,600 sq km 28,248,000	24 per sq km 2.1%	*Bogotá 4,486,200	69% 81%	coffee, fuel oil, cotton	$40,000 $1,435
Costa Rica	51,000 sq km 2,693,000	44 per sq km 2.6%	*San José 867,800	43% 93%	coffee, bananas, sugar	$3,300 $1,390
Cuba	114,478 sq km 9,995,000	86 per sq km 1.1%	*Havana 1,924,886 '81	67% 96%	sugar, copper & nickel ores, shellfish, tobacco	$14,900 $1,534
Dominica	790 sq km 74,000	94 per sq km 0.2%	*Roseau 20,000	27% 95%	bananas, coconut oil & other oils, cocoa	$57 $883
Dominican Republic	48,692 sq km 6,416,000	130 per sq km 2.7%	*Santo Domingo 1,313,172	52% 68%	sugar, nickel, coffee, tobacco, cocoa	$7,600 $1,400
Ecuador	274,540 sq km 9,091,000	33 per sq km 3.1%	*Quito 918,900 1,278,900	43% 84%	crude oil, bananas, coffee, cocoa, fish & fish products	$13,300 $1,507
El Salvador	21,400 sq km 4,829,000	222 per sq km 3.0%	*San Salvador 440,000	41% 65%	coffee, cotton, sugar	$3,600 $700

§Growth Rate: Includes natural increase and migration.

†GNP: Gross National Product, or the total value of all the goods and services produced by the people of a country within one year.

N/A: Data not available

National Profiles: *Latin America*

Independent Nation	Area/ Population	Population Density/ §Growth Rate	*Capital/ Largest City and Population	Urban Population/ Literacy Rate	Chief Exports (in order of Importance)	†GNP (in millions of U.S. dollars)/ Per Capita Income
Grenada	344 sq km 113,000	328 per sq km 1.3%	*Saint George's 7,500	15% N/A	cocoa, nutmeg, bananas, mace	$70 $850
Guatemala	108,880 sq km 7,956,000	73 per sq km 3.1%	*Guatemala City 1,307,300	36% 50%	coffee, cotton, sugar, bananas	$8,600 $1,114
Guyana	214,970 sq km 837,000	4 per sq km 0.5%	*Georgetown 170,000	30% 85%	bauxite, sugar, rice, alumina shrimp, molasses	$430 $539
Haiti	27,713 sq km 5,803,000	212 per sq km 2%	*Port-au-Prince 763,000	21% 23%	coffee, light industrial goods, bauxite	$1,500 $300
Honduras	112,150 sq km 4,424,000	39 per sq km 3.4%	*Tegucigalpa 502,500	31% 47%	bananas, coffee, beef, ores	$2,800 $710
Jamaica	11,422 sq km 2,388,000	197 per sq km 1.7%	*Kingston 684,000	37% 76%	alumina, bauxite, sugar	$3,000 $1,360
Mexico	1,978,800 sq km 77,659,000	39 per sq km 2.6%	*Mexico City 9,191,300	65% 74%	cotton, coffee, minerals, shrimp, petroleum	$168,000 $2,273
Nicaragua	147,900 sq km 2,914,000	20 per sq km 3.6%	*Managua 552,000	40% 66%	cotton, coffee, chemicals, meat	$2,500 $846
Panama	75,650 sq km 2,101,000	28 per sq km 2.1%	*Panama City 655,000	50% 85%	petroleum products, bananas, sugar	$3,945 $1,934
Paraguay	406,630 sq km 3,623,000	9 per sq km 2.7%	*Asunción 455,517	42% 81%	cotton, soybeans, timber, oilseeds	$5,800 $1,411
Peru	1,284,640 sq km 19,157,000	15 per sq km 2.6%	*Lima 3,100,000	73% 72%	copper, fish & fish products, other metals, coffee	$19,200 $1,018
St. Christopher and Nevis	261 sq km 44,000	167 per sq km −0.3%	*Basseterre 14,700	N/A 80%	sugar	$416 $920
St. Lucia	616 sq km 120,000	194 per sq km 1.1%	*Castries 45,000	17% 78%	bananas, cocoa	$121 $980
St. Vincent and the Grenadines	389 sq km 138,000	354 per sq km 3.2%	*Kingstown 23,200	N/A 82%	bananas, arrowroot, copra	$69 $539
Surinam	142,709 sq km 370,000	3 per sq km 1.8%	*Paramaribo 67,700	44% 80%	alumina, bauxite, aluminum, rice, wood	$1,044 $2,916
Trinidad and Tobago	5,128 sq km 1,168,000	227 per sq km 1.6%	*Port-of-Spain 47,300	49% 89%	crude oil, petroleum products	$7,316 $6,651
Uruguay	186,998 sq km 2,926,000	16 per sq km 0.3%	*Montevideo 1,255,600	83% 94%	wool, hides, meat, clothing, textiles	$9,400 $3,201
Venezuela	911,680 sq km 18,552,000	20 per sq km 3.1%	*Caracas 2,700,000	75% 86%	crude oil, petroleum products, iron ore	$69,300 $4,716

A chain of volcanic highlands runs through much of Central America. In the Colombian lowlands (right), the Amazon River has flooded a thick forest.

Mountains. Look closely at the landforms of the Western Hemisphere on the world graphic-relief map (pages 12–13). Both North and South America have high, rugged mountains on their western sides. In Mexico, the Sierra Madres split into the Western and Eastern ranges. Find these ranges on the map on page 447. Between these mountains lies a high plateau. What is the name of the mountain range forming South America's western edge? This range forms the longest mountain chain in the world.

Throughout history, the Andes Mountains have been a barrier to communication among regions of South America. Look at the graphic-relief map on page 29. What are the heights of Mounts Huascaron, Chimborazo, and Illimani? Mt. Aconcagua, at 22,834 feet (6,850 m) above sea level, is the highest peak in the Western Hemisphere. Travel over these rugged mountains is difficult.

Highlands. In Mexico, the highland plateau bordered by the Sierra Madres fills most of the country's width. In southern Mexico, another highland region near Guatemala forms the foothills of the mountains that stretch through Central America.

There are also large areas of highlands in South America. On the map on page 29, find and name the highland region in Venezuela. The Brazilian Highlands in eastern South America are made up of low mountains, older and more worn than the Andes in the west. The highlands in Patagonia in the south of the continent have somewhat higher hills. Because Patagonia receives very little rain, the hills there are less eroded than in wetter highland areas. Sheep grazing is common in Patagonia.

Plains and rivers. Most of Mexico's lowland plains lie along the coast of the Gulf of Mexico. There are also stretches of plains along Central America's Caribbean coast. Look at the map of land use on page 30. How do Central Americans along the Caribbean coast use their land?

Most of the rivers in Mexico, Central America, and the West Indies are short, swift, and generally unnavigable. The Rio Grande, on its course along the U.S.-Mexican border, runs through deep, narrow canyons. Only in some places in the east does the river valley widen enough to allow some irrigated farming.

In South America, on the other hand, there are several great river systems, each flowing through wide stretches of lowlands. The Amazon, the second longest river in the world, is navigable for about 2,000 miles (3,200 km). But the Amazon is surrounded by a huge, untamed, tropical rain forest. Very few people live in this hazardous, isolated area. Since it flows through such a sparsely populated and undeveloped region, the Amazon has few practical uses for irrigation.

Another great river system, the Río de la Plata, is in central South America. Find the Río de la Plata on the map on page 29. It is actually an estuary, a finger of the sea, rather than a river. What rivers flow into the Río de la Plata? These rivers, navigable for hundreds of miles, were very important trade routes during the colonial period. The ports of Buenos Aires and Montevideo have become important centers for trade today.

The port of Buenos Aires lies at the edge of a vast area of grassland plains called the Pampas. Find the Pampas on the map on page 29. Unlike most other regions of Latin America, the Pampas enjoy a combination of flat lands, rich soils, and a favorable climate.

A third important river in South America is the Orinoco in Venezuela.

AVERAGE TEMPERATURE: BELEM AND QUITO

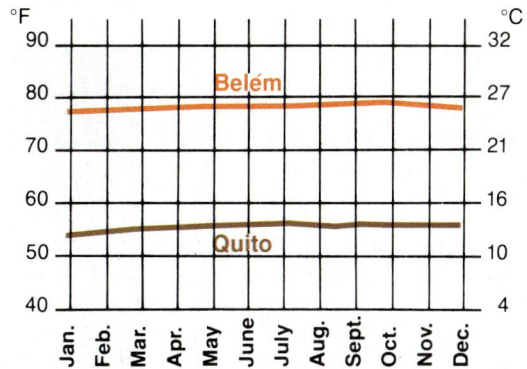

Find Belém, Brazil, and Quito, Ecuador, on the map on page 447. What do you think accounts for the difference in their temperatures?

Look at the map on page 29. Into what body of water does the Orinoco flow? North of the Orinoco are the llanos (län' ōs), flat grasslands that cover about one third of Venezuela's area. For many years, the sparsely populated llanos were used mainly as grazing lands. Herders there, however, were always at the mercy of the region's frequent floods and droughts. In the 1950s, the Venezuelan government began to build a series of dams to regulate the region's supply of water. In addition to making herding easier, water from these dams has made large areas of once-dry lands suitable for farming.

Climates. The map on page 200 shows the diversity of climates within Latin America. Where are the areas with the coldest temperatures?

Most of Latin America lies within the low latitudes, where high temperatures are common. But high altitudes in much of the region keep temperatures fairly cool.

Another feature that moderates Latin American climate is the Peru Current

A bulldozer crew lays out a route through the Amazon rain forest for a highway connecting Brasília with Belém on Brazil's coast.

along South America's west coast. The waters of the Peru Current keep temperatures between northern Chile and the equator cooler than average for the low latitudes.

The Peru Current also helps create the Atacama Desert on South America's western coast. Winds crossing the cold current from the west lose all their moisture over the water. Some people in the Atacama have never seen rain.

Population. Latin America's 393 million people come from a variety of racial and cultural backgrounds. People of European descent make up the largest group. Another group, the Indians, numbers over twenty million. A third group is the mestizos, people of mixed Indian and European backgrounds. There are also many Latin Americans with mixed black and white ancestry.

Only about 40 percent of Latin America's land is suitable for farming. Some areas are too steep, too swampy, too cold, or too dry to encourage settlement. In more favorable areas, Latin Americans have tended to settle in fairly isolated clusters, separated by sparsely populated regions. Recently, modern means of transportation and communication have brought people in different clusters into closer contact. But there are still many isolated communities in Latin America.

Latin America's population is growing quickly, at a rate of about 2.3 percent a year. The population in Latin American cities is growing at an even faster rate, close to 5 percent a year. Such a rapid urban growth rate makes it almost impossible for cities to keep up with demands for decent housing and sanitation, good schooling, and jobs that pay fair wages. Slum areas have spread out in many of Latin America's major cities. Still, many people moving to the cities feel that even life in an urban slum is better than the poor farming life they left behind.

Keeping Facts in Focus

1. What mountain range forms South America's western edge?
2. What river drains most of Venezuela?
3. How does the Peru Current help create the Atacama Desert on South America's western coast?
4. What are mestizos?
5. What percentage of Latin America's land is suitable for farming?
6. Why are some areas of Latin America unable to support many people?

The People's Heritage

Between 20,000 and 40,000 years ago, humans first began to settle in the Western Hemisphere. These early peoples, the ancestors of American Indians, probably came from Asia across the Bering Strait into what is now Alaska. Gradually, these early hunters spread southward. By about 6,000 to 10,000 years ago, human settlements had reached the southern tip of the South American continent.

Some of the early Indian tribes were nomads; others developed stable farming methods. Early farmers in Mexico were the first to domesticate maize. Throughout Latin America, farmers also grew squash, beans, potatoes, and tobacco.

Several advanced cultures grew up in Latin America. By 600 B.C., the Maya had built a highly developed civilization in Central America. Their efficient farming methods and stable government allowed the Maya time to create great buildings and beautiful works of art. They also developed an accurate calendar and achieved a good understanding of astronomy.

Another great Indian culture, that of the Aztec, developed in what is now central Mexico. And the civilization of the Inca grew up in the steep valleys and high plateaus of the central Andes. The Maya, Aztec, and Inca all became wealthy from the gold and silver they mined in their regions.

Europeans in the Americas. Soon after the voyage of Columbus to the Americas in 1492, other explorers, both Portuguese and Spanish, began to ar-

Gauchos, the earliest cowboys in the Pampas, left a rich folklore in Argentina and Uruguay.

rive in Latin America. Early European explorers convinced their governments that the "new world" they had discovered was full of riches. Before long, both Spain and Portugal had claims over large parts of Latin America. European conquerors, armed and on horseback, proved too strong for the Indians, and the ancient empires crumbled.

As Europeans began setting up plantations and mining for metals, they tried to force the Indians to work for them. Many Indians died from overwork and from diseases Europeans brought with them. Those who survived gradually adopted the customs, language, and religion of their European conquerors. Some European men married Indian women, blending the races as well as the cultures. Well-planned towns sprang up, each with a government building and Catholic church in its center.

The dwindling Indian population became too small to provide the labor for European plantations and mines. In the early 1600s, Europeans in Latin America began to bring in black African slaves for labor. From Africa, many slaves were shipped to the West Indies. There they were taught about Christianity and made to learn the basics of

Catholics near Cuenca, Ecuador, carry branches in observance of Palm Sunday.

some European language. Once "seasoned," they were sold all over the Americas.

The struggle for independence. For nearly 300 years, Latin American lands were colonies of European powers. Indians, blacks, and people of mixed races resented their second-class lives. They had to work hard on land they could never hope to own and pay heavy taxes to their European governments. Rebellions broke out in the early 1800s. Under such leaders as Simón Bolívar, José de San Martín, Miguel Hidalgo, and Bernardo O'Higgins, many Latin American lands finally achieved political independence.

A continuing struggle. Independence did not solve all the problems. Tensions between members of the rich, upper class and the poor, landless workers continued. Frequent revolutions and violent border disputes made governments unstable. Some people began to feel that the Catholic church enjoyed too many privileges. Throughout the 1800s and into the 1900s, Latin Americans fought for reforms. Gradually, democracy became more widespread, more people were able to own land, and the church gave up much of its power. Industry helped create a growing middle class in Latin America, as well as a stronger economic base.

The struggles continue. Military dictators control some Latin American countries. Others are in upheaval. Social unrest and poverty are widespread. Land redistribution and other programs of the Organization of American States (OAS) and the Latin America Free Trade Association (LAFTA) are of some help. Latin American nations face major challenges as growing populations increase the difficulties of achieving stability.

Keeping Facts in Focus

1. Where did the Inca civilization develop? Where did the Aztec live?
2. What were some achievements of the Maya?
3. What problems did Latin American nations face after independence?
4. What answers would you give to the key questions on page 446?

Working with Ideas

In what ways is the physical and cultural development of Latin American countries similar to that of the countries of Sub-Saharan Africa? In what ways is it different?

Lesson 2 Mexico and Central America

Reading Focus

1. In this lesson, the discussion of agriculture in Mexico fills eleven paragraphs. The details in this discussion are important. But the general idea they are supporting is even more important. The information in these eleven paragraphs can be summed up in this way: "Land reforms have distributed Mexico's scarce arable land fairly among the farmers. But there is still a shortage of land. In the north, lands are too dry for much farming. The highland slopes and lowland swamps also have farming problems. The southern part of the central plateau is the best farming region. It is also the most densely populated." This summary leaves out some important details. But it does help focus the main idea of the discussion. As you finish each main part of the lesson, try to summarize the information you have just read to form a general idea that is easy to remember.

2. As you read this lesson, look for the answers to these key questions:
   a. In what ways is the Mexican government trying to improve its nation's farming conditions?
   b. What are the major export crops of Central America?
   c. What features have slowed industrial growth in Central America?
   d. What are Mexico's chief industrial products?

Vocabulary Focus

hacienda	*henequen*	*fluorite*
ejido	*maguey*	

Mexico and the countries of Central America together form a region called the Central American Mainland. The countries in this region, most of which were colonies of Spain, share many similarities in history, culture, and landforms. But there are also important differences, especially between Mexico, the largest country in this region, and the Central American countries. Use the map on page 447 to find and name the seven independent nations of Central America.

The Central American Mainland is, on the whole, a rugged, mountainous region. Within the mountains, however, there are many plateaus, some made fertile by rich volcanic ash. A few wide basins and river valleys also relieve the rugged landscape.

Only a few areas are lowlands, and these are not ideal for human settlement. They are generally very hot, and north of the Tropic of Cancer they are also very dry. South of the Tropic, rains are much heavier, but swamps and dense forests make the land hard to adapt for human use.

The physical features of the Central American Mainland do create some problems for the 100 million people living there. Only a small percentage of the land is suitable for farming. Even the suitable lands require much effort to maintain. The mountain slopes make soil erosion a problem for many farmers. The mountains also make trade and travel difficult, isolating many people. On the other hand, mountain streams supply the region with sources of waterpower and irrigation. And the region's many areas of high altitude enjoy a cool and healthful climate. In Mexico, the mountains provide still another important benefit: minerals. Mexico, the most industrialized country on the Central American Mainland, owes much of its wealth to its abundant supply of minerals.

Agriculture: Mexico and Central America

Most people in Mexico and Central America are farmers. The region's wide range of altitudes and climates allows farmers to grow many different kinds of crops. In most areas, farmers work small plots of land and grow only enough food for their own use. In some areas, however, especially in the low-lands, there are large plantations for commercial agriculture.

Farming and grazing in Mexico. For many years after the Spanish conquest, most of Mexico's farmlands were owned by only a few wealthy landlords. Landless farmers worked on these large estates, called haciendas ((h)äs·ē·en′ dəz), for little or no pay. Often they received only a small share of the crops or a small plot of ground for their own use in exchange for their labor. The hacienda system remained even after Mexico achieved its independence from Spain in 1821. In fact, it was another hundred years before real progress in the form of government land reforms began to improve farmers' lives in Mexico.

Under the land-reform programs of the 1920s, the government broke up the large haciendas into smaller units called ejidos (ā·hē′ thōs). Several families, sometimes even whole communities, own and work the ejidos. On some ejidos, farmers work only their own plot of land. On cooperative ejidos, farmers share the work of the entire farm and also share all profits.

Today much of Mexico's land is back in the hands of the farmers themselves. But there are still many farming problems. One of the biggest is the shortage of arable land. Only about 12 percent of Mexico's land is used for agriculture, or about one and a half acres (.6 ha) per person. The government hopes to increase the amount of arable land to 15 percent of the country's total area. Toward this goal, it is building irrigation dams, clearing swamps, and educating farmers in soil conservation methods.

Northern grazing lands. Much of the northern part of Mexico is dry. Little land can be cultivated so most people use these dry lands to raise cattle. About three fourths of Mexico's total pasture land is located in this northern region. Here ranches are huge. Cattle must graze over great stretches of land to find enough grass for feed.

Irrigation has made some farming possible in this dry zone. A dam on the Colorado River in Baja California has made the once-dry land suitable for growing cotton, vegetables, olives, and grapes. Streams rising in the Sierra Madres are also used to irrigate lands on the central plateau and in the desert region east of the Gulf of California. Commercial fertilizers and modern machinery help some of these irrigated farmlands produce high yields of wheat, cotton, corn, and vegetables.

Farming on highlands. Another difficult farming area is the highland slopes of the Sierra Madre. On the steep highland slopes, most people are subsistence farmers who grow maize and rice. In the few areas of flat valleys, farmers grow cacao and coffee, two commercial crops. Although the soils are fertile in these volcanic highlands, usable land is very scarce.

Farming on coastal lowlands. A rainy hot climate, swamps, and malaria make the Gulf Coastal lowlands difficult to farm. For many years, these conditions kept all but a few people from settling in this region. In recent years, however, the government has built dams, drained swamps, and controlled insects that carry disease in the lowlands. Now peo-

Government irrigation programs have greatly increased agricultural output in Baja California.

ple from all over Mexico are moving into this region to grow sugar, coffee, bananas, and other tropical crops. In the sparsely populated Yucatán Peninsula, henequen is the chief product. Fiber from the henequen plant is used to make rope and twine.

The central plateau. Only the southern part of the central plateau, Mexico's most densely populated region, has excellent conditions for farming. Averaging about 5,000 feet (1,500 m) above sea level, this region enjoys a temperate climate, hot sunshine, sufficient rains, and fertile soils.

In some parts of this region, as in most places in Mexico, farmers still use the traditional methods of their ancestors. Most use only wooden plows, oxen, and sharp sticks to cultivate the land and plant corn, squash, and beans.

Costa Rica was the first Central American country to begin growing coffee. Today coffee accounts for 18 percent of the nation's foreign earnings. The large plantations are in the highlands, where volcanic soils enrich the flavor of the beans. Workers pick the beans by hand and gather to unload them for shipment to the processing plant (right).

Even in this good farming region, harvests are sometimes poor. Since the region is overcrowded, most farming plots are very small. Some of the land is worn out from overuse. Farmers must find ways, therefore, to use all their available land. They graze sheep, goats, and pigs in the high mountain pastures. In dry, rocky fields, farmers can grow maguey (mə·gā′). The sap from this plant is made into a drink popular among the Indians.

Although some people in this region are subsistence farmers, there is also much commercial farming. Near Mexico City, large, irrigated fields grow corn, cotton, and wheat. Commercial farmers often use modern methods and machinery. Improved roads help these farmers transport their goods to the many food-processing industries in the area.

Agriculture in Central America. About two thirds of Central America's people earn their living by farming. As in Mexico, people in Central America cluster in the highlands, avoiding the humid coastal areas. Many are subsistence farmers who grow maize, rice, and

The red outer layer of the beans is removed before the beans are washed and dried. After roasting, the beans turn a dark brown color.

cassava. They often use traditional but effective farming methods that help reduce soil erosion on the slopes.

Commercial farming in the highlands centers on coffee. Large coffee plantations cover much of the highlands and plateaus in Central America, especially in Guatemala, Costa Rica, and El Salvador. Many subsistence farmers add to their small incomes by working during harvest on these plantations.

Ten percent of the world's coffee comes from Central America. Much of it is sold to the United States and West Germany. Money from the sale of coffee beans buys Central America's much-needed imports, such as manufactured goods and some food products. When coffee crops are large, the countries can afford to import what they need. But in a bad year, when frost or other problems ruin coffee crops, the whole region suffers. To reduce their dependence on coffee, some governments are trying to encourage farmers to grow other products for export. Cotton has become an important cash crop in recent years, as has sugarcane, rice, and henequen. The cattle industry is also growing rapidly. Nicaragua is the leading cattle country in Central America.

Bananas are another important export crop in Central America, especially in the coastal lowlands. Bananas from

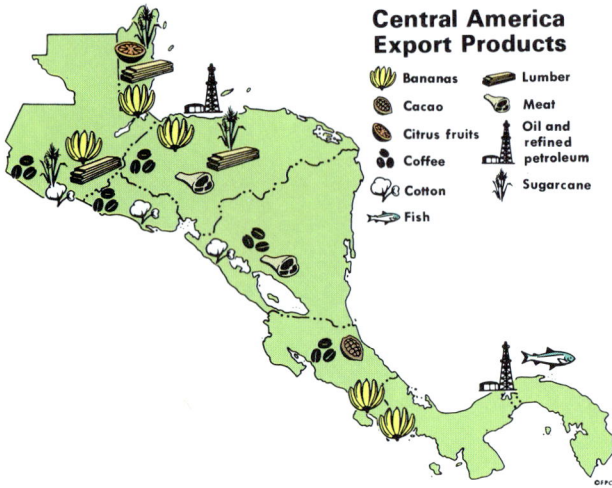

Central America Export Products

- 🍌 Bananas
- 🥥 Cacao
- 🍊 Citrus fruits
- ☕ Coffee
- Cotton
- 🐟 Fish
- Lumber
- Meat
- Oil and refined petroleum
- Sugarcane

Central America make up about 13 percent of the world's total supply. Central America also has many thick forests that provide income for some people, especially in Guatemala, Belize, and Honduras. Trees from these forests provide the raw materials for such products as medicines, perfumes, soap, chewing gum, and kapok, a cottonlike fiber.

Industry in Central America

To strengthen their economies, the Central American countries need to grow a wider variety of crops. They also need to develop manufacturing industries to reduce their dependence on imports. Generally lacking in mineral wealth, the Central American countries must look for other industrial resources to help balance their primarily agricultural economies.

Only Nicaragua and Honduras have significant mineral deposits. In 1968, gold, silver, lead, and zinc were discovered in northern Nicaragua. Mining is therefore playing a growing part in this nation's economy. Honduras discovered its supply of minerals, especially silver, much earlier. By now, however, the silver deposits are seriously depleted. Honduras is turning to smaller industries, such as sugar refineries and factories that produce cement, beer, and soft drinks.

Poor transportation in both Honduras and Nicaragua has slowed their industrial growth. To improve transportation, Nicaragua is now opening its San Juan River to commercial traffic. It is also building an inland canal system. A hydroelectric station on the San Juan River powers Nicaragua's sugar refineries, textile mills, and cement plants.

Manufacturing in El Salvador, the most industrialized Central American

Tropical diseases, such as malaria and yellow fever, were great obstacles to the building of the Panama Canal. Only after the swamps were cleared and mosquitoes controlled could the work on the canal be completed.

Industries of Mexico

Coal
C Copper
General Manufacturing
I Iron Ore
L Lead
M Manganese
Oil
Pottery
Shipping
S Silver
Steel Products
Su Sulphur
Textiles
Z Zinc
Tourism

Five Major Industries of Mexico: 1980s
(billions of U.S. dollars)

	0	5	10	15	20	25	30	35	40
Agriculture, forestry, and fishing	██████████████								
Manufacturing	████████████████████████████████████								
Mining and metallurgy	███████████								
Petroleum	████████████████████████								
Trade, restaurants, and hotels	██████████████████████████████████								

country, accounts for about 17 percent of the gross national product. A power station on the river Lempa fuels El Salvador's textile and food-processing plants and other light industries. Costa Rica, second only to El Salvador in industrial development, also has plentiful supplies of waterpower. Its main industrial products are textiles, chemicals, and leather goods.

Panama's economy is unique. Much of its income depends on services it provides to people using the Panama Canal. Find this canal on the map on page 447. What bodies of water does it link? This seaway, completed in 1914 by the United States, is vital for both commercial and military traffic. Until recently, the United States had continuing control over the Panama Canal. But new treaties between the United States and Panama, signed in 1977, grant the Panama government complete control over the canal beginning in the year 2000. The Panama government will then be able to keep all the tolls collected from ships using the canal. This increased revenue will help Panama develop its other sources of income, which include oil and sugar refineries, paper mills, and cement and food-processing plants.

Industry in Mexico

Mexico is far ahead of its Central American neighbors in industrial development. It is the world's largest producer of fluorite, used in glassmaking, and graphite. It is also well supplied with other minerals, including silver, lead, copper, and zinc. Petroleum is also abundant along Mexico's Gulf Coast. Foreign investors developed many of these coastal oil reserves in the early 1900s. But by 1938, the Mexican government had taken over control of producing and marketing all Mexican oil. Look at the graph on this page. How much money do the petroleum, mining, and metal industries earn for Mexico? Recently, lower oil prices have caused problems for Mexico, especially in bringing in enough revenue to pay back large loans in foreign banks.

Monterrey (mänt′ə·rā′), the nation's third largest city, is near deposits of coal, iron ore, and limestone. All of these are used in Monterrey's modern steel mills. Monterrey's industries produce over half of Mexico's iron and steel. Textile mills, cement plants, and other industries in Monterrey have attracted many rural people seeking jobs.

Some of the copper processing in Mexico is still done on a small scale. Where is the nation's major deposit of copper?

Police officers direct traffic in Mexico City. The population of Mexico City's urban area may be greater than 30 million in the year 2000.

Mexico City, with an urban population of 17 million is the largest urban area in the world. About 60 percent of Mexico's industrial workers live in or near Mexico City. Textile industries, food processing, paper, and wood-pulp industries employ thousands of workers. But many people seeking jobs in Mexico City are unskilled. Some cannot find jobs, and slums are a growing problem.

Lingering Problems

While industry has helped create huge, modern cities, many Mexicans, especially Indians, still live traditional, isolated lives. As in Central America, illiteracy is still a problem. The governments of Mexico and Central American countries do require all students to attend school. But they lack the funds to enforce this requirement and provide enough schools and teachers for all students. Even where there are schools, many children do not attend. Instead, they must find jobs to help support their families.

Land reform in Mexico has increased the number of landowners. In Central America, however, there are still too few people who own the land they work. As a result, poverty is widespread, and political unrest has increased. In Nicaragua and El Salvador, for example, more and more people are demanding changes in their governments and a fair share of their countries' wealth.

Keeping Facts in Focus

1. Why are the Gulf Coast lowlands difficult to farm?
2. What are ejidos?
3. What is the most industrialized country on the Central American Mainland?
4. In or near what city do about 60 percent of Mexico's industrial workers live?
5. What answers would you give to the key questions on page 455?

Working with Ideas

1. How will diversification of agriculture aid many of the countries on the Central American Mainland?
2. In what ways might Mexico be considered an industrialized nation? A developing nation?

462

Lesson 3 The West Indies

Reading Focus

1. Throughout this book, some of the words listed under the **Vocabulary Focus** are names of minerals and other raw materials. Some minerals, such as gold and silver, are valuable as precious metals. Others are valuable as the raw materials from which certain industrial products can be made. Knowing the possible uses of a certain mineral will help you understand its value. The names of three minerals are listed in this lesson's **Vocabulary Focus.** Look these words up in the Glossary. If there were local manufacturing industries near deposits of silica, gypsum, and asphalt, what kinds of goods do you think they would produce? What important product can be made from bauxite? In each lesson, make sure you understand the industrial uses that the region's raw materials have. With such an understanding you will be able to predict what products, if any, are produced locally or determine why the mineral is a valuable export.

2. As you read this lesson, look for the answers to these key questions:
   a. What island grouping has the largest islands?
   b. What jobs does tourism create for the people of the West Indies?
   c. In general, what is the standard of living in the West Indies?
   d. In what ways is Puerto Rico different from its neighbors? In what ways is Cuba different?

Vocabulary Focus

mulatto	*gypsum*	*asphalt*
commonwealth	*silica*	

The region called the West Indies is a chain of islands stretching between North and South America. It divides into three main island groupings: Greater Antilles, Lesser Antilles, and The Bahamas. Look at the map on page 464. Which island grouping contains the largest islands and countries? In which grouping are the islands that form Trinidad and Tobago? Which island grouping is made up of only one country?

In all, there are over 7,000 islands in the West Indies. Some of them are little more than rocks rising above the surface of the sea. Like the islands of Oceania, West Indian islands were formed in several different ways. Some of the large islands, such as Cuba and Jamaica, are the tops of sunken mountain ranges. Many islands in the Lesser

Major Island Groups

Bahamas

Greater Antilles

Lesser Antilles

©FPC

Island Types

Coral Limestone

Limestone and Granite

Coral Limestone

(Tops of mountain range)

Volcanic

Coastal

©FPC

Population

Persons Per Square Mile

- Less than 300
- 300 to 700
- Over 700

BAHAMAS

CUBA

JAMAICA

HAITI

DOMINICAN REPUBLIC

PUERTO RICO

TRINIDAD AND TOBAGO

©FPC

Many islanders in The Bahamas make a living selling handmade goods to tourists.

Residents and Tourists

Unlike their Mexican and Central American neighbors, most people of the West Indies are blacks or mulattoes. Mulattoes are people of mixed black and white backgrounds. Even in Cuba, where whites make up the majority of the population, about a quarter of the people are blacks or mulattoes.

But the African heritage is only one of the cultural influences that have shaped the West Indies. People of Spanish and Indian descent in the Dominican Republic, for example, have kept many of the traditions of their ancestors. Neighboring Haiti has a mixed French and African culture. While most Haitians are

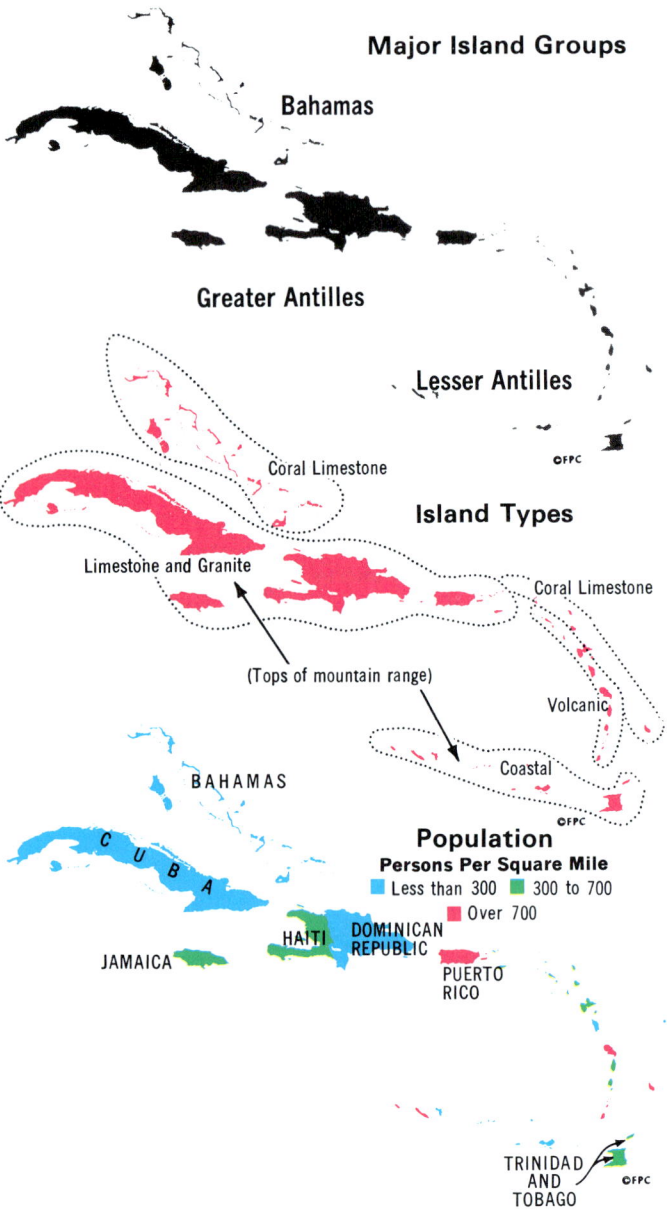

Antilles were formed by volcanoes that built up layers of lava that eventually rose above the water's surface. And most islands in The Bahamas are coral islands, formed by a buildup of limestone from the skeletons of coral. The illustration on this page shows the different island types in the West Indies.

Nassau, on New Providence Island in The Bahamas, is a favorite destination of luxury cruisers. Large, modern hotels line the shores of many islands in The Bahamas.

blacks, their official language is French. French influence is also strong in Guadaloupe and Martinique. And in Puerto Rico, a commonwealth of the United States, U.S. influence is strong.

The millions of tourists who come to the West Indies every year also influence ways of life on the islands. Modern hotels and restaurants now line the white, sandy beaches of many West Indian cities. The warm climate and beautiful mountains, forests, and beaches attract many visitors, especially in the winter. Tourists spend over 300 million dollars every year in The Bahamas alone. Jamaica and Puerto Rico also attract huge numbers of tourists. Many islanders earn their livings by working in hotels and resorts. Others sell souvenirs and hand-crafted products to the eager tourists. For most of these islands, tourism adds to the national income produced by the exports of their major products.

Agricultural Products

In addition to tourism, agricultural products are another important source of income to the West Indies. Some of these products are listed on the graph on page 466. What is the most important export crop?

The Dominican Republic, where the world's largest sugar mill is located, and Cuba are two of the world's leading producers of sugarcane. The sugarcane harvest lasts about six months. During the harvest season, workers on the sugar plantations can earn enough to support their families. But when the work in the fields is over, about one fifth of all workers in Cuba and the Dominican Republic find themselves unemployed. Many flock to the cities, where they live in slums. People call these months the dead season. In Cuba, the only communist country in the Western Hemisphere, discontent during the dead season was one factor that led to the communist takeover.

Plantations in the West Indies also grow large crops of tobacco and coffee. Tobacco thrives in the valleys, while coffee grows well on hillsides and mountain slopes. Bananas and cacao are also plantation crops. Most plantations in the West Indies belong to foreigners or wealthy island landlords. Laboring islanders work the plantations for low

VALUE OF AGRICULTURAL EXPORTS
(in millions of U.S. dollars)

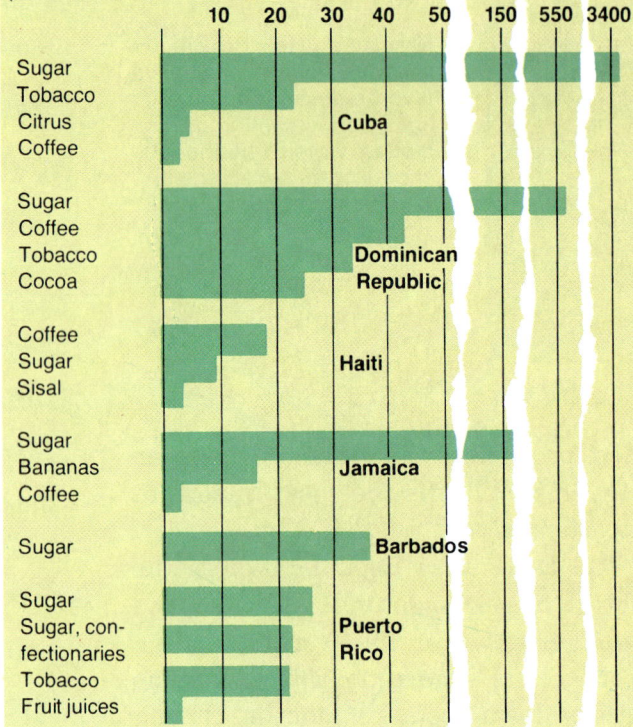

	10	20	30	40	50	150	550	3400
Cuba								
Sugar								
Tobacco								
Citrus								
Coffee								
Dominican Republic								
Sugar								
Coffee								
Tobacco								
Cocoa								
Haiti								
Coffee								
Sugar								
Sisal								
Jamaica								
Sugar								
Bananas								
Coffee								
Barbados								
Sugar								
Puerto Rico								
Sugar								
Sugar, confectionaries								
Tobacco								
Fruit juices								

wages. Only a few own their own small plots of land. Those who do often use traditional farming methods to grow only small yields of corn, beans, rice, fruits, and vegetables. Some also graze cattle. Especially in Haiti, many subsistence farmers grow barely enough food to feed their families. Poverty and illiteracy are widespread.

Industrial Products

The West Indies do not have enough mineral resources, electrical power, or highly-trained workers to support much heavy industry. As a result, most industries in the West Indies revolve around agricultural products grown on the islands. Tobacco crops go to industrial plants that manufacture cigars, cigarettes, and other tobacco products. Other industrial plants prepare coffee, cacao, and tropical fruits for shipment abroad. Food processing is a major industry in The Bahamas. Fishing is also an important industry in The Bahamas, and it is growing in Jamaica and Cuba as well.

Mining is important in only a few West Indian countries. Jamaica is the world's second largest producer of bauxite, used in making aluminum. It also has good supplies of gypsum and silica, from which building materials can be made. In recent years, Jamaica has attracted foreign businesses, many from the United States, that are eager to develop Jamaica's mineral resources.

San Juan, the capital of the U.S. Commonwealth of Puerto Rico, is a major manufacturing center in the West Indies. Many U.S. firms have plants on the densely populated island.

With livestock as their only source of power, many Haitians must use a slow process to press the juices from sugarcane.

Mining is also growing in Cuba. The government-owned mines and mineral reserves have good supplies of nickel ore and limestone. Oil is important to the economies of Trinidad and Tobago and the islands of Curaçao (k(y)o͞or′ə·sō′) and Aruba. The country of Trinidad and Tobago is also a major world supplier of natural asphalt.

Industry plays an especially strong role in the economy of Puerto Rico. In the late 1960s, most Puerto Ricans were farmers. Today, over half of all Puerto Rican workers are employed in industry. As U.S. citizens, Puerto Ricans have received much industrial assistance from the United States. Manufactured goods now account for over three quarters of Puerto Rico's gross national product. Over 2,600 factories in Puerto Rico produce such goods as chemicals, food products, clothing, electrical machinery, and petroleum and coal products.

Puerto Rico's industries have helped it achieve the highest standard of living in the West Indies. With help from the United States, Puerto Rico provides such public benefits as free schooling and low-cost housing to its crowded population. Many people in other West Indian countries, especially Haiti, live very poor lives, in sharp contrast to the luxury surrounding the region's tourists. Unemployment and overcrowding are common problems.

Cubans have become communistic to help ease some of their problems. Although the government is a dictatorship, some progress has been made. Literacy has risen. The country has more skilled workers, and increased trade has improved the country's economy.

Patterns of Government in Latin America

All Latin American countries share a common colonial history. Many also share another common feature: instability. Throughout the region, especially in Central and South America, constitutional governments are prey to military leaders who can organize enough power to overthrow them. Once in power, these military governments are often harsh. As they violate human rights, unrest grows in the country. Sometimes the unrest leads to a revolt that forces a return to civilian rule. But the civilian leaders are often either too weak or too harsh themselves to rule effectively. Before long, the military forces again take power, and the cycle continues.

Some Latin American countries do not follow this cycle. Mexico, for example, and several West Indian countries, are now enjoying stable, democratic governments. Cuba, although under military rule, also has a stable government. The chart below describes both the generally stable governments of Latin America and the more extreme military regimes.

CIVILIAN REPUBLICS	MILITARY REPUBLICS
Chief executive, either president or prime minister, popularly elected from a number of political parties. Representative assembly also popularly elected. Some countries, such as Guyana and Trinidad and Tobago, are independent nations of the British Commonwealth. Government generally tries to protect rights and freedoms of people. Mixed economy: some free enterprise and some government-controlled businesses. Extent of government-provided public services varies according to country's wealth. Most can afford only limited programs of social welfare. Where practiced (selected countries): The Bahamas, Colombia, Costa Rica, Guyana, Mexico, Venezuela.	Constitution often suspended when military comes to power. Military leaders rule with almost total power. Cabinet of Ministers advises military president. Assembly and political parties often dissolved. Little protection of human rights. Opposition not tolerated; political prisoners often tortured. Government often censors press. Extent of government-provided public services varies from country to country. In communist Cuba, the government provides all public services and controls all business. Where practiced (selected countries): Cuba, Nicaragua, Chile, Paraguay, Surinam.

Keeping Facts in Focus

1. What are the names of the three island groupings in the West Indies? What island types are found in each group?
2. What answers would you give to the key questions on page 463?

Working with Ideas

1. Why is tourism so important in the West Indies?
2. Compare Cuba and Puerto Rico. How do the two countries differ in their attempts to solve problems?

Lesson 4 South America

Reading Focus

1. Effective writers are careful to make their purposes clear. Without actually saying "The purpose of these paragraphs is . . . ," they still try to focus their ideas on one main point that serves a specific purpose. Sometimes the purpose is to introduce a topic. At other times, the writer's purpose is to prove a point. And sometimes the author's purpose is to summarize or conclude a discussion. What main purpose do the first two paragraphs of this lesson serve? What main point do they make? Try to look for the author's purpose whenever you read so that you can decide whether or not the purpose is successfully achieved.

2. As you read this lesson, look for the answers to these key questions:
   a. How do reform programs in South America compare to those in Mexico?
   b. Where are most of South America's huge agricultural plantations?
   c. In what areas is mineral wealth concentrated in South America?
   d. What problems face South America today?

Vocabulary Focus

llama	*wilderness*	*sodium nitrate*	*nationalize*
coca leaves	*tungsten*	*capital*	*inflation*

Twelve independent countries and the colonial territory of French Guiana make up the vast continent of South America. The countries vary greatly in size. Surinam, covering roughly the same amount of land as Wisconsin, is dwarfed by Brazil, which covers an area larger than the mainland United States. South America's population is also varied. In several countries, including Peru, Bolivia, and Paraguay, half or more of the people are Indians. Throughout the continent, there are also many mestizos. Along South America's northeast coast, many people are blacks or mulattoes. Many of their ancestors were African slaves sent to South America in the 1600s to work the large sugarcane plantations. In Argentina, Uruguay, and southern Brazil, all but about 10 percent of the people are of European descent, especially Spanish and Portuguese. After World War II, immigrants from Germany, Japan, and China also came to this region.

South America also reveals economic diversity. Most of its 269 million people are very poor farmers who battle hunger and disease every day. Only a few South Americans are extremely wealthy.

The mountains of Ecuador make transportation difficult. As a result, many people in Quito, the capital city, must meet their needs by buying and exchanging locally produced goods.

These people are the owners of huge businesses, including banks, industries, and agricultural estates. In some South American countries, such as Venezuela, a middle class of professional workers is also on the rise.

Agricultural Patterns in South America

South America is well known for its several important agricultural exports, including coffee, meats, sugar, and bananas. These exports come from large estates. But in most South American countries, estates cover less land than farms that produce maize, beans, rice, and cassava for local use. Despite the many acres used for subsistence farming, each farmer's plot and yields are small.

Life for subsistence farmers. Many poor farmers have never been exposed to modern changes. They often live in isolated clusters in mountain basins or along mountain streams. They use only simple tools, such as sticks for planting and carts for hauling. In the higher mountain regions, pack animals such as the donkey and llama provide the only source of transportation. Children suffering from lack of food chew on coca leaves to stop their hunger pains. Many die before they reach the age of five. Reform programs have usually neglected the subsistence farmers.

Commercial farming. Most of the huge commercial estates are in South America's eastern countries. In the western countries, the estates are generally smaller. Commercial crops grown in the lowlands of Guyana, Colombia, Ecuador, Peru, Bolivia, and Chile include sugarcane, cotton, and bananas. Many South American countries depend on one agricultural product for their chief export. From the charts on pages 448-449, name the countries whose chief export is an agricultural product.

Irrigated vineyards in some parts of Chile and Argentina produce grapes. Columbia, Ecuador, and Venezuela also produce cacao. Cacao also grows in Brazil's coastal lowlands, along with tobacco, bananas, and sugarcane.

Most of Brazil's commercial farming, however, centers on coffee. Brazil and Colombia are the world's leading producers of coffee. Brazil alone supplies almost half the world's coffee. The United States is the biggest single customer of Brazilian coffee.

The llanos in northeastern Colombia are well suited to cattle raising. Cowherders lead livestock into a corral to be marked for ownership.

Most of Brazil's coffee plantations are in the eastern part of the country, north and west of its largest city, São Paulo. The rich, red upland soils of this region are ideal for coffee. On most plantations, large landowners hire tenant farmers to work some land for a few years. The tenants' job is to clear the land and plant and care for the coffee trees. During the few years before the trees start bearing beans, the tenants can grow their own food in between the rows of coffee or on small plots nearby. Once the beans appear, the tenants must move on and look for work on other plantations. Landowners then hire migrant workers to harvest the coffee beans.

Reforms of this hacienda system have begun in some South American countries. In some places, farmers are paid in wages so they are not tied to the plantation. Plantation workers in some countries, such as Peru, have formed unions to help them raise their pay and improve their working conditions. While they are still too poor to own much land of their own, many plantation workers make a much better living than the remote subsistence farmers.

Farming and grazing in the Pampas. The fertile Pampas of Argentina and Uruguay support several different kinds of agriculture. Near Buenos Aires is a large region devoted to truck, fruit, and dairy farming. Maize is the chief crop near Rosario, while wheat grows well in areas further west. In the areas too cool or dry for grain, pasture grasses thrive. Huge livestock ranches help produce thousands of tons of meat exports. Argentina is a world leader in beef exports and depends on them for about a third of its income. The National Profiles chart on page 449 shows that Uruguay depends on animals and animal products as its chief exports.

Livestock raising is important in other parts of South America as well. In

South America Export Products

- 🍌 Bananas
- B Bauxite
- 🟤 Cacao
- ☕ Coffee
- C Copper
- 🧵 Cotton
- I Iron ore
- 🥩 Meat
- ⚙️ Machinery
- N Nitrates
- 🛢️ Oil
- 🌿 Sugarcane
- T Tin
- 🌽 Wheat
- 🪵 Wood
- Wool

©FFC

some mountain basins in the Andes, livestock raising supplements subsistence farming. Livestock grazed in the northeastern region of Colombia is an important part of that nation's economy. In the irrigated central valley of Chile west of the Andes, there are also many large livestock ranches. Patagonia in southern Argentina is a large sheep-grazing area. Argentina is one of the world's leading producers of wool.

Natural Resources

South America has many areas of wilderness, some still unexplored. One of the largest areas of wilderness is the Amazon Basin, a huge tropical rain forest. It was not until the mid 1970s that the Brazilian government explored and mapped this vast region. The explorers found a long river formerly unknown, as well as signs of rich mineral deposits. Many such areas of the continent remain to be explored and developed. Until they are, all of South America's natural resources cannot be known.

Many of South America's more accessible forests provide income for some countries. Thirty percent of Paraguay's economy depends on lumber. Forests of mahogany and cedar supply wood for Colombia's lumber industry. Northeastern Brazil also has huge forests whose trees supply materials for many valuable products.

The waters surrounding South America are another valuable resource. Fishing for ocean trout, tuna, and sardines is an important coastal industry. Chile has a growing fishing industry, and Peru's fishing industry is the largest in the Southern Hemisphere.

Mountains of minerals. The largest area of mineral deposits in South America is a zone in the Andes that runs from Colombia to central Chile. Colombia has large gold deposits, as well as most of the world's supply of emeralds. Ecuador also has large gold deposits, although mining employs only one percent of Ecuador's workers. Peru is among the world's leading producers of lead, copper, zinc, and silver. It also has good reserves of iron ore. Poor transportation, however, makes it very costly to ship these minerals from their sources high in the mountains. Bolivia is second only to Malaysia in tin production, sup-

Chile is the world's largest exporter of copper. Copper from the vast mines of Chuquicamata is processed in nearby smelting plants. A hydroelectric station in the Andes Mountains powers the region's copper mines and refineries.

Vast oilfields lie under Lake Maracaibo in Venezuela. Since the first drilling there in 1917, Venezuela has become a major oil producer.

plying more than one tenth of the world's total. Bolivia also has copper, silver, zinc, and tungsten. Chile, at the southern end of this zone, exports more copper than any other country. It also has the world's only deposits of natural sodium nitrate. Foreign companies control a great part of this region's mining activities.

Another source for minerals is the Guiana Highlands. Find this region on the map on page 29. Gold and diamonds lie within these highlands in Venezuela. There are also large reserves of iron ore, developed mainly by American capital. Guyana and Surinam are leading world producers of bauxite.

A third large area rich in minerals is the Brazilian Highlands. For many years, gold and diamonds were the chief mineral products from this area. In 1910, scientific reports revealing the area's huge reserves of iron ore and manganese brought many more miners to Brazil. Although poor transportation

from the highlands is still a problem, Brazil has become a leading supplier of iron ore and manganese.

Fuels in the lowlands. Coal is generally lacking in South America. Small amounts of low-grade coal appear in southern Brazil, central Chile, and Colombia. Oil, however, is abundant in several countries. The huge petroleum reserves in and near Lake Maracaibo in Venezuela make this nation, a member of OPEC, the eighth largest oil producer in the world. Argentina has recently become the second largest oil producer in South America. Ecuador, also an OPEC member, is the third largest oil producer in South America. The valley of the Magdalena River in Colombia is another important source of petroleum. Petroleum has also been discovered in Argentina, Peru, and Bolivia.

South America also has fields of natural gas. But they are often far from the cities where gas could be used commercially. In Argentina, a pipeline over

Rio de Janeiro, regarded by many as one of the most beautiful cities in the world, was the capital of Brazil until 1960.

1,000 miles (1,600 km) long transports natural gas from its source in Patagonia to Buenos Aires. There it is widely used for power.

Rapidly Growing Industries

Most South American countries still depend on the export of raw materials, such as food and metals, for most of their income. In some countries, such as Argentina, Uruguay, Peru, and Brazil, food processing, meat packing, and textile plants have developed around these raw materials. In other South American countries, such as Ecuador and Paraguay, a lack of good transportation, fuels, and trained workers has limited manufacturing to a few small shops in the cities. Here craft workers make such items as pottery, tools, hats, and leather goods.

But elsewhere on the continent, manufacturing industries are growing quickly. Steel mills, chemical and cement factories, and wood processing plants are growing in the cities of Argentina, Brazil, and Chile. Huge hydroelectric stations are either planned or under construction in many South American countries. Transportation facilities are being expanded and improved. Industry is playing a greater and greater part in South American economies. There are now about 47,000 manufacturing plants in Colombia. In Chile, industry makes up about one fourth of the gross national product.

Industrial production in Brazil has grown at an especially high rate, but slowed in recent years. The area around São Paulo and Rio de Janeiro is a huge manufacturing district. Brazilian petroleum products, machinery, and such goods as automobiles, furniture, tires, radios and televisions, and clothing are finding eager markets both at home and overseas.

Brazil, like other South American countries, was not highly industrialized before World War II. After the war, South America, as a developing region, began to receive various kinds of foreign economic aid. Loans from the World Bank, a United Nations organization, went for such projects as road building in Colombia and the steel mill at Volta Redonda in Brazil. The United States also gave money to developing South American nations. Under a plan called the Alliance for Progress, the United States agreed to supply 20 percent of the cost of certain development projects in Latin America. Programs such as these speeded industrial growth in South

Brazil's new capital, Brasilia, rises amid the scrub of the country's vast interior. Brazil has the largest debt to U.S. and other banks of any country in the world. Several other South American countries also face heavy loan obligations.

America. But more important than these were the private investments of industrialists from the United States, European countries, and Japan.

The investment of private foreign capital skyrocketed South American industrial production. But it has created many serious problems. Since much of South America's wealth is in foreign hands, governments cannot use it to provide much needed public services and equal opportunities. In some countries, industries originally developed by foreigners have been taken over by the governments. The Peruvian government, for example, nationalized several U.S.-owned industries, including an oil company and a copper mine. To continue developing its resources, however, Peru must still invite foreign investors. American capital continues to be used in developing a new copper mine in Peru. When completed, this mine will increase Peru's copper production by two thirds. But in this case, Americans must share the profits with the mine workers. Over the years, mine workers will come to own half of the company. In most cases of foreign investment, however, there are no such profit-sharing programs.

Many South American problems can be eased only through costly development projects. To solve the serious overcrowding in their cities, for example, South American countries must develop their wilderness areas and try to spread their populations more evenly throughout their lands. Brazil has already done much to develop its interior. It created Brasília, now its capital, far from the crowded coast, to attract people into the unsettled interior. But this project was very expensive. South American governments, already fighting a serious inflation problem, must continue to seek foreign capital.

Keeping Facts in Focus

1. What are South America's chief commercial crops? Which one is Brazil's chief crop?
2. What different kinds of farming take place in the Pampas?
3. Which country supplies most of the world's emeralds? Which leads the world in the export of copper?
4. Which South American countries have excellent supplies of petroleum?
5. How did the Alliance for Progress help South America develop its resources?
6. What answers would you give to the key questions on page 469?

Working with Ideas

What advantages and disadvantages result when much of a country's wealth is in foreign hands? How does nationalization of an industry affect foreign investors?

Unit 14 REVIEW WORKSHOP

Test Your Geographic Knowledge

A. Fill in the word or words that best complete each statement.

1. _____ is the largest country in Latin America.
2. In Mexico, a high _____ lies between the western and eastern ranges of the Sierra Madres.
3. The _____ Mountains stretch along South America's west coast.
4. The South American river that is really an estuary is called _____.
5. The flat grasslands north of the Orinoco River are called the _____.
6. Although most of Latin America lies within the low latitudes, high _____ in many regions keep temperatures fairly cool.
7. _____ are people of mixed Indian and European backgrounds.
8. Latin America's population is growing at a rate of about _____ to _____ percent a year.
9. The best farming region in Mexico is the southern part of the central _____.
10. The most important farm crop in the Central American highlands is _____.
11. Much of Panama's income depends on the _____.
12. Industries in the city of _____ produce over half of Mexico's iron and steel.
13. The Dominican Republic and Cuba are two of the world's largest producers of _____.

B. Tell whether each statement is true or false.

1. Puerto Ricans are citizens of the United States.
2. Communist Cuba has been able to achieve economic independence.
3. Commercial farms cover more land in South America than subsistence farms.
4. Argentina and Chile are the world's leading producers of coffee.
5. Patagonia is important for its many sheep ranches.
6. The largest area of mineral deposits in South America is the Guiana Highlands.
7. The Brazilian Highlands have vast reserves of iron ore and manganese.
8. Venezuela is the only South American country that is a member of OPEC.
9. South American countries depend on the export of manufactured goods for most of their income.
10. The Alliance for Progress is a program through which Latin American countries receive money from European countries for certain development projects.
11. The city of Brasília was created in part to attract people into the sparsely populated interior of Brazil.

Apply Your Reading Skills

A. Match each of the products listed below to the country that is its biggest producer in Latin America.

Products	Countries
1. Petroleum	a. Chile
2. Sugarcane	b. Venezuela
3. Bauxite	c. Colombia
4. Tin	d. Jamaica
5. Emeralds	e. Cuba
6. Copper	f. Bolivia

B. Use library resources to help you write a report on ONE of the following topics.

1. The history of the Cuban revolution and the problems and accomplishments of the Communist government.
2. The activities of the Catholic Church in promoting human rights and freedom in Latin America.
3. Descendants of the Inca living in South America today.

Apply Your Geographic Skills

Look closely at the graph on page 477. Remember that the graph provides information only on the *percentage* of workers in the labor force, not on actual *numbers* of workers. Then decide which of the following questions you can answer with the information provided by the graph. On a separate sheet of paper, write the numbers 1–10. If the graph provides the answer, write the answer next to the number of the question. If you cannot answer the question with information from the graph, put an X next to the number of the question.

1. In which region are about 36% of the workers employed in agriculture, hunting, forestry, and fishing?
2. Which region employs a greater number of workers in construction?
3. Which region employs a greater percentage of its laborers in construction?
4. Which region has more mines and quarries?
5. Which region has more advanced transportation, communications, and utilities?
6. Which region employs a greater percentage of its laborers in service industries?
7. In which region is about 12% of the labor force employed in commerce?
8. Which region produces the most food?
9. In what occupation are about 13% of South America's laborers employed?
10. Which region employs a greater number of workers in manufacturing?

Percent of Labor Force Employed in Certain Occupations*

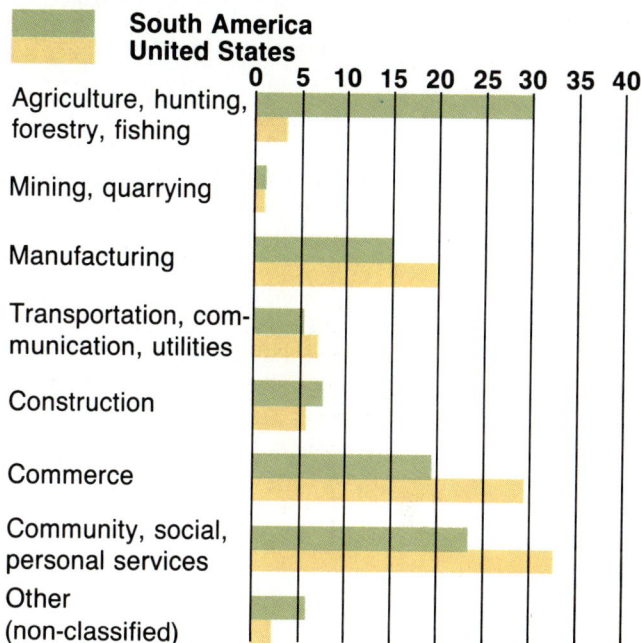

*Excluding French Guiana, Guyana, and Surinam; percentages do not add to 100.00 due to rounding.

Apply Your Thinking Skills

Imagine that you are the owner of each of the following businesses. Refer back to the lessons to determine from which Latin American countries you could buy goods and materials to use in your business. What goods and materials could you buy from these countries?

1. Glass-making factory
2. Cement factory
3. Meat-processing plant
4. Wool refinery and mill
5. Wine-making plant
6. Petrochemical plant
7. Aluminum plant
8. Furniture factory
9. Heavy machinery factory
10. Candy company

Discuss These Points

When asked in 1978 what South America's greatest problem was, Bolivian president General Hugo Banzer Suarez replied that it was the "ever-increasing population. This forces the use of funds for education and infant health which could otherwise be used to create new jobs." Do you agree that this is South America's greatest problem? What measures can South American governments take to keep pace with the growing population?

Expand Your Geographic Sights

Carter, William E. *South America* rev. ed. (Watts, 1983).

Fincher, E. B. *Mexico and the United States: Their Linked Destinies* (Harper and Row, 1983).

Fisher, John K. *Latin America: From Conquest to Independence* (Day, 1972).

May, Charles P. *Peru, Bolivia, Ecuador: The Indian Andes* (Nelson, 1969).

Sherlock, Philip. *The Land and People of the West Indies* (Lippincott, 1967).

Smith, Eileen L. *Mexico: Giant of the South* (Dillon, 1983).

Williams, Byron. *Continent in Turmoil: Background Book for Young People on Latin America* (Parents Magazine Press, 1971).

Unit 15   The United States and Canada

Niagara Falls, on the Niagara River connecting Lake Erie and Lake Ontario, is one of the most spectacular sights in the world. Tourists can view the roaring falls from either the United States or the Canadian side.

479

Lesson 1 Exploring the United States and Canada

Reading Focus

1. One of the vocabulary words in this lesson is *intermountain*. *Inter-* is a prefix that means *between*. What do you think the word *intermountain* means? Can you think of other words that begin with the prefix *inter-*?

   Like prefixes, suffixes affect the meanings of words. But suffixes come at the ends of words. This lesson, for example, contains the word *strongest*. The suffix *-est* means *the most*. *Strongest*, therefore, means *the most strong*. Think of some other words with the *-est* suffix. Another suffix you will find in this lesson is *-able*. This suffix has the same meaning as the word *able*. What do you think *navigable* means? As you read, use prefixes and suffixes to help you understand the meanings of new words.

2. As you read this lesson, look for the answers to these key questions:
   a. What are the eight natural regions of the United States and Canada? What physical features does each contain?
   b. What is the pattern of population distribution in the United States and Canada? What natural features help explain this pattern?
   c. How did the early Indians in North America make their living?
   d. What are some problems facing the United States and Canada today?

Vocabulary Focus

continental divide	*permafrost*	*bison*
intermountain	*surplus*	

The vast and varied lands of the United States and Canada cover about four fifths of the North American continent. Within their broad expanses lie fertile plains, thickly forested mountains, and rich mineral deposits. Long, navigable rivers cross the region, and the coastal waters brim with fish. Few other countries in the world enjoy the variety and abundance of natural resources that the United States and Canada possess.

The people's heritage is as rich and varied as the resources of their lands. From their colonial ties with Great Britain, the people of the United States and Canada have inherited their democratic traditions and their language. The customs of North American Indians, French Canadians, black Americans, and millions of immigrants from all over the world have also helped enrich the heritage of the United States and

Canada. The varied peoples of the United States and Canada have used their good resources to build two of the world's most prosperous nations.

The Physical Setting

The United States and Canada occupy a physical setting that has generally

National Profiles: *United States and Canada*

Independent Nation	Area/ Population	Population Density/ §Growth Rate	*Capital/ Largest City and Population	Urban Popula- tion/ Literacy Rate	Chief Exports (in order of Importance)	†GNP (in) millions of U.S. dollars)/ Per Capita Income
Canada	9,971,500 sq km 25,142,000	3 per sq km 1.0%	*Ottawa 695,000 Toronto 2,850,000	76% 99%	motor vehicles, metal ores, machinery, newsprint, grain	$288,800 $11,725
United States	9,371,829 sq km 236,413,000	25 per sq km 0.9%	*Washington, D.C. 638,432 New York 7,071,639	79% 99%	machinery, chemicals, motor vehicles, agricultural products	$3,363,300 $12,530

§Growth Rate: Includes natural increase and migration.

†GNP: Gross National Product, or the total value of all the goods and services produced by the people of a country within one year.

favored their development into powerful countries. Their location, size, resources, and climate have helped them achieve strong economies as well as an important place in world affairs.

A key location. The United States and Canada lie in the Northern and Western hemispheres. With Asia to the west, Europe to the east, and Latin America to the south, the United States and Canada have good advantages for trade. At the same time, the wide oceans to the east and west have helped protect the countries from attack.

Enormous size. Look at the graph on this page. What is the only country larger than Canada? How many countries are larger than the United States? What is the approximate area of the United States and Canada combined?

The vast area of the United States and Canada spans a variety of climate

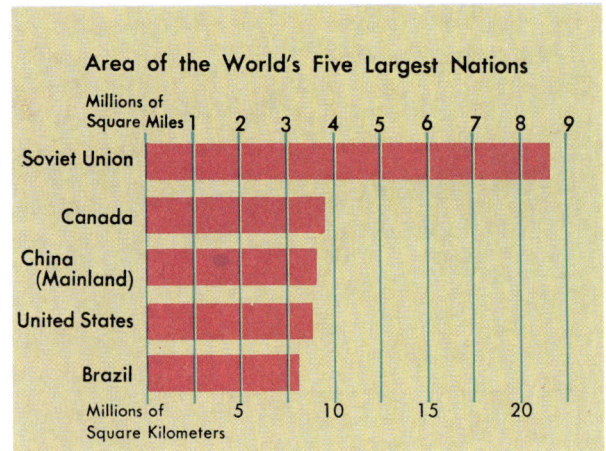

Area of the World's Five Largest Nations

zones and natural regions. In the far north, Canada's Ellesmere Island reaches to within 500 miles (800 km) of the North Pole. Parts of the United States mainland, on the other hand, reach almost as far south as the Tropic of Cancer. Look at the map on page 10. What is the latitude of Hawaii?

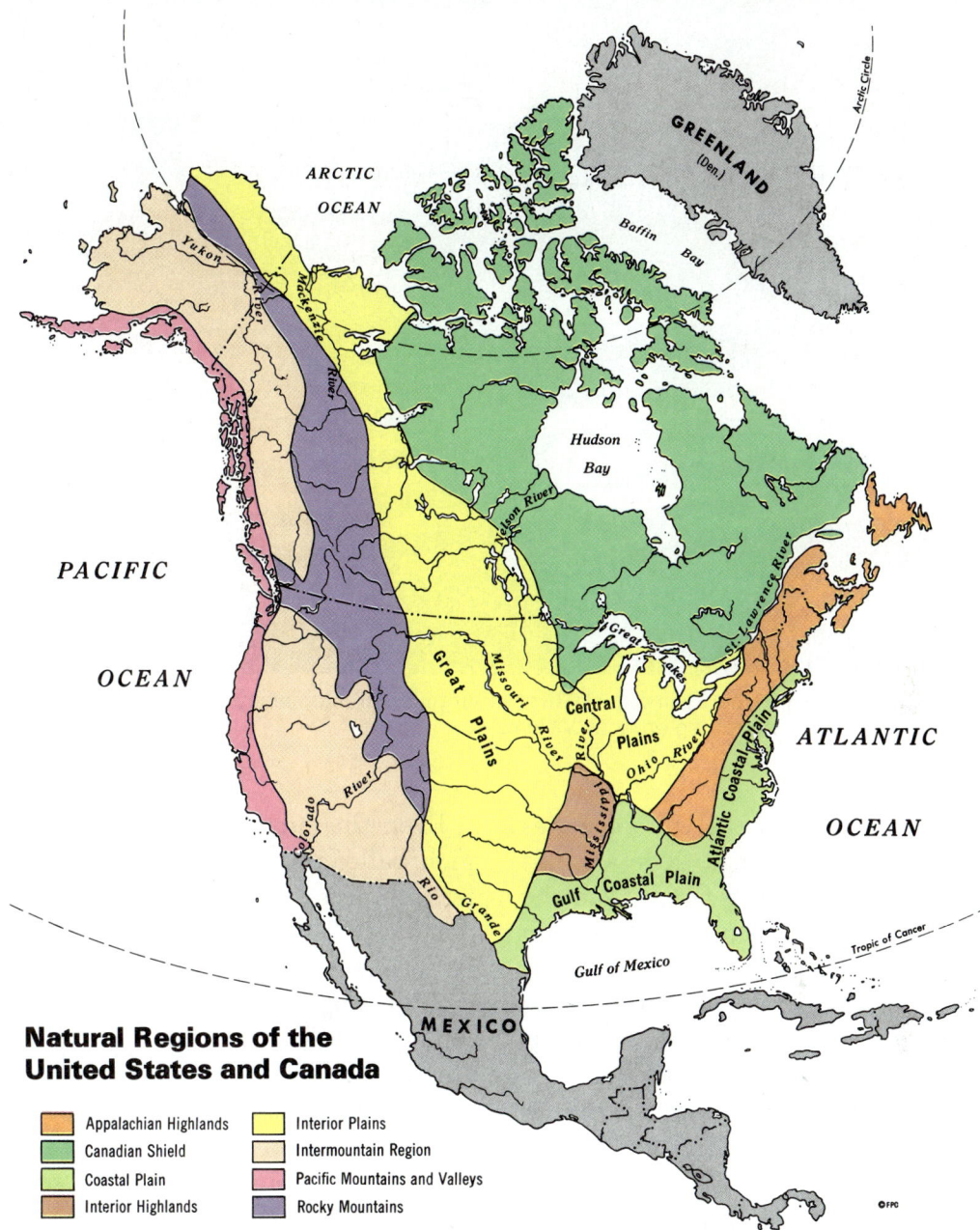

Natural Regions of the United States and Canada

	Appalachian Highlands		Interior Plains
	Canadian Shield		Intermountain Region
	Coastal Plain		Pacific Mountains and Valleys
	Interior Highlands		Rocky Mountains

A variety of natural regions. The map on this page shows the eight natural regions of the United States and Canada. Which six regions spread through both countries?

The oldest land region of the North American continent is the *Canadian Shield*. Long ago, the glaciers that covered much of North America scraped off

the Shield's topsoil. As the glaciers melted, they left nearly two million square miles (5 million sq km) of swamps, lakes, and forests. Because of the lack of soil, the Shield is unsuitable for farming, but it is rich in minerals.

Spacious plains. Find the *Interior Plains* and *Coastal Plain* on the map on this page. These plains cover about

The wide, fertile valley of the Shenandoah River in Virginia attracted many early settlers. In what natural region does Virginia lie?

half the land area of the United States and Canada. Their broad, generally flat lands make a sweeping curve from the Atlantic coast of the United States up to the Arctic Ocean.

The Interior Plains border the Canadian Shield. This vast region contains two sections, the Central Plains and the Great Plains. The Great Plains, forming the western section of the region, lie over 2,000 feet (600 m) above sea level.

The Interior Plains have some of the world's most fertile soil. At one time, forests and grasslands covered these plains. Over the years, however, farmers have cleared these lands and turned them into one of the most productive food-growing areas in the world.

The Coastal Plain is a wide arc of land that slopes gradually to the shore and continues under the ocean. The fertile soils of the Coastal Plain support such crops as grains, cotton, and tobacco.

Even the submerged part of this region, called the continental shelf, is rich in resources. The waters above it abound in fish. The many drowned river valleys, formed when the level of the coastal waters rose and filled in the valleys, make good natural harbors. And in the Gulf of Mexico, the continental shelf itself contains large amounts of oil, natural gas, and other minerals.

Majestic highlands. There are several areas of sprawling highlands in the United States and Canada. The rugged highlands make farming and transportation difficult. But they are rich in other resources, especially forests and minerals.

Find the *Appalachian Highlands* and *Interior Highlands* on the map on page 483. The broad valley of the Mississippi River separates these two highland regions. Both highlands are the eroded remains of ancient mountains. The heavily forested Appalachians rise to about 6,000 feet (1,830 m) above sea level. The Interior Highlands are lower, reaching a height only about 2,800 feet (654 m) above sea level.

Most people in the Appalachian region live in the fertile valleys between the mountains. Lumbering and mining are important Appalachian industries. Many people of the Interior Highlands farm small plots of land.

The mountains in the west are younger and more rugged than the eastern highlands. Some peaks in the *Rocky Mountains* reach heights above 14,000 feet (4,370 m). The beautiful scenery of the Rockies attracts visitors from all parts of Canada and the United States. The mountains are also a rich source of wood products, copper, silver, oil, and other raw materials.

Many rivers have their source in the Rockies. The Rockies form the continental divide, separating the drainage basins of rivers flowing eastward from those of rivers flowing to the west. Look at the map on pages 34–35. Name two rivers on the western side of the divide.

The highlands in the *Pacific Mountains and Valleys* region are also young

and rugged. Some peaks in the coastal ranges rise even higher than peaks in the Rockies. Mount McKinley in the Alaska Range is over 20,000 feet (6,100 m) above sea level. It is the highest point in North America.

Low valleys separate the various mountain ranges in this region. Many of these valleys, such as the San Joaquin (San Wä·kēn') and the Willamette, are covered with fertile soil washed down from nearby mountains.

Between the Pacific ranges and the Rockies is the *Intermountain Region*. Much of this region is a high, very dry plateau. Several small but rugged mountain ranges and deep canyons cut through this region. The most famous canyon is the Grand Canyon, which is over one mile (1.6 km) deep.

Lakes and waterways. The United States and Canada have an abundance of water resources. Their many lakes and rivers supply fresh water and valuable transportation routes to people deep in the continent's interior.

The Great Lakes hold more fresh water than any other system of lakes in the world. They lie between the United States and Canada and form part of the border between these nations. Find the five Great Lakes—Superior, Michigan, Huron, Erie, and Ontario—on the map on page 27. Which is the most northerly? Which lake lies entirely within the United States? A number of huge industrial cities, including Chicago, Cleveland, and Toronto, have developed along the shores of the Great Lakes.

In the 1950s, Canada and the United States cooperated in building the St.

ARCTIC OCEAN

Hudson Bay

CONTINENTAL DIVIDE

PACIFIC OCEAN

ATLANTIC OCEAN

©FPC

Death Valley, California, in the Intermountain Region, is the lowest spot in North America, lying 282 feet (85 m) below sea level.

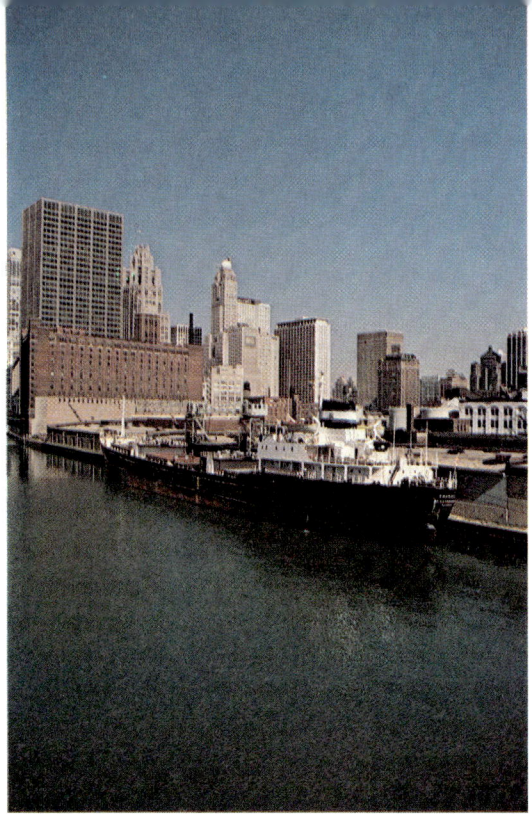

A freighter enters the Ogden Slip in Chicago, Illinois. On which Great Lake is Chicago located?

Lawrence Seaway. This system of locks and canals connects the Great Lakes with the St. Lawrence River and the Atlantic Ocean. Except during the winter, when ice closes the lakes, ocean ships can use the Seaway to reach such cities as Toronto, Chicago, or Duluth.

One of the busiest waterways of North America is the Mississippi River. Hundreds of barges carry grain, salt, and other freight to the cities along the river and its tributaries. Look at the map on page 26. Trace the course of the Mississippi from its source near St. Paul to its mouth at the Gulf of Mexico. What large river flows into the Mississippi from the east? From the west?

There are many other important rivers in North America. The Rio Grande forms much of the border between the United States and Mexico. The Colorado River provides irrigation and drinking water for much of the dry southwestern United States. The Columbia, Snake, and Fraser rivers are important sources of hydroelectric power in both the United States and Canada. Look again at the map on page 26. What large river flows through northern Canada into the Arctic Ocean? What large river flows through central Alaska into the Bering Sea?

Contrasts in climate. Turn to the world climate map on page 200. What two climate zones cover most of northern Canada and Alaska? Both of these regions are so cold that the ground in many places is always frozen. Construc-

Vancouver, in the Canadian province of British Columbia, has long been a center for lumbering and shipping. How does the location of Vancouver favor the development of shipping and trade?

tion workers must use special building methods in these areas to keep houses and roads from melting this permafrost and sinking into the ground.

Much of the western half of the United States has a very dry climate. Compare the climate map with the political map (pages 32-33). What states lie within the desert climate zone?

Many regions of the United States and Canada enjoy a more moderate climate than the dry western areas and cold north. The eastern half of the United States and parts of Canada have long, warm summers with plenty of moisture. In the southeastern and Pacific coastal regions, even the winter temperatures are mild. These areas attract many vacationers as well as year-round residents.

Land use. Favorable climates, rich soils, and flat lands combine in several parts of the United States and Canada to create excellent conditions for farming. Modern farming methods and machinery help farmers in the United States and Canada produce much more food than their countries can use. Both countries export much of their surplus corn, wheat, soybeans, and other food products.

The map on page 30 shows other ways people of the United States and Canada use their land. In what part of the United States is the land used widely for grazing? What are the main land uses near the city of Seattle?

Resources for industry. The resources of the United States and Canada support thriving urban industries as well as such rural industries as farming and forestry. Busy mills, plants, and factories use the region's coal, iron ore, natural gas, copper, nickel, and other raw materials to produce a wide variety of goods. The United States and Canada also have abundant human resources for industry. Many excellent schools and universities help provide training in advanced technology. Highly skilled workers, a large labor force, and a good supply of natural resources have helped the United States and Canada become leaders in the industrialized world.

487

People and cities. Look at the population map on page 30. Which has more people, the eastern half of the United States or the western half? Where do most Canadians live, in the northern part of their country or the southern part? What physical features help explain the population distribution in the United States and Canada?

As highly industrialized countries, the United States and Canada have many large cities. About three fourths of the people in these countries live in urban areas. People from many different backgrounds take advantage of the cities' opportunities for jobs, education, and recreation.

In general, the people of the United States and Canada enjoy a high standard of modern living. The governments of both countries offer many public services. Both countries, however, are fighting inflation and unemployment problems. Both countries are also troubled by dwindling energy supplies and growing energy needs. Along with other industrialized countries, the United States and Canada are looking for ways to balance the needs of their people with the limited resources of the earth.

Keeping Facts in Focus

1. What are the five highland regions in the United States and Canada? Which are old and eroded? Which are young and rugged?
2. What are some of the large lakes and river systems in the United States and Canada? What are some of the ways people use these water resources?
3. Where are the most populated areas of the United States and Canada? Where are the areas of extreme climates?

The Eskimos in Alaska are descendants of early migrating Asians. Why do you think many early people migrated further south?

The People's Heritage

The earliest people in the lands of the United States and Canada were the ancestors of the American Indians. These people began to arrive from Asia between 20,000 and 40,000 years ago. At first, they were nomadic hunters. But by about 2,000 years ago, some Indians in the southwest, the Pacific coast region, and the east were living in villages. Some villagers grew crops, especially corn. Others hunted small animals or fished. Some Indian groups on the Great Plains continued to hunt herds of bison and other large animals.

Explorers and settlers. Europeans began to arrive in North America in the 1500s. These early explorers, whose expeditions included a number of black Africans, were looking for such riches as gold, silver, and furs. Many explorers traded with the Indians, but they did little to disturb the Indians' traditional ways. When European farmers, merchants, and other settlers began to arrive, however, many Indians were

Paul Cuffee, a black from Massachusetts, used much of his shipping fortune to help other blacks in the U.S. return to Africa.

forced off their lands. Many people lost their lives in the frequent battles between Indians and settlers.

Despite the hardships in settling new lands, Europeans continued to move to North America. People from France settled in the fertile St. Lawrence River Valley of Canada. Their customs and language still dominate this region today. People from England settled up and down the Atlantic coast from Newfoundland to Georgia. The English settlers brought ideas of democracy, individual freedom, and fair government that are still at work today in both the United States and Canada.

Fishing, farming, trapping, and trading were important activities of the early European settlers. In the southern colonies, farming was especially important. By the late 1700s, southern colonists were growing tobacco and cotton on large plantations.

Since the time of the first European settlers, many other people have come to the United States and Canada for a variety of reasons. Some people came to own and farm the fertile lands. Some searched for gold or other valuable minerals. Others sought political or religious freedom. Still others came for good jobs in growing industries. Not all people, however, came with hopes of a better life. Large numbers of black Africans, for example, were brought to work as slaves, especially in the southern colonies. It was not until nearly a hundred years after the United States became independent that all slaves were freed.

The United States gained its independence from Britain after a bloody revolution that ended in 1783. Canada became independent peacefully, but gradually. The British gave the Canadian provinces limited independence in 1867. But it was not until 1931 that Canada became a completely independent nation within the Commonwealth.

A new wave of immigrants. In the mid-1800s, famines and political conflicts in Ireland, Britain, and Germany caused millions of people to leave their homes and flee to the United States and Canada. In the second half of the 1800s, the wave of immigration grew even stronger. Millions more immigrants came from eastern and southern Europe and Asia. Each group of people brought its own traditions that enriched the developing cultures of the United States and Canada.

489

IMMIGRATION TO THE UNITED STATES SINCE 1820

Years	Number of Immigrants	Selected countries of origin
1820-1840	750,949	Ireland, Great Britain, France
1841-1860	4,311,465	Ireland, Germany, Great Britain, Sweden
1861-1880	5,127,015	Germany, Ireland, Canada, China
1881-1900	8,934,177	Germany, Austria-Hungary, Russia, Italy
1901-1920	14,531,197	Italy, Russia, Greece, Japan, Austria-Hungary
1921-1940	4,635,640	Poland, Czechoslovakia, Canada, Mexico
1941-1960	3,550,518	Germany, Philippines, Mexico, Korea
1961-1980	7,815,000	Mexico, West Indies, Philippines, Vietnam
Total	49,655,000	

Entering the machine age. During the 1800s, the industrial revolution was beginning to change the United States and Canada from rural, farming countries to urban, industrial societies. People built miles of railroads, canals, and highways to help move raw materials from the mines and forests to the busy new factories of the growing cities. By the beginning of the 1900s, these factories were producing such goods as steel, machinery, wood pulp and paper, chemicals, and motor vehicles.

During the early 1900s, the industrial strength as well as the population of the United States and Canada continued to grow. A serious depression in the 1930s slowed economic growth. But by the late 1940s, the nations were recovering. Many of the benefits that both governments now offer their people were designed to help curb the kind of poverty that was widespread during the hard depression years.

The nations today. While still strong today, the United States and Canada are facing some difficult problems, especially inflation and shortages. Their growing industries and cities need more energy resources than the countries can supply. Both the United States and Canada must import much of their oil, mainly from the Middle East. The high price of imported oil has driven up the price of U.S. and Canadian goods. People in both countries are seeking new energy sources in order to reduce the need for oil imports.

Despite their problems, the United States and Canada are still strong and productive. Their lands produce much of the world's food. Their industries produce goods used by countries around the globe. And their good relations with each other show that two powerful nations can grow side by side in peace.

Keeping Facts in Focus

1. What were some of the activities of the early European settlers in the United States and Canada?
2. How did the immigrants affect life in the United States and Canada?
3. How does the energy shortage in the United States and Canada affect inflation?
4. What answers would you give to the key questions on page 480?

Working with Ideas

1. Of what benefit to the United States and Canada are surplus agricultural products?
2. What do you consider the most pressing problems facing the United States and Canada today?

Lesson 2　The Northeastern United States

Reading Focus

1. The last five lessons of this book examine many ways of life in the United States and Canada. Within these lessons, you will learn about the region in which you live and the regions nearby. You can get even more information about your own region from the daily newspapers published in your area. As you study the rest of this unit, refer often to your local newspapers. Look for stories that might give you insight into your region's agricultural activities, industries, and cities or towns. Read these stories with a geographer's eye, always looking for relationships between people and their physical environment. Newspapers will help give you up-to-the-minute information about the changing conditions of your local environment.

2. As you read the lesson, look for the answers to these key questions:
   a. What difficulties do Northeastern farmers face?
   b. How did the location and physical setting of the Northeastern states favor the growth of trade and manufacturing?
   c. What types of industry are important in the Northeast?
   d. What is urban sprawl? What problems has it created?

Vocabulary Focus

fall line　　　　　*urban sprawl*

megalopolis　　　*urban renewal*

In 1787, leaders from the thirteen original states drafted the Constitution of the United States. Since that time, the population of the United States has grown from 4 million to more than 236 million. There are now fifty states, and they cover nearly a third of the North American continent. The vast area of the United States makes it the fourth largest country in the world.

Differences in climate, landforms, resources, and ways of living separate the fifty states into several distinct regions.

The map on page 492 shows the four main regions of the United States. In the following lessons, you will learn about life in each of these regions. You will also learn what special contributions each region makes to the United States as a whole.

New England and the Middle Atlantic States

Find and name the Northeastern states on the map on page 492. Which

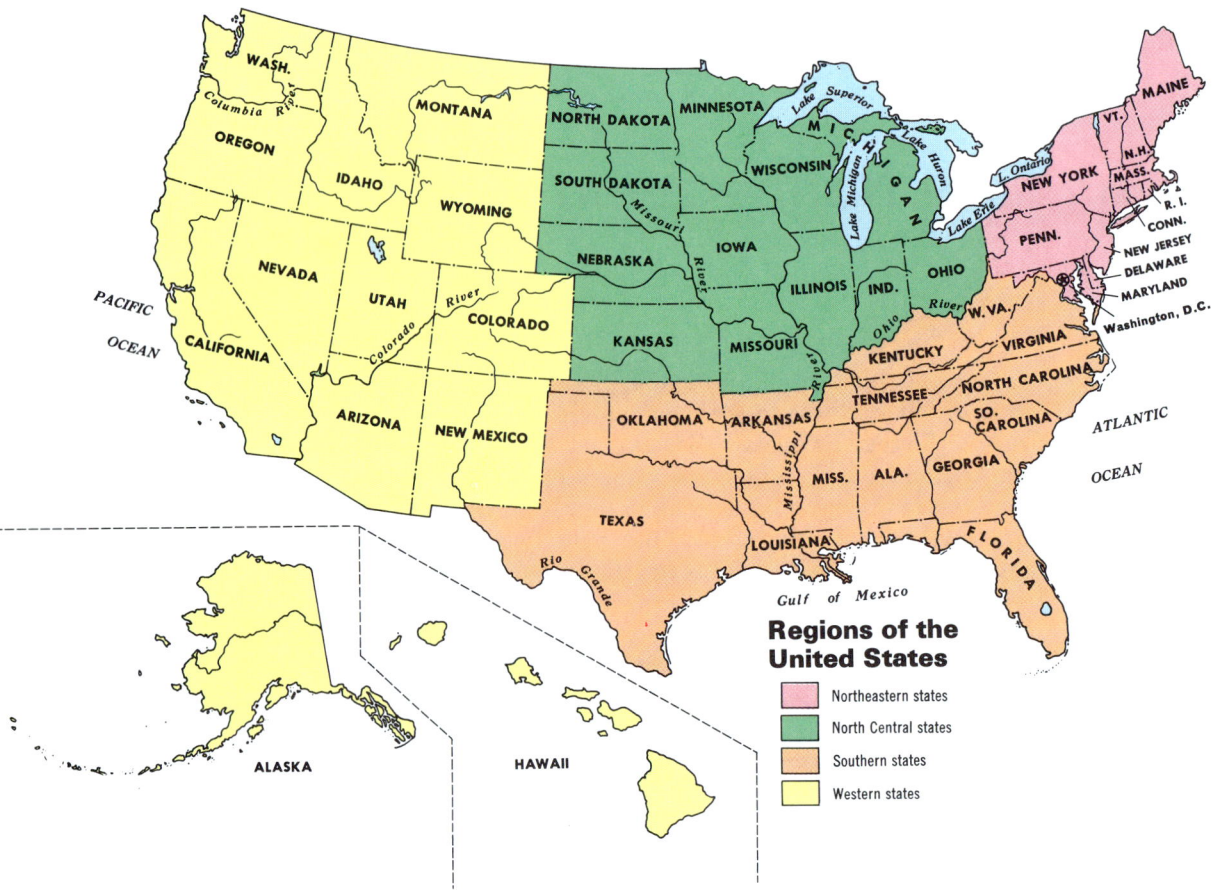

Regions of the United States

- Northeastern states
- North Central states
- Southern states
- Western states

of these states share a border with Canada? Which lie along the shores of a Great Lake? What is the only land-locked state in the Northeast?

The Green Mountains of Vermont divide the Northeastern states into two smaller sections. The section east of the mountains is New England. New England is a region of rocky uplands, broad river valleys, and rocky coastlines. Many people in New England work as seafarers, small farmers, or merchants.

The states to the west of the Green Mountains are the Middle Atlantic states. The Appalachian Highlands run through the center of this region. They separate the Atlantic coastal plain from the lowlands along the Great Lakes. Trade, government, transportation, and other services employ over two thirds of the workers in this region. The Middle Atlantic states are also part of the great manufacturing heartland of the United States. Their busy factories produce over one quarter of the total goods manufactured in the United States.

Rocky Lands and Rich Waters

The rocky lands of both New England and the Middle Atlantic states have made farming difficult through the years. Even when the land is cleared, the soils in some regions are not very fertile. Many of the region's farms are small, averaging fewer than 200 acres (81 ha) each. The poor soil needs large amounts of fertilizer. The hilly land,

Rockport, Massachusetts, is typical of the fishing harbors that dot the New England coastline.

The Potomac River reaches the fall line near the nation's capital city of Washington, D.C.

especially in New England, limits the use of modern farm machinery. And the short growing season in much of the Northeast limits the types of crops the farmers can grow.

Potatoes are one crop that the region's farmers can grow successfully. Many of the french fries you eat are probably made from Maine potatoes. In the parts of the Middle Atlantic states with a longer growing season, farmers can also grow grains, fruits, and vegetables.

Dairy and poultry farming is better suited to much of the Northeast than crop growing. Much of the grain to feed the animals, however, must come from the Middle West or other parts of the United States.

Fishing. The region's rich coastal fishing grounds help make up for the poor soils and rocky lands. Northeastern fishers catch large numbers of clams, lobsters, cod, herring, and flounder. Seafood from Chesapeake Bay and the waters off Cape Cod accounts for about one fourth of the total catch in the United States.

The Growth of Trade and Manufacturing

The poor farming conditions in the Northeast led the early European settlers to seek some other ways to make a living. They soon found that their location and physical setting favored the growth of trade and manufacturing.

The fall line. Many sheltered bays and inlets dot the Atlantic coast of the Northeast. Wide, navigable rivers flow into many of these bays. The early colonists found that ships could sail up these rivers for many miles, until they reached the fall line.

The fall line is the boundary between the Piedmont, a plateau east of the Appalachians, and the Atlantic Coastal Plain. Eastward flowing rivers fall over the edge of the plateau onto the plains at this point. Since ships cannot navigate the falls, the colonial sailors had to stop at the fall line to unload their cargoes. Workers then transferred goods to wagons and other means of transportation and carried them inland.

493

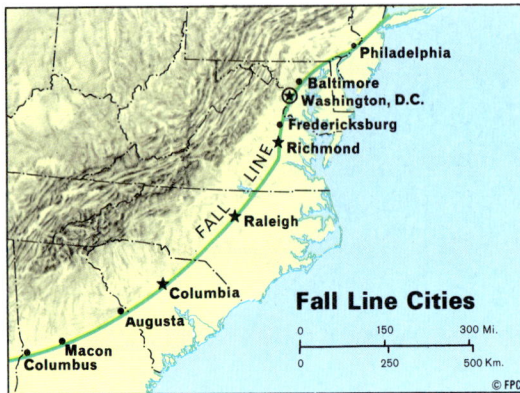

Fall Line Cities

Philadelphia
Baltimore
Washington, D.C.
Fredericksburg
Richmond
Raleigh
Columbia
Augusta
Macon
Columbus

FALL LINE

0   150   300 Mi.
0   250   500 Km.
© FPC

Many cities sprang up along these fall line transfer points in both the Northeastern and Southeastern states. The map on this page shows some of these cities. Which three cities lie within the Northeastern states?

At first, fall line cities were mainly trading centers. But when the industrial revolution began in the 1700s, people realized that the falls were also a cheap source of energy. Industrialists soon built waterwheels and dams to harness the waterpower. Many new factories sprang to life.

Both trade and manufacturing continued to grow as the Northeastern states developed. The region's many forests provided timber for shipbuilding. By the early 1800s, merchant ships from the Northeast were trading on every continent of the world.

Trade within the United States also expanded. The Erie Canal, completed in the early 1800s, linked the Mohawk River with Lake Erie. The canal opened a direct trading route between New York City and the rapidly growing West. Later in the 1800s, railroads through the Appalachians made travel faster and trade more convenient than ever before.

In the mid 1800s, a new method of making steel helped Northeastern manufacturing grow rapidly. This new method required large amounts of coal, iron ore, and limestone. The area of western Pennsylvania near Pittsburgh had many of these raw materials nearby. Pittsburgh soon became one of the world's leading steel-producing cities.

Today's Industries

Manufacturing employs more Northeasterners than any other single industry. Factories in the Northeast produce a wide range of goods, from heavy machinery and metal products to such consumer goods as clothing, books, and magazines.

Thriving trade. Western Pennsylvania is well supplied with raw materials for its industries. But other parts of the Northeast lack many needed materials.

The Allegheny and Monongahela rivers meet in the city of Pittsburgh, Pennsylvania, to form the Ohio River.

Named for the Indians who first inhabited it, Manhattan today forms the heart of the country's largest city, New York.

Both the Declaration of Independence and the Constitution of the United States were formally approved in Independence Hall in Philadelphia.

For this reason, trade is still vital to the Northeastern states. Some of the region's factories must import iron ore from Canada, crude oil from the Middle East and South America, and other raw materials. The region's dairy farmers and food processors must import much of their grain from the Midwestern states.

Export trading is also important in the Northeast. Once the raw materials are manufactured into finished products, they are shipped to other parts of the United States and countries around the world. The region has many busy ports to handle both incoming materials and outgoing products.

People serving people. In addition to manufacturing and trade, service industries play an important role in the economy of the Northeast. About one fifth of the region's workers are employed in such service industries as entertainment, recreation, and housekeeping. In the capital of the United States, Washington D.C., almost half the workers hold government service jobs. Commercial industries are also important. New

York and Connecticut are the insurance centers of the United States. The Wall Street district of New York City is in many ways the financial capital of the world. Billions of dollars change hands in the district's banks and stock exchanges every business day.

An Urban Region

The region's many industries have helped make the Northeast one of the most highly urbanized areas of the world. About 80 percent of the region's people live in urban areas.

Cities of the Northeast. Along the Atlantic coast between Boston and Washington D.C., there is one large city after another. About 40 million people live in this urbanized area. Geographers call this urban strip a "megalopolis," or "great city." Find some of the cities of this megalopolis on the map on pages 32–33.

New York is by far the largest city in both the United States and Canada. Over nine million people live in this sprawling urban area. Philadelphia,

Washington, D.C., became the capital of the United States in 1880. The White House and Washington Monument draw many visitors each year.

with close to five million people, is the second largest urban area in the Northeastern states.

Today, more Northeasterners live in suburbs surrounding the cities than in the cities themselves. The suburbs are expanding outward into areas that were once rural. This urban sprawl has led to problems in both the central cities and the spreading suburbs.

As wealthier people move to the suburbs, their taxes no longer go to the central city for public services. As industry moves to the suburbs, fewer jobs are available in the cities. And as suburban areas grow, they take up valuable farmland. In some places, city problems such as crowding, crime, and pollution are moving to the suburbs along with the people and industries.

Cities everywhere are facing financial problems. Welfare services, quality education, and police and fire protection are all very costly. Because the cities in the Northeast are the oldest in the country, their buildings often need frequent repair. The cost of urban renewal, so important in the older Northeast, is also

sky-high. Some people predict that Northeastern cities face a grim future unless they can draw people and industries back into the cities to help boost their tax bases.

But many people enjoy city living and have no desire to move away. They like the opportunities cities offer for education, jobs, and recreation. They like the rich mixture of races and cultures that adds variety to city neighborhoods. To many people, cities are the centers of social, cultural, and historic traditions of the Northeast.

Keeping Facts in Focus

1. What are some of the agricultural products of the Northeast?
2. What are the two largest urban areas in the Northeast?
3. What answers would you give to the key questions on page 491?

Working with Ideas

Compare living in a large central city with living in a smaller suburban area or with living in a small rural village. Which would you prefer to live in? Why?

Lesson 3 The Southern United States

Reading Focus

1. Before reading this lesson, skim through the pages and look for familiar names and places. Maybe you live near some of these places or have visited them. Or perhaps you have seen pictures of them in books or magazines or on television. Try to remember everything you know about these places. Compare your memories with what you learn as you read. By calling personal experiences to mind, you can enrich your reading and gain a fuller understanding of the subject.

2. As you read the lesson, look for the answers to these key questions:
   a. What are some important rivers in the Southern states?
   b. How has Southern agriculture changed since the 1930s?
   c. What is the Tennessee Valley Authority? How has it helped the South?
   d. How has the population distribution of the South changed since the early 1900s?

Vocabulary Focus

floodplain	*menhaden*
silt	*synthetic*

Within the past thirty years, the South has grown into a highly productive region. Especially since World War II, people in the South have developed the rich resources of their lands and waters. Southern farms, once threatened by erosion, worn-out soils, and flooding, are now producing good yields of a variety of crops. Modern cities in the South have grown into bustling centers of trade, industry, and tourism. The people of the South are working together to make their region one of the fastest-developing areas of the United States.

Southern Lands and Resources

Find and name the Southern states on the map on page 492. Which states border the Atlantic Ocean to the east? What body of water lies to the south?

The Mississippi River divides the Southern states into two sections. The states to the east of the Mississippi are the Southeastern states. The four states to the west of the river are the South Central states.

Landforms and climate. Use the map on page 483 to find the five natural regions of the South. As in the Northeastern states, the fall line separates

497

What are the leading coal-mining states in the nation?

highland areas from the coastal plain. Find the Southern cities located near the fall line on the map on page 494.

Most of the South has a humid subtropical climate. Rainfall is heavy throughout the year except in western Texas and Oklahoma. In the northern parts of the region and in the highlands, winters can get quite cold. But in Florida and along the Gulf Coast, winter temperatures rarely fall below freezing. Summers throughout the South are very hot.

Many useful rivers. Because rainfall is so plentiful, many rivers cross the Southern states. Some rivers, such as the Tennessee and its tributaries, are important sources of hydroelectricity. Southerners use other rivers, such as the Arkansas and Rio Grande, for irrigation. But the most useful and important river of all is the Mississippi.

The Mississippi and its major tributary, the Missouri, form the third longest river system in the world. Only the Nile and the Amazon are longer. The floodplain of the Mississippi is covered with rich, fertile silt, good for farming. The river has also been a major highway for trade in the United States since the early 1800s. New Orleans, near the mouth of the Mississippi, is the second busiest port in the United States.

Thick forests. Forests cover much of the Appalachians as well as many of the swampy coastal areas of the South. Timber from these forests accounts for about one third of all lumber produced in the United States. Southern forests supply the raw materials for much of the paper, turpentine, and other wood products made in the United States. Wood from these forests also supplies materials for the South's large furniture industry.

Mineral treasures. The Southern states are a vast storehouse of mineral wealth. Beneath their lands lie large deposits of coal, petroleum, natural gas, and other important minerals.

One of the richest coalfields in the world lies in the Appalachians. About half of all the coal mined in the United States comes from the five Appalachian states—Virginia, West Virginia, Kentucky, Tennessee, and Alabama. West Virginia and Kentucky are the country's leading coal-mining states.

While the Southeastern states are rich in coal, the South Central states have rich deposits of petroleum and natural gas. Fuels from these four states and Mississippi make up about 70 percent of the oil and 80 percent of the natural gas produced in the United States. In the past, these states ac-

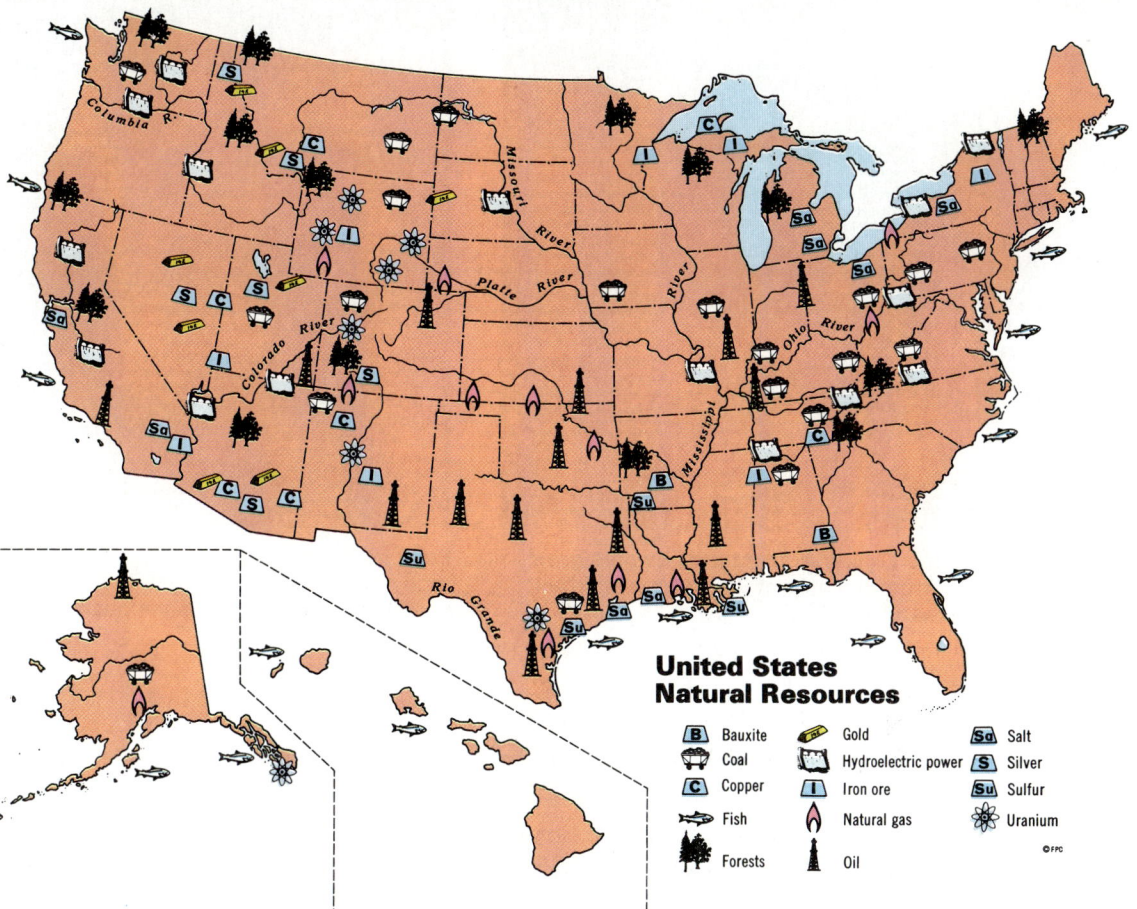

United States Natural Resources

B Bauxite	**Gd** Gold	**Sa** Salt
Coal	Hydroelectric power	**S** Silver
C Copper	**I** Iron ore	**Su** Sulfur
Fish	Natural gas	Uranium
Forests	Oil	

©FPC

counted for an even larger share of the total production of oil and natural gas. Texas is still the leading oil producing state.

Several other minerals are also important in the South. Alabama and North Carolina have good supplies of tungsten. Arkansas, Alabama, and Georgia contain all the bauxite in the United States. Salt and sulfur are in good supply in the coastal areas of Texas and Louisiana. And there are several deposits of iron ore in Alabama.

Farming and Fishing in the South

The Southern states are favored with wide plains, warm temperatures, and plenty of rainfall. Yet in the years before World War II, Southern farmers faced serious problems. Years of planting one crop—tobacco in the north and cotton further south—had taken a heavy toll on the land. Once-fertile soils were exhausted. Erosion removed whatever good soil was left. Many farmers could make only poor livings. Others went bankrupt and lost their farms completely.

Far-reaching reforms. In the 1930s, the United States government began programs to help the Southern farmers. The government hired jobless people to plant trees, build dams and flood-control projects, and do other work to reclaim the land. The government also encouraged farmers to use such methods as

Before the Civil War, plantation owners grew wealthy from the sale of their crops and built stately mansions.

Peanuts are an important crop in the South today, especially in Georgia. Much of the peanut crop is used to make peanut butter.

crop rotation, fertilizing, and terracing to protect and enrich the soil. Smaller farms were combined to make larger farms, each about 300 acres (120 ha). On these larger fields, farmers could use the latest farm machinery to increase their yields. New crops, such as soybeans, also helped bring new life to the farmlands of the South.

Southern farmers today grow a large variety of crops. They grow rice along parts of the Gulf Coast in Texas and Louisiana and along the Mississippi River in Arkansas. Sugarcane and citrus fruits grow well in Florida's hot, humid climate. Cotton is an important crop in Mississippi, Arkansas, and Texas. And tobacco is a major crop in Kentucky, North Carolina, and South Carolina.

Huge livestock ranches. Livestock raising has also become important in the modern South. Large ranches and farms that are run like factories produce millions of cows, pigs, sheep, chickens, and turkeys. They also produce a wide range of dairy products. Besides raising the livestock, these large farms also process, store, and distribute the meat and dairy products. Many small farmers cannot compete with these huge farms. Since the large farms do not have to pay others to store, process, or distribute their goods, they are better able to absorb the increasing costs of fuel and feed than the smaller farms.

Thriving fisheries. Fishing has always been profitable in the Southern states. The warm waters off both the Atlantic and Gulf coasts brim with many varieties of fish. In fact, the South has the nation's leading fisheries. Seafood from Southern waters makes up more than half the total catch of the United States. Menhaden, which is used for fish meal and animal feed, is the most common catch. In the Atlantic, crabs are an important catch. Fishers from Texas and Louisiana catch shrimp in the Gulf of Mexico. And the catfish from Southern rivers is a favorite treat for many people.

Industry in the Southern States

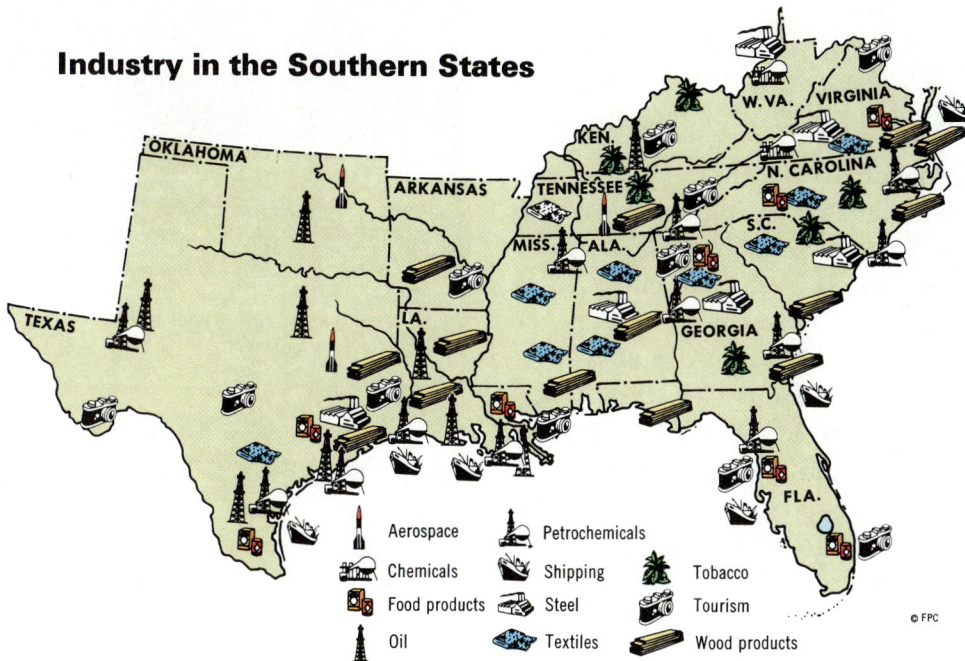

Map legend:
- Aerospace
- Chemicals
- Food products
- Oil
- Petrochemicals
- Shipping
- Steel
- Textiles
- Tobacco
- Tourism
- Wood products

© FPC

A Land of Booming Industries

Southern industry as well as agriculture has developed rapidly since the 1930s. Today's industries take full advantage of the region's many valuable resources.

The Tennessee Valley Authority. One of the most important events in the history of the South was the creation of the Tennessee Valley Authority in 1933. This government-owned corporation has helped control floods, develop river travel, and harness waterpower. TVA dams use the region's swift-flowing rivers to provide large amounts of low-cost electricity. Besides bringing electricity to the region's homes, this cheap power helped attract new industry to the area. The South can now boast thriving industries based on mining, processing, tourism, and government research.

Mining. Mining employs more people in the South than it does in any other region of the country. In the South Central states, the large reserves of oil and natural gas have given rise to other industries as well. These include oil refining and the production of chemicals. Oil-related industries have helped such cities as Dallas, Fort Worth, Houston, and Tulsa to grow rapidly.

In Alabama, coal and iron ore deposits near Birmingham have helped that city grow into the "Pittsburgh of the South." Birmingham is the South's leading steel-producing city.

Processing. The revival of Southern agriculture has helped the region's industry grow as well. Food products such as canned and preserved fruits and vegetables, sugar, soft drinks, and vegetable oils are among the most important products of the South.

Southern textile mills use both cotton and synthetic fabrics to make cloth, carpets, and clothing. The South leads the nation in textile production. Georgia, South Carolina, and North Carolina all have large textile mills.

Other industries. Tourism has long been an important industry in the South. The year-round sunshine in some parts of the South attracts many vacationers, especially in the winter.

The South also has many government-related industries. Oak Ridge, Tennessee, is a center for nuclear research. Huntsville, Alabama, is the home of a large missile-building plant. The Lyndon B. Johnson Space Center of NASA is in Houston, Texas. And Cape Canaveral, Florida, is the launching pad for all U.S. spacecraft.

A Growing Urban Region

As late as 1900, only 15 percent of the people in the South lived in urban areas. Today, about two thirds of all Southerners live in the industrial cities and suburbs, and their numbers continue to grow.

Dallas and Fort Worth form the largest urban center of the South and the eighth largest in the country. Dallas is the headquarters of many oil firms. It is also an important manufacturing, banking, and insurance center. Fort Worth has many manufacturing industries. It also serves as a distribution center for western Texas.

Outlook for the Future

The South faces the same problems as many other regions—providing jobs for the unemployed, reducing overcrowding in its cities, and protecting the environment from the effects of pollution. As it strengthens its economy by

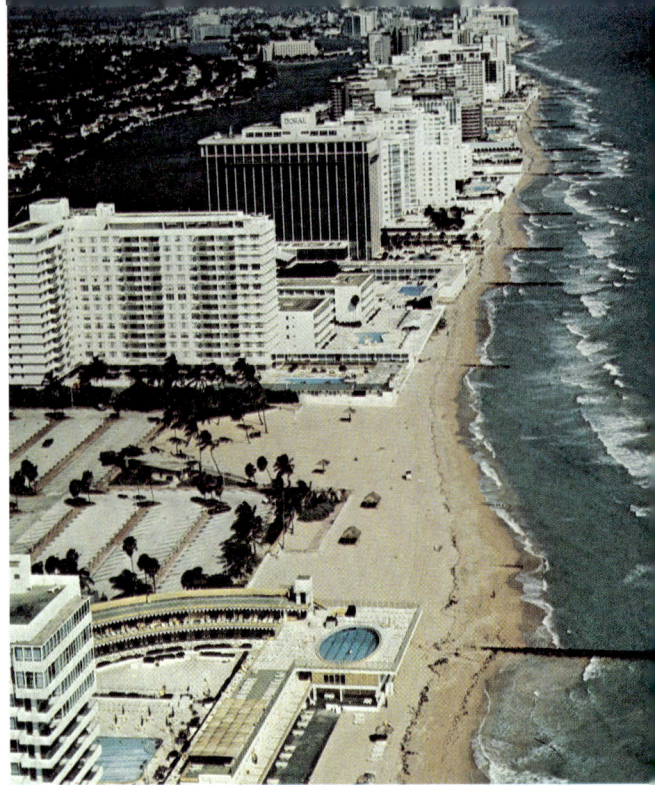

The warm waves of the Atlantic Ocean and the luxurious beachside hotels attract visitors from all over the country to Miami, Florida.

the development of private industries, the South hopes to ensure a stable future and a good life for its people.

Keeping Facts in Focus

1. What problems faced Southern farmers in the years before World War II? How were these problems solved?
2. What are the major agricultural products of the South?
3. What minerals are in good supply in the South? Why is Birmingham, Alabama, called the "Pittsburgh of the South"?
4. What industries are important in the South?
5. What answers would you give to the key questions on page 497?

Working with Ideas

Why might a textile manufacturer choose to build a new plant in the South rather than in the Northeast?

Lesson 4 The North Central United States

Reading Focus

1. The North Central states make up one of the leading food-producing regions in the world. The map on page 505 shows the three main farming belts in the region. Before reading the lesson, become familiar with the North Central states by making an agricultural chart like the one shown below. Use the three farming belts as headings. Below each heading, list the states that lie within that belt. Some states lie within two belts, so list those states under both proper headings.

Corn Belt	Wheat Belt	Hay-and-Dairy Belt

2. As you read this lesson, look for the answers to these key questions:
   a. How have the land and climate in the North Central region helped it become the world's breadbasket?
   b. What are the three main farming regions of the North Central states?
   c. What are some large cities in the North Central states?
   d. What kinds of transportation facilities help make the North Central region an important crossroads?

Vocabulary Focus

domestic	*combine*	*waterborne*
tornado	*harvester*	*milling*

The North Central states, also called the Middle West, form the heartland of the nation in many ways. Their farms help feed people in all parts of the United States—and some foreign lands as well. Their factories make a variety of goods for both domestic use and export. And their location in the center of productive lands helps make this region the crossroads of the North American continent.

The Natural Setting

Find the North Central states on the map on page 492. Which states border the Great Lakes? Through which states does the Missouri River flow?

The Mississippi River divides the North Central states into two sections. The five Great Lakes states to the east have busy factories and huge cities. The Plains states to the west form the most productive farming region in the world.

United States Agricultural Products

Symbol	Product	Symbol	Product	Symbol	Product
🐄	Cattle and hogs	🐔	Poultry	🌸	Sugar beets
🍊	Citrus fruits	⚬	Rice	🌿	Tobacco
🌽	Corn	🌾	Sorghum grain	🧺	Vegetables
☁	Cotton	S	Soybeans	🌾	Wheat
🥔	Potatoes	🎋	Sugarcane		

© FPC

Amid bustling activity, traders at the Chicago Board of Trade buy and sell corn, wheat, soybeans, and other products of Midwestern farms.

The lands. Most of the Middle West lies within the wide Interior Plains. Glaciers leveled these lands long ago, leaving rich soil behind. Tall prairie grasses grew on these rich plains. Humus from decayed prairie grass further enriched the soil. In the 1800s, farmers cleared the prairies and replaced the grasses with acres and acres of sown wheat, corn, hay, and other crops.

The weather. The North Central states have a continental climate, with cold winters and hot, wet summers. Blizzards often sweep across the plains during the winter. Thunderstorms and tornadoes come with the warmer spring

North Central States Farming Belts

Corn belt

Dairy and hay belt

Wheat belt

★ State capitals

● Other cities

© FPC

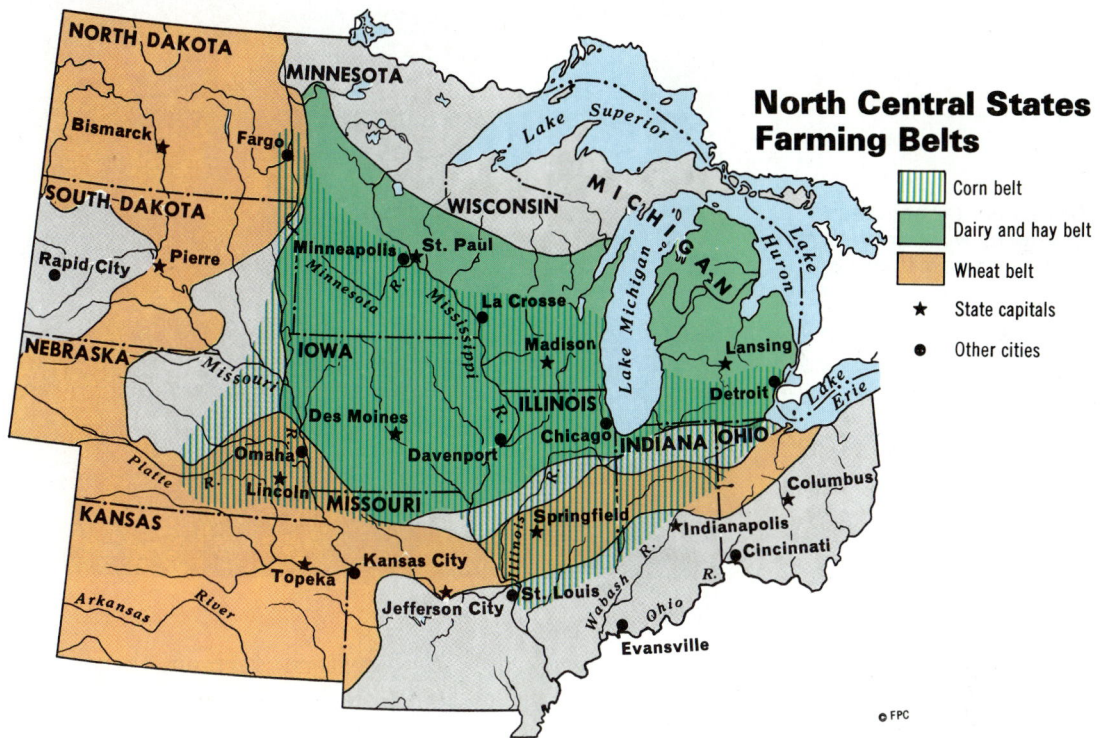

and summer. These storms are sometimes destructive. But they also provide moisture for the area's extensive farmland.

Food Producer for the World

Flat plains, fertile soils, and humid weather have helped make the Middle West the world's breadbasket. Midwestern farms are large, averaging about 360 acres (144 ha). The flat lands and large farms allow widespread use of the most modern farm machinery, including combines and harvesters. Midwestern farmers also use specially developed breeds of corn, wheat, and other plants. These sturdy breeds resist disease and help increase the farmers' yields.

Midwestern farms produce almost 40 percent of the world's corn, half the world's soybeans, and about 10 percent of the world's wheat. About half the country's wheat and soybeans and one

fourth of its corn is for export. Midwestern farms have made the United States the world's largest exporter of grains.

The corn, wheat, and dairy belts. The North Central states grow some fruits, vegetables, sugar beets, and sorghum. But corn, wheat, and dairy farming are the region's most important agricultural activities. The map on this page shows what states lie within the wheat, corn, and dairy belts.

Corn grows well in the five-month growing season and humid climate of Nebraska, Iowa, Illinois, Indiana, and Ohio. To enrich and protect their soils, cornbelt farmers use many fertilizers. They also rotate the planting of soybean and corn crops in their fields. Many farmers in the corn belt also raise cattle and hogs for meat. Iowa has more hogs and pigs than any other state, and only Texas has more cattle.

Wheat needs less water than many other crops. It grows well in the dry

505

When their corn is ripe, Midwestern farmers must sometimes work around the clock to harvest their crops.

Plains states. Wheat belt farmers grow two types of wheat. In the areas of mild climate, they plant winter wheat in the autumn and harvest it in the spring. Where the weather is more severe, farmers grow spring wheat. These farmers plant their wheat in the spring and harvest it in the autumn. Kansas and North Dakota are the two leading wheat-producing states in the country.

Hay-and-dairy farming needs plenty of rainfall but only a short growing season. Look at the map on page 505. What states lie within the hay-and-dairy belt? Farmers in these states use the hay they grow as feed for their dairy cattle. Wisconsin, sometimes called "America's Dairyland," has more milk cows than any other state.

Problems of Midwestern farmers. In spite of high yields, some Midwestern farmers make only small profits. The prices they get for their products have not kept pace with their costs. The cost of fuel, fertilizer, and farm equipment has risen sharply. As in the South, small farmers are sometimes forced to sell out to huge business farms. In recent years, some Midwestern farmers have sold their lands to European and other foreign corporations. While less than 1 percent of U.S. farmland is foreign-owned, many people in the United States fear a growing foreign influence over one of the nation's most valuable resources.

To increase their profits, some small farmers have formed cooperatives. In farm cooperatives, several farmers share the costs of producing and storing farm goods. They also share the combined profits of their farms. The U.S. government has also helped farmers make a better profit. The government guarantees a reasonable price for the farmers' goods.

Minerals and Waterways

The Midwest is not only a leading farming region, but it is the nation's leading industrial region as well. Rich mineral resources and an excellent system of inland waterways have helped the Midwest grow into an industrial giant.

This open-pit iron ore mine is in the Mesabi Range near Hibbing, Minnesota. What problems of open-pit mining are visible in the picture?

Rich mineral deposits. Almost every state in the Middle West produces some minerals. Iron mines near Lake Superior have been operating since the 1800s. As a result, most of the high-grade ore has already been removed. But the mining industry has developed ways to use the low-grade iron ore that remains. Minnesota, Michigan, and Wisconsin produce almost 90 percent of the iron ore mined in the United States.

Coal is plentiful in Illinois, Indiana, and Ohio. Oil, natural gas, limestone, sand, lead, clay, zinc, and sulfur are other minerals of the Midwest.

Inland waterways. Busy inland waterways help transport minerals and other materials to the Midwest's many facto-

ries. Nearly half of the waterborne commerce in the United States is on the Mississippi River system or the Great Lakes. The St. Lawrence Seaway, linking much of the Midwest to ocean trade routes, is also very busy. Canals at several points even connect the Seaway to the Mississippi system.

Giant Cities

The cities of the Middle West are part of the greatest industrial region in the world. The Great Lakes area especially has many large cities. Between Milwaukee and Gary, a megalopolis has grown along the shores of Lake Michigan. Another megalopolis runs along the western and southern shores of Lake Erie from Detroit to Youngstown. Each of these huge urban strips has about 10 million people. The photo on page 61 shows what these areas look like from a space satellite.

Industrial products. Factories in the five Great Lakes states produce over one fourth of the goods manufactured in the United States. Iron and steel are the most important industries. Chicago and Gary form the busiest steel-making center in the world. Cleveland, Youngstown, and Detroit are also important steel producers.

Midwestern steel has helped build other local industries. The most important of these is the automobile industry. Detroit is the world's largest automobile producer. Factories in other cities, such as Akron, make auto parts.

Farm machinery is a very important product in the cities of Moline and

The production of automobiles employs thousands of workers in the Midwest. This woman works in a plant in Kenosha, Wisconsin.

The colleges and universities of the North Central states serve as centers of research and training for thousands of young Americans.

Milwaukee. Other Great Lakes products include musical instruments, processed foods, and chemicals.

Heavy industry is less important in the Plains states than food processing. Omaha and Kansas City have large meatpacking industries. Minneapolis, St. Louis, Des Moines, and Kansas City are important centers for milling, or grinding grain into flour.

Bringing goods to market. The railways, highways, and waterways of the Middle West are always busy transporting large amounts of goods. Chicago is the world's largest railroad city. It also has the world's busiest airport and miles of modern highways. Located on two major canals that link the Great Lakes with the Mississippi, Chicago is also the nation's busiest inland port. St. Louis, Missouri, is another busy port. Look at the map on pages 34–35. What two rivers come together near St. Louis?

Future Prospects

The Middle West will continue to be a leading farming and industrial region. In many ways, Midwesterners owe their prosperity to the resources of their region. With careful use and protection of these resources, the Midwest will remain the nation's heartland.

Keeping Facts in Focus

1. What are some important inland waterways in the Midwest?
2. What are the main industrial products of the North Central states?
3. What Midwestern cities are especially important transportation centers?
4. What answers would you give to the key questions on page 503?

Working with Ideas

How have the Great Lakes influenced the development of industry in the North Central States? Of agriculture?

Lesson 5 The Western United States

For hundreds of years, the West was populated mainly by American Indians. Many of the Western tribes depended on hunting and gathering. Indians in the Southwest also farmed, and fishing was important to Indians along the Pacific coast. Even when Spanish settlements began in the 1700s, the population of the West remained small. Spaniards from Mexico set up missions in the southwest and grazed livestock over the region's dry, rugged lands.

But in the mid 1800s, rich deposits of minerals, especially gold and silver, drew many more people to the West.

Despite the dry climate and rough lands, the population of the West grew quickly as people moved westward to mine, raise cattle and sheep, and grow crops.

Mining and farming remain important in the West today. But the West also has many booming new industries. Most people who move West today hope to find jobs in these growing industries.

The Last Frontier

The Western states cover an area of about 1,748,000 square miles (4,527,320

American Indians built these cliff houses in southwestern Colorado nearly 1,000 years ago. Today the deserted city attracts tourists from all over North America.

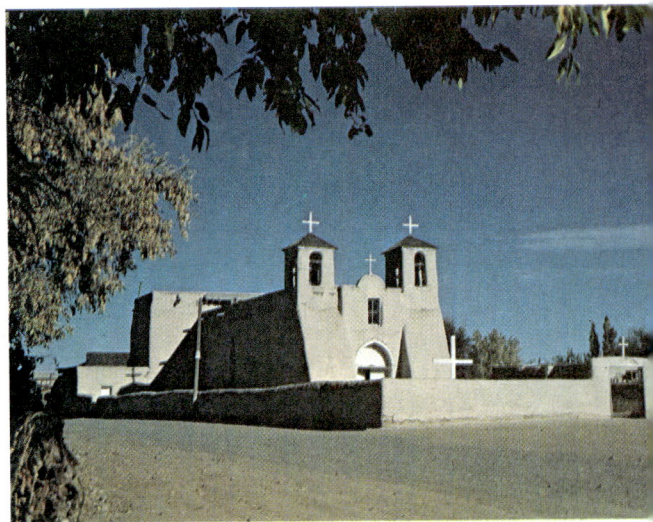

Spanish settlers built this New Mexico mission in 1772. How did the dry climate influence the types of building materials used in the Southwest?

sq km). They take up about half the land area of the United States. Look at the map on page 492. Washington, Oregon, and California are called the Pacific Coast states. The eight states to the east of these are called the Rocky Mountain states.

The distant states of Alaska and Hawaii are also considered Western states. Find these states on the map. Alaska is by far the largest state in the nation. It covers more area than Texas, California, and Montana combined. Hawaii is a group of volcanic islands in the Pacific Ocean. How many main islands make up Hawaii?

Mountains, valleys, and plateaus. Look at the map on page 483. Within what four natural regions do the Western states lie? What are the four natural regions of Alaska? Most of the people of the West live in the Pacific Mountains and Valleys region.

A variety of climates. The varying altitudes within the Western states help create a variety of climates. The Pacific Ocean also influences climate, especially in the coastal states.

All three Pacific Coast states have a mild climate. The ocean moderates the area's temperatures, keeping them comfortable the year round. Westerly winds from over the ocean bring plenty of rain to the coastal areas of Washington, Oregon, and northern California. Look at the climate map on page 200. How do summers in southern California differ from those in coastal regions further north?

In the mountainous regions, altitude has an important influence on climate. The higher the altitude, the lower the temperature. The mountains also keep much of the rain from reaching the plains, plateaus, and basins.

Water—A limited resource. The people of the dry Western states have made extensive use of their limited water resources. Almost every large river in the region has been dammed for hydropower, irrigation, drinking, or recreation. Locate the Colorado, Snake, Missouri, and Platte rivers on the map on pages 34–35. What other rivers can you find in the Western states?

The scarcity of water in parts of the West has led to some problems. Each Western river must serve the water needs of many people. In some areas, people are overusing their water supplies. For example, people are drawing so much water from the Colorado River for drinking and irrigation that the river dwindles into a salty marsh as it reaches its mouth. Careful use of the region's limited water supplies is an important concern, especially to the area's farmers.

A Land of Agricultural Variety

The varying climates and landforms in the West support a wide variety of agricultural products. There are regions where wheat, potatoes, fruits, vegetables, and cotton can each be sown successfully. There are also many areas well suited to livestock grazing.

The Great Plains. Eastern Montana, Wyoming, and Colorado are part of the wheat belt of the dry Great Plains. Some Plains farmers also grow sugar beets and alfalfa, and many graze cattle and sheep.

The Intermountain Region. Farmers in the drier Intermountain Region must use irrigation to support their crops. Cotton, potatoes, hay, wheat, barley, sugar beets, and tomatoes are the region's most common crops. The sale of cattle and dairy products, however, earns the region's farmers even more money than the sale of crops. Because the region has only a sparse covering of grass, herds must be able to graze over large areas. Ranches in the Rocky

These workers are harvesting flowers in southern California. Some workers migrate from place to place in the Western states, helping to harvest the region's many different crops.

511

A small tourist train winds its way through the spectacular Colorado Rockies.

Loggers in Oregon use small tugs to help increase their efficiency.

Mountain states are the largest in the country, averaging over 2,000 acres (819 ha) each. Ranchers must often use jeeps and helicopters to round up their widely scattered herds.

The Pacific Coast. Farmers in the fertile Pacific Coast valleys must also use irrigation. The Willamette Valley of Oregon and the Central Valley of California are especially productive. Farmers in these valleys grow rice, fruits, vegetables, and cotton. Livestock grazing is as important in the coastal region as it is in the mountain areas.

The distant states. Hawaii's tropical climate is ideal for growing sugarcane, the islands' most important product, and pineapples. These crops flourish in the islands' year-long growing season and rainy weather.

Alaska's growing season is very short. The few farms in Alaska's southern region do not grow enough to feed the state's people. Alaskans must import much of their food from warmer regions.

Minerals and Forests

When a California ranch worker discovered gold in 1848, the news spread quickly. Thousands of people from the eastern United States rushed westward hoping to make a quick fortune. Within the next few years, prospectors discovered silver as well as gold in Nevada, Colorado, Montana, and Alaska. In fact, almost every Western state proved to have at least a small deposit of these valuable minerals.

Mineral riches. Since the "Gold Rush" days, people have discovered many other minerals in the Western states. Some of these have proven to be as important as gold and silver.

Two of the most valuable resources of today's world are oil and natural gas. The West has large deposits of these energy sources in California, New Mexico, Wyoming, Colorado, Montana, and Alaska. The West also has large deposits of oil shale, rocks from which crude oil can be distilled. Many scientists believe that Wyoming and Colorado might contain the largest reserves of oil shale in the world.

Deposits of coal and iron ore in the Rocky Mountain states have helped the steel industry grow in Colorado, Utah, and California. Uranium from the West helps supply nuclear power plants

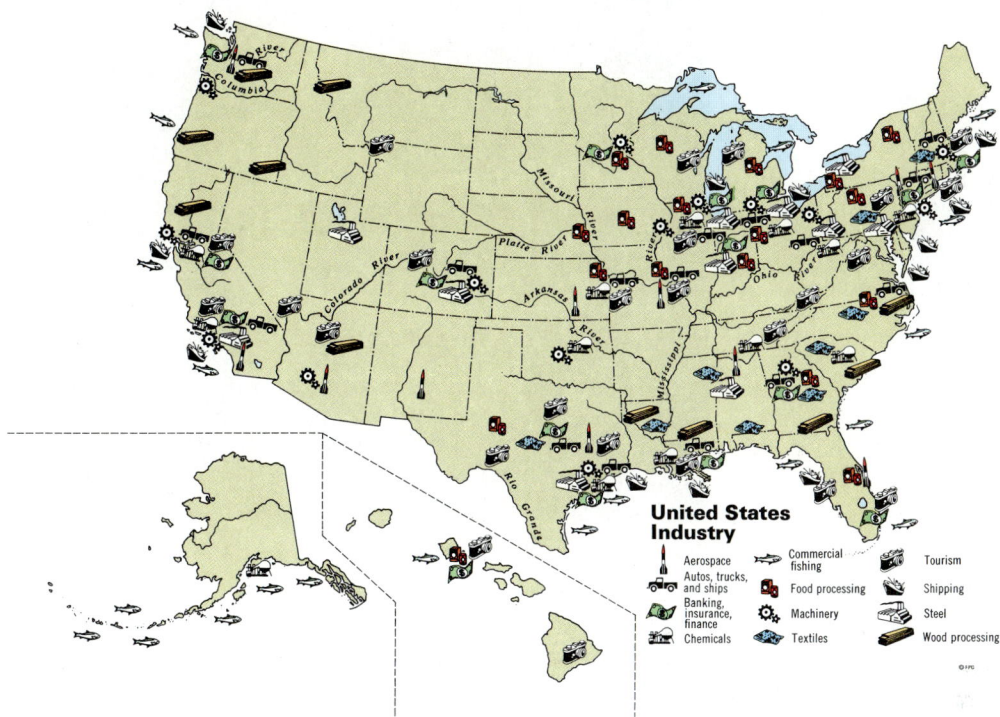

United States Industry

Symbol	Industry	Symbol	Industry	Symbol	Industry
	Aerospace		Commercial fishing		Tourism
	Autos, trucks, and ships		Food processing		Shipping
	Banking, insurance, finance		Machinery		Steel
	Chemicals		Textiles		Wood processing

throughout the country. Western states also supply the nation's industries with nickel, mercury, and copper.

Acres and acres of trees. Forestry is important throughout the Western states. Millions of acres of fir, spruce, pine, and cedar cover the region's mountains. California's redwood trees are the largest trees in the world. California, Oregon, Washington, and Alaska supply almost half the nation's timber.

Fast-Growing Cities and Industries

The warm climate and rich resources of the West continue to attract large numbers of people today. In fact, the West is the fastest-growing region of the United States.

An urban region. Four out of every five Westerners live in urban areas. Like other parts of the country, the West has a sprawling megalopolis. About 15 million people live along the Pacific coast between Santa Barbara and the Mexican border.

Many Western cities have had their greatest growth since 1900. At the beginning of this century, the population of Los Angeles was about 100,000. Today, Los Angeles is the center of an urban area with about 7.5 million people.

Western cities differ in some ways from the older cities of the Northeast and Middle West. For example, the generally flat lands of many Midwestern and Northeastern cities have made the building of public transportation routes fairly easy. But some Western cities have grown up in mountainous areas where people live in valleys and on the hillsides. Rapid transit systems are hard to construct in these regions. Although San Francisco has a good public transit system, most Westerners depend heavily on automobiles. California has about twice as many cars as any other state, and it has traffic and pollution problems to match.

A curious bear looks on during construction of the Alaska pipeline. Completed in 1977, the pipeline carries crude oil from the Arctic Coast to the Gulf of Alaska.

Industrial prosperity. The West is the fastest-growing manufacturing region in the United States. Even the mountain states, which face high transportation costs and limited water supplies, have growing industries. Denver, Pueblo, Salt Lake City, and Phoenix have important steelmaking, food processing, forestry, and aerospace industries.

Most Western manufacturing, however, is centered in the three Pacific Coast states. California is one of the most productive manufacturing states in the country. Plants in southern California make ships, airplanes, petroleum products, cars, and other goods. The entertainment industry is also important in southern California. In Oregon and Washington, wood processing is a leading industry. Look at the map on page 513. What are some important industries near Seattle, Washington?

Western trade and commerce have grown along with the region's industry. Seattle, San Francisco, Portland, and Los Angeles all have fine harbors that encourage the area's flourishing trade. Much of the region's foreign trade is with Japan and other Asian countries.

Wilderness tug-of-war. There are many wilderness areas in the Western states that have been set aside by the government as national parks or forests. Some industrialists want to tap wilderness resources, believing they would ease present-day shortages of raw materials and energy. Environmentalists want these areas preserved for future generations. Which side of the question would receive your support? Why?

Keeping Facts in Focus

1. What natural regions are in the West?
2. What type of climate does Hawaii have? How has this affected its agriculture? How does Alaska's climate affect its agriculture?
3. What are some important industries in the Pacific Coast states?
4. What answers would you give to the key questions on page 509?

Working with Ideas

If you had to live in a different section of the United States from the section you now live in, which section would you choose? Why would you choose it? How might you choose to earn a living in that section? What problems might you have to help solve?

Lesson 6 Canada

Reading Focus

1. In any large society, the opinions and viewpoints of one group of people often differ from those of another. In Canada, for example, most local government leaders support programs that will give the most benefit to their own province. But sometimes Canada's central government opposes these programs. The central government, which has to consider the needs of the whole country, often sees the programs from a different point of view. Balancing the two viewpoints usually involves talking over the differences and agreeing on a compromise. As you read this lesson, look for ways in which the viewpoint of one group of Canadians may differ from other viewpoints. Try to think of possible compromises that could help resolve these differences.

2. As you read this lesson, look for the answers to these key questions:
   a. In what ways is Canada different from the United States? In what ways is it similar?
   b. What region of Canada has most of the nation's farmlands?
   c. How do Canada's wood pulp production and fish exports rank in the world market?
   d. What are some problems in Canadian industry today?

Vocabulary Focus

province   *flax*
autonomy   *newsprint*

Life in Canada is similar in many ways to life in the United States. Both countries have productive farmlands, great factories, and many well-educated people living in large modern cities. Both countries have been built mainly by immigrants, especially from Europe, who sought better lives in North America. People in both countries speak English, although French is also an official language in Canada. And people in both the United States and Canada are trying to balance the comforts and drawbacks of their complex societies to face the future with confidence.

But Canada is also unique in many ways. Its central government in Ottawa has less power over the provinces than Washington has over the states. Canada's land area is greater than that of the United States, and its population much smaller. Although the United States has

The Canadian Parliament meets in Ottawa (above). The Canadian National Tower (right) rises above Toronto's skyline.

had a great influence on Canada, that influence has not always been welcome. Strong national feelings are growing as Canadians are seeking to develop their unique place in the world economy.

Natural and Political Regions of Canada

Canada is the second largest country in the world, covering 3,851,809 square miles (9,976,139 sq km). Within its borders lie six different natural regions. The largest of these is the wide Canadian Shield. Look at the map on page 483. What are the other five natural regions of Canada? What large Canadian island is within the Appalachian region? What mountain range lies to the west of the Interior Plains?

Climate. The world map on page 200 shows Canada's five climate zones. In the north, summers are cool and winters are very cold. The growing season in most of the north is shorter than three months.

In the more southerly regions, temperatures are milder and growing sea-

sons somewhat longer. About 90 percent of Canada's 25 million people live in these southern regions, mainly in cities.

Canada's three largest cities, Toronto, Montreal, and Vancouver, have attracted over one fourth of the country's population. Look at the map on page 517. Find and name some of Canada's other large cities. Although there are some

Glaciers and ice-covered lakes are common in the Columbia Icefields in Alberta's Jasper National Park.

Map labels (Canada Graphic-Relief Map):

Yukon R. · Fairbanks · ALASKA · Dawson · YUKON TERRITORY · Whitehorse · Juneau · Rocky Mountains · BRITISH COLUMBIA · Coast Mountains · Kitimat · Vancouver Island · Vancouver · Victoria · Seattle · WASH. · OREG. · IDAHO · Fraser R. · Columbia R. · Peace R. · Athabasca R. · ALBERTA · Edmonton · Calgary · Medicine Hat · MONTANA

Polar Ice Pack · ARCTIC OCEAN · North Magnetic Pole · Banks I. · Tuktoyaktuk · Victoria Island · Mackenzie River · Great Bear Lake · Port Radium · Yellowknife · Great Slave Lake · NORTHWEST TERRITORIES · Arctic Circle 66½° N. · Baker Lake · Lake Athabasca · SASKATCHEWAN · Saskatoon · Regina · Winnipeg · Brandon · Saskatchewan R. · Lake Winnipeg · MANITOBA · Flin Flon · Thompson · Nelson R. · Churchill R. · Churchill · Severn R. · Lake of the Woods · N. DAKOTA · S. DAK. · NEBR.

Greenland (Denmark) · Davis Strait · Baffin Bay · Baffin Island · Foxe Basin · Frobisher Bay · Hudson Strait · Ungava Peninsula · Hudson Bay · James Bay · Scheffervile · NEWFOUNDLAND · Labrador · Goose Bay · QUEBEC · Sept-Iles · Moosonee · ONTARIO · Gander · Gulf of St. Lawrence · Prince Edward Island · Newfoundland · Charlottetown · Gaspé Pen. · St. Lawrence River · NEW BRUNSWICK · Fredericton · St. John · Sydney · NOVA SCOTIA · Halifax · Bay of Fundy · Quebec · Trois-Rivières · Montreal · Timmins · Sudbury · Sault Ste. Marie · Soo Canals · Thunder Bay · L. Nipigon · L. Superior · Duluth · MINN. · WIS. · IOWA · Chicago · ILL. · IND. · OHIO · MICH. · L. Michigan · L. Huron · Ottawa · Toronto · Hamilton · London · Windsor · Detroit · Buffalo · L. Ontario · L. Erie · N.Y. · PA. · N.J. · New York · CONN. · R.I. · MASS. · Boston · N.H. · VT. · ME. · St. John · ATLANTIC OCEAN

Legend box:

HEIGHT OF LAND
OVER 13,000 FEET
6,600 TO 13,000
3,300 TO 6,600
1,650 TO 3,300
650 TO 1,650
0 TO 650 FEET

DEPTH OF WATER
0 TO 600 FEET
BELOW 600 FEET

Title box:

Canada
Graphic-Relief Map
⊛ National Capital   ★ Other Capitals
● Other Cities
SCALE
One inch—about 500 miles
Miles 0 200 400 600 800 1000
Kilometers 0 200 400 600 800 1000 1200 1400 1600

cities in the cold north, their populations are much smaller than those of the south.

The provinces. Like the United States, Canada is divided into smaller political units. There are ten provinces and two territories. The map on this page shows the provinces, territories, and capital cities.

The Atlantic provinces, in the extreme southeast of Canada, are Newfoundland, Nova Scotia, Prince Edward Island, and New Brunswick. Labrador, on the mainland, is part of the province of Newfoundland. Find these provinces on the map above.

To the west of the Atlantic provinces is Quebec. Most of the people in Quebec are descended from the early French settlers in Canada. Many of these people still speak French and follow French customs. In 1969, French became Canada's second official language.

Ontario, Canada's most populous province, is west of Quebec. What lakes border Ontario on the south?

The provinces of Manitoba, Saskatchewan, and Alberta are called the Prairie provinces. The country's westernmost province is British Columbia. What ocean borders British Columbia? What is the capital of this province?

Patterns of Government in the United States and Canada

Both the United States and Canada are federations, or unions of several political units. In the United States, these units are the fifty states. In Canada, they are the ten provinces and two territories. A federation allows the central, or federal, government to make many decisions that affect the whole country. In general, the provinces of Canada have somewhat more power than the states of the United States. But in both countries, the federal government has the power to decide the law of the land.

The British crown is the crown ruler of Canada as well. The governor general is the crown's representative in Canada. But the prime minister, the leader of the majority party in the House of Commons, actually directs the government. The Cabinet members, also chosen from the House of Commons, head the various government departments and advise the prime minister. Canadian senators are appointed by the governor general. Members of the House of Commons are elected by popular vote. The House of Commons has more lawmaking power than the Senate.

In the United States, the president is the chief executive. The vice-president assists the president and also presides over the Senate. The president appoints the Cabinet members. Members of both the Senate and the House of Representatives are elected by popular vote. The U.S. Senate has more lawmaking power than the Canadian Senate.

In both the United States and Canada, a Supreme Court heads the judiciary. Many legal cases that start in lower courts reach the Supreme Court for a final decision.

The Structure of the Federal Governments in the United States and Canada

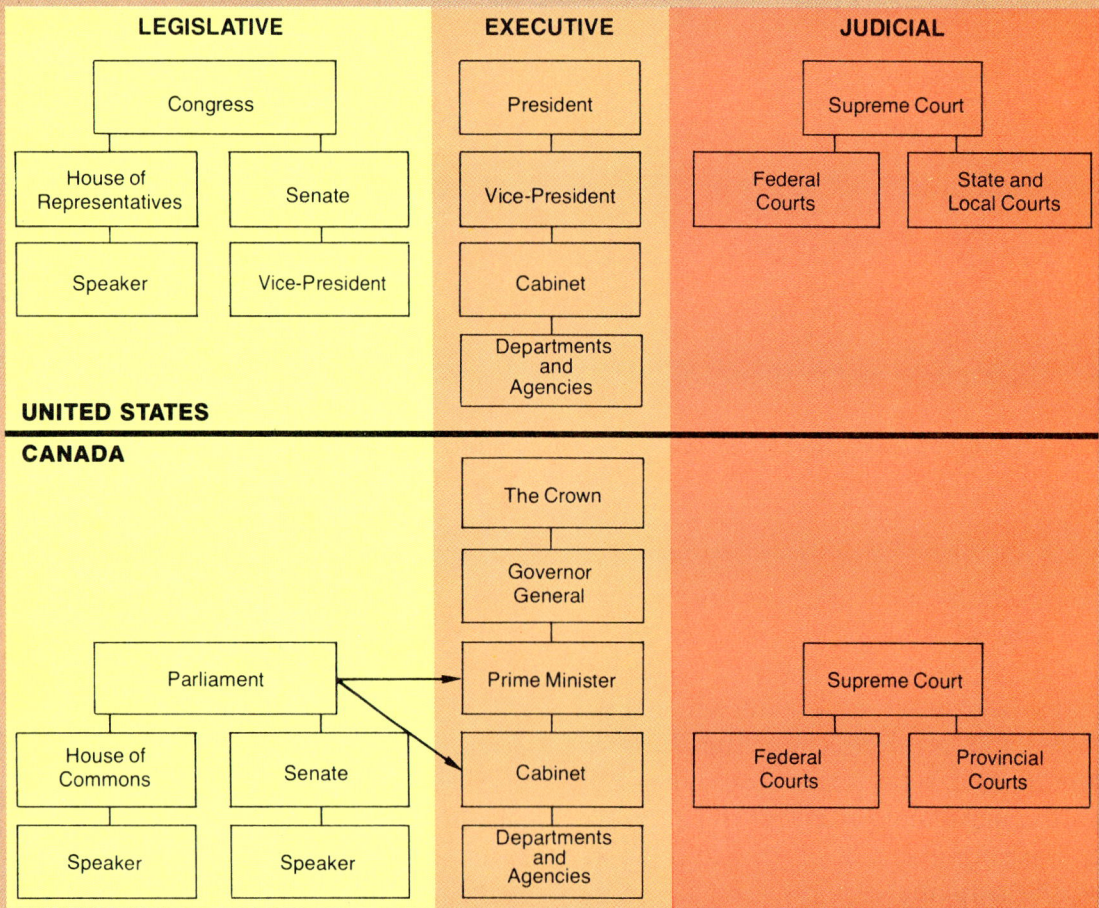

UNITED STATES

LEGISLATIVE: Congress → House of Representatives → Speaker; Senate → Vice-President

EXECUTIVE: President → Vice-President → Cabinet → Departments and Agencies

JUDICIAL: Supreme Court → Federal Courts; State and Local Courts

CANADA

LEGISLATIVE: Parliament → House of Commons → Speaker; Senate → Speaker

EXECUTIVE: The Crown → Governor General → Prime Minister → Cabinet → Departments and Agencies

JUDICIAL: Supreme Court → Federal Courts; Provincial Courts

Canada's most northerly lands are not part of any province. They divide into two territories: Yukon and Northwest Territories. The combined population of these two territories is only about 60,000 people.

Canada's constitution gives the provinces much more independent power than the U.S. Constitution gives the states. As a result, each province is able to decide its own policies in several important areas. Sometimes the policies and goals of one province conflict with those of the central government or other provinces. French-speaking Canadians in Quebec, for example, have often been at odds with Canada's English-speaking majority. In fact, many people in Quebec want their province to become an independent country.

Other provinces are also seeking autonomy, but in economic rather than political matters. The constitution gives the provinces full ownership of all resources within their borders. With today's shortages, these resource rights have helped several of the provinces develop somewhat greater wealth than other parts of the country.

The Use of Natural Resources

Like the United States, Canada has abundant resources to support a variety of industries and occupations. Canada also has many highly skilled workers who know how to make good use of their resources.

Fertile lands. Canada is mainly an industrial country. But agriculture still accounts for about 12 percent of the

Wheat grown on fertile fields in Western Canada has helped Canada become the world's second-leading grain-exporting nation.

nation's exports. The number of farmers has declined in recent years, but production has increased. Machines and modern farming methods help Canada's farmers achieve very high yields.

Over 80 percent of Canada's farmlands are in the Prairie provinces of the West. Wheat is the main crop in these provinces. The Prairie provinces are part of the same wheat belt that extends into the United States. Canada is one of the world's leading wheat exporters. It uses only about one fourth of its wheat for its own needs. Other Prairie crops include barley, oats, and flax. On many farms, grain cropping is combined with livestock raising.

Dairy farming is especially important in Quebec and Ontario. Cows from these provinces supply nearly three fourths of Canada's milk. Ontario also has many farms for special crops, such as sugar beets, tobacco, and fruits.

Canada's forests. Forests cover nearly half of Canada's land. Using timber from these abundant forests, Canada produces more newsprint than any other country. It is also the world's second largest producer of wood pulp.

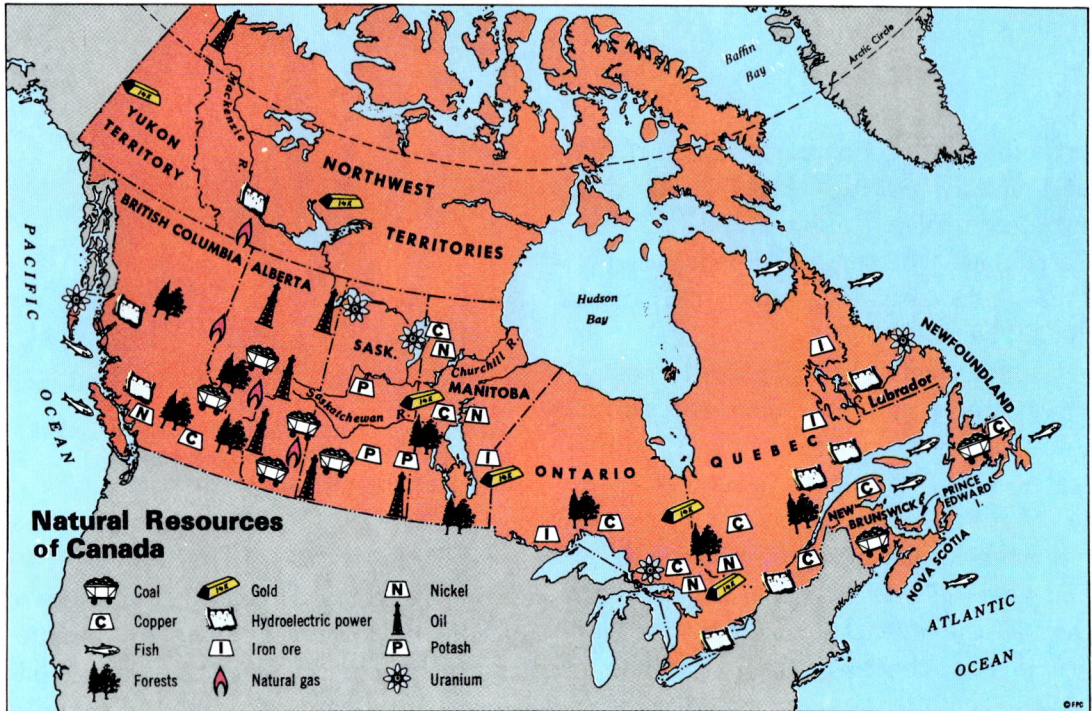

Natural Resources of Canada

Symbol	Resource
Coal	Coal
C	Copper
Fish	Fish
Forests	Forests
Gold	Gold
Hydroelectric power	Hydroelectric power
I	Iron ore
Natural gas	Natural gas
N	Nickel
Oil	Oil
P	Potash
Uranium	Uranium

Fish. Canada has long been a leading exporter of fish. The value of its fish exports ranks fourth highest in the world. The coastal waters of Newfoundland and Nova Scotia are some of the best fishing grounds in the world. Ships from all over the world come to catch cod, herring, mackerel, lobsters, and scallops from these waters. In recent years, however, overfishing has been a problem. The Canadian government has limited foreign fishing in these waters to protect the supply of fish.

Waterways. Canada's many waterways are used for both trade and hydropower. The St. Lawrence Seaway is a major trade route, handling more freight than the Panama and Suez canals combined. Niagara Falls, between Lakes Erie and Ontario, has long been used for electricity. Dams on the St. Lawrence, Ottawa, Churchill, Nelson, Peace, Fraser, and Columbia rivers also provide electricity for thousands of homes and industries.

Mining and Other Industries

Canada has large amounts of important minerals, including oil, iron ore, coal, natural gas, nickel, copper, gold, and uranium. The map on this page shows where many of these mineral deposits lie.

Canada also has many manufacturing industries. Most of Canada's factories are located in southern Ontario along the Great Lakes. Montreal is also an important manufacturing center. Canadian factories produce paper products, cars, metal products, iron and steel, and food products. About one fifth of Canada's exports are communication and transportation equipment.

It is over mining and industrial development that Canada's provinces have

A refinery in Montreal processes crude oil that has been transported by pipeline from Portland, Maine. What are some other examples of cooperation between Canada and the United States?

had the sharpest conflicts with the central government. Many of these conflicts arise from the high degree of U.S. control in Canada's industries.

United States companies control over a third of Canada's paper and pulp industry and over half of its natural gas and oil industry. Almost half of Canada's mining, smelting, and manufacturing industries are also U.S.-owned. In addition, the United States is by far Canada's biggest trading partner. Canada's central government is looking for ways to reduce the nation's economic dependence on U.S. capital and trade.

But some provinces still seek U.S. investment to help them develop their resources. Alberta, for example, which already produces 85 percent of Canada's oil, has many untapped reserves. These reserves are estimated to be greater than those of Saudi Arabia. Alberta welcomes U.S. investment to help it tap these resources.

In Saskatchewan, conflicts have arisen over some of the world's richest potash mines. The provincial government of Saskatchewan is buying out these mines from their American owners. Even though this takeover is reducing

foreign investment, the central government has opposed the plan. It wants a share in the wealth to use for the whole country's benefit. Similar conflicts have grown out of British Columbia's zinc and copper development projects.

In recent years, Canada has begun to make trade agreements with the European Community (Common Market) and with Communist countries around the world. While reducing its dependence on U.S. trade, Canada is keeping close ties with the U.S. in other ways.

Keeping Facts in Focus

1. How do Canadian provinces differ from U.S. states?
2. What are the major agricultural products of Canada?
3. What are Canada's major mineral resources?
4. How is Canada seeking to reduce its dependence on the United States?
5. What answers would you give to the key questions on page 515?

Working with Ideas

Why are most of Canada's major cities and its population concentrated in the southern part of the country? Is this chiefly an advantage or a disadvantage for Canada? Why?

UNIT 15 REVIEW WORKSHOP

Test Your Geographic Knowledge

A. Choose the letter of the item that correctly completes each statement.

1. All the following natural regions are found in both the United States and Canada except the (a) Rocky Mountains, (b) Coastal Plain, (c) Canadian Shield.
2. Most of the interior of the United States and Canada is made up of (a) fertile plains, (b) forested highlands, (c) arid plateaus.
3. The Missouri River flows from the Rocky Mountains to the (a) Gulf of Mexico, (b) Mississippi River, (c) Great Lakes.
4. In the late 1800s, millions of immigrants came to the United States and Canada from (a) Africa, (b) Latin America, (c) southern and eastern Europe.
5. Glaciers helped level the land in (a) Alaska, (b) Texas, (c) the Midwest.
6. Canada's most populous province is (a) Ontario, (b) Alberta, (c) Quebec.
7. The fall line is the boundary between (a) the Rocky Mountains and the Great Plains, (b) the Piedmont and the Coastal Plain, (c) the United States and Canada.
8. The major agricultural products of the Western states are (a) corn and soybeans, (b) cattle and sheep, (c) apples and sugarcane.
9. Over one fourth of the goods manufactured in the United States are produced in (a) the Great Lakes states, (b) the Northeast, (c) California.
10. Nearly half of Canada's land is covered by (a) glaciers, (b) farmland, (c) forests.

B. Tell whether each statement is true or false.

1. Irrigation for farming is needed only in the Western states.
2. Birmingham, Alabama, is often called the "Pittsburgh of the South."
3. French is Canada's official language.
4. The United States and Canada have more energy resources than are needed by their industries and cities.
5. Midwestern farmers grow about half the world's soybeans.
6. Pikes Peak is the highest point in North America.
7. The Lyndon B. Johnson Space Center of NASA is at Cape Canaveral, Florida.
8. Chicago has the world's busiest airport.
9. Labrador is part of the province of Quebec.
10. The St. Lawrence Seaway connects the Great Lakes with New York City.

Apply Your Reading Skills

Choose the letter of the phrase that correctly defines the italicized word in each sentence.

1. Extreme cold and *permafrost* have made it difficult for Canadians to use the many resources of their northern lands. (a) vast distance (b) deep snow cover (c) permanently frozen ground
2. Unless they conserve their *domestic* resources, people in the United States might have to increase their dependence on imports. (a) mass-produced (b) own country's (c) energy
3. The development of *synthetic* fabrics has decreased the demand for cotton. (a) imported (b) human-made (c) colorful
4. During the Gold Rush, there were many *prospectors* in San Francisco. (a) poorly built houses (b) ships (c) people who search for valuable minerals
5. The *floodplain* of the Sacramento River is covered by rich soil. (a) lands that are flooded when a river overflows (b) delta (c) wall built to protect people against an overflowing river

Apply Your Geographic Skills

Use the map on page 523 to find the answers to these questions.

1. In which states is aerospace an important industry?
2. What industries are important in the area around Los Angeles?
3. In what two regions of the country is the textile industry concentrated?

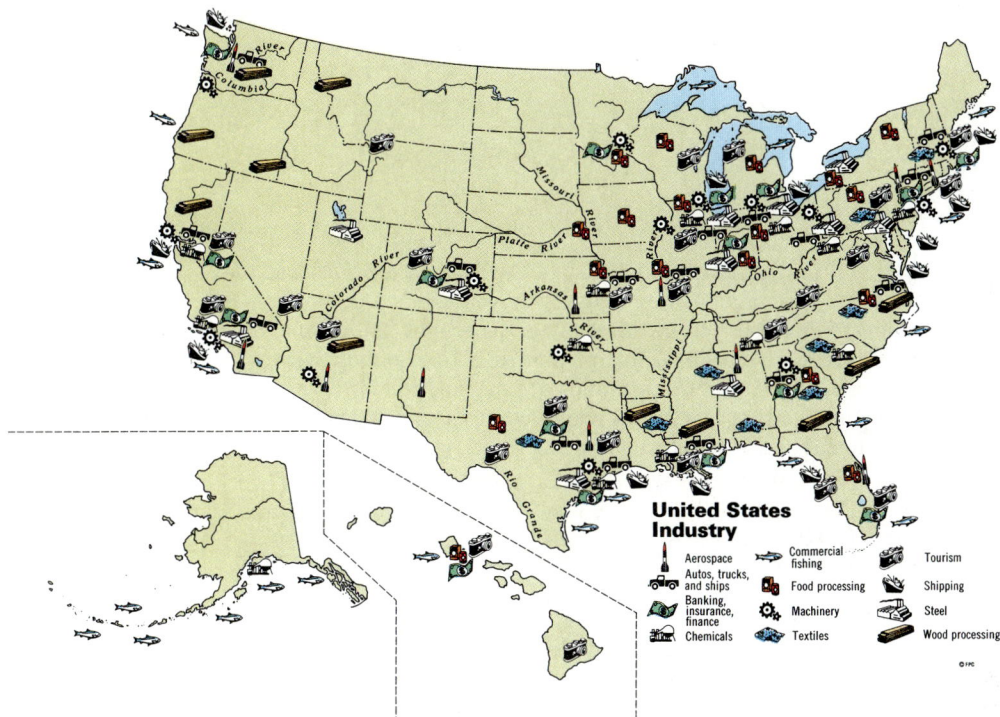

United States Industry

Aerospace
Autos, trucks, and ships
Banking, insurance, finance
Chemicals
Commercial fishing
Food processing
Machinery
Textiles
Tourism
Shipping
Steel
Wood processing

4. What industry is important throughout the Plains states?
5. What two regions of the country have the greatest concentration of industry?
6. What industries are especially important along the Gulf Coast?

Apply Your Thinking Skills

Read the following five descriptions of cities in the United States and Canada. Use the maps, charts, and text from the lessons to identify each city.

1. The location of this city near the mouth of the busiest river in the United States has helped it become the second busiest port in the country.
2. This Canadian city is the second largest French-speaking city in the world. It is also a major trade and manufacturing center.
3. Located near the junction of the two longest rivers in the United States, this city is a leading grain-milling center.
4. The capital of the leading manufacturing state in the United States, this city is located in the center of the most productive agricultural valley in the West.
5. The growth of this fall-line city was due not to trade, but to the fact that it is the national capital.

Discuss These Points

Both pollution and inflation are serious problems in the United States and Canada. Poisonous chemicals and other wastes in the environment can affect people's health and well-being. At the same time, antipollution laws require manufacturers to buy new equipment. This added equipment increases the cost of producing goods, thus forcing prices to rise even higher. What do you think is more important, a clean environment or a stable economy? How do you think people in the United States and Canada can balance these two conflicting needs?

Expand Your Geographic Sights

Acheson, Patricia C. *Our Federal Government: How It Works* (Apollo, 1970).

Neidhardt, Wilfred, *Canadian-American Relations* (Viking, 1975).

Peck, Anne M. *Pageant of Canadian History* (McKay, 1963).

Ross, Frances A. *The Land and People of Canada* (Lippincott, 1964).

Sabin, Louis. *Canada* (Troll Assoc., 1985).

Sterling Editors. *Canda in Pictures* (Sterling, 1966).

The United States (Time-Life, 1984).

Glossary

In this Glossary, you will find all the terms listed in the *Vocabulary Focus* at the beginning of each lesson. You may use the Glossary as you would a dictionary. But only the meanings most helpful to you in the study of geography are given. Almost all of the single word terms have been spelled phonetically—by sounds (within parentheses following the word), and these sounds are marked to help you pronounce them correctly. The Pronunciation Key below shows what the marks mean and gives examples of how to pronounce the sounds.

Pronunciation Key

a *at*	e m*e*t	ī h*i*ke	o͞o g*oo*d
ā f*a*de	ē m*e*	ō *o*pen	o͞o s*oo*n
ã *ai*r	ə *a*bout; aft*e*r; *u*p	ô f*o*r; *aw*ful	ou l*ou*d
ä f*a*r; f*a*ther; h*o*t	i *i*t	oi *oi*l	zh vi*si*on

Aboriginals (ab′ə·rij′nəlz) The earliest people to live in Australia and their descendants.

acacia (ə·kā′shə) A tropical tree that yields gum arabic, a substance used in making candy, gum, and glue.

accessible (ak·ses′ə·bəl) Easily reached.

aerospace (ãr′ō·spās) Having to do with air travel and space flight.

affluent (af′lo͞o·ənt) Wealthy enough to meet basic needs comfortably and still have money left for nonnecessities.

agricultural (ag′ri·kəlch′ə·rəl) Having to do with caring for the soil, producing crops, and raising domestic animals.

air mass A body of air in which weather conditions are uniform.

air pressure The weight or force of air.

alloy (al′oi) A metal made by melting together two or more metals or a metal and a nonmetal.

alluvial (ə·lo͞o′vē·əl) Formed from the deposits of flowing water.

aluminum (ə·lo͞o′mə·nəm) A very light, silver-white metal that has many uses in manufacturing.

antimony (ant′ə·mō′nē) An element used to strengthen metals and to produce various plastics, chemicals, and medicines.

apartheid (ə·pär′tāt) A strict policy of separation of the races, practiced in South Africa.

arable (ãr′ə·bəl) Well-suited for growing crops.

armaments (är′mə·mənts) Military arms and equipment, such as guns and tanks.

artesian well A well in which water rises to the surface from internal pressure.

asphalt (as′fôlt) A tar-like substance occurring in natural deposits or as a by-product of oil refining.

astronomy (ə·strän′ə·mē) The study of the heavenly bodies and their motions.

atmosphere (at′mə·sfir) The layers of air that surround the earth.

autonomy (ô·tän′ə·mē) Self-government.

axis (ak′səs) An imaginary rod on which the earth rotates.

baobab (bou′bab) A tropical fruit-bearing tree whose bark may be used to make cloth and rope.

bilharzia (bil·här′zē·ə) A human disease caused by the presence of certain worms in the body and resulting in loss of blood and damage to tissue.

bioconversion (bī′ō·kən·vər′zhən) The changing of organic matter, such as plants or garbage, into fuel.

birth rate The number of births each year for every 1,000 people in a population.

bison (bīs′ən) A type of buffalo that once roamed the North American plains in large herds.

boundary (boun′d(ə)rē) A line that divides or marks the limits of a political or geographic unit.

breadfruit A tropical fruit that looks and feels like bread when baked.

broadleaf Pertaining to trees that have wide flat leaves, such as oaks, elms, and maples.

calorie (kal′ə·rē) A unit for measuring the heat or energy produced by food in the human body.

capital (kap′ət·əl) Plant and machinery owned by a business; also, money used for investment, in order to bring in more money.

capitalist (kap′ət·əl·əst) Pertaining to a system in which production of most goods and services is controlled by individuals rather than the government.

carbohydrate (kär′bō·hī′drāt) A combination of carbon, hydrogen, and oxygen that is found in sugars and starches and is needed in the human diet.

carbon dioxide A colorless, odorless gas found in the air, without which plants cannot live.

cash Money readily available for use.

cassava (kə·säv′ə) A small tropical shrub, the roots of which can be eaten or used to make starch or tapioca.

cellulose (sel′yə·lōs) The solid or fibrous part of plants, useful for making such products as paper and cotton cloth.

chemical (kem′i·kəl) (*noun*) A substance that can act on other substances to create a new one or cause permanent changes; (*adjective*) having to do with the changes that take place in the makeup of substances.

citrus (si′trəs) A type of tree or shrub grown in warm climates that bears such fruit as oranges or lemons.

climate (klī′mət) The average conditions of weather in a region over a period of many years.

coca leaves The leaves of a South American shrub containing a pain-killing drug.

combine (käm′bīn) A machine that cuts and threshes grain.

commercial (kə·mər′shəl) Intended to be sold; having to do with business.

commonwealth (käm′ən·welth′) An independently governed area that is part of the United States by a special arrangement—Puerto Rico, for example.

communal (kə·myōōn′əl) Shared, used, or particpated in by all members of a community.

commune (käm′yōōn) An organized community in which almost all property, work, and profits are managed by the group or its leaders.

communism (käm′yōō·niz′əm) An economic system in which there is a high degree of state ownership of and control over the means of production, resulting in an absence of private profits.

commuter (kə·myōōt′ər) A person who travels back and forth regularly between two places.

competition (käm′pə·ti′shən) The effort engaged in by two or more people or groups acting independently to achieve the same goal.

composite (käm·päz′ət) A type of volcano that has alternating layers of lava and ashes.

condense (kən·dens′) To change from a gas into a liquid.

conservation (kän′sər·vā′shən) Careful saving and protection of something, such as natural resources.

continental divide A mountain ridge that separates rivers flowing in opposite directions across a continent.

cooperative (kō·äp′ə·rət·iv) Large farm on which the farmers either work together on shared lands or work the land in return for wages paid by the government; an organized group of people who work together in a business, sharing expenses and profits.

copra (kō′prə) Coconut meat that has been dried.

coral (kôr′əl) Tiny sea animals whose skeletal remains build up a deposit of limestone that can form a reef.

core (kôr) The central part of the earth, consisting of solid metal surrounded by a layer of hot liquid metal.

cost (kôst) The amount paid to obtain or produce something.

cottage industry A business in which people produce goods in home workshops.

crater (krāt′ər) A deep depression forming the opening of a volcano.

crust (krəst) The thin outermost shell of the earth, composed mainly of basaltic and granitic rock.

current (kər′ənt) A part of a body of water that flows in one direction; the flow of water in one direction.

cyclone (sī′klōn) A violent tropical storm with strong winds and heavy rains that forms in the Indian Ocean.

death rate The number of deaths each year for every 1,000 people in a population.

deciduous (di·sij′ə·wəs) Pertaining to plants or trees that are bare of leaves during one season of the year.

deforested (dē·fôr′ə·stəd) Cleared of forests.

demand (di·mand′) The amounts of goods or services that people are willing and able to buy.

demographer (di·mäg′rə·fər) A person who studies the characteristics of a population and the causes and effects of population changes.

developing nation A country that depends on agriculture for most of its income.

dew point The temperature to which air must be cooled before it becomes saturated, or filled, with moisture.

diagram (dī′ə·gram) A drawing that shows how things are arranged, how parts go together, or the steps in a process.

dictator (dik′tāt·ər) A ruler who has absolute power and complete control over a country.

distributary (dis·trib′yə·ter′ē) A stream that flows out of rather than into a main river.

diversity (də·vər′sət·ē) Wide variety.

domestic (də·mes′tik) Made at home; having to do with home.

domesticate (də·mes′ti·kāt) To tame or adapt in ways that are helpful to people.

dormant (dôr′mənt) Temporarily inactive.

doubling time The number of years in which a population is expected to double its present size, based on the current growth rate.

duster (dəs′tər) A device that applies pesticides to crops.

ecology (i·käl′ə·jē) The study of the relationships among living things and their surroundings.

ejido (ā·hē′thō) A large Mexican farm that is owned and worked by several families or a community.

element (el′ə·mənt) Any of the more than 100 simple substances into which all matter can be broken down.

elevation (el′ə·vā′shən) The height above sea level.

emigration (em′ə·grā′shən) The action of leaving one country to enter another with the intention of permanent residence.

empire (em′pīr) A group of colonial territories ruled by the same government.

energy (en′ər·jē) The ability or capacity to make things move, change, or do work.

equinox (ē′kwə·näks) Either of the two times of the year when the direct rays of the sun strike the equator and day and night are of equal length everywhere on the earth.

erosion (i·rō′zhən) The process by which land is worn down and the debris is carried away by either wind, water, or glacial action.

estuary (es(h)′chə·wer′ē) A river valley flooded by the sea.

ethnic (eth′nik) Having to do with groups who have common racial or cultural characteristics or national backgrounds.

evaporate (i·vap′ə·rāt) To change from a liquid into a gas.

evergreen Pertaining to plants or trees that remain green the year round.

exploit (ik·sploit′) To use something to its fullest advantage.

export (*verb*) (ek·spôrt′) To send a product to another country for sale.

export (*noun*) (ek′spôrt) A product sent to another country.

extinct (ik·stinkt′) No longer active.

fall line The points at which the land drops sharply and rivers generally become unnavigable.

fallow (fal′ō) Characteristic of arable land left idle during a growing season.

fat An oil that is necessary in the human diet.

fault (fôlt) A break in the earth's crust caused by the land moving up, down, or sideways on either side.

fertile (fərt′əl) Suited for growing crops.

fiord (fē·ôrd′) A narrow inlet of the sea formed between steep cliffs, usually by a glacier.

fishing bank A shallow place in the ocean over a continental shelf where many fish live and can easily be caught.

fission (fish′ən) The splitting of atoms, resulting in the release of large amounts of energy.

flax (flaks) A plant that yields fibers used to make linen thread or cloth and seeds used to make linseed oil.

floodplain Level land built up by deposits from a river that overruns its banks.

fluorite (floor′īt) A mineral used in industrial processes, such as glassmaking.

forecast (fôr′kast) A prediction of a future condition or happening.

fossil fuel A substance derived from the remains of living things that releases energy when it is burned.

front A narrow band where two air masses meet, interact, and often produce stormy weather.

frost A coating of thin ice crystals formed when moisture-laden air close to the ground is chilled below its dew point.

fusion (fyoo′zhən) The uniting of atoms in a way that produces a huge amount of energy.

futurist (fyoo′chə·rist) One who studies present conditions in order to anticipate the possibilities of the future.

gasification (gas′ə·fə·kā′shən) A process that converts materials, such as low-grade coal, into a gaseous fuel.

geologist (jē·äl′ə·jəst) A scientist who studies the solid matter of the earth.

geothermal (jē′ō·thər′məl) Produced by the heat of the earth.

glacial (glā′shəl) Having to do with glaciers, bodies of moving ice on the earth.

glaze A thin coating of ice formed when rain freezes on contact with cold surfaces.

graph (graf) A drawing that uses bars, circles, points and lines, or symbols to present sets of facts.

graphic An illustration, map, photograph, diagram, graph, or chart used to present information.

graphic-relief map A map that uses shading and colors to show the physical features of a region.

great circle The equator, all meridians, and any other imaginary line that divides the earth into two equal parts.

grid Two sets of lines that cross each other, as on a map.

groundwater Water that lies below the surface of the earth.

gully A ditch or channel made by running water.

gum arabic A substance obtained from acacia trees, used in making candy, gum, and glue.

gypsum (jip′səm) A white or yellowish mineral used to make plaster, paint, cement, and other products.

hacienda ((h)äs·ē·en′də) A large family estate that includes homes, gardens, mills, and other necessities for owners and workers on the estate.

harvester (här′vəst·ər) A machine that picks crops.

heavy industry An industry in which heavy materials are handled.

helium (hē′lē·əm) A gaseous element, lighter than air, that makes up much of the sun and other stars.

hemisphere (hem′ə·sfir) Half of the earth.

henequen (hen′i·kən) A tropical plant that yields strong fibers used to make rope and twine.

horizontal (hôr′ə·zänt′əl) Characterized by a side-to-side direction, one parallel to the horizon.

humidity (hyoo·mid′ət·ē) Moisture or wetness in the air.

humus (hyoo′məs) The fertile part of soil made from dead plants and leaves.

hurricane (hər′ə·kān) A tropical storm that forms over the Atlantic or Pacific oceans off the coasts of Mexico.

hydroelectric (hī′drō·i·lek′trik) Having to do with electricity produced by water rushing over a dam or waterfall.

hydrogen (hī′drə·jən) The lightest element, one of the most plentiful on earth, and a main element of the sun.

hydropower (hī′drō·pou(ə)r) The energy of rushing water.

igneous (ig′nē·əs) A characteristic of rocks formed by the cooling and solidifying of molten materials.

immigration (im′ə·grā′shən) The action of entering one country from another with the intention of permanent residence.

import (*verb*) (im·pôrt′) To buy and bring a foreign product into a country.

import (*noun*) (im′pôrt) The product bought from a foreign country.

income The amount of money received, often through investment or labor.

industrialist (in·dəs′trē·ə·ləst) One who owns or manages a large-scale business or manufacturing company.

industrialization (in·dəs′trē·ə·lə·zā′shən) The large-scale production of goods by machinery rather than by hand.

industrialized (in·dəs′trē·ə·līzd) Having factories, machinery, and large-scale production of goods.

inflation (in·flā′shən) An economic condition in which more and more money is required to buy the same amount of goods.

intensive farming Highly efficient farming methods that produce large yields from small plots of land.

interior (in·tir′ē·ər) Land located away from coasts or borders.

intermittent (int′ər·mit′ənt) A kind of volcano in which eruptions are followed by long periods of inactivity.

intermountain Between two or more mountains.

invest To spend money for something expecting that it will bring about future profits.

ionosphere (ī·än′ə·sfir) The layer of the atmosphere beginning about twenty-five miles above the earth's surface containing charged particles, or ions.

iron sands Sands containing iron.

irrigation (ir′ə·gā′shən) The act of supplying water for dry land to enable people to grow crops.

jet stream A high-speed, high-altitude westerly wind that influences weather in the middle latitudes.

kibbutz (kib·ŏŏts′) A cooperative farm in Israel on which the members share all property.

kinetic energy The capacity of moving objects to do work.

kola nut The seed of a tropical evergreen tree, used in making soft drinks and medicines.

land grant Land given to someone by the government, usually for the purpose of encouraging settlement and development of that land.

landlocked Without an outlet to an ocean or sea.

land reform A policy of breaking up large landholdings in order to divide the land more equally among the people.

latitude (lat′ə·tōōd) Distance north or south of the equator, measured in degrees.

leaching (lēch′ing) The process by which water seeping downward through layers of soil dissolves and removes minerals.

leeward (lē′wərd) Facing away from the wind.

levee (lev′ē) A riverbank higher than the surrounding land.

lichen (lī′kən) A flat, leafless plant growing on rocks or tree trunks.

life expectancy The average number of years a person at birth is expected to live.

light industry An industry in which lightweight materials are handled.

literacy (lit′ə·rə·sē) The ability to read or write.

llama (läm′ə) An animal of South America, similar to a camel but smaller and having no humps.

llano (län′ō) The Spanish word for large grassy plain, usually used in reference to plains in South America that are north of the Amazon River.

loess (les) Fertile soil deposited by the wind, often yellowish-brown.

longitude (län′jə·tōōd) Distance east or west of the prime meridian, measured in degrees.

luxury (lək′shə·rē) Something costly that is enjoyed rather than needed.

maguey (mə·gā′) A plant of southern Mexico that yields fibers used in weaving and in making several kinds of drinks, including tequila.

mainland Land on a continent as opposed to the land on offshore islands.

maize (māz) Indian corn.

malaria (mə·ler′ē·ə) A disease transmitted to humans by the bite of certain mosquitoes.

malnutrition (mal′nōō·trish′ən) Severe lack of important nutrients, such as protein, minerals, and vitamins, that results in illness and poor health.

mantle (man′təl) The middle layer of the earth, consisting of hot, heavy, and mostly solid rock.

map key An explanation of the meaning of the symbols on a map. The map scale is often included in the key.

market (mär′kət) A geographic area in which goods are bought and sold.

megalopolis (meg′ə·läp′ə·ləs) A heavily populated area in which the suburbs of large cities extend far enough to meet the suburbs of other large cities.

menhaden (men·hād′ən) A fish used for fertilizer and animal feed, found mostly off the Atlantic coast of the United States.

meridian (mə·rid′ē·ən) A north-south line of the global grid; also called a line of longitude.

merino (mə·rē′nō) A type of sheep originating in Spain that yields a high-quality wool.

mestizo (me·stē′zō) A person having parents or ancestors of both American Indian and European origins.

metamorphic (met′ə·môr′fik) A characteristic of rocks formed originally from sediments or molten materials and later changed by heat and pressure.

migrant (mī′grənt) A person who moves often from one place to another in order to work, especially at harvesting crops.

migration (mī·grā′shən) The movement of people from continent to continent or from place to place within a continent.

millet (mil′it) A grain grown in hot, dry areas, especially in Africa and Asia.

milling Grinding grain into flour.

mineral (min′ə·rəl) An element or combination of elements found in nature, usually as a solid, such as coal, copper, and granite.

moderate (*adjective*) (mäd′ə·rət) Calm, not violent or extreme.

moderate (*verb*) (mäd′ə·rāt) To keep from extremes.

molten (mōlt′ən) Made liquid by heat.

monarchy (män′ər·kē) A form of government headed by a king or queen.

monsoon (män·sōōn′) A seasonal shift in prevailing wind patterns.

mortgage (môr′gij) To use property as security for a loan. If the loan is not paid back, the property belongs to the lender.

moshav (mō·shäv′) A cooperative farm in Israel on which families farm their own land but join with other members to buy supplies and sell crops.

mulatto (m(y)o͞o·lät′ō) A person having parents or ancestors of both black and white races.

mutton (mət′n) Meat from fully grown sheep.

nationalism (nash′nəl·iz′əm) The belief that people should live together as a separate, independent nation within the boundaries of their own country; strong loyalty that persons have for their nation and culture.

nationalize (nash′nəl·īz) To put control or ownership in the hands of the government.

natural environment Surroundings that are not artificial, or produced by people.

navigable (nav′i·gə·bəl) The condition of a waterway that is deep and wide enough to permit passage of ships.

needleleaf Pertaining to trees that have needle-shaped leaves, such as firs, pines, and spruces.

newsprint Low quality paper made chiefly from wood pulp and used mostly for newspapers.

nitrogen (nī′trə·jən) A gas without color, odor, or taste that makes up about 78 percent of air.

nodule (nä′jo͞ol) A small lump of minerals.

nomadic (nō·mad′ik) Moving from place to place.

nuclear (no͞o′klē·ər) Having to do with the use of the nucleus, or central part, of an atom to generate energy.

oasis (ō·ā′səs) A small area in a desert region that has water and fertile soil.

oil shale A sedimentary rock that produces oil after it is crushed and heated.

one-crop economy A system in which a nation depends largely on the production and sale of one crop for its income.

orbit (ôr′bət) The path of one object as it travels around another.

oxygen (äk′si·jən) A colorless, odorless, tasteless gas that makes up about 21 percent of air and is a necessary part of many life processes.

ozone (ō′zōn) A form of oxygen present in the stratosphere or produced in the troposphere when sunlight reacts with chemicals.

Pampas (päm′pəz) The Spanish word for grassy plains, used in reference to plains in South America that are south of the Amazon River.

pan-Africanism The belief that African nations should unite to promote their own interests and to keep out foreign influences.

parallel (pâr′ə·lel) An east-west line of the global grid; also called a line of latitude.

permafrost (pər′mə·frôst) Permanently frozen soil.

persimmon (pər·sim′ən) A tree bearing a red-orange fruit that is edible when very ripe.

pesticide (pes′ti·sīd) A chemical used to kill plant and animal pests.

phosphorus (fäs′f(ə)rəs) A nonmetallic mineral found in such foods as eggs, milk, and fish and extracted from natural deposits for fertilizer or other uses.

plankton (plank′tən) The tiny plants and animals that fish feed on in ocean water.

plantain (plant′ən) A green, bananalike fruit that is a staple food in tropical regions.

plantation (plan·tā′shən) A large commercial farm usually devoted to one crop.

plate (plāt) A section of the earth's crust that moves.

plutonium (plo͞o·tō′nē·əm) An element used to make nuclear energy.

political map A map that shows boundaries of nations and other political units.

pollutant (pə·lo͞ot′nt) A harmful waste product scattered or dispersed on land or in water or air.

population density The average number of people living on a square unit of land in a given area.

population growth rate A figure computed by adding the rate of natural increase (births minus deaths per 1,000 persons) to the net migration rate (immigration minus emigration divided by the total population) and expressed as a percent.

porous (pôr′əs) Having the ability to let water or other liquids pass through.

potash (pät′ash) A chemical compound used to make fertilizer and glass.

prairie A large plains region with tall grasses, usually found in the United States or Canada.

precipitation (pri·sip′ə·tā′shən) Moisture that falls to earth—such as rain, snow, hail, and sleet.

prevailing wind A horizontal motion of the air that comes from the same direction most of the time.

production (prə·dək′shən) The making, creating, or bringing forth of something.

productive (prə·dək′tiv) Successful at making, creating, or bringing forth large quantities.

professional (prə·fesh′ən·əl) A person who works at a job requiring special education.

profit (präf′ət) Money that is left over after all expenses have been paid.

prospector (präs′pek·tər) A person who searches for mineral deposits.

protein (prō′tēn) A substance needed to build body cells.

province (präv′ən(t)s) One of the political units into which a country may be divided.

radiation (rād′ē·ā′shən) Energy sent out from an object in the form of waves or particles.

radioactivity (rād′ē·ō·ak·tiv′ət·ē) A property of objects that send out particles or waves of energy as their atoms break down.

rapids (rap′ədz) A generally shallow part of a river in which the current moves swiftly over rocks.

rate of natural increase The difference between the birth rate and the death rate, usually expressed as a percent.

reclaim (rē·klām′) To make land suitable for human use.

refinery (ri·fīn′(ə)rē) A factory for purifying metals, oil, sugar, or other raw materials.

refrigerated (ri·frij′ə·rāt′əd) Kept cold, often for the purpose of preserving food.

regulate (reg′yə·lāt) To bring under control.

relative humidity The amount of moisture in the air at a given temperature compared with the amount of moisture the air could possibly hold at that temperature.

relief (ri·lēf′) The changes in elevation from one point to another in a given stretch of land.

renewable resource A useful natural material of which new reserves are constantly supplied, such as water, air, lumber, or soil.

reservoir (rez′ərv·wär′) A lake where water is stored for future use, often formed by a dam.

retirement (ri·tī(ə)r′mənt) A worker's generally permanent withdrawal from the labor force.

revenue (rev′ə·n(y)o͞o) Money that goes into the government treasury, often through taxation.

revolution (rev′ə·lo͞o′shən) One complete orbit of the earth around the sun.

ridge (rij) A narrow chain of raised land either on a landmass or on the ocean floor.

river system A river and all its branches.

rotation (rō·tā′shən) One complete turn of the earth on its axis.

sago flour A powdered starch made from a sago palm and used to thicken foods and stiffen cloth.

savanna (sə·van′ə) A level, tropical grassland with scattered trees.

scale (skāl) The length used to represent a larger unit of measurement in making a map, drawing, or model.

sediment (sed′ə·mənt) Fine particles of sand, mud, clay, and other materials.

sedimentary (sed′ə·ment′ə·rē) A characteristic of rocks formed by the cementing together of small pieces of rock and sand.

seismic (sīz′mik) Having to do with vibrations from earthquakes.

self-sufficient Able to meet all needs and wants without the help of others.

shale (shāl) Rock that splits easily into thin layers.

silica (sil′i·kə) The most abundant material of the earth's crust, used in making glass and radios.

silica sands Sands containing silica.

silicon (sil′i·kən) A nonmetallic element that combines readily with other elements of the earth's crust.

silt Fine soil deposited by flowing water.

smelt To take pure metal out of an ore, usually by a melting and separating process.

sodium nitrate A type of salt that is a valuable fertilizer and is also used in preserving meat.

soil The uppermost layer of land, most favorable to plant growth when it contains both mineral matter and humus.

solar (sō′lər) Having to do with the sun.

solar system The sun and the planets that revolve around it.

solstice (säl′stəs) Either of the two times in the year when the direct rays of the sun are farthest north or south of the equator.

source region A place where air masses form and take on generally uniform characteristics.

sphere (sfir) An object shaped like a ball.

station (stā′shən) A livestock ranch in Australia or New Zealand.

statistic (stə·tis′tik) A numerical figure in a collection of such figures.

steppe (step) A vast area of dry, level grassland, usually found in Asia.

strategic (strə·tē′jik) Necessary to the outcome of a planned action.

stratosphere (strat′ə·sfir) The layer of the atmosphere above the troposphere that is free of water vapor and strong vertical air movements.

strip mining A method of taking minerals from the ground by removing all the soil and plants that cover the deposit.

subarctic (səb′ärk′tik) Having to do with the regions immediately south of the Arctic Circle.

subcontinent (səb′känt′ən·ənt) A large part of a continent that forms a section of its own.

subsidy (səb′səd·ē) Money given by the government to support various persons or organizations, including farmers, homeowners, businesses, and industries.

subsistence (səb·sis′tən(t)s) Producing or having barely enough to live.

subtropical (səb′träp′i·kəl) Having to do with the regions bordering the tropics.

supply (sə·plī′) The amount of goods and services people are willing and able to provide.

surplus (sər′pləs) An amount over and above what is immediately needed.

symbol (sim′bəl) An object, sign, or picture used to represent something.

synthetic (sin·thet′ik) Made or manufactured, rather than occurring naturally.

table A list of related facts.

tariff (tār′əf) A tax on imports and, in some countries, also on exports.

taro (tär′ō) A tropical plant whose roots yield a starch that can be eaten.

technology (tek′näl′ə·jē) Scientific advancement in such areas as tools, weapons, transportation, and all methods of providing for human life and comfort.

temperature (tem′pə(r)·chər′) The warmth or coldness of any given thing measured in degrees on the Fahrenheit or Celsius scale.

temperature range The difference between the highest and lowest temperatures of an area for a given period of time.

terraced (ter′əst) Made into levels with each level higher than the previous one.

theory (thē′ə·rē) An explanation based on either incomplete evidence or on reasoning and guesswork.

thermal (thər′məl) Having to do with heat.

threshing Separating the heads of grain from the stalks of the plants.

tide The regular rising and falling of the oceans and the waters connected with them.

tilling Working the land—plowing, planting, and raising crops.

titanium (tī·tān′ē·əm) A light but strong metallic element, often used to make steel more resistant to heat.

tornado (tôr·nād′ō) A funnel-shaped cloud with violent winds that drops from a storm cloud, destroying almost everything it touches on land.

transition (tranz·ish′ən) The passage from one place or condition to another.

tremor (trem′ər) A slight vibrating or shaking motion produced by an earthquake.

trench A long, narrow opening on the ocean floor, formed by movements of the earth's plates.

trend An indication of a future condition or development.

troposphere (trōp′ə·sfir) The layer of the atmosphere closest to the earth's surface, where day-to-day weather occurs.

tsetse fly An insect that carries a disease deadly to humans and animals.

tsunami (tsoo·näm′ē) The Japanese word for a fast-moving ocean wave caused by an earthquake or volcanic eruption.

tungsten (təng′stən) A metallic element that remains strong at high temperatures, used in making lights, electronic equipment, and steel.

turbine (tər′bən) An engine with parts that move in a circular motion.

typhoon (tī·foon′) A tropical storm occurring over the Pacific Ocean between the international dateline and the coasts of Asia and Australia.

uranium (yō·rā′nē·əm) Extremely rare radioactive element found as an ore and used to produce nuclear energy.

urbanization (ər′bə·nə·zā′shən) The movement of people from rural to urban areas; the effect of such movement.

urban renewal A program to rebuild or restore buildings in a city.

urban sprawl Housing, shopping, and industrial developments on the outskirts of a city.

vertical (vərt′i·kəl) Characterized by a top-to-bottom or bottom-to-top direction.

vitamin (vīt′ə·mən) Any of certain types of substances that are essential to the human diet.

wadi (wäd′ē) A usually dry riverbed that fills with rushing water during a cloudburst, often found in desert regions.

waterborne Carried by water.

water cycle The movement of water from earth to air and back to earth again by the processes of evaporation, condensation, and precipitation.

water distillation A process by which impure water is made suitable for use.

water table The top part of a layer of water-filled soil, usually lying beneath the earth's surface.

water vapor Water in a gaseous state.

wave The up and down movement of water caused by the transfer of energy across its surface.

weathering (weth′ər·ing) Any process occurring at or near the earth's surface that causes chemical changes in rocks or that physically breaks rocks into smaller pieces.

wetland An area—such as a swamp, marsh, or bog—where the soil is very wet or under shallow water, often found near the edges of larger bodies of water.

wilderness (wil′dər·nəs) A region in which the natural environment has not been disturbed by human activity.

windward Facing the wind.

winnowing (win′ō·ing) A process that uses the wind to separate kernels of grain from their outer coats.

Index

Acknowledgments

Artwork
Cunningham David, 77, 78, 83,
  85, 88, 101, 105, 110, 119,
  121, 125, 142, 143, 155,
  157, 161, 169, 175, 179,
  197, 253
Darcy, Tom; NEWSDAY, 283
Harris, Sidney J., 189
Clyde Wells; AUGUSTA
  CHRONICLE, 279

Maps
Follett Cartographers

Photographs
Page 7: National Aeronautics
  and Space Administration
38: General Electric Company,
  Beltsville Photographic
  Engineering Laboratory in
  cooperation with the
  National Geographic Society
  and the National
  Aeronautics and Space
  Administration
40: The Christian Science
  Publishing Society
61: top, Robert Frerck; middle,
  Tom Stack &
  Associates/Keith Gunnar;
  bottom, National
  Aeronautics and Space
  Administration
62: Dr. Georg Gerster
67: Wm. Parker
70: top, Tom Stack &
  Associates/Warren and
  Genny Garst; bottom, Milt
  and Joan Mann
71: top left, Peter Arnold,
  Inc./Clyde Smith; top right,
  Peter Arnold, Inc./James
  Karales; center right, Tom
  Stack & Associates/Fredrick
  H. Kerr; bottom right, Tom
  Stack & Associates/John
  Neel
74: Kennecott Copper
  Corporation
79: Woods Hole Oceanographic
  Institution
84: Hawaii Natural History
  Association and the National
  Park Service
85: U.S. Department of the
  Interior/K. Segerstrom
86: Peter Arnold, Inc./W. H.
  Hodge

87: Shostal Associates/
  Russell B. Lamb
89: Ward's Natural Science
  Establishment, Inc.
92: Dimensions/Walter A.
  Frerck
93: Soil Conservation Service
94: top, Tom Stack &
  Associates/Jim Holland;
  bottom, Dimensions/Robert
  A. Asher
95: Tom Stack & Associates/Bill
  N. Kleeman
96: Historical Pictures Service,
  Inc.
99: Wm. Parker
108: top, Robert Frerck; bottom,
  Tom Stack &
  Associates/John R. Neel
109: Instituto Italiano di Cultura
114: Peter Arnold, Inc./Stephen
  J. Krasemann
117: Van Cleve, Inc./Michael
  Bertan
118: Bruce Coleman, Inc./
  T. Daniel
124: left and right, Van Cleve,
  Inc./Michael Bertan
128: AP Laserphoto
130: Terry McClellan
133: Peter Arnold, Inc./Klaus D.
  Francke
135: Van Cleve, Inc./G. Lucas
136: top left, Van Cleve,
  Inc./Leonard Lee Rue IV; top
  right, Peter Arnold; center,
  Ward's Natural Science
  Establishment, Inc.
137: Van Cleve, Inc./Ed Bryan
141: Van Cleve, Inc./Barbara
  Van Cleve
144: Van Cleve, Inc./Barbara
  Van Cleve
145: National Aeronautics and
  Space Administration
147: Peter Arnold, Inc./Kahana
  Film Productions
152: Van Cleve, Inc./Nancy
  Haynes
156: Tom Stack &
  Associates/Bryon Crader
157: The Foto Place
159: top and bottom, The Foto
  Place/Tony Castelvecchi
162: Van Cleve, Inc./Peter Fronk
163: Robert Frerck
165: left, Central Scientific Co.;
  right, Van Cleve, Inc./Nancy
  Simmerman

174: Milt and Joan Mann
176: Van Cleve, Inc./Bruce Fritz
180: Tom Stack & Associates/Rick Myers
181: top, Van Cleve, Inc./U.S. Weather Bureau; left, Van Cleve, Inc./Stephen Gwin, U.S. Weather Bureau
186: Robert Frerck
191: top, Bruce Coleman, Inc./Giorgio Gualco; bottom left, Bruce Coleman, Inc./Robert E. Pelham; bottom right, Bruce Coleman Inc./C. Haagner
193: Peter Arnold, Inc./ W. H. Muller
199: Florida Department of Commerce, Division of Tourism
205: Bruce Coleman, Inc./Peter Ward
208: Editorial Photocolor Archives
209: Bruce Coleman, Inc./Dale and Marion Zimmerman
210: left, Bruce Coleman, Inc./Jen and Des Bartlett; right, M.K. Parker
211: Bruce Coleman, Inc./ W. E. Ruth
214: Robert Frerck
219: Robert Frerck
220: Robert Frerck
221: left, Bruce Coleman, Inc./Bill Brooks; right, Bruce Coleman, Inc./M. Timothy O'Keefe
222: Robert Frerck
223: World Food Programme/ T. Fincher
225: top, Peter Arnold, Inc./Jacques Jangoux; center left, Robert Frerck; center right, Victor Englebert
228: all, Noel Vietmeyer
229: Victor Englebert
231: Bruce Coleman, Inc./Larry R. Ditto
234: top left, Peter Arnold, Inc./Jacques Jangoux; top right, Bruce Coleman, Inc./Nicholas DeVore III; center left, Bruce Coleman/Norman Myers
237: Terry McClellan
240: Amoco Torch Magazine, Standard Oil Company (Indiana)
244: left and right, Globe-Union, Inc.
249: left and right, Roloc Color Slides

251: Wm. Parker
252: National Aeronautics and Space Administration
253: G. R. Robert
254: Wm. Parker
255: Chicago Sun Times/John H. White
262: Tom Stack & Associates/R.A. Guy
265: Bruce Coleman, Inc./Nicholas DeVore III
266: Peter Arnold, Inc./Jacques Jangoux
267: Peter Arnold, Inc./Jacques Jangoux
275: top, Peter Arnold, Inc./Jacques Jangoux; center, Peter Arnold, Inc./Harvey Lloyd; bottom, Milt and Joan Mann
279: Peter Arnold, Inc./Harvey Lloyd
280: Milt and Joan Mann
282: Food and Agriculture Organization/F. Botts
286: Mel Wilson
294: Bruce Coleman, Inc./ W. Grant
295: Bruce Coleman, Inc./Heinz Herfort
297: left, Bruce Coleman, Inc./Jonathan T. Wright; right, Bruce Coleman, Inc./ J. Messerschmidt
300: Wm. Parker
302: Milt and Joan Mann
303: Milt and Joan Mann
307: Tom Stack & Associates
309: Bruce Coleman, Inc./ W. Ferchland
310: Bruce Coleman, Inc./ F. Jackson
311: Bruce Coleman, Inc./ J. Messerschmidt
315: Bruce Coleman, Inc./ J. and M. Williamson
316: Peter Arnold, Inc./Tom Pix
317: left, Robert Frerck; right, Bruce Coleman, Inc./ J. Messerschmidt
320: Peter Arnold, Inc./Jacques Jangoux
321: Tom Stack & Associates/ M. Gilson
322: Howard Sochurek
328: Tom Stack & Associates/Bryon Crader
333: top and bottom, Robert Frerck
337: Mel Wilson
338: top left, Root Resources/Mike Mitrakul; top right, Robert Frerck;

center right, Tom Stack & Associates/Bryon Crader
341: Robert Frerck
343: Robert Frerck
344: Black Star/Harry Redl
346: left and right, Robert Frerck
349: left, Milt and Joan Mann; right, Root Resources/John LaDue
351: Robert Frerck
356: Root Resources/Jane Shepstone
357: left, Tom Stack & Associates/Gary Stallings; right, Sygma/J.P. Laffont
359: Peter Arnold, Inc./Stephanie FitzGerald
360: top, Wide World; center, Milt and Joan Mann
362: left, Milt and Joan Mann; right, Norma Morrison
366: Rapho Guillumette/Minoru Aoki
367: Bruce Coleman, Inc./ M. Tonooka
370: Bruce Coleman, Inc./ F. Prenzel
376: top right, Bruce Coleman, Inc./ N. DeVore III; center left, Tom Stack & Associates/Ben Cropp
377: Tom Stack & Associates/Warren Garst
379: Tom Stack & Associates/Warren Garst
380: Bruce Coleman, Inc./ F. Prenzel
381: Bruce Coleman, Inc./ F. Prenzel
382: left and right, Bruce Coleman, Inc./F. Prenzel
385: Bruce Coleman, Inc./ N. DeVore III
387: left, Tom Stack & Associates/Jane Downtown; center, Tom Stack & Associates/ Bill Kleeman; left, Bruce Coleman, Inc./ N. DeVore III
390: Peter Arnold, Inc./Klaus D. Franke
393: Bruce Coleman, Inc./Giorgio Gualco
397: Bruce Coleman, Inc./Giorgio Gualco
399: Sygma/Milner
401: Bruce Coleman, Inc./Giorgio Gualco
403: left, Peter Arnold, Inc./Harvey Lloyd; right, Peter Arnold, Inc./Dan Porges